**Surrogate Modeling**
**and Optimization**

# Surrogate Modeling and Optimization

Theories, Applications, and Limitations

*Nam-Ho Kim*
*University of Florida*
*Gainesville, Florida, USA*

WILEY

Published by JohnWiley & Sons, Inc., Hoboken, New Jersey.



***Library of Congress Cataloging-in-Publication Data Applied for:***

Hardback ISBN: 9781394245819

Cover Design: Wiley
Cover Image: © agsandrew/Getty Images

Set in 9.5/12.5pt STIXTwoText by Straive, Pondicherry, India

*Dedicated to Professor Raphael "Rafi" Tuvia Haftka, a great mentor, colleague, and friend*

# Contents

# Preface

Surrogate modeling is a powerful engineering tool to approximate a quantity of interest (QoI) as a function of input variables. The QoI can be an outcome of computer simulation or experimental measurement, which may or may not include noise. Input variables can be any parameter that affects the QoI, but in this book, they represent design variables that can be changed during optimization. The functional relationship between the QoI and input variables is unknown in many engineering applications because it is often governed by integro-differential equations. Computer simulation and experimental measurement can provide the value of QoI for selected input variables, but the behavior of the QoI over the entire input space is generally unknown. Surrogate modeling approximates this complex relationship in the form of simple analytical/statistical functions using a set of samples, which were obtained by computer simulations or experiments at the different values of input variables. The main goal of surrogate modeling is to obtain an accurate approximate function using a minimum number of samples. Accuracy in this context means the prediction capability of a surrogate model at unsampled locations, not the error between samples and surrogate predictions. The main use of a surrogate model is to replace expensive computer simulations and experiments during design optimization, sensitivity analysis, and/or uncertainty quantification.

There are many books on surrogate modeling and optimization in the market. Many of them focus on how to build surrogate models and how to use them for optimization. There are also many commercial and public-domain programs for surrogate modeling. However, in reality, it requires significant technical expertise to use surrogate models properly. This is partly because building surrogate models is one thing and understanding them is another. An important goal of this book is to explain the meaning of built surrogates and to provide an in-depth understanding of such surrogates. This book emphasizes when and how a surrogate model can fail in approximating complex functions. One of the best ways to understand the theory of surrogate models is to implement it using a computer program. Throughout this book, numerous Matlab codes are provided as practice tools.

This book is organized into three parts and ten chapters. Part 1 introduces the basics of surrogate modeling. This book considers the polynomial response surface (PRS) and Kriging surrogates as two pillars of surrogate modeling, as they approach the surrogate approximation in completely different directions. Chapter 2 introduces the traditional PRS surrogate. Even if PRS is a basic surrogate model, profound theories and experiences are available in the literature. The basic assumption of PRS is that the model form is correct, but that the samples have uncorrelated noise. Because of this assumption, the fitting process of PRS is regression, where the surrogate converges to the mean of the samples' trend. Various error measures and ways of evaluating the goodness of fit are introduced in this chapter. The statistical view of regression can provide a good contrast with the Kriging surrogate in Chapter 7. Since the quality of surrogates is significantly affected by the number and location of samples, Chapter 3 presents various designs of experiments. Many standard designs of experiments are developed for linear and quadratic PRS in box-like design space. For general surrogates, various optimal designs are introduced, as well as space-filling designs of experiments. The main questions of design of experiments are (i) how many samples are good enough and (ii) where to locate the samples. This chapter focuses on answering these two questions for various methods of design of experiments.

Part 2 introduces various theories and algorithms of design optimization. Chapter 4 presents engineering optimization problems and their formulations. Optimality criteria are considered the most important optimization theories. Optimality criteria for both unconstrained and constrained optimization problems are discussed. In general, numerical algorithms for solving optimization problems can be categorized into gradient-based and gradient-free algorithms. Chapter 5 summarizes gradient-based optimization algorithms. Most algorithms require determining a search direction and step size. In this process, the gradient and Hessian information play a critical role, which assumes that the objective and constraints are a continuous and smooth function of design variables. Chapter 6 summarizes gradient-free optimization algorithms. They are often referred to as global optimization algorithms, but in this book, they are referred to as global search algorithms. This group of algorithms does not require gradient or Hessian information, and the objective and constraints do not need to be continuous. Some global search algorithms are deterministic, such as the Nelder-Mead and DIRECT algorithms, while others are stochastic, such as genetic algorithms and particle swarm algorithms. But there is no algorithm that can guarantee finding the global optimum design.

Part 3 presents advanced topics in surrogate modeling, which include the Kriging surrogate, neural network models, multi-fidelity surrogates, and efficient global optimization. Kriging surrogates as described in Chapter 7 are the second pillar of this textbook, where the basic assumption is that the model form is uncertain, but samples are accurate with distance-dependent correlation. This yields an interpolating approximation, not regression as is the case in PRS. Although Kriging surrogate is considered one of the most advanced surrogates in many engineering fields, it is challenging to set up the surrogate model form properly and estimate its hyperparameters properly, which will be the focus of Chapter 7. Neural network models in Chapter 8 have gained a lot of popularity due to advances in computing power and deep learning algorithms. This book focuses on uncertainty associated with neural network training and predictions. In the surrogate perspective, neural network models work well in the interpolation region but perform poorly in the extrapolation region. Since obtaining many samples is expensive, multi-fidelity surrogates in Chapter 9 have attracted a lot of attention recently. The basic idea is to combine a small number of high-fidelity samples with many low-fidelity samples (or low-fidelity model itself) to improve prediction accuracy. In practice, however, multi-fidelity surrogates tend to save computational time for obtaining high-fidelity samples but are not able to improve prediction accuracy significantly. Chapter 10 presents an interesting approach to optimization using surrogates and adaptive sampling, which is referred to as efficient global optimization. The basic idea is to use prediction uncertainty information to find the location of an additional sample to search for the best design. The main mechanism is a smart mixture of exploring design space and improving a good design locally (exploitation).

The substances of this book come from portions of the Approximation and Optimization in Engineering course that I taught with Prof. Raphael "Rafi" Haftka for several years at the University of Florida. His unique perspectives toward surrogate modeling are refined into this book, including the viewpoint of error as uncertainty, issues relating to extrapolation in exploring design space, detailed understandings of assumptions in different surrogates, and situations in which a surrogate is more accurate than samples. A personal goal of this textbook is to leave such unique perspectives in an archived form.

Many examples and exercise problems were derived from students' assignments when I taught this substance at the University of Florida. I am thankful to the students who took the course for their valuable suggestions, especially regarding the example and exercise problems.

March 2025

Nam-Ho Kim

## Acknowledgment

This work was supported by the National Research Foundation of Korea (NRF) grant funded by the Korea government (MSIT) (No.RS-2024-00341872).

## About the Companion Website

This book is accompanied by a companion website:

**www.wiley.com/go/kim/surrogate_modeling**

Scan this QR code to visit the companion website

**This website include:**

- Solution manual
- PPT
- Other

# Part I

# Basics of Surrogate Modeling

# 1

# Introduction to Surrogate Models

## 1.1 What Is Surrogate Modeling?

In engineering design, a quantity of interest (QoI) is a function of many input parameters or design variables. For example, the maximum stress in a cantilevered beam is a function of the beam length, cross-sectional dimensions, and applied loads. Such a functional relationship is often governed by complex physics, and it may not exist in a simple analytical form. Many engineering mechanics problems are governed by integro-differential equations, which often require computationally intensive numerical methods to find the relationship. Finite element methods, computational fluid dynamics, and rigid-body dynamics comprise a short list of examples of numerical methods used to establish the relationship between QoI and input parameters or design variables. Even if modern technology has enhanced computational power significantly, engineers still anticipate a significant amount of computational time to calculate the QoI.

Figure 1.1(a) shows an example of dispersed multiphase flow, where numerous particles interact with shock waves generated from explosives [1]. Physics of shock interaction with a cloud of particles has many interesting applications, such as explosive volcanic eruptions, dust explosions in coal mines, and supernovae [2]. Simulation of shock interaction with distributed particles, as shown in Figure 1.1(b), creates many computational challenges. Problems involving distributed particles present a wide range of length scales, from hundreds of nanometers (shock thickness) to meters (size of the domain) and an associated wide range of time scales from microseconds to minutes. Given this wide range of length and time scales, fully resolved particle simulations of dispersed multiphase flows involving a large number of particles are practically impossible, where the complete details of both the flow around the particles and the particle motion are captured. Instead, Euler–Lagrange formations are often adopted with point-particle force and heat transfer models to couple the particle momentum and energy with those of the surrounding fluid. While the governing Navier–Stokes equations are solved in the region outside the particles with a sufficiently high resolution in the Eulerian frame of reference, the motion of each particle is tracked by solving its equations of motion in the Lagrangian frame of reference. Even such an approximate model pushes the limit of current computational resources. The simulation in Figure 1.1(b), for example, took 15,000 seconds with 2,400 clusters in a high-performance computing facility at US National Laboratories.

In many engineering processes, multiple simulations are frequently required by changing input parameters or design variables. To find the optimal airfoil shape, for example, an engineer simulates the airflow around a wing for different design variables, such as length, curvature, aspect ratio,

*Surrogate Modeling and Optimization: Theories, Applications, and Limitations*, First Edition. Nam-Ho Kim.
 Published 2026 by John Wiley & Sons, Inc.
Companion website: www.wiley.com/go/kim/surrogate_modeling

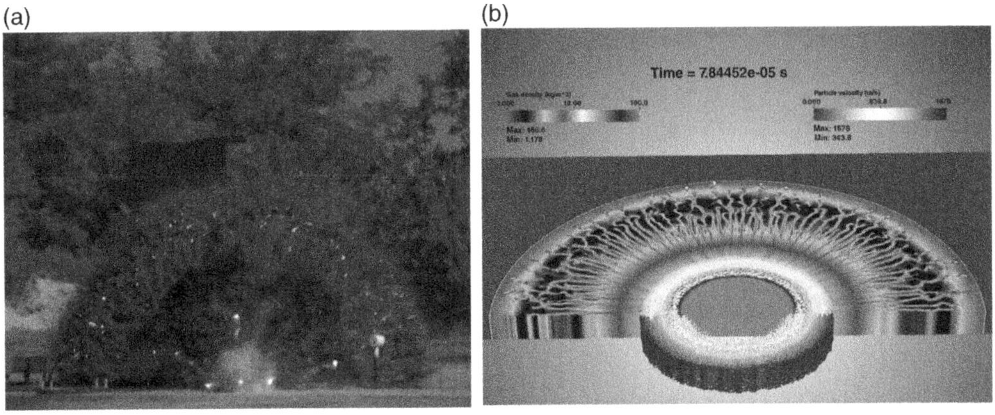


**Figure 1.1** Experiment and simulation of dispersed multiphase flow: shock wave interaction with a particle bed. (a) Experiment at Eglin Air Force Base and (b) simulation at the University of Florida.

chord length, and even material. During design optimization, design variables are updated at each iteration to improve the objective function while satisfying constraints. Commonly, the design variables are updated more than 100 times before reaching an optimum design. Since the objective function and constraints need to be recalculated with each updated design, numerous simulations are required during the design optimization process. Other examples are design space exploration, sensitivity analysis, what-if analysis, reliability analysis, and uncertainty quantification. All of these examples require numerous simulations by changing input parameters or design variables. As shown in the previous example, if a single simulation is already computationally expensive, it becomes easily impractical to run numerous simulations.

Even if the process of calculating QoI is governed by complex physics and expensive numerical simulations, the relationship between the QoI and input parameters can be simplified. In the case of a cantilevered beam shown in Figure 1.2, for example, the maximum stress can be written as

$$\sigma_{\max} = \frac{6FL}{bh^2} \tag{1.1}$$

where $F$ is the applied force at the tip, $L$ is the length of the beam, $b$ is the width, and $h$ is the height of the cross-section. Even if in this simple example the QoI (maximum stress) is expressed in an analytical form, we may consider that it is from complex physics and expensive numerical simulation. However, the relationship between maximum stress and applied force is linear, $\sigma_{\max} = \alpha F$, and it is inversely proportional to the height of the beam, $\sigma_{\max} = \beta h^{-2}$. Therefore, without considering complex physics, it is possible to express the relationship using simple functions,

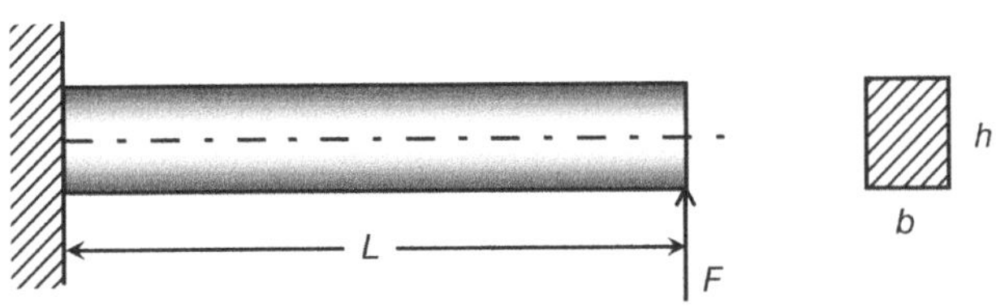


**Figure 1.2** Bending of a cantilevered beam.

such as $\sigma_{max} = f(F, L, b, h)$. However, the issues of determining the functional form and calculating the coefficients remain.

Surrogate modeling is a powerful engineering tool when a QoI cannot be easily calculated (or measured), so an approximation of the QoI is used instead. It mimics (i.e., approximates) the behavior of the simulation model as closely as possible while being computationally cheap in evaluating the QoI. It is a so-called data-driven approach where it does not concern physics or underlying theory. Instead, it only concerns input–output behavior, and the relationship is represented using mathematical functions. This approach is also known as a response surface model [3], meta-model [4], reduced model [5], model emulator [6], proxy model [7], lower-fidelity model [8], behavioral modeling, or black-box modeling in different communities. When only a single input parameter is involved, the modeling process is known as curve-fitting.

## 1.2 Surrogate Models

The idea of surrogate modeling is to represent the relationship between the QoI and input parameters using known functional forms with unknown function coefficients. Different surrogate models can be defined based on how to select the functional forms and how to determine unknown coefficients. However, all of them share some common characteristics. As will be discussed in more detail later, the most important characteristic relates to samples. All surrogate models utilize samples to determine the unknown coefficients. Samples in a surrogate model mean data between input parameters and output QoIs. The dimension of a surrogate model refers to the number of input variables. Figure 1.3(a) shows an example of samples in a two-dimensional surrogate model. Each sample consists of the paired values of input parameters and output QoI. Let input parameters be $x_1$ and $x_2$ and output QoI be $y$. Then, a sample can be represented by $S = (x_1, x_2, y)$. In general, the $k$-th sample of $N$-dimensional surrogate can be written as $S^{(k)} = (\mathbf{x}^{(k)}, y^{(k)})$, where $\mathbf{x} = \{x_1, x_2, ..., x_N\}^T$ is the vector of input sample locations. In this text, a sample refers to $S^{(k)}$, while a sample location refers to $\mathbf{x}^{(k)}$. The example in Figure 1.3(a) has 21 samples. For a given input sample location, $\mathbf{x}^{(k)}$, it is necessary to evaluate the output QoI sample $y^{(k)}$, which requires expensive numerical simulation or a physical experiment. The major cost of the surrogate model occurs at this stage; that is, each sample requires simulation/experimentation. Once $m$ numbers of samples, $\mathbf{S} = \{S^{(1)}, S^{(2)}, ..., S^{(m)}\}$,

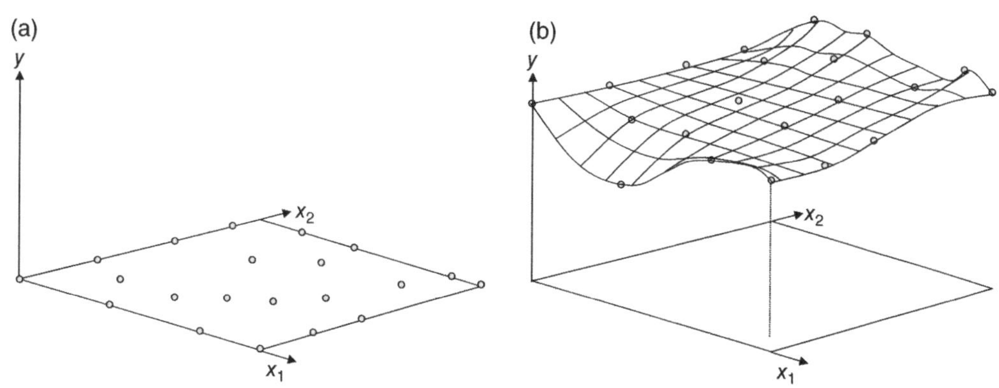


**Figure 1.3** An example of samples in a 2D domain and the corresponding surrogate model. (a) Sample locations and (b) samples and fitted surrogate model.

are obtained, Figure 1.3(b) shows a surrogate model that fits using these samples. The surrogate model can be simple polynomials or complicated statistical functions. However, it is assumed that the cost of constructing the surrogate model and evaluating the output QoI using the surrogate model is assumed to be negligible compared to that of evaluating the output QoI samples $y^{(k)}$. Under this assumption, the cost of a surrogate model can be determined based on the number of samples to construct it. In general, the accuracy of a surrogate model is proportional to the number of samples. Therefore, there is a trade-off between the accuracy of a surrogate model and the cost of constructing it.

Surrogate modeling began in field experiments where the trend of a response change was measured by varying input parameters [3]. Normally, the input parameters were varied by two or three levels, and the change in the response was modeled using linear or quadratic polynomials. Since measured data from an experiment normally contained noise, it was important that the surrogate model filters out the measurement noise. The polynomial response surface (PRS) was designed for this particular purpose, where the functional form was assumed to be a linear combination of monomial bases with unknown coefficients. These coefficients were determined to minimize the error between the surrogate prediction and measured data. The errors between the surrogate prediction and measured data are considered random noise. Therefore, the surrogate prediction is different from the data due to the presence of noise. The fundamental assumption of PRS is that the form of the surrogate model is correct, but the data have random noise. Figure 1.4(a) shows an example of a 1D PRS surrogate where four samples were used to fit a quadratic function. As expected, the PRS surrogate does not pass through the sample points. Instead, the surrogate prediction distributes positive and negative errors, such that the mean of the errors becomes zero (unbiased). Indeed, this is the process of determining the unknown coefficients of the PRS surrogate. Figure 1.4(a) also shows confidence intervals (grey area). We will discuss the PRS surrogate in more detail in Chapter 2.

As computer simulation was greatly advanced due to the development of powerful computers, people started to construct a surrogate model out of computer simulation results rather than physical experiments, which are often called numerical experiments. The data generated from computer simulations have different characteristics from that of physical experiments. Even if a computer

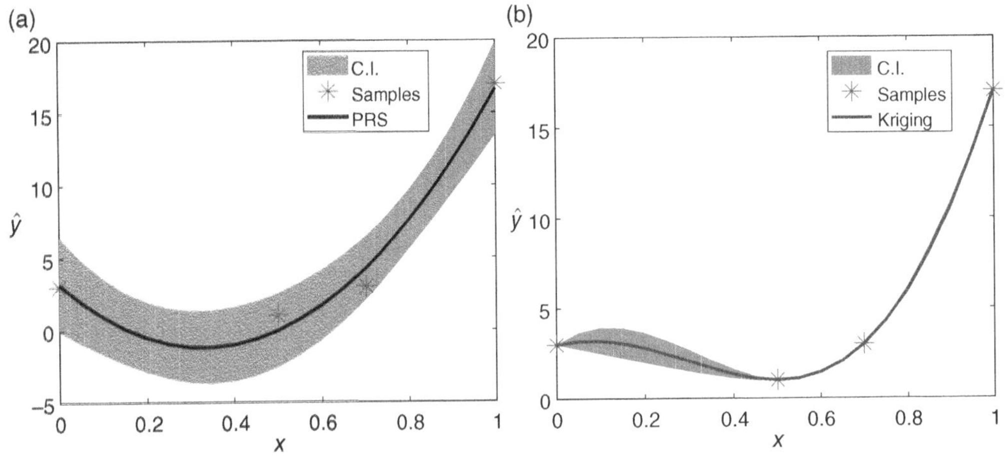


**Figure 1.4** Polynomial response surface versus Kriging surrogate models. (a) Polynomial response surface and (b) Kriging surrogate.

simulation has random noise, it is often from numerical truncation errors, which is negligible compared to other types of errors. On the other hand, simulation data would not be exact due to various assumptions in modeling and numerical errors. These assumptions and errors tend to generate bias-like predictions, i.e., predictions are shifted from the true data. However, since the goal of surrogate modeling is to approximate simulation results, not the true function, the simulation data are considered exact, and the difference between surrogate prediction and simulation data is considered a prediction error, not noise. Thus, surrogates that predict the simulation data exactly are preferred over ones that cancel noise at data points. For example, the Kriging surrogate model, which is also called the Gaussian process model, is popular for computer simulation. The fundamental assumption of the Kriging surrogate is that data are inherently exact, but the model form has uncertainty. Due to this assumption, the Kriging surrogate predicts the simulation data exactly, but the uncertainty increases as a prediction point is farther away from data points. Figure 1.4(b) shows an example of a Kriging surrogate, where the same four samples were used to fit the surrogate. Since the Kriging surrogate considers that there is uncertainty in the model, the prediction is represented by the mean curve (red-colored curve) with 95% confidence intervals (curves with the green area). Notably, that the prediction uncertainty disappears at sample locations because the Kriging surrogate considers data to be exact, i.e., that there is no uncertainty at data points. We will discuss the Kriging surrogate in more detail in Chapter 7.

As will be discussed in subsequent chapters, there are many different surrogates, including polynomial response surface, Kriging, radial basis function, support vector machine, space mapping, artificial neural network, polynomial chaos expansion, and Bayesian network. Which surrogate model is the best? This is a question without an answer because often the comparison is invalid and meaningless. As discussed previously, the PRS and Kriging surrogates have different assumptions. Therefore, it is obvious that one will be better than the other when one's own set of assumptions is satisfied. For example, if the PRS surrogate has the exact model form and the data have Gaussian noise, it may perform better than the Kriging surrogate. On the other hand, if the data are accurate but the PRS surrogate does not have an exact model form, the Kriging surrogate may perform better than the PRS surrogate. When the nature of the true function and its data are unknown, it is not clear which surrogate model will be the most accurate. However, if more information is available, it is possible to find a good surrogate that matches the available information. For example, when the nature of the true function is known, physics-based surrogates such as space-mapping-based models might be most efficient. When a set of data are given, it is difficult to suggest the "best" surrogate. In fact, there is no consensus on how to obtain the most reliable estimates of the accuracy of a given surrogate. Instead of selecting the best surrogate, it might be possible to eliminate bad surrogates. Goel et al. [9] suggested using an ensemble of multiple surrogates to reduce the risk of using a bad surrogate. This is a particularly attractive idea because we assume that the cost of constructing a surrogate is negligible compared to obtaining data. Therefore, it is relatively easy to build multiple surrogates and remove those predictions far away from other surrogates.

## 1.3 Design of Experiments: Sampling

As the purpose of surrogate modeling is to approximate a QoI using a set of samples, the goal is to make the surrogate model as accurate as possible using as few samples as possible. To achieve this goal, the first task is to determine how many samples to use and where to locate the samples. The methodology of selecting samples is called "design of experiments (DoE)" or "experimental design." This terminology was developed because PRS was initially designed to approximate responses from experiments. Since this is an important topic in surrogate modeling, Chapter 3 is dedicated to discussing the design of experiments. It is generally accepted that DoE is as important as the surrogate itself because the accuracy of the surrogate depends on the number and location of samples. In addition, there is no single DoE scheme that is the best for all surrogates. Various DoE techniques cater to different sources of errors, in particular, errors due to noise in the data or errors due to an improper surrogate model.

Samples do not need to be generated simultaneously. After starting from a small number of samples, additional samples can be added sequentially to improve the accuracy of surrogate prediction. The strategy of adding additional samples is called adaptive sampling, sequential design, or active learning. Usually, adaptive sampling strategies require a criterion to determine the next sample locations. Many criteria have been proposed, such as D-optimal design and minimum bias design. These criteria are often based on a trade-off between minimizing bias (i.e., error) and/or minimizing variance (i.e., uncertainty).

When a surrogate model is sufficiently flexible, a better surrogate can be obtained with more samples. In addition, it is better to locate the samples such that they are more or less uniformly distributed to the entire sampling space, which is called "space-filling design." To have a space-filling design, some sampling strategies minimize the maximum distance between samples. Others uniformly divide the sampling space and put a sample in each segment.

## 1.4 Interpolation Versus Extrapolation

Although it is better to distribute samples uniformly to the entire sampling space, it is often impossible to generate enough samples. This is particularly the case when the dimension of input variables is high. To illustrate, consider two-dimensional space, as shown in Figure 1.5(a). First, the range of each input variable is halved so that there are four quadrants. Assume that a sample is

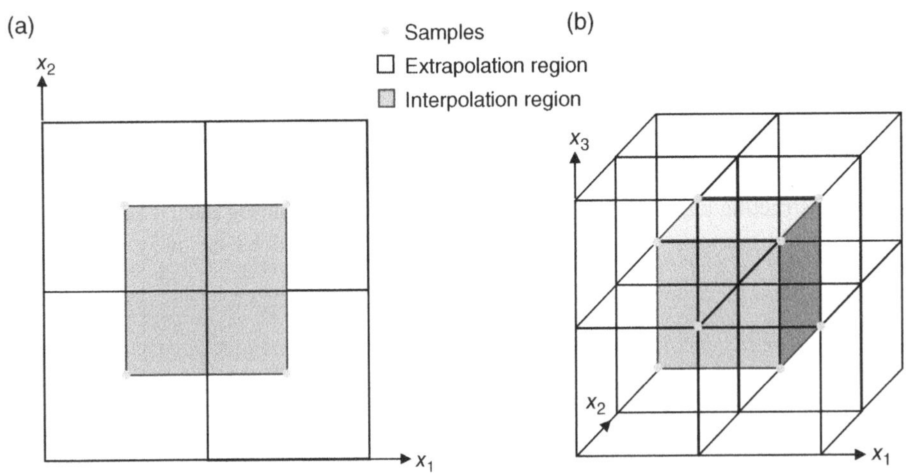


**Figure 1.5** Interpolation versus extrapolation region in the sampling space. (a) 2-dimensional space and (b) 3-dimensional space.

located at the center of each quadrant. Therefore, in two-dimensional space, there are four samples. Figure 1.5(b) shows the same practice in three dimensions, now with eight samples—one sample at each octant. If the dimension of input variables increases, the number of samples increases rapidly. For example, in the case of 10-dimensional space, 1,024 samples are required to locate a sample to each orthant (orthant is hyperoctant, which is the analog in $n$-dimensional Euclidean space of an octant in three dimensions). Therefore, the number of samples increases exponentially, proportional to the number of dimensions, which is called the "curse of dimensionality."

A more disruptive fact is that one sample at each orthant is not enough for space-filling purposes. To explain, consider the difference between "interpolation" and "extrapolation." Even if there are several different definitions, this book considers that interpolation means that the prediction point is within the convex hull of samples. If the prediction point is out of the convex hull, it is referred to as extrapolation. In general, the prediction in the interpolation region is more accurate than that of the extrapolation region. Therefore, it is natural to generate samples such that they cover the input space as much as possible, i.e., maximizing the interpolation region. The challenge is that in general, the extrapolation region is much larger than the interpolation region. For example, in the two-dimensional case in Figure 1.5(a), the interpolation region is in the grey area, while the extrapolation is outside of that. In the case of a 2-dimensional space, the interpolation region only covers 25% of the entire input space. This portion rapidly decreases as the dimension increases. For example, in the 3-dimensional space in Figure 1.5(b), the interpolation region covers 12.5%, and the 10-dimensional space covers less than 0.1% of the input space. Therefore, it would be hard to rely on a prediction that comes from 99.9% extrapolation.

The issue of extrapolation is a major challenge in surrogate modeling. This is a major problem in both one- and multi-dimensional design space. To illustrate the issue dramatically, consider the following one-dimensional function:

$$y(x) = x + 0.5\sin(5x) \tag{1.2}$$

The function is defined in the domain $x \in [0, 10]$. To fit the function, 21 equally-spaced samples were generated in the region $x \in [1, 9]$. Therefore, the interpolation region is $x \in [1, 9]$, while the extrapolation region is $x \in [0, 1] \cup [9, 10]$. That is, 80% of the domain is the interpolation region, while 20% is the extrapolation region. The following Matlab code is used to generate samples, fit the surrogate, and plot the surrogate and true functions:

```
x=linspace(1,9,21);
y=x+0.5*sin(5*x);
net=newrb(x,y);
xf=linspace(0,10,101);
yf=xf+0.5*sin(5*xf );
ysim=sim(net,xf );
plot(xf,yf ); hold on;
plot(xf,ysim,'ro');
```

For this particular example, the true function in Eq. (1.2) is used to generate samples; that is, there is no error or noise in the samples. Figure 1.6 shows the true function and surrogate prediction along with the interpolation and extrapolation regions. It turns out that surrogate predictions are accurate in the interpolation region. In fact, the surrogate predictions were exact in this region. In the extrapolation region, however, the surrogate predictions showed significant errors. Considering that the extrapolation region is only 20% of the entire domain and the true function shows a repeated pattern, the surrogate model completely loses its prediction capability.

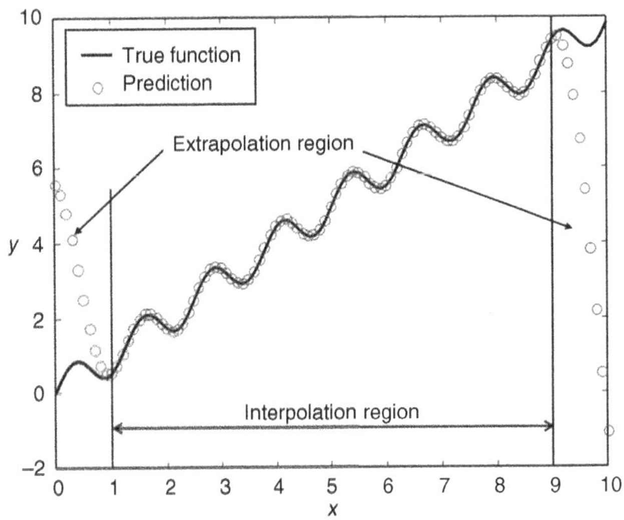


**Figure 1.6** Fitting a function using a radial-basis neural network model in Matlab.

A surrogate with a good fit in the interpolation region can fail miserably in predicting the extrapolation region. This is one of the biggest challenges associated with surrogate modeling. In general, it is difficult to make the entire input domain the interpolation region, i.e., samples are located in all vertices of the input domain. As shown in Figure 1.5, this will increase the number of samples exponentially, along with the dimension of input parameters. It is also possible that an interpolation region may have significant error if the distance between samples is large. For example, consider the two-dimensional space in Figure 1.5(a). If samples are located at all corners, the maximum distance between samples is $\sqrt{2}$, assuming that all input variables are normalized between [0, 1]. This distance increases to $\sqrt{3}$ for three dimensions, and in general, $\sqrt{n}$ for $n$ dimensions. As the distance between the samples increases, the prediction accuracy decreases, which is similar to the behavior seen in the extrapolation region.

## 1.5 Flowchart of Surrogate Modeling

Since we discussed some important tasks of surrogate modeling individually, it might be beneficial to discuss the flowchart of general surrogate modeling. Figure 1.7 shows a flowchart of surrogate modeling. Although there are many different ways of describing the surrogate modeling process, we use four steps in this text. The first step is the design of experiments (DoE), where the number and location of samples is planned in the input variable space. Normally, the input variables are normalized, and thus, the input variable space is a hyper-cube. Then, DoE distributes samples in this domain. Since there are infinitely many possibilities for locating samples, it is important to assess the goodness of different DoEs. Also, since each sample comes from an expensive computer simulation or experiment, the number of samples is limited.

The second step is to run simulations or experiments at selected DoE locations. As previously mentioned, it is assumed that numerical simulations or experiments are expensive compared to fitting or evaluating the surrogate. When computer simulations are used to evaluate the QoI at sample locations, it is possible that parallel processing can be used to save computational time. It is also noted that the obtained QoI at sample locations might include random noise or error.

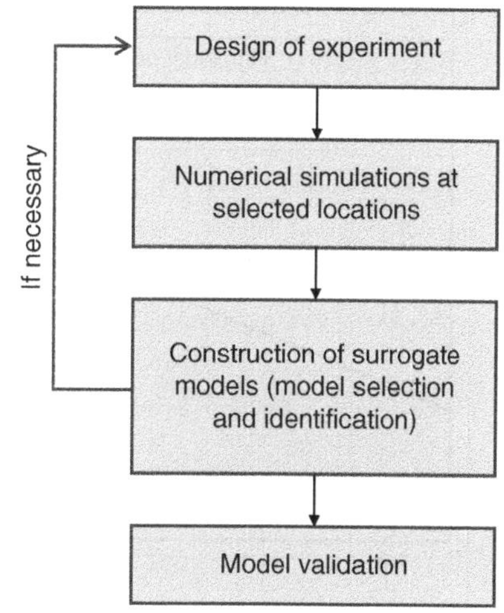


**Figure 1.7** Flowchart of surrogate modeling.

The third step is to build a surrogate using samples. This step can be broken down into two stages. The first is to select a surrogate model form, and the second is to determine unknown coefficients using samples, which are also called model identification. The model selection is closely related to the number and location of samples. A flexible model usually comes with greater numbers of unknown coefficients, which requires more samples. It is always a good strategy to start with a simple model and progressively increase its complexity. Model identification is the process of determining unknown model parameters by minimizing the errors between the surrogate predictions and samples. Depending on how to define the error, different outcomes can be obtained.

The last step is model validation, which evaluates the accuracy of the constructed surrogate model. There is a philosophical challenge in the validation step. Samples are used to fit the surrogate model, which means the surrogate model is constructed to minimize the errors between the surrogate predictions and samples. Therefore, it is highly likely the errors are small at sample locations. However, this does not mean that the surrogate is a good predictor at unsampled locations. For example, Figure 1.8 shows a surrogate that passes all sample points but has poor prediction capability. The five samples were generated from a linear function $y = x$, $x \in [-2, 2]$, while the PRS was assumed to be a quintic polynomial, which has six unknown coefficients. Since the number of unknown coefficients is greater than the number of samples, the fitting process is indefinite, and it is possible that there are infinitely many combinations of coefficients whose curves pass through all the sample points. Figure 1.8 illustrates one particular case where $y = -2x + 3.75x^3 - 0.75x^5$. In such a case, the surrogate predictions are exact at all sample points. However, the predictions at unsampled locations are very poor in this example. If only samples are used to evaluate surrogate, it may decrease the accuracy of the surrogate.

One way of resolving this issue is to save some samples for the purpose of validation. That is, out of $m$ samples, only $k < m$ samples are used for fitting the surrogate, and $m - k$ samples are used for validation. For example, the implementation of the neural network in Matlab uses 70% of the samples for fitting the network while the remaining 30% is used for testing and validation. However, one may get a better surrogate using $m$ samples than using $k$ samples. Therefore, it is good to save $m - k$ samples for the purpose of validation, at the cost of deteriorating the accuracy of the surrogate. One possibility is that $m - 1$ samples are used to fit the surrogate, and the other sample is used to measure the prediction error. This process can be repeated $m$ times by dropping one sample at a time to obtain $m$ prediction errors. These $m$ errors can represent the prediction capability of the

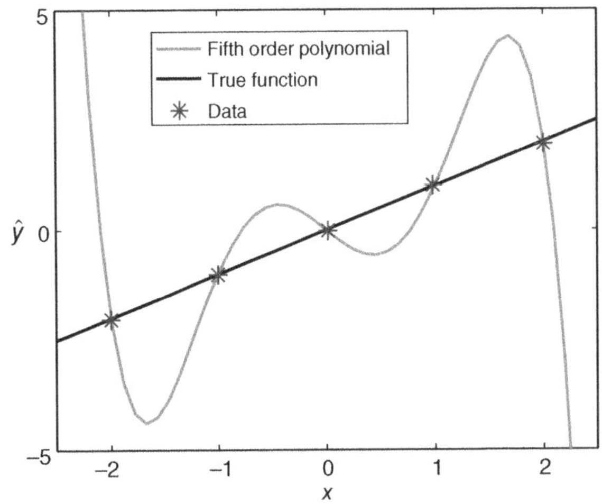


**Figure 1.8** A surrogate model that is perfect at sample locations but has a large prediction error at unsampled locations.

surrogate. This process is called "cross-validation." The advantage of cross-validation is that no samples are wasted for the purpose of validation.

If the model is not sufficiently accurate, it is necessary to add additional samples (adaptive sampling) and reconstruct the surrogate model. If the model form limits accuracy, it can also be modified. For example, if a linear PRS surrogate cannot represent the trend of data, a higher-order PRS can be used to match the data trend. This remodeling process requires several iterations until the validation error becomes less than a threshold.

## 1.6 Overview of Surrogate Modeling

With the flowchart of surrogate modeling in Figure 1.7 in mind, Figure 1.9 shows the overview of surrogate modeling. The simulation-based model, $f(\mathbf{x})$, represents an expensive numerical simulation or physical experiment. Even if the functional form of the simulation model is hidden or it is too complicated to be used, it is possible to evaluate the simulation model at a given sample location $\mathbf{x}_i$, $f_i = f(\mathbf{x}_i)$; this process is referred to as the analysis problem. The outcomes of the analysis problem, $f_i$, $i = 1, \ldots, m$, are called either samples or data. In general, since the analysis problem is considered computationally expensive, it is expected to have a small number of samples. Once $m$ samples are available, they are used to build the estimated model, $\hat{f}(\mathbf{x})$, which is also called an approximate model or a surrogate model. The estimation process can be understood as a nonlinear inverse problem. That is, we want to determine a continuous function, $\hat{f}(\mathbf{x})$, as a function of input variables from a limited amount of data, $\mathbf{f} = \{f_1, f_2, \ldots, f_m\}^T$. For example, the function can be determined by minimizing the error between the data and prediction as

$$\min_{\hat{f} \in H} \hat{f} = \frac{1}{m} \sum_{i=1}^{m} L\left[f_i - \hat{f}(\mathbf{x}_i)\right] \tag{1.3}$$

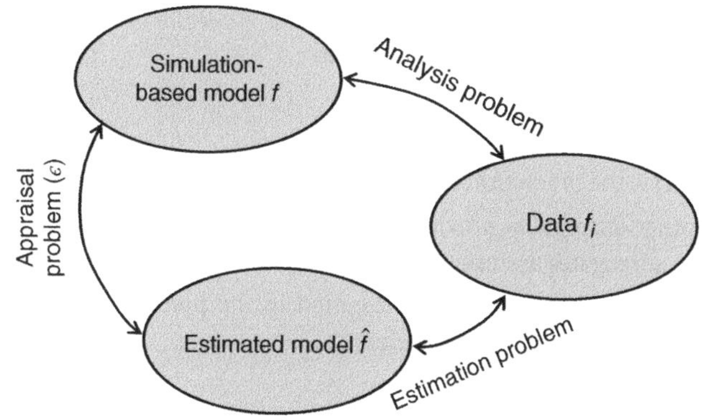


**Figure 1.9** Overview of the surrogate modeling process.

where $L[\cdot]$ is a loss function used to quantify the empirical error, and $H$ is the family of surrogate models under consideration. The loss function increases if there is a discrepancy between the data and prediction. Different outcomes are expected when different definitions of the loss function are used.

In addition to the nonlinear inverse estimation problem, the data may be exact or may include noise. Therefore, the estimation problem usually does not have a unique solution. That means, it is possible to have multiple estimated models. Such possibility of multiple models is referred to as "prediction uncertainty" in this text. For example, if the multiple estimated models are represented by a Gaussian distribution, the surrogate prediction can be expressed in terms of the mean and variance. Once the surrogate model is constructed, it is required to evaluate the accuracy of the model, which is called the appraisal problem. The appraisal problem uses several definitions of error, $\epsilon$, to assess the goodness of the estimated model against the simulation-based model. As discussed in Figure 1.7, if the appraisal error is not satisfactory, additional samples are added, or the model form is changed to improve the prediction accuracy.

An introductory level of understanding surrogate modeling would be to check if the surrogate is accurate or not at sample locations. The next level would be to understand the accuracy at unsampled points (i.e., prediction points). The accuracy at prediction points indeed represents the prediction capability of the surrogate. In order to assess the accuracy at prediction points, the concept of "prediction uncertainty" is often used. Consider the one-dimensional surrogate model (blue-colored curve) that fits five samples in Figure 1.10. The surrogate has interpolation capability such that the surrogate passes through all sample points; that is, the prediction error is zero at sample locations. However, there is no proof that the surrogate is accurate at un-sampled

**Figure 1.10** Prediction uncertainty in a surrogate.

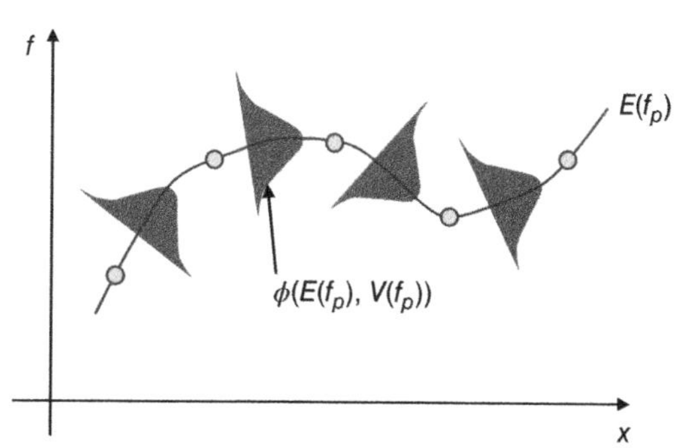

locations. Therefore, it is natural to assume that there is a prediction error at unsampled locations. A rigorous method of expressing surrogate prediction can be written as

$$f_p(\mathbf{x}) = \hat{f}(\mathbf{x}) + \epsilon(\mathbf{x}) \tag{1.4}$$

where $\hat{f}(\mathbf{x})$ is the surrogate model and $\epsilon(\mathbf{x})$ is the prediction error. Some surrogate models only provide information regarding $\hat{f}(\mathbf{x})$, but other surrogates provide additional information regarding $\epsilon(\mathbf{x})$. For example, both PRS and Kriging surrogates assume that the surrogate prediction is uncertain with $\hat{f}(\mathbf{x})$ being the mean prediction. The uncertainty is presented in the prediction error, which is often represented by a normal distribution with zero mean:

$$\epsilon(\mathbf{x}) \sim N(0, \sigma^2(\mathbf{x})) \tag{1.5}$$

where $\sigma^2(\mathbf{x})$ is called the prediction variance. In this perspective, the surrogate prediction is a distribution with the expected value $E(f_p) = \hat{f}(\mathbf{x})$ and variance $V(f_p) = \sigma^2(\mathbf{x})$.

Considering the mean and variance of surrogate prediction, the quality of a surrogate can be evaluated in two different aspects. The first is based on the accuracy of the mean prediction, while the second is based on the prediction uncertainty. Even if a mean prediction shows a good accuracy compared with samples, a large prediction uncertainty may indicate that it is possible that the true QoI can be far away from the mean prediction. Therefore, a good surrogate should have a good mean prediction as well as a small prediction uncertainty.

## 1.7 Smoothness and Loss Function

As mentioned previously, surrogate modeling can be considered a nonlinear inverse problem where a continuous function is identified by using a finite number of samples. Since the inverse problem does not have a unique solution, it is possible to have infinitely many surrogates that satisfy the inverse problem. For example, Figure 1.11 shows that two surrogate models pass through

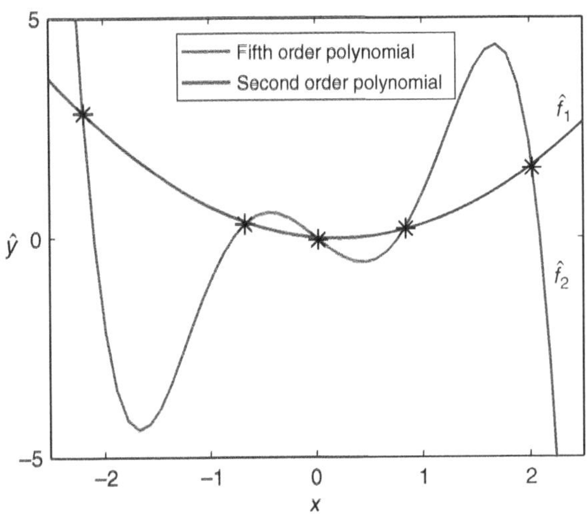


**Figure 1.11** Different surrogate that fits the same samples.

the five samples. Since both surrogates pass through the samples, the fitting errors of both surrogates are zero. Therefore, it is unclear which surrogate is better.

If multiple surrogates have the same fitting error, a smoother one is considered a better one. In the case of the two surrogates in Figure 1.11, the second-order polynomial is considered to be smoother than the first-order one. Although both surrogates are a smooth function of input variable $x$, the blue surrogate changes the sign of the slope less than that of the red one. Therefore, smoothness is related to the change in the sign of the slope, i.e., gradient. When a PRS surrogate is used, a surrogate with lower polynomial orders is considered smoother than the one with higher polynomial orders. For example, a cubic polynomial function has one inflection point, while a quartic polynomial function has two. Therefore, a polynomial function with a smaller number of inflection points is considered smoother.

The nonlinear inverse estimation problem in Eq. (1.3) can be modified such that a smoother model can be preferred. Knowing that smoothness is related to the change in the sign of the slope, a function with high-order derivatives can be penalized during the estimation process. That is, the following formulation can be used instead of Eq. (1.3):

$$\min_{\hat{f}\in H} \hat{f} = \frac{1}{m}\sum_{i=1}^{m} L\left[f_i - \hat{f}(\mathbf{x}_i)\right] + \lambda \int \left\| D^k \hat{f} \right\|_H d\mathbf{x} \tag{1.6}$$

where the error is represented in the loss function, while the smoothness is represented using $D^k\hat{f}$, which is $k$th-order derivative of $\hat{f}$. When a surrogate has higher-order derivatives, it is penalized by the regularization parameter $\lambda \geq 0$. With a large $\lambda$, the estimation process will choose a smooth function even if the function has a relatively large error, which causes a biased error. With a small $\lambda$, the process will choose the most flexible model to fit the data.

In Eqs. (1.3) and (1.6), the loss function $L[\cdot]$ is used to determine the closedness of the model to the data. The loss function is supposed to increase as the error term, $e_i = f_i - \hat{f}(\mathbf{x}_i)$, increases. There are many different forms of loss function. Figure 1.12 shows examples of loss functions. First, the quadratic loss function penalizes a large error using a quadratic function. This is equivalent to the $L_2$-norm in mathematics, the square sum of errors. This is the most commonly used loss function as

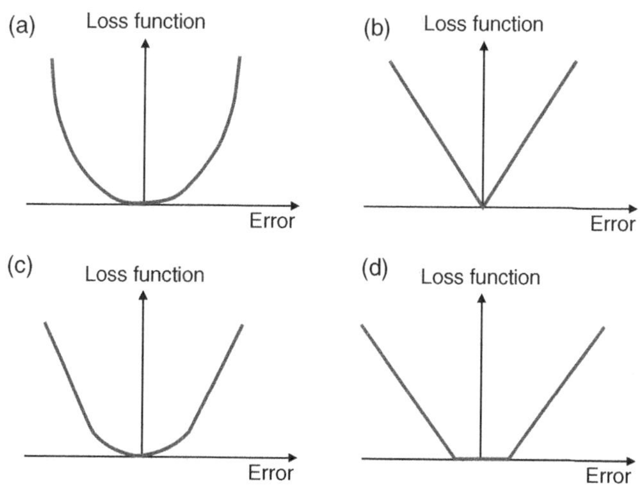


**Figure 1.12** Loss functions. (a) Quadratic loss function ($L_2$-norm), (b) linear loss function ($L_1$-norm), (c) Huber loss function, and (d) $\epsilon$-loss function.

it is a smooth function of error. The quadratic loss function is easy to estimate its parameters because its derivatives become a linear system of equations. However, this function is sensitive to a small number of outliers (i.e., large errors) and tends to ignore the effect of small errors. Second, the linear loss function is also called the Laplace loss function, which is equivalent to $L_1$-norm in mathematics. This basically uses the absolute value of the errors such that the contribution is linearly proportional to the magnitude. The linear loss function is not smooth at the origin. Third, the Huber loss function in Figure 1.12(c) tries to take advantage of both the linear and quadratic loss functions. It is quadratic for small errors while linear for large ones. Therefore, the Huber loss function is smooth at the origin and not too sensitive to large errors. Last, the $\epsilon$-loss function in Figure 1.12(d) is popular in support vector regress surrogates, where errors whose magnitude is less than $\epsilon$ are considered to be zero. This loss function is particularly useful when data includes small random noise, which should not affect the decision-making process for selecting a surrogate.

# 2

# Polynomial Response Surfaces

## 2.1 Introduction

Most optimization algorithms that are in use for solving analytical engineering optimization problems are sequential in nature. That is, the objective function and constraints are evaluated at one point at a time, and the values at that point, as well as previous design points, contribute to a decision on where to move for the next evaluation in the design space.

When the objective functions and (or) the constraints are evaluated by experiments rather than by analytical/computational evaluations, there is usually an incentive to perform the experiments in batches rather than sequentially. One reason for batching the experiments is that most require setup time, advanced planning, and reservations of experimental facilities or technicians. Another reason is that experimental errors make it difficult to interpret the results of a single experiment. When a batch of experiments is performed, errors in one or two experiments tend to stand out. Duplicating experiments for identically nominal conditions permits us to estimate the magnitude of experimental scatter due to errors and variability in the properties of the tested designs. Finally, some of the experimental scatters can be averaged out by performing a large number of experiments.

Because of these advantages of running experiments in batches, experimental optimization has followed a different route from analytical optimization. The standard approach is to use an optimization strategy that is based on the results of a batch of experiments. On the basis of the experiments, we construct approximations to objective functions and/or constraints and perform optimization based on these approximations. In most cases, the optimum obtained is then tested, and if satisfactory results are obtained, the design procedure is terminated. In some cases, the optimum is used as the central design point for a new batch of experiments, and the process is repeated once or twice. This process is sometimes called sequential approximate optimization. However, because of the cost and time associated with conducting experiments, it is rare that the process is iterated to convergence.

When analytical calculations were mostly based on closed-form solutions or numerical models that required minimal modeling and computations, the difference between analysis and experiments was very clear. However, today numerical evaluations of objective functions and constraints often share many of the properties of experimental evaluations. First, numerical models such as finite element structural models require a substantial investment of time to set up and debug. Furthermore, the evaluation of such models may require large computational resources, so that the cost of numerical simulation may be comparable to the cost of experiments. Second, with

*Surrogate Modeling and Optimization: Theories, Applications, and Limitations*, First Edition. Nam-Ho Kim.

Published 2026 by John Wiley & Sons, Inc.
Companion website: www.wiley.com/go/kim/surrogate_modeling

analytical simulations based on complex numerical models, many sources of noise are often found in the results of numerical simulations. These include round-off errors as well as errors due to incomplete convergence of iterative processes. Additionally, numerical simulations are usually based on the discretization of continua, and the accuracy of this discretization depends on the shape of the domains being discretized. For example, when stress analysis is performed in an elastic body using finite element discretization, the discretization error does not change smoothly with the shape of the body, because the number of finite elements can vary only in integer increments.

These growing similarities between analytical simulations and experiments create incentives to run analytical simulations in batches and use sequential approximate optimization. Additionally, the growing availability of parallel computers also provides incentives for running analytical simulations in batches. Finally, numerical simulations are often run with software packages that are difficult to connect directly to optimization programs. Approximate sequential optimization provides a mechanism for running these software packages in a stand-alone mode and connecting the optimization programs to approximation.

The polynomial response surface (PRS) is the oldest form of a surrogate model, where it approximates discrete experimental data (output) with different test parameters (input) using simple polynomials. PRS was invented in the 1920s to characterize crop yields in terms of inputs such as water and fertilizer. It was called the response surface approximation. The algebraic function to fit data is called a surrogate, metamodel, or approximation. The term "surrogate" captures the purpose of the fit: using it instead of experiments for prediction. It is often important when data is expensive and noisy, especially for optimization and uncertainty quantification.

The process of identifying the relationship between the input parameters and the output quantity of interest (QoI) is statistically referred to as regression. The term comes from the paper by Galton [10], where regression was used that happened to be about a phenomenon called regression toward the mean. He found that children of tall parents tended to be shorter than their parents, while children of short parents tended to be taller than their parents.

Through regression, the seemingly complicated behavior of QoI is approximated by simple functions. Normally, linear or quadratic polynomials are used for approximation because the main goal is to find the trend of the QoI as test parameters change [3]. Since most experimental data include measurement noise, the PRS surrogate is designed to compensate for the noise. Therefore, the PRS surrogate normally does not pass through the data points. Instead, the differences between PRS predictions and measured data are assumed to be randomly distributed. In fact, the differences between PRS predictions and measured data are used to estimate the level of noise in the measurement.

In the modern form of PRS, the surrogate approximates the trend of data using a linear combination of polynomials. In this form, each monomial term is referred to as "basis," and the constants that are multiplied by the bases are called unknown coefficients. The major process of building a PRS surrogate is to find the unknown coefficients by minimizing errors between the data and the surrogate predictions at the data points. In statistics, this is called linear regression, where the output QoI is a linear function of unknown coefficients—the basis functions can be nonlinear. In particular, when a single QoI is modeled as a linear combination of basis functions, it is called simple linear regression, compared to multiple linear regression, where multiple QoIs are modeled simultaneously [11]. This is different from multivariate linear regression, where multiple "correlated" QoIs are predicted, rather than a single QoI [12]. In the following discussion, we will only consider simple regression.

In the regression process, it is assumed that the numbers and orders of basis functions are predetermined, and only the unknown coefficients are to be determined. Therefore, the following questions naturally arise: (i) how to determine the numbers and orders of basis functions, (ii) how to determine the unknown coefficients, and (iii) how much data is required to build the surrogate model.

The second question of how to determine the unknown coefficients is the easiest part of surrogate modeling, which will be discussed in Section 2.2 in detail. In general, the unknown coefficients are determined by minimizing errors between the given data and surrogate predictions. It is obvious that the outcome will be different depending on how to define the errors and how to minimize them. Sezction 2.3 will discuss how to estimate the unknown coefficients using linear regression.

The first question of how to determine the numbers and orders of basis functions is subjective and requires domain knowledge of the specific application. Since the success of PRS surrogate modeling depends on the selection of basis functions, this step is critically important, and yet, it is hard to choose an appropriate set of basis functions. For example, when real physics behaves like a sinusoidal function, it would be hard to approximate the function using a set of polynomials. If the domain of input variables is not too wide, the Taylor series expansion can be used to justify the polynomial-based approximation. In general, when the model form of PRS is unknown, stepwise regression can be used, which will be discussed in Section 2.5. This method starts with enough numbers and orders of basis functions and gradually removes those basis functions that are not important or significant for the approximation.

The last question of how much data is required relates to the complexity of the surrogate model and the accuracy of prediction. Since the main usage of samples is to determine the unknown coefficients, the number of samples should be larger than that of unknown coefficients. Theoretically, $m$ numbers of samples should be good enough to determine $m$ numbers of coefficients. However, this corresponds to solving a linear system of $m$ equations, not regression. In general, the required number of samples is at least two- or threefold of that of the unknown coefficients. Not only the number of samples but also the location of samples is important in building a good surrogate model. It is always good that the sample locations are populated in the entire input space while maintaining a small distance between them. However, as shown in Figure 1.5, it is difficult to cover the entire input space with a small number of samples, and it is inevitable to deal with a large extrapolation region, which is indeed the major challenge in surrogate modeling. Since selecting samples is an important topic, Chapter 3 is dedicated to the design of experiments. Therefore, in this chapter, only the first two questions will be addressed.

## 2.2 Curve Fitting

Before discussing the details of PRS, it would be beneficial to discuss curve fitting first, since it is the simplest version of PRS. Consider that experimental data $y_i$ are measured at different values of parameter $x_i$. Then, we want to fit a curve that matches the data best in some sense. When we fit a curve to data, it is necessary to ask the following questions: (i) what is the error metric for the best fit? and (ii) what is more accurate, the data or the fit? Although the answer to the first question is a matter of choice, the second one is subtle and needs to be understood thoroughly. In the following example, we start with the following assumptions: (i) data are noisy, (ii) the functional form of the true function is known, and (iii) data are dense enough to allow us to filter noise in data.

In order to explain the accuracy between the data and the fit, let us generate samples from a linear function $y = x$ at $x = 1, 2, \ldots, 30$. In order to make the data noisy, we can add random noise from a normal distribution $\sim N(0, 1^2)$ to the data. Using randomized data, we can fit a linear polynomial using the `polyfit` function of Matlab, which determines the unknown regression coefficients. Once the `polyfit` function calculates the unknown coefficients, `p`, they can be used in the `polyval` function to evaluate the surrogate at any prediction point, `x`. The following Matlab script generates samples, fits the samples with a linear polynomial, and calculates the root-mean-square (RMS) error of the data and that of the fit.

```
noise=randn(1,30);
x=1:1:30; y=x+noise;
[p,s]=polyfit(x,y,1);
yfit=polyval(p,x);
plot(x,y,'+',x,x,'r',x,yfit,'b');
legend('Data','True function','Fitted function');
RMS_data=sqrt(sum((x-y).^2)/30)
RMS_fit=sqrt(sum((x-yfit).^2)/30)
mean(noise)
```

Due to the randomness of noise, the above Matlab code yields different results at different trials. In this particular case, the fitted polynomial function turned out to be $\hat{y}(x) = -0.2954 + 0.9997x$. Figure 2.1 shows the data (plus markers), fitted function (solid line), and the true function (dashed line). By visual inspection of the figure, the fitted function seems more accurate than the data. This can be checked numerically by calculating the errors of the fit at the data points using either the RMS error or the maximum error. The following equation is used for the RMS error:

$$e_{\mathrm{RMS}} = \sqrt{\frac{1}{m}\sum_{i=1}^{m}\left(y(x_i) - \hat{y}(x_i)\right)^2} \tag{2.1}$$

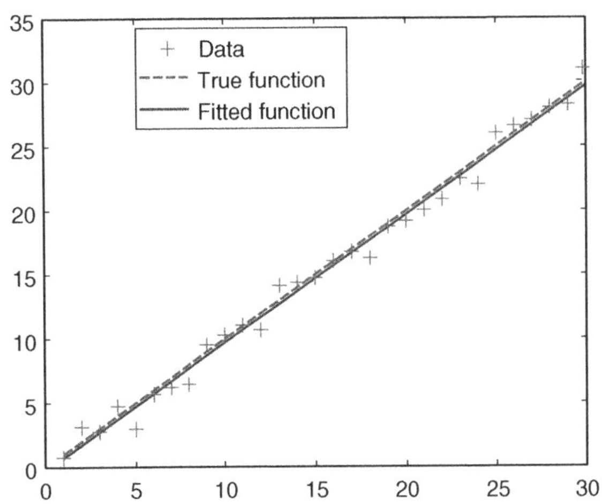


**Figure 2.1** Fitting noisy data with a linear polynomial function.

where $y(x_i)$ is the value of the true function at $x_i$, and $\hat{y}(x_i)$ is that of the fitted function. The RMS error of the data can be calculated by replacing $\hat{y}(x_i)$ with $y_i$. For the given example, $e_{RMS}$ of data was 0.9367, which is close to the standard deviation of noise $\sigma_{noise} = 1$. On the other hand, $e_{RMS}$ of the fitted function was 0.3006. Therefore, the fitted function is more accurate than the data. This happened because the fitting process is based on minimizing the RMS error between the data and the fitted function. When the functional form is accurate, regression serves to filter out noise with dense data.

Since the synthetic data are generated by adding a random noise $\sim N(0, 1^2)$ to the true function, it is questionable why the fitted function is not exact if regression is supposed to filter out noise. This can be explained by the imperfection of the noise samples. Even if the noise samples are randomly generated from the population distribution, $N(0, 1^2)$, the actual samples may not follow it. In fact, the mean of the noise samples was $\mu_{noise} = -0.3005$. This is the culprit of $e_{RMS} = 0.3006$ of the fitted function. That is, the fitted function passes through the mean of the noise. If the same fitting process is repeated with the zero-mean noise, `noise = noise - mean(noise)`, $e_{RMS}$ of data is close to the standard deviation of the samples and $e_{RMS}$ of the fitted function is close to zero. Since the distribution of samples converges to that of the population as the number of samples increases, it is expected that the fitted function becomes accurate as the number of samples increases.

In this elementary example, it is assumed that the true function $y = x$ is known, and the synthetic data are generated by adding random noise. In reality, however, the true function is unknown, and only data are given without knowing the distribution of noise. Therefore, the RMS error of the data and that of the fitted function are not available. The only available information is the error between the data and the fitted function. For the given samples, $e_{RMS}$ between the data and the fitted function was 0.8872, which is the combined effect of the noise and the error in the fitted function. This happened to be smaller than $e_{RMS}$ of data because the mean of the noise was negative. This discrepancy would be reduced as the mean of the noise converges to zero, which can happen as the number of samples increases.

The fact that the fitted function converges to the true function and $e_{RMS}$ converges to the standard deviation of the noise is true only when the functional form of the fitted function is the same as that of the true function. The difference in the functional form between the true function and the fitted function is called the model-form error. When a model-form error is present, the fitting process treats the model-form error as if it is noise because there is no way to distinguish the model-form error from the noise. Therefore, it is critically important to choose a proper model form for successful curve fitting.

In order to investigate the effect of model-form error, 31 equally spaced samples are generated in $x \in [0, 3]$ from a true function of $y(x) = x^2$. The measurement environment is simulated by adding normally distributed noise $\sim N(0, 0.3^2)$ to the samples. The bias in the noise samples is removed by making the mean of the noise samples to be zero. These samples are fitted using a linear PRS with two unknown coefficients, $\hat{y}(x) = a_0 + a_1 x$ using `polyfit` function in Matlab. In this particular case, the fitted function turned out to be $\hat{y}(x) = -1.4131 + 2.9754x$. Figure 2.2 compares the data samples (plus markers), the true function (red curve), and the fitted function (blue curve). It is clear that the regression process determines its coefficients such that the fitted linear function approximates the quadratic trend the best. In the presence of model-form error, however, the error metrics can be difficult to explain. For example, $e_{RMS}$ of data is 0.2932, which is close to the standard deviation of noise, $\sigma_{noise} = 0.3$. However, $e_{RMS}$ of the fitted function is 0.7148. Therefore, in the presence of model-form error, it is possible that the data can be more accurate than the fit. Without knowing the true function, $e_{RMS}$ between the data and the fitted function is 0.7963, which is a combined effect of model-form error and noise in the data.

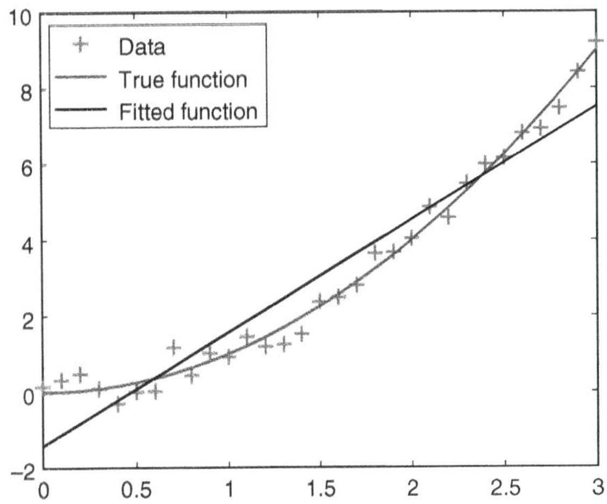


**Figure 2.2** Fitting data with model-form error.

```
noise=0.3*randn(1,31); noise=noise-mean(noise);
x=linspace(0,3,31); ytrue=x.^2; y=ytrue+noise;
[p,s]=polyfit(x,y,1);
yfit=polyval(p,x);
plot(x,y,'+',x,ytrue,'r',x,yfit,'b');
legend('Data','True function','Fitted function');
RMS_data=sqrt(sum((ytrue - y).^2)/31)
noise_std=std(noise)
RMS_fit=sqrt(sum((ytrue - yfit).^2)/31)
RMS_disc=sqrt(sum((y - yfit).^2)/31)
```

Even if it is difficult to identify the model-form error without knowing the true function, there is still a way to understand the presence of model-form error. Since the nature of the noise is random, the errors between the samples and the fitted function are supposed to be randomly distributed. In Figure 2.2, for example, the errors between the data and the true function are indeed randomly distributed. Some errors are positive, while others are negative, and their locations are well-mixed. On the other hand, however, the errors between the data and the fitted function are grouped. That is, most errors are positive when $x \leq 0.5$ or $x \geq 2.5$, while they are mostly negative when $0.5 \leq x \leq 2.5$. This indicates that the fitted function underestimates the data in the ranges of $x \leq 0.5$ and $x \geq 2.5$, while overestimating in the range of $0.5 \leq x \leq 2.5$. Therefore, when a cluster of errors is in the same sign, it is a good indicator that there is a model-form error in the fitted function.

The major source of error in the fitted function in Figure 2.2 is due to the fact that a lower-order polynomial is used for the fitted function. However, increasing the degree of the polynomial fit using `polyfit` does not always result in a better fit. This is because high-order polynomials can be oscillatory between the sample points, leading to a poorer prediction capability between samples. A good example was shown in Figure 1.11, where a quadratic polynomial fits better than a quintic polynomial even if both models pass through all samples. If the level of error is similar, a lower-order polynomial is considered better because it tends to be smoother between samples.

One interesting question would be that if a higher-order polynomial includes the lower-order polynomial, will the fitting results be the same? That is, consider the following quintic polynomial function:

$$\hat{y}^{(5)}(x) = a_0 + a_1x + a_2x^2 + a_3x^3 + a_4x^4 + a_5x^5 \tag{2.2}$$

The question is, if samples are generated from a quadratic polynomial, will the higher-order coefficients, $a_3$, $a_4$, and $a_5$, be zero after the fitting process? Unfortunately, the answer to this question is no because (i) since samples have noise, the quadratic polynomial function may have a larger error than the quintic polynomial function, and (ii) in general, more degrees-of-freedom can yield a smaller error in the fitting process. Therefore, it would be necessary to introduce a fitting strategy to penalize the higher-order polynomials against the lower-order ones.

Polynomials are unbounded and oscillatory functions by nature. Therefore, they are not well-suited to extrapolating bounded data or monotonic (increasing or decreasing) data. The curve fitting process shown in Figure 2.2 only takes into account the accuracy within the range of samples. However, if the accuracy is measured beyond the range of samples, the error in PRS can be significantly increased. In Figure 2.2, for example, it can be easily estimated that the error between the true function and the fitted function can significantly increase when $x < 0$ or $x > 3$.

## 2.3 Linear Regression

Selecting sample locations in the design space where experiments are to be performed is possibly the most important part of obtaining a good approximation to QoI that may be used in optimization and uncertainty quantification. However, this sample selection turns out to be a difficult optimization problem itself and will be discussed in the next chapter. In this section, the second question of how to determine the unknown coefficients of PRS is discussed first.

### 2.3.1 Polynomial Response Surface

In the following explanations, the goal is to approximate the true function $y(\mathbf{x})$ using PRS. In order to determine the unknown coefficients, it is assumed that the functional form of PRS is already determined, and the number of samples $n_y$ are available. The samples are prepared in the form of pairs of input variables and output QoI: $(\mathbf{x}_i, y_i)$, $i = 1, \ldots, n_y$, where $\mathbf{x}_i = \{x_{1i}, x_{2i}, \ldots, x_{ni}\}^T$ is the vector of input variables at the $i$th sample and $y_i$ is the measured/simulated QoI at $\mathbf{x}_i$; i.e., $y_i = y(\mathbf{x}_i)$. For example, in the case of the cantilevered beam problem in Figure 1.2, the input variable is $\mathbf{x} = \{F, L, b, h\}^T$ with $n = 4$ and the output QoI is $y = \sigma_{\max}$.

We want to approximate the true QoI $y(\mathbf{x})$ using a polynomial function $\hat{y}(\mathbf{x}, \boldsymbol{\beta})$, where $\boldsymbol{\beta} = \left\{\beta_1, \beta_2, \ldots, \beta_{n_\beta}\right\}^T$ is the vector of parameters that need to be determined. The relationship between the true function and the approximate function can be given as

$$y(\mathbf{x}) = \hat{y}(\mathbf{x}, \boldsymbol{\beta}) + \epsilon(\mathbf{x}) \tag{2.3}$$

where $\epsilon(\mathbf{x})$ is the approximation error. In this book, we refer to $\hat{y}(\mathbf{x}, \boldsymbol{\beta})$ as a surrogate model. Different surrogate models can be defined depending on how to define the functional form of $\hat{y}(\mathbf{x}, \boldsymbol{\beta})$ and how to determine the unknown model parameters. The vector of unknown parameters $\boldsymbol{\beta}$ often does not have any physical meaning. Rather, we select a functional representation for $\hat{y}(\mathbf{x}, \boldsymbol{\beta})$, with $\boldsymbol{\beta}$ representing some coefficients to be determined so as to fit the data well.

The goal of surrogate modeling is to determine the unknown coefficient $\boldsymbol{\beta}$ so that the approximation error $\epsilon(\mathbf{x})$ is minimized. Therefore, natural questions are (i) what is the form of the approximation function $\hat{y}(\mathbf{x}, \boldsymbol{\beta})$ and (ii) what measure is used to minimize the error $\epsilon(\mathbf{x})$. To address these questions, it is assumed that the true function $y(\mathbf{x})$ is unknown, but we can evaluate it as discrete points. In experiments, for example, even if the functional relationship between input variables and output QoI may not be known, we can still perform experiments to measure QoI by changing input variables. The same is true for complex numerical simulations, where multiple simulations can be performed by changing input variables. Therefore, the first step in surrogate modeling is to perform $n_y$ numbers of experiments to obtain samples: $(\mathbf{x}_i, y_i)$, $i = 1, \ldots, n_y$. Then, Eq. (2.3) can be written at each sample location as

$$y_i = \hat{y}(\mathbf{x}_i, \boldsymbol{\beta}) + e_i, \quad i = 1, \ldots, n_y \tag{2.4}$$

where $e_i = \epsilon(\mathbf{x}_i)$ is the error at the $i$th sample location. Then, the requirement of minimizing error $\epsilon(\mathbf{x})$ is related to minimize the errors at the sample locations.

Two simple examples of the approximate function $\hat{y}(\mathbf{x}_i, \boldsymbol{\beta})$ are

$$\hat{y}(x, \boldsymbol{\beta}) = \beta_1 + \beta_2 x + \beta_3 x^2 \tag{2.5}$$

$$\hat{y}(x, \boldsymbol{\beta}) = \frac{\beta_1}{x + \beta_2} \tag{2.6}$$

The functional form in Eq. (2.5) is a quadratic PRS, while that of Eq. (2.6) is a rational function. It is noted that the approximate function in Eq. (2.5) is a linear function of coefficients, while that in Eq. (2.6) is a nonlinear function of coefficients. Identifying unknown coefficients that are in a linear relationship with the approximate function is called linear regression, as in Eq. (2.5), while that in a nonlinear relationship is called nonlinear regression, as in Eq. (2.6). In this chapter, we will only handle those functional forms that yield linear regression. In particular, when all basis functions are monomial, the surrogate model is referred to as a polynomial response surface (PRS).

In general, the vector of unknown coefficients, $\boldsymbol{\beta}$, is found by minimizing the error so that the approximate function is the best fit. As mentioned before, the approximation error is defined at every sample location. In order to determine the best fit, it is necessary to define a scalar measure of error from the individual errors at sample locations. The following three error measures can be used:

$$e_{\text{RMS}} = \sqrt{\frac{1}{n_y}\sum_{i=1}^{n_y} (y_i - \hat{y}(\mathbf{x}_i, \boldsymbol{\beta}))^2} \tag{2.7}$$

$$e_{\text{av}} = \frac{1}{n_y}\sum_{i=1}^{n_y} |y_i - \hat{y}(\mathbf{x}_i, \boldsymbol{\beta})| \tag{2.8}$$

$$e_{\max} = \max_{n_y} |y_i - \hat{y}(\mathbf{x}_i, \boldsymbol{\beta})| \tag{2.9}$$

The RMS error is most popular because it is a smooth function of $\boldsymbol{\beta}$, but it tends to ignore small errors and emphasize large errors too much. The average error weighs small and large errors equally, while the maximum error considers the largest error only. The average error and maximum error are non-smooth functions of $\boldsymbol{\beta}$, which is the main reason that they are not used often. The errors in Eqs. (2.7)–(2.9) are in fact norms in mathematics. For example, they are $L_2$-norm, $L_1$-norm, and $L_\infty$-norm, respectively. It is noted that the RMS error in Eq. (2.7) is different from the one in Eq. (2.1). The RMS error in Eq. (2.1) is the true error between the true function and the surrogate model, while the one in Eq. (2.7) is the discrepancy between the samples and surrogate model. Unfortunately, the true RMS error cannot be found because we do not know the true function in most cases.

When the approximate function is given as a linear combination of unknown coefficients (e.g., the one in Eq. (2.5)), we can rewrite the surrogate model by focusing on the coefficients as

$$\hat{y}(\mathbf{x}, \boldsymbol{\beta}) = \sum_{i=1}^{n_\beta} \beta_i \xi_i(\mathbf{x}) = \boldsymbol{\xi}(\mathbf{x})^T \boldsymbol{\beta} \tag{2.10}$$

where $\boldsymbol{\xi}(\mathbf{x}) = \left\{\xi_1(\mathbf{x}), \xi_2(\mathbf{x}), \ldots, \xi_{n_\beta}(\mathbf{x})\right\}^T$ is the vector of basis functions. For example, the quadratic polynomial function in Eq. (2.5) has three basis functions: $\xi_1(x) = 1$, $\xi_2(x) = x$, and $\xi_3(x) = x^2$. In general, the number of terms $n_\beta$ and all basis functions are assumed to be given. Therefore, unknown coefficients, $\boldsymbol{\beta}$, are the only undetermined terms in Eq. (2.10).

The exact coefficients, $\boldsymbol{\beta}$, can be found when the number of sample points $n_y$ goes to infinity. That is, when sample points are defined at all the points in the design space, which is impossible because the input variables are real numbers. With a finite number of samples, we can only find the estimate of $\boldsymbol{\beta}$, which will be referred to $\mathbf{b}$ in the following derivations. From the expression $\hat{y}(\mathbf{x}_i, \boldsymbol{\beta}) = \boldsymbol{\xi}(\mathbf{x}_i)^T \mathbf{b}$, Eq. (2.4) can be rewritten in matrix–vector notation as

$$\begin{Bmatrix} e_1 \\ e_2 \\ \vdots \\ e_{n_y} \end{Bmatrix} = \begin{Bmatrix} y_1 \\ y_2 \\ \vdots \\ y_{n_y} \end{Bmatrix} - \begin{bmatrix} \xi_1(\mathbf{x}_1) & \xi_2(\mathbf{x}_1) & \cdots & \xi_{n_\beta}(\mathbf{x}_1) \\ \xi_1(\mathbf{x}_2) & \xi_2(\mathbf{x}_2) & \cdots & \xi_{n_\beta}(\mathbf{x}_2) \\ \vdots & \vdots & \ddots & \vdots \\ \xi_1(\mathbf{x}_{n_y}) & \xi_2(\mathbf{x}_{n_y}) & \cdots & \xi_{n_\beta}(\mathbf{x}_{n_y}) \end{bmatrix} \begin{Bmatrix} b_1 \\ b_2 \\ \vdots \\ b_{n_\beta} \end{Bmatrix} \tag{2.11}$$

$$\mathbf{e} = \mathbf{y} - \mathbf{X}\mathbf{b}$$

In Eq. (2.11), the $(n_y \times n_\beta)$ design matrix $\mathbf{X}$ is defined using the vectors of basis functions at all sample locations. Also, it is often called "regressor" in the statistics community.

Once the errors at sample locations are defined, the regression process determines the unknown coefficients, $\mathbf{b}$, by minimizing these errors. Since $\mathbf{b}$ is determined through the regression process, it is often referred to as regression coefficients. For that purpose, we need a scalar measure of error, instead of the vector of errors. Among the scalar measures of errors given in Eqs. (2.7)–(2.9), we will use the RMS error here. It is noted that the magnitude of $e_{\text{RMS}}$ is not the focus here. Instead, the coefficients $\mathbf{b}$ that minimizes $e_{\text{RMS}}$ is what we want. Using the definition of $e_{\text{RMS}}$ in Eq. (2.7) and that of errors in Eq. (2.11), the RMS error is redefined as

$$e_{\text{RMS}} = \sqrt{\frac{1}{n_y}\sum_{i=1}^{n_y} e_i^2} = \sqrt{\frac{1}{n_y}\mathbf{e}^T\mathbf{e}} \tag{2.12}$$

Therefore, minimizing $e_{\text{RMS}}$ is equivalent to minimizing $\mathbf{e}^T\mathbf{e}$. The square sum of errors can be expressed as

$$\mathbf{e}^T\mathbf{e} = (\mathbf{y} - \mathbf{X}\mathbf{b})^T(\mathbf{y} - \mathbf{X}\mathbf{b}) = \mathbf{y}^T\mathbf{y} - \mathbf{y}^T\mathbf{X}\mathbf{b} - \mathbf{b}^T\mathbf{X}^T\mathbf{y} + \mathbf{b}^T\mathbf{X}^T\mathbf{X}\mathbf{b} \tag{2.13}$$

In the above equation, it is noted that $\mathbf{y}^T\mathbf{X}\mathbf{b} = \mathbf{b}^T\mathbf{X}^T\mathbf{y}$ because both are scalar.

The square sum of errors in Eq. (2.13) is a quadratic function of $\mathbf{b}$. Therefore, if the matrix $\mathbf{X}^T\mathbf{X}$ is positive definite, the function is convex, and a local minimum will be the global minimum. In elementary polynomial functions with a single variable, this corresponds to a quadratic function $ax^2$ with $a > 0$. The minimum of a quadratic function can be found from the condition that the derivative (i.e., gradient) becomes zero. In the same way, the coefficient vector $\mathbf{b}$ that satisfies the first-order Kuhn–Tucker optimality condition will be the optimal coefficients:

$$\begin{aligned} \frac{d}{d\mathbf{b}}\left(\mathbf{e}^T\mathbf{e}\right) &= -2\mathbf{X}^T\mathbf{y} + 2\mathbf{X}^T\mathbf{X}\mathbf{b} = 0 \\ \mathbf{X}^T\mathbf{X}\mathbf{b} &= \mathbf{X}^T\mathbf{y} \\ \mathbf{b} &= \left(\mathbf{X}^T\mathbf{X}\right)^{-1}\mathbf{X}^T\mathbf{y} \end{aligned} \tag{2.14}$$

Equation (2.14) is often referred to as the normal equation of linear regression equation, and $\mathbf{X}^T\mathbf{X}$ is referred to as the information matrix or moment matrix. Solving Eq. (2.14) is computationally efficient because the dimension of the moment matrix is $(n_\beta \times n_\beta)$, and the number of unknown coefficients $n_\beta$ is much smaller than the number of samples $n_y$. This can be considered as the major advantage of linear regression, where the global optimum coefficients can be obtained by solving a single matrix equation. In the case of nonlinear regression, a nonlinear optimization problem needs to be solved iteratively.

The normal equation has a unique solution $\mathbf{b}$ when the moment matrix $\mathbf{X}^T\mathbf{X}$ is positive definite. When monomial basis functions are used, the moment matrix is ill-conditioned for a large $n_\beta$, which means that the moment matrix is almost singular. To avoid some of the effects of the ill-conditioning, we can formulate the regression problem in a slightly different form. Consider the system of equations:

$$\mathbf{X}\mathbf{b} = \mathbf{y} \tag{2.15}$$

If we could solve this equation exactly, then from Eq. (2.13) we see that the RMS error will be zero. However, this system has $n_y$ equations for $n_\beta$ unknowns, with $n_y$ in general larger than $n_\beta$, so that in general, we cannot find an exact solution. That is, any vector $\mathbf{b}$ will not satisfy Eq. (2.15) exactly, but instead there will be a vector of residuals (differences between the left-hand side and the right-hand side of the equation). The solution of the normal equation is the vector $\mathbf{b}$ that minimizes the sum of the squares of the residuals. However, instead of solving the normal equations, there are numerical methods, such as the QR decomposition, that solve for Eq. (2.15) directly for the least-squares solution, and these are usually more numerically stable than directly solving the normal equations. To improve numerical stability, it is also recommended to translate and scale all variables so that each variable changes in the range (−1, 1) or (0, 1) [13].

**Example 2.1 Linear PRS with three samples**

Using three samples $(x_i, y_i) = (0, 0), (1, 1), (2, 0)$, fit a linear PRS $\hat{y}(x, \mathbf{b}) = b_1 + b_2x$. Use different error metrics in Eqs. (2.7)–(2.9) and compare them.

**Solution:**

The three samples are applied to the linear PRS model to calculate the error in Eq. (2.11) as

$$\left.\begin{aligned} e_1 &= y(0) - b_1 - b_2 \cdot 0 \\ e_2 &= y(1) - b_1 - b_2 \cdot 1 \\ e_3 &= y(2) - b_1 - b_2 \cdot 2 \end{aligned}\right\} \rightarrow \underbrace{\begin{Bmatrix} e_1 \\ e_2 \\ e_3 \end{Bmatrix}}_{\mathbf{e}} = \underbrace{\begin{Bmatrix} 0 \\ 1 \\ 0 \end{Bmatrix}}_{\mathbf{y}} - \underbrace{\begin{bmatrix} 1 & 0 \\ 1 & 1 \\ 1 & 2 \end{bmatrix}}_{\mathbf{X}} \underbrace{\begin{Bmatrix} b_1 \\ b_2 \end{Bmatrix}}_{\mathbf{b}}$$

Then, the matrix and vector that are required in the regression equation in Eq. (2.14) can be obtained as

$$\mathbf{X}^T\mathbf{X} = \begin{bmatrix} 1 & 1 & 1 \\ 0 & 1 & 2 \end{bmatrix} \begin{bmatrix} 1 & 0 \\ 1 & 1 \\ 1 & 2 \end{bmatrix} = \begin{bmatrix} 3 & 3 \\ 3 & 5 \end{bmatrix}, \quad \mathbf{X}^T\mathbf{y} = \begin{bmatrix} 1 & 1 & 1 \\ 0 & 1 & 2 \end{bmatrix} \begin{Bmatrix} 0 \\ 1 \\ 0 \end{Bmatrix} = \begin{Bmatrix} 1 \\ 1 \end{Bmatrix}$$

Therefore, the regression equation, $\mathbf{X}^T\mathbf{X}\mathbf{b} = \mathbf{X}^T\mathbf{y}$, can be solved for regression coefficients as

$$\begin{cases} 3b_1 + 3b_2 = 1 \\ 3b_1 + 5b_2 = 1 \end{cases} \rightarrow \begin{cases} b_1 = \dfrac{1}{3} \\ b_2 = 0 \end{cases}$$

Therefore, the fitted linear PRS becomes a constant function, i.e., $\hat{y}(x, \mathbf{b}) = 1/3$. Figure 2.3(a) compares the fitted PRS with the three sample points when the RMS error is used. It is clear that the fitting process selected the coefficients such that the positive error and the two negative errors are equally distributed. At all sample locations, the errors are $e_1 = e_3 = -1/3$ and $e_2 = 2/3$, respectively. Then the RMS error becomes

$$e_{\text{RMS}} = \sqrt{\frac{1}{3}\left[\left(-\frac{1}{3}\right)^2 + \left(\frac{2}{3}\right)^2 + \left(-\frac{1}{3}\right)^2\right]} = 0.47$$

In general, it is difficult to find the PRS that minimizes the maximum error and the average error. For this simple example, however, it can be easily done. We expect that, like the line that minimizes the RMS error, the lines that minimize the other two errors would be horizontal lines of the form $\hat{y}(x, \mathbf{b}) = b_1, 0 \le b_1 \le 1$. This is expected as samples are symmetric with respect to $x = 1$. To minimize the maximum error, it is obvious that we must have $b_1 = 0.5$ such that all errors are of the same magnitude. This results in a maximum error as

$$e_{\max} = \max(|0 - b_1|, |1 - b_1|, |0 - b_1|) = 0.5$$

In this case, this is also the average error and the RMS error because all three points have the same 0.5 error. To minimize the average error, we note that for the range $b_1 \in [0, 1]$, the average error will be

$$e_{\text{av}} = \frac{1}{3}(|0 - b_1| + |1 - b_1| + |0 - b_1|) = \frac{1 + b_1}{3}$$

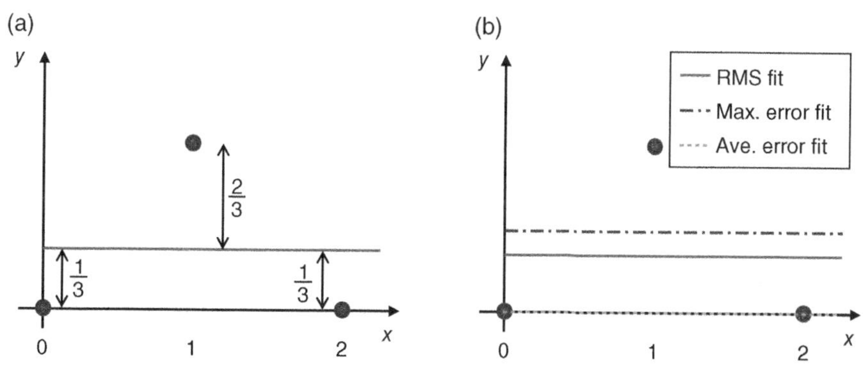


**Figure 2.3** Fitting a linear PRS with three samples: (a) with RMS error and (b) with average error and maximum error.

**Table 2.1** Comparison of the three linear PRS fits with the three different error metrics.

| | RMS fit<br>$\hat{y} = 1/3$ | Ave. error fit<br>$\hat{y} = 0$ | Max. error fit<br>$\hat{y} = 0.5$ |
|---|---|---|---|
| $e_{RMS}$ | 0.471 | 0.577 | 0.5 |
| $e_{av}$ | 0.444 | 0.333 | 0.5 |
| $e_{max}$ | 0.667 | 1.0 | 0.5 |

The average error becomes minimum when $b_1 = 0$, at which the minimum average error becomes $e_{av} = 1/3$. Figure 2.3(b) compares all three linear PRSs. As can be seen in the figure, even if the same samples are used, the fitted surrogate models can be different with different error metrics.

At first glance, it looks like the maximum and RMS error metrics yield a better fit than the average error metric, but the quality of a fit depends on the error metric used as well. For example, Table 2.1 evaluates the three PRS fits with the three different error metrics. It turns out that an error is a minimum when a surrogate is fitted with the same error metric.

### 2.3.2 Polynomial Response Surface in Multiple Dimensions

So far, we have only discussed a PRS in one-dimensional input variables. The `polyfit` and `polyval` Matlab functions in Section 2.2 can only be used for one-dimensional input variables with a fixed degree of polynomials. However, linear regression is flexible enough such that it can be applied to multi-dimensional input variables with an arbitrary degree of polynomials. In fact, the basis functions do not have to be monomials. As long as the basis functions are linearly independent and the QoI has a linear relationship with the regression coefficients, the linear regression is well-defined. In Matlab, the multiple linear regression function `regress` is used for this purpose.

**Example 2.2 Profit surrogate model of a soft can**

A soft-drink company wants to determine the size of a can such that it can maximize its profit. The dimensions of the can are determined by input variables: diameter $D \in [1.5, 3.5]$ in. and height $H \in [3.0, 7.0]$ in. The true profit is an explicit function of these two variables:

$$p(D, H) = 0.2361\pi D^2 H - 3.858 \times 10^{-4}\pi^2 D^4 H^2 - 0.1\pi\left(0.5D^2 + DH\right)$$

However, it is assumed that the true profit is unknown. Instead, the company conducted field tests and obtained nine samples, as shown in the table.

| Sample no. | $D$ | $H$ | $p$ | $p_{true}$ |
|---|---|---|---|---|
| 1 | 1.8 | 3.6 | 5.9 | 5.6 |
| 2 | 2.4 | 4.8 | 13.1 | 13.1 |
| 3 | 3.0 | 6.0 | 22.0 | 21.9 |
| 4 | 1.5 | 4.6 | 4.7 | 4.7 |
| 5 | 2.1 | 5.8 | 12.0 | 12.0 |
| 6 | 2.7 | 7.0 | 20.9 | 20.9 |
| 7 | 1.4 | 5.6 | 4.9 | 4.9 |
| 8 | 2.0 | 6.8 | 12.5 | 12.5 |
| 9 | 2.6 | 8.0 | 21.4 | 21.4 |

(a) Fit a quadratic polynomial to the profit per can using the experimental nine samples in column p and plot the profit contours with samples. (b) Repeat the fitting process using the true samples that are generated from the true profit equation (use the last column of the table). Compare the two PRS and explain the reason for the difference.

**Solution:**

The quadratic PRS with two input variables has the following form with six unknown coefficients:

$$\hat{p} = b_1 + b_2 D + b_3 H + b_4 D^2 + b_5 DH + b_6 H^2$$

By comparing with the true function, it is obvious that the quadratic PRS cannot be exact because the true function includes higher-order polynomial terms of input variables. And yet, it is difficult to use a higher-order PRS because then the number of unknown coefficients can be more than that of the number of samples. It is also possible that the PRS can include specific terms, such as $D^2$, $DH$, $D^2H$, and $D^4H^2$. However, it is only possible if the model form is already known.

The design matrix can be calculated using all nine samples as

$$\mathbf{X} = \begin{bmatrix} 1 & 1.8 & 3.6 & 3.24 & 6.48 & 12.96 \\ 1 & 2.4 & 4.8 & 5.76 & 11.52 & 23.04 \\ 1 & 3.0 & 6.0 & 9.00 & 18.0 & 36.0 \\ 1 & 1.5 & 4.6 & 2.25 & 6.9 & 21.16 \\ 1 & 2.1 & 5.8 & 4.41 & 12.18 & 33.64 \\ 1 & 2.7 & 7.0 & 7.29 & 18.9 & 49.0 \\ 1 & 1.4 & 5.6 & 1.96 & 7.84 & 31.36 \\ 1 & 2.0 & 6.8 & 4.00 & 13.6 & 46.24 \\ 1 & 2.6 & 8.0 & 6.76 & 20.8 & 64.0 \end{bmatrix}$$

a) The unknown coefficients can be calculated from the regression equation in Eq. (2.14). Then the PRS approximate of the profit can be written as

$$\hat{p} = -7.1827 + 2.8835D + 0.3022H + 0.3576D^2 + 1.0024DH - 0.0707H^2$$

b) In comparison, if the true values in the last column of the table are used, the PRS approximate becomes

$$\hat{p}_{\text{true}} = -7.9801 + 2.5672D + 0.6756H + 0.2956D^2 + 1.0898DH - 0.1150H^2$$

Note that four of the six coefficients are fairly close, whose difference is less than 10% of each other. However, the coefficients of $H$ and of $H^2$ are quite different, indicating that these coefficients are less important for fitting the samples. This does not mean, however, that these coefficients may not affect predictions at other points besides the sample points. We are thus warned that these coefficients may need special treatment. The profit-per-can plot based on the experimental samples is shown in Figure 2.4: one from a quadratic PRS using experimental samples and the other using true samples. Both PRSs are obviously quite similar to each other. The following Matlab code is used for plotting Figure 2.4.

```
% Quadratic PRS for profit-per-can
x=[1.8 3.6; 2.4 4.8; 3.0 6.0; 1.5 4.6; 2.1 5.8; 2.7 7.0; 1.4 5.6; 2.0
6.8; 2.6 8.0];
p=[5.9 13.1 22.0 4.7 12.0 20.9 4.9 12.5 21.4]';
```

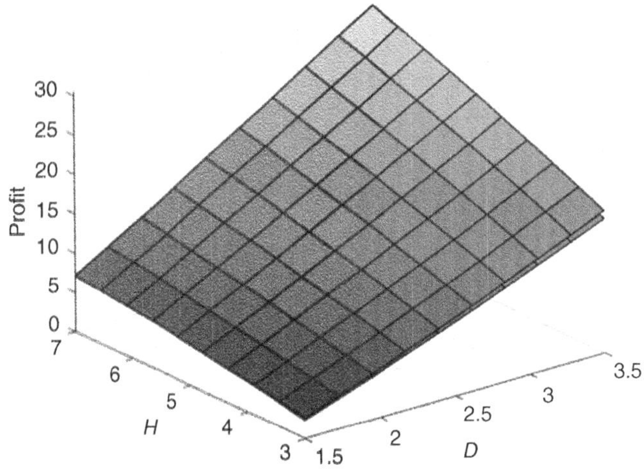


**Figure 2.4** Quadratic polynomial response surface model for the profit-per-can.

```
D=x(:,1); H=x(:,2);
ptrue=0.2361*pi*D.^2.*H-3.858E-4*pi^2*D.^4.*H.^2-0.1*pi*(0.5*D.^2+D.*H);
%
X=[ones(9,1) D H D.^2 D.*H H.^2];
A=X'*X;
Btrue=X'*ptrue;
B=X'*p;
btrue=A\Btrue;
b=A\B;
%
d=linspace(1.5,3.5,10);
h=linspace(3.0,7.0,10);
[D,H]=meshgrid(d,h);
P=b(1)+b(2)*D+b(3)*H+b(4)*D.^2+b(5)*D.*H+b(6)*H.^2;
Ptrue=btrue(1)+btrue(2)*D+btrue(3)*H+btrue(4)*D.^2+btrue(5)*D.*H
+btrue(6)*H.^2;
surf(D, H, P); hold on; surf(D, H, Ptrue);
```

### 2.3.3 Curse of Dimensionality

As shown in the previous example, the number of regression coefficients for the quadratic PRS is six for two-dimensional input variables. The number of regression coefficients rapidly increases as the order of PRS increases and the dimension of the input variable increases. Considering the fact that the required number of samples is about two- or threefold of the number of regression coefficients, the required number of samples rapidly increases along with the dimension of input variables and the order of PRS, which is referred to as the curse of dimensionality. In the case of a quadratic and a cubic PRS, the numbers of regression coefficients are, respectively,

$$n_\beta^{(2)} = \frac{1}{2}(n+1)(n+2) \tag{2.16}$$

$$n_\beta^{(3)} = \frac{1}{6}(n+1)(n+2)(n+3) \tag{2.17}$$

where $n$ is the dimension of input variables. For example, when $n = 10$, $n_\beta^{(2)} = 66$, and $n_\beta^{(3)} = 286$. Therefore, in 10-dimensional input variables, we need more than 800 samples to fit the PRS, which becomes impractical for many applications with expensive simulations or experiments.

Although the rapid growth of the number of required samples is a part of the curse of dimensionality, the major difficulty comes from the fact that when the dimensionality increases, the volume of the design space increases so fast that the available samples become sparse. In order to obtain a reliable result, the number of samples needed often grows exponentially with dimensionality. For example, let us assume that all input variables are normalized by a unit length. If we generate 10 evenly spaced samples in a one-dimensional domain, then the distance between the samples would be 0.1. An equivalent sampling of a 10-dimensional unit hypercube with a lattice that has a spacing of 0.1 between adjacent samples would require $10^{10}$ sample points, which are practically impossible to use. Therefore, in a high-dimensional design space, samples are sparse and the distances between samples are quite large. When the distance between samples is large, it would be difficult to capture the trend of the function accurately. In addition, sparse sampling inevitably causes a large extrapolation region, as shown in Figure 1.5, where the volume covered by samples (interpolation region) is dramatically reduced as the dimension increases.

Based on the above discussion, it would be challenging to build an accurate and reliable surrogate model with more than 10 design variables. Unfortunately, it is common that many engineering applications have more than 10 design variables. Instead of trying to make a surrogate model that is accurate in high-dimensional input variables with a small number of samples, it would make more sense to reduce the number of variables. It is hard to imagine that a QoI strongly depends on many variables simultaneously; such a system would be very sensitive and volatile to a small change in variables. Instead, many engineering systems strongly depend on only a handful of variables. Therefore, it would make sense to identify those design variables that significantly affect the QoI and fix all other variables that are not significant. Then, the surrogate model is built using only those significant variables. This process of identifying significant input variables and fixing insignificant variables is called sensitivity analysis [14]. If 100 samples are affordable, it would be better to build an accurate surrogate with four significant input variables, rather than building an inaccurate surrogate with all 10 variables.

It is also possible to combine multiple variables together instead of using individual variables, which is called dimensional analysis. A good example is nondimensionalization in fluid mechanics, where multiple variables are combined together to describe a physically important phenomenon [15]. Another method of dimension reduction is the transformation of data from a high-dimensional space into a low-dimensional space so that the low-dimensional representation retains some meaningful properties of the original data, ideally close to its intrinsic dimension. Principal component analysis [16] or proper orthogonal decomposition (POD) is one of the dimension reduction, which performs a linear mapping of the data to a lower-dimensional space in such a way that the variance of the data in the low-dimensional representation is maximized. In practice, the covariance matrix of the samples is constructed, and the eigenvectors on this matrix are computed. The eigenvectors that correspond to the largest eigenvalues (the principal components) can now be used to reconstruct a large fraction of the variance of the original data.

### 2.3.4 Assumptions in Linear Regression

Although the process of regression analysis in Eq. (2.14) is simple, the accuracy analysis of regression is quite comprehensive. Before we discuss regression accuracy, it would be useful to summarize the assumptions in the regression analysis.

1) The regression analysis assumes that the input variables can be fully controllable; that is, we can select the input variables without any errors associated with them. This might sound obvious but in practice, it could be hard to control the input variables precisely during experiments. When a sample is given in the form of $(\mathbf{x}_i, y_i)$, the regression analysis assumes that there is random noise in $y_i$, but $\mathbf{x}_i$ is precise.
2) The functional form of PRS is a linear combination of unknown coefficients and basis functions. Therefore, the only limitation is that the relationship between the QoI and unknown coefficients is linear. This is flexibility rather than limitation. Because of this flexibility, PRS often shows "too much power," in that it tends to overfit the data. Using a too-small number of samples, too many basis functions and/or too-high order of monomial basis functions can cause overfitting. There are several techniques available to prevent overfitting, such as ridge regression, lasso regression, and Bayesian linear regression.
3) The noise in each sample is independent. More specifically, the noise in samples is statistically uncorrelated. For example, the noise in two samples is nothing related to the distance between the two samples. In fact, linear regression allows us to have samples at the same location, i.e., $\mathbf{x}_i = \mathbf{x}_j$, but the output QoI samples may not be the same, i.e., $y_i \neq y_j$ because the noise in the two samples is independent. When we generate noise samples from $\sim N(0, \sigma^2)$, we automatically apply the independence assumption.
4) The variance of the noise is assumed to be constant throughout the entire design space. That is, the noise is normally distributed with a zero mean and a constant standard deviation $\sigma$. Based on this assumption, minimizing the RMS error can provide an unbiased fit. This seems a reasonable assumption, but in practice, the noise in experiments often depends on the value of measurement. When the QoI does not vary rapidly over the design space, this assumption can be reasonable. When the QoI varies over orders of magnitude, however, this assumption may not represent the reality. For example, when we measure stress in a component, it might be reasonable to say that the measurement error is 10 MPa when the level of stress is 500 MPa, which corresponds to a 2% error. However, the same measurement error of 10 MPa may not be acceptable when the level of stress is 10 MPa because that would imply the actual stress could vary anywhere between 0 and 20 MPa. In many engineering experiments, the measurement noise is often proportional to the magnitude of the data. In such a case, it makes more sense to assume that the coefficient of variation (CoV) remains the same, not the standard deviation. If we want to check if samples satisfy this assumption, the residuals at all sample locations can be plotted. If residuals are randomly scattered about the horizontal midline at 0 with a similar bound, it means that this assumption is satisfied. If samples do not satisfy this assumption, it will result in an overall "average" estimate of variance being used instead of one that takes into account the true variance structure. This leads to less precise parameter estimates and biased standard errors, resulting in misleading tests and interval estimates.
5) The moment matrix $\mathbf{X}^T\mathbf{X}$ is positive definite such that the regression analysis yields a unique solution $\mathbf{b}$. In order to have a positive definite matrix $\mathbf{X}^T\mathbf{X}$, the design matrix $\mathbf{X}$ must have a full column rank of $n_\beta$. Rank deficiency can happen when a linear relationship exists between two or more basis functions or when a smaller number of samples is used than the number of

coefficients to be estimated. It is noted that monomial basis functions are linearly independent. In order to prevent a near-singular matrix, it is recommended that the number of samples should be two- or threefold larger than that of the coefficients.

## 2.4 Goodness of Fit

The goal of the surrogate fitting process, in particular the linear regression in the previous section, is to find the best fit that can approximate the true function. In order to determine the quality of the approximation, it would be necessary to define the measure of goodness. In the case of the curve fitting in Section 2.2, the RMS error at sample locations was used as an error measure, which was minimized to determine regression coefficients. As shown in Section 2.3, different error measures would yield different fitting results. Different from the error measures to fit a surrogate, the measure of goodness can be considered post-processing in a way. After fitting a surrogate by minimizing an error measure, the measure of goodness characterizes the fidelity of the surrogate for predicting the behavior in future simulations. It is possible that a measure can be local at a point or in a region, but in this section, we limit ourselves to global measures, which are a single number that characterizes the overall fidelity of the surrogate model in the entire design space.

However, achieving this goal is often doomed because of a simple reason: we do not know the true function, and therefore, we cannot compare the accuracy of the surrogate model against the true function. Although samples represent the true function, as we discussed in the previous section, they include random noise (or, at least we assume it). Therefore, fitting the samples exactly does not guarantee that the surrogate model is accurate compared to the true function. A good example is the one shown in Figure 1.8, where the quintic polynomial passes through all samples with zero error, but it is not an accurate surrogate model except for the sample locations. Therefore, the biggest challenge in assessing the accuracy of a surrogate model is how to evaluate its accuracy without knowing the true function. All information that we have (or we assume) is that the noise is normally distributed with respect to the true function, but we still do not know the variance of the distribution. Therefore, the goal of assessing the accuracy of a surrogate model to the true function is difficult to achieve.

Instead, it is possible to evaluate the accuracy of a surrogate model against the samples. That is, how well the surrogate fits the samples, which is referred to as "goodness of fit." In evaluating the goodness of fit, it is important to remember that errors are minimized at sample locations during the fitting process. For example, if the number of samples is the same as the number of regression coefficients, i.e., $n_y = n_\beta$, theoretically, it is possible that the regression equation can be solved such that the surrogate model can pass through all sample points. That is, the errors at all sample points are zero. However, this does not mean that the surrogate model is accurate at unsampled points. Therefore, the goodness of fit does not mean simply measuring the error at sample points.

In general, the goodness of fit is evaluated in two ways: (i) equivalence between the surrogate and samples and (ii) prediction accuracy of the surrogate. Equivalence measures include the coefficient of multiple determination and the adjusted coefficient of multiple determination. They measure the equivalence between the surrogate and the samples in terms of variability. The first provides the fraction of the variability in the samples captured by the surrogate. The second adjusts it in an attempt to estimate the fraction that will be captured by using the surrogate to predict values at other points. Good fidelity will be reflected in these coefficients being close to one. The second category of prediction accuracy measures estimates what will be the RMS error in predictions based on the surrogate and includes cross-validation error and standard error. The cross-validation error is a measure that can be applied to any surrogate, while the standard error applies only to linear regression with specific assumptions on the noise in the samples. A surrogate model is considered to be good when these error measures are small compared to the average value of the samples.

### 2.4.1 Estimation of Noise in Samples

In most applications, the PRS approximation we construct based on given data is intended for the prediction of QoI at other design points, typically for improving the design. Therefore, the ultimate test of the PRS is how well it predicts the QoI at other points of interest. However, if the PRS does not approximate the response well even at the data points, we cannot expect it to approximate well other points in the design space. The most immediate measures of the accuracy of the fit to the samples are the various errors discussed earlier, the RMS error, the average error, and the maximum error.

Since the samples are assumed to include both the true function value and noise, it is important to estimate the level of noise accurately so that the surrogate can approximate the true function, not the noise. After fitting a surrogate $\hat{y}(\mathbf{x}, \mathbf{b}) = \xi(\mathbf{x})^T\mathbf{b}$, the model predictions at sample points become $\hat{y}(\mathbf{x}_i, \mathbf{b}) = \xi(\mathbf{x}_i)^T\mathbf{b}, i = 1, \dots, n_y$. These $n_y$ predictions can be written in matrix–vector notation as $\hat{\mathbf{y}} = \mathbf{X} \cdot \mathbf{b}$, and the regression errors at the sample points become $\mathbf{e} = \hat{\mathbf{y}} - \mathbf{y}$. Then, the RMS error in Eq. (2.7) can be rewritten as

$$e_{\mathrm{RMS}} = \sqrt{\frac{SSe}{n_y}} \tag{2.18}$$

where the square-sum-error is defined as $SSe = \mathbf{e}^T\mathbf{e} = \mathbf{y}^T\mathbf{y} - 2\mathbf{b}^T\mathbf{X}^T\mathbf{y} + \mathbf{b}^T\mathbf{X}^T\mathbf{X}\mathbf{b}$. However, this calculation of RMS error is quite misleading if we want to use it to assess the accuracy of PRS. This becomes clear if we note that the number of sample points $n_y$ is equal to the number of coefficients, $n_\beta$, then the PRS will pass through the sample points, and the error will be zero. We certainly do not expect that the error will be zero at other points, not included in the data. In fact, fitting a PRS to the same number of points equal to the number of coefficients (known as a saturated design) is known to often lead to poor approximation, especially when there is noise in the data.

An impressive body of theoretical work has been done for the case where the noise in the samples is random with a normal distribution with zero mean and a standard deviation of $\sigma$, and where the noise at one point is uncorrelated with the noise at other data points (e.g., Myers and Montgomery [3]).

Since the surrogate fitting process minimizes the RMS error, it is minimum at sample points but underestimates the error at unsampled (prediction) points. Therefore, it is not a good measure to assess the accuracy of the surrogate model. However, the square-sum-error can be used to estimate the variance of noise in samples. The variance of random samples $y_i$ is defined as

$$\mathrm{Var} = \frac{1}{n_y - 1}\sum_{i=1}^{n_y} (y_i - \mu_y)^2 \tag{2.19}$$

where $\mu_y$ is the mean of the samples, and the denominator $n_y - 1$ is used for unbiased variance, which represents the number of degrees-of-freedom. It is reminded that PRS assumes that the model is accurate, but the samples have a random noise. The surrogate $\hat{y}(\mathbf{x}, \mathbf{b})$ is considered a mean prediction, and the difference between the samples and the mean predictions is considered noise.

Therefore, in the viewpoint of noise, its mean is zero, and the unbiased estimate of the variance of the noise from the samples is

$$\hat{\sigma}^2 = \frac{SSe}{n_y - n_\beta} \tag{2.20}$$

The square root of Eq. (2.20), $\hat{\sigma}$, is called the standard error of noise.

Since the surrogate model has $n_y$ samples with $n_\beta$ model parameters, the denominator $n_y - n_\beta$ represents the number of degrees-of-freedom. If this estimate of the RMS error is larger than what we can tolerate for predicting values of the response at candidate design points, we conclude that the PRS is inadequate. In this case, we must change the form of the response surface to try to fit the samples better. The simplest approach is to add terms to the polynomial approximation. If we used a linear polynomial to start with, we may want to go to a quadratic. If we used a quadratic, we may want to use a cubic.

Unfortunately, while we can always improve the fit to the samples by increasing the number of terms in the PRS, it is not clear that these gains will translate into gains in predicting the PRS at other points. As we add coefficients, we run the danger of "overfitting" the samples. This danger is particularly acute when the samples contain a substantial amount of noise. As we increase the number of coefficients, we may be fitting the noise rather than the underlying response. This danger is captured by Eq. (2.20). As we add more terms, we expect to decrease the numerator, but the denominator will also decrease.

There are several assumptions under the noise estimate in Eq. (2.20). First, it is assumed that the true function is described by the model form of the surrogate. If there is a model-form error, it is embedded in the estimated noise. Second, the samples include noise that is normally distributed with a zero mean and the same standard deviation at every sample. Lastly, the noises in different samples are not correlated. Under these assumptions, the standard error, $\hat{\sigma}$, is an estimate of the standard deviation of the noise. The standard error will be used to estimate the prediction error in Section 2.6. That is the error between the true function and the surrogate prediction.

**Example 2.3 Quadratic PRS with five samples**

The following five equally spaced samples are generated from a true function $y = x$ and added noise $\sim N(0, 1^2)$: $(x_i, y_i)$ = (0, 1.5326), (2.5, 1.7303), (5.0, 5.3714), (7.5, 7.2744), (10.0, 11.1174). Fit the samples using a quartic PRS (five coefficients) and see if the surrogate fits the trend or the noise.

**Solution:**

The five samples can be fitted using `polyfit` Matlab function. Since the quartic PRS has five coefficients, the number of samples and the number of coefficients is the same, i.e., $n_y = n_\beta$. Therefore, the regression equation is equivalent to solving a linear system of equations. In this case, the linear system of equations has a unique solution that makes the errors zero. The fitted quartic PRS becomes

$$\hat{y}(x) = 1.5326 - 2.1864x + 1.3397x^2 - 0.1970x^3 + 0.0095x^4$$

Figure 2.5 compares the quartic PRS with samples and the true function. It is noted that the quartic PRS passes through all samples with zero errors. However, there is a significant error at unsampled locations. In this case, the PRS fits noise rather than the trend of the true function. This happens when the number of samples is not enough to determine the mean of the trend of the samples. In order to have a meaningful regression performance, it is often recommended that the number of samples should be two- or threefold larger than that of the regression coefficients. The following Matlab code can be used to plot Figure 2.5.

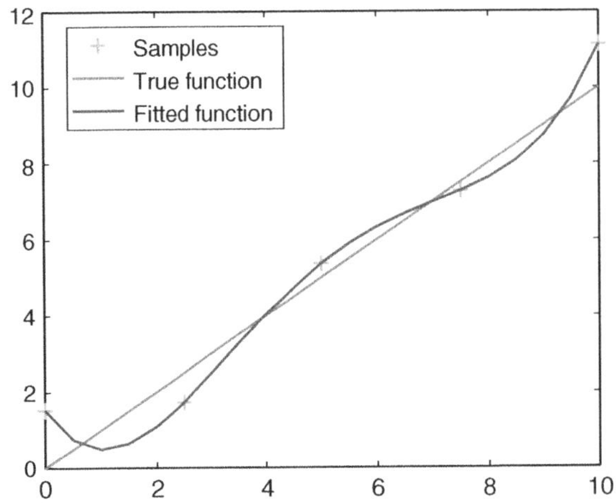


**Figure 2.5** Fitted quartic polynomial response surface of noisy samples from $y = x$.

```
x=[0 2.5 5 7.5 10]';
y=[1.5326 1.7303 5.3714 7.2744 11.1174]';
xp=0:0.5:10; ytrue=xp;
[b,s]=polyfit(x,y,4);
yfit=polyval(b,xp);
plot(x,y,'+',xp,ytrue,'r',xp,yfit,'b');
legend('Samples','True function','Fitted function');
```

As mentioned before, the PRS assumes that the model form is correct, but samples have noise. Because of this assumption, the estimate of noise variance in Eq. (2.20) is reasonable when the model form is correct. If the model form has an error, it will be included in the estimated noise variance. The following example shows how the model-form error can contribute to the noise variance estimate.

**Example 2.4 Estimation of noise using linear and quadratic PRS with 20 samples**
Generate 20 equally spaced samples in $x \in [0, 10]$ from a true function $y = x^2$ and added noise $\sim N(0, 1^2)$. Fit the samples using linear and quadratic PRS and estimate the noise from the two surrogates.

**Solution:**
In order to prevent different random samples, `rng default` Matlab command is used, which will reset the random number generator sequence. In addition, the noise samples are shifted to have a zero mean by the following command: `noise=noise-mean(noise)`. In this specific case, the standard deviation of noise samples is 1.4797, which is different from the population distribution $\sim N(0, 1^2)$. The two fitted surrogate models become

$$\hat{y}_L(x) = -16.2473 + 10.0916x$$

$$\hat{y}_Q(x) = -0.5748 + 0.1657x + 0.9926x^2$$

The estimated standard deviations from the two surrogate models are, respectively,

$$\hat{\sigma}_L = \sqrt{\frac{SSe_L}{20-2}} = 8.7153, \quad \hat{\sigma}_Q = \sqrt{\frac{SSe_Q}{20-3}} = 1.5336$$

Compared to the actual noise standard deviation of 1.4797, the estimate from the quadratic PRS is close to the actual one, but the estimate from the linear PRS is quite different from the actual one. This is because the linear PRS has a significant level of model error. Figure 2.6(a) shows the two PRS predictions. The quadratic PRS follows the trend of samples well, while the linear PRS approximates the trend linearly. As shown in Figure 2.6(b), the errors of the quadratic PRS are randomly distributed with respect to the zero line, while the errors of the linear PRS are all positive in $x < 2$ or $x > 8$ and all negative in $2 < x < 8$. This is a good indicator of the presence of model-form error. The following Matlab code is used to fit the surrogate and plot the results.

```
rng default;
x=linspace(0,10,20);
ytrue=x.^2;
noise=randn(1,20); noise=noise-mean(noise);
y=ytrue+noise;
bL=polyfit(x,y,1);
yL=polyval(bL,x);
sigmaL=sqrt((y-yL)*(y-yL)'/(20-2))
bQ=polyfit(x,y,2);
yQ=polyval(bQ,x);
sigmaQ=sqrt((y-yQ)*(y-yQ)'/(20-3))
plot(x,y,'+',x,x.^2,'r',x,yL,'b',x,yQ,'k');
legend('Samples','True function','Linear PRS','Quadratic PRS');
figure(2);
plot(x,y-yL,'+',x,y-yQ,'o');
legend('e_L','e_Q');
```

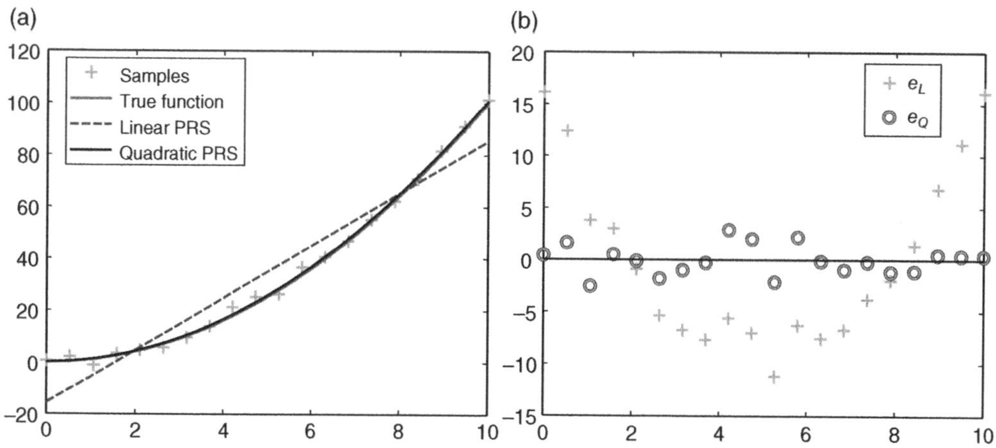


**Figure 2.6** Fitted linear and quadratic polynomial response surface of noisy samples from $y = x^2$. (a) Linear and quadratic PRS and (b) errors of the two surrogates.

### 2.4.2 Coefficient of Multiple Determination

As mentioned before, the accuracy at sample points is different from the accuracy at prediction points. Therefore, instead of measuring accuracy at sample points, it would be possible to measure the equivalence between the samples and PRS predictions in terms of variability. That is how well a surrogate captures the variability in samples. The coefficient of multiple determination ($R^2$) is a numerical index that reflects the degree to which variation in the samples is accounted for by the predictions. If a surrogate passes through all sample points, then the variability of the predictions at sample points would be the same as that of the samples. In that case, the entire variability in the samples is explained by the variability of the model predictions. On the other hand, if the model prediction is a constant at the mean of the samples, then none of the variability of the samples can be explained by the model predictions. In such a case, the entire variability becomes the square-sum-error $SS_e$ in Eq. (2.18).

Since we do not know the true function, the coefficient of multiple determination is defined as the variation with respect to the mean of samples, which is defined as

$$\bar{y} = \frac{1}{n_y}\sum_{i=1}^{n_y} y_i \tag{2.21}$$

The vector version of the mean can be defined as $\bar{\mathbf{y}} = \{\bar{y}, \bar{y}, ..., \bar{y}\}^T$ whose dimension is $n_y \times 1$. The variation of samples from the mean is defined as

$$SS_y = \sum_{i=1}^{n_y} (y_i - \bar{y})^2 \tag{2.22}$$

This is also known as the total variation. Similarly, the variation of the PRS $\hat{y}$ from the same mean is defined as

$$SS_r = \sum_{i=1}^{n_y} (\hat{y}_i - \bar{y})^2 \tag{2.23}$$

where $\hat{y}_i = \hat{y}(\mathbf{x}_i)$. This is also known as the explained variation. That is, the variation can be explained by the fitted model. If samples were generated by adding random noise to a true function, and if the fitted PRS model is close to the true function, then $SS_y$ is larger than $SS_r$ because of the variation from the random noise in the samples. In fact, the variation from the random noise corresponds to the sum-of-square errors in Eq. (2.18), which is also known as the unexplained variation. In linear regression, these three variations have the following relationship:

$$SS_y = SS_r + SS_e \tag{2.24}$$

That is, the total variation in samples is the sum of the explained variation by the model and the unexplained variation by noise. The relationship in Eq. (2.24) holds if and only if $\mathbf{y}^T\bar{\mathbf{y}} = \hat{\mathbf{y}}^T\bar{\mathbf{y}}$. Since $\bar{\mathbf{y}}$ is a constant, the condition means that the sum of samples equals to the sum of predictions at sample points, or equivalently, the sum of residuals $e_i = y_i - \hat{y}_i$ is zero. Since the first column of the design matrix $\mathbf{X}$ is all ones, the first element of $\mathbf{X}^T\mathbf{e}$ is the sum of residuals. From the regression analysis, it can be shown that

$$\mathbf{X}^T\mathbf{e} = \mathbf{X}^T\left[\mathbf{I} - \mathbf{X}(\mathbf{X}^T\mathbf{X})^{-1}\mathbf{X}^T\right]\mathbf{y} = \left[\mathbf{X}^T - \mathbf{X}^T\right]\mathbf{y} = \mathbf{0} \tag{2.25}$$

Therefore, the condition $\mathbf{y}^T\bar{\mathbf{y}} = \hat{\mathbf{y}}^T\bar{\mathbf{y}}$ holds for linear regression of PRS. The proof of Eq. (2.24) is left as an exercise problem.

The ratio of Eq. (2.23) over Eq. (2.22), denoted by $R^2$, measures the fraction of the variation in the samples is captured by the PRS:

$$R^2 = \frac{SS_r}{SS_y} = 1 - \frac{SS_e}{SS_y} \tag{2.26}$$

$R^2$ is a measure of the goodness of fit of a PRS model [17], often interpreted as the proportion of sample variation "explained" by the model prediction. If the fitted model $\hat{y}$ comes from linear regression, it is expected that $0 \leq R^2 \leq 1$. An $R^2 = 1$ indicates that the fitted model explains all the variability in the samples $\mathbf{y}$, while $R^2 = 0$ indicates no "linear" relationship between the samples and predictions. An intermediate value such as $R^2 = 0.7$ means that 70% of the variability in the samples may come from the surrogate predictions, and the remaining 30% may come from the noise in the samples. Theoretically, it is not possible to have a negative value of $R^2$ if the constant basis is used in the regression and $R^2$ is calculated using the same samples with the fitting process. Having said that, it is possible to have a negative $R^2$ when (i) different sets of samples are used for testing the goodness of a surrogate than the fitting samples, (ii) the surrogate is fitted using different methods other than linear regression, and (iii) the basis functions do not include the constant term, such that the condition $\mathbf{y}^T\bar{\mathbf{y}} = \hat{\mathbf{y}}^T\bar{\mathbf{y}}$ does not satisfy. In such cases, it is possible that the condition in Eq. (2.24) may not satisfy and $SS_e$ can be larger than $SS_y$.

It is interesting to note that $R^2$ monotonically increases as the PRS includes additional basis functions, i.e., increasing the number of coefficients $n_\beta$. This can be considered a major drawback of $R^2$ to be used as a criterion for assessing the goodness-of-fit of a surrogate model. A good example was shown in Figure 2.5, where the order of PRS is increased to a quintic polynomial, the PRS passes through all the samples, and thus, $R^2 = 1$. However, this does not mean that the PRS is accurate in prediction. Therefore, $R^2$ may not represent the accuracy of a surrogate, especially when the number of coefficients is close to the number of samples.

This leads to the alternative approach of looking at the adjusted $R^2$, namely $R_a^2$, which penalizes $R^2$ as extra basis functions are included in the model. The penalization is based on the degrees-of-freedom of the estimate of the variance around the mean. First, the square-sum-error $SS_e$ is penalized by the degrees-of-freedom $n_y - n_\beta - 1$ because it uses $n_y$ samples but $n_\beta$ coefficients have been used. Also, the mean of the samples is considered as an additional coefficient. In the same way, the variance of samples $SS_y$ is penalized by the degrees-of-freedom $n_y - 1$ because the mean is fixed. After using these penalizations on the original definition of $R^2$, the adjusted coefficient of multiple determination $R_a^2$ is given as

$$R_a^2 = 1 - \frac{SS_e/\left(n_y - n_\beta - 1\right)}{SS_y/\left(n_y - 1\right)} = 1 - \left(1 - R^2\right)\left(\frac{n_y - 1}{n_y - n_\beta - 1}\right) \tag{2.27}$$

If the adjusted value, $R_a^2$, decreases as we increase the number of coefficients, it is a warning that we may be fitting the samples better, but losing predictive capability.

$R_a^2$ can be negative and always less than or equal to $R^2$. When adding an additional basis, $R_a^2$ increases only when the basis contributes to prediction accuracy. Therefore, $R_a^2$ can be a good indicator if an additional basis should be introduced or not. Let us assume that a series of additional basis functions is available in the order of their significance. Then, each basis function is introduced in the linear regression, and $R_a^2$ can be calculated each time. The level at which $R_a^2$ reaches a maximum, and decreases afterward, would be the regression with the ideal combination of having the best fit without fitting the noise. Therefore, $R_a^2$ is more appropriate when evaluating the PRS fit and comparing alternative models.

**Example 2.5 Linear PRS with four samples**

A QoI $y$ is a function of a single variable $x$, and has eventually been identified to follow a simple relationship $y_{\text{true}} = x$. However, the first set of measurements that were taken is given as four $(x, y)$ sample pairs: $(-2, -1.5)$, $(-1, -1.5)$, $(1, 1.25)$, $(2, 1.75)$. From theoretical considerations, you also know that $y(0) = 0$. Fit a linear function and a quadratic function to the samples, compare the two fits and see how they model the true function.

**Solution:**

Linear PRS: The PRS is of the form $\hat{y}(x) = b_1 + b_2x$, and using the point (0, 0) as another sample point, we can define the following matrices and vectors for regression:

$$\mathbf{X} = \begin{bmatrix} 1 & -2 \\ 1 & -1 \\ 1 & 0 \\ 1 & 1 \\ 1 & 2 \end{bmatrix}, \quad \mathbf{y} = \begin{Bmatrix} -1.5 \\ -1.5 \\ 0 \\ 1.25 \\ 1.75 \end{Bmatrix}, \quad \mathbf{X}^T\mathbf{X} = \begin{bmatrix} 5 & 0 \\ 0 & 10 \end{bmatrix}, \quad \mathbf{X}^T\mathbf{y} = \begin{Bmatrix} 0 \\ 9.25 \end{Bmatrix}$$

By solving the normal equation in Eq. (2.14), the unknown regression coefficients are identified as $b_1 = 0$ and $b_2 = 0.925$. Therefore, the following linear PRS and residuals at sample locations can be obtained:

$$\hat{y}^{(1)}(x) = 0.925x, \quad \mathbf{e} = \begin{Bmatrix} 0.35 \\ -0.575 \\ 0.0 \\ 0.325 \\ -0.1 \end{Bmatrix}$$

The coefficient of multiple determination can be calculated from Eqs. (2.21) to (2.27) as

$$\bar{y} = 0, \quad SS_y = 9.125, \quad SS_r = 8.5563, \quad R^2 = 0.9377, \quad R_a^2 = 0.8754$$

Therefore, based on the coefficient of multiple determination, the linear PRS performs well in predicting the trend of samples. In fact, the linear PRS appears to be satisfactory in terms of error measures as well:

$$e_{\text{RMS}} = 0.337, e_{\text{av}} = 0.27, e_{\text{max}} = 0.575$$

Lastly, the estimate of the standard deviation of noise in Eq. (2.20) becomes

$$\hat{\sigma}^2 = \frac{SS_e}{n_y - n_\beta} = 0.1896, \quad \Rightarrow \quad \hat{\sigma} = 0.4354$$

which is close to the true standard deviation of noise $\sigma = 0.3953$, which can be obtained by comparing it with $y_{\text{true}}$ and the samples.

Quadratic PRS: The PRS is of the form $\hat{y}(x) = b_1 + b_2x + b_3x^2$, and using point (0, 0) as another sample point, we can define the following matrices and vectors for regression:

$$\mathbf{X} = \begin{bmatrix} 1 & -2 & 4 \\ 1 & -1 & 1 \\ 1 & 0 & 0 \\ 1 & 1 & 1 \\ 1 & 2 & 4 \end{bmatrix}, \quad \mathbf{y} = \begin{Bmatrix} -1.5 \\ -1.5 \\ 0 \\ 1.25 \\ 1.75 \end{Bmatrix}, \quad \mathbf{X}^T\mathbf{X} = \begin{bmatrix} 5 & 0 & 10 \\ 0 & 10 & 0 \\ 10 & 0 & 34 \end{bmatrix}, \quad \mathbf{X}^T\mathbf{y} = \begin{Bmatrix} 0 \\ 9.25 \\ 0.75 \end{Bmatrix}$$

By solving the normal equation in Eq. (2.14), the unknown regression coefficients are identified as $b_1 = -0.1071$, $b_2 = 0.925$, and $b_3 = 0.0536$. Therefore, the following quadratic PRS and residuals at sample locations can be obtained:

$$\hat{y}^{(2)}(x) = -0.1071 + 0.925x + 0.0536x^2, \quad \mathbf{e} = \begin{Bmatrix} 0.243 \\ -0.521 \\ 0.107 \\ 0.379 \\ -0.207 \end{Bmatrix}$$

The coefficient of multiple determination can be calculated from Eqs. (2.21) to (2.27) as

$$\bar{y} = 0, \quad SS_y = 9.125, \quad SS_r = 8.5964, \quad R^2 = 0.9421, \quad R_a^2 = 0.7684$$

We see that there is only a small improvement in $R^2$, and that $R_a^2$ is actually poorer, which is an indication that we have not gained any predictive capabilities by adding the quadratic terms. The same conclusion can be obtained based on the error metrics as

$$e_{\mathrm{RMS}} = 0.325, \quad e_{\mathrm{av}} = 0.291, \quad e_{\max} = 0.521$$

These errors have not changed much compared to the linear model, with small improvements in the RMS and maximum errors, and a small increase in the average error. The estimate of the standard deviation of the noise in the samples from Eq. (2.20) becomes

$$\hat{\sigma}^2 = \frac{SS_e}{n_y - n_\beta} = 0.2643, \quad \Rightarrow \quad \hat{\sigma} = 0.5141$$

which is a small increase over the linear case. This increase is another indication that the quadratic approximation is not better than the linear approximation.

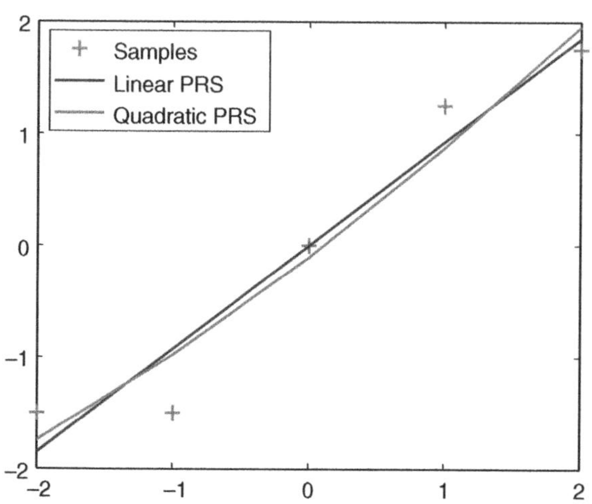


**Figure 2.7** Compares the linear and quadratic PRS along with five samples. There is no significant difference between the two surrogates, but indicates that the linear PRS is slightly better than the quadratic one.

We discussed before that PRS starts from the assumption that the model form is accurate, but samples have random noise that follows a normal distribution with zero mean. In reality, however, we do not know the correct model form. We do not know if the true functional form is polynomial either. From Taylor's expansion theorem, it is possible that any continuous function can be approximated using polynomials. Therefore, we can accept the idea that the true function is in the form of polynomials. Then the remaining question is what orders of polynomials should be used. As shown in the previous example, a higher-order polynomial does not mean better approximation. Therefore, in order to find an appropriate model form, we start from lower-order polynomials and gradually increase the order until the approximation becomes worse, or we start from higher-order polynomials and gradually remove basis functions that do not contribute significantly. In Section 2.5, we will discuss the second approach, which is called backward elimination.

**Example 2.6 Issue of the coefficient of multiple determination**

Compare the average error and the coefficient of multiple determination of two PRS surrogates that are generated from two true functions $y^{(1)}(x) = x$ and $y^{(2)}(x) = 0.1x$. In order to build the surrogates, generate 31 equal-interval samples from the design space $x \in [0, 30]$ and add random noise from the normal distribution $\sim N(0, 1^2)$. Check if $e_{\text{av}}$ and $R^2$ can be used to assess the accuracy of surrogates.

**Solution:**

In order to generate the same sequence of random numbers, the `rng default` command is used before generating random numbers in Matlab. After generating samples of true functions $y^{(1)}_{\text{true}}$ and $y^{(2)}_{\text{true}}$, the same random noises are added to the true functions. Since the input variable is one-dimensional, `polyfit` and `polyval` Matlab functions are used for building the linear PRS. The two PRS surrogates are, respectively,

$$\hat{y}^{(1)}(x) = 0.5728 + 0.9993x$$

$$\hat{y}^{(2)}(x) = 0.5728 + 0.0993x$$

Both PRSs have the same average error $e_{\text{av}} = 0.5628$. However, the coefficients of multiple determination are quite different. $R^2$ for $\hat{y}^{(1)}$ is 0.9806, while that of $\hat{y}^{(2)}$ is 0.3325. Therefore, even if the average error of the two surrogates is the same, $R^2$ of the first surrogate looks much better than the second. This happens because even if the noises are the same, $y^{(1)}$ varies much faster than $y^{(2)}$. As shown in Figure 2.8(a), the difference between samples and PRS surrogate seems small, while the difference looks significant in Figure 2.8(b). Therefore, $R^2$ does not reflect the accuracy of a surrogate model. It shows the ratios of the variation of PRS surrogate to the variation of samples. The following Matlab code is used to plot Figure 2.8.

```
rng default;
x=linspace(0,30,31); noise=randn(1,31);
y1true=x; y1=y1true + noise;
y2true=0.1*x; y2=y2true + noise;
b1=polyfit(x,y1,1);
b2=polyfit(x,y2,1);
y1p=polyval(b1,x);
y2p=polyval(b2,x);
eva1= sum(abs(y1p-y1true))/31;
eva2= sum(abs(y2p-y2true))/31;
meany1=mean(y1);
meany2=mean(y2);
```

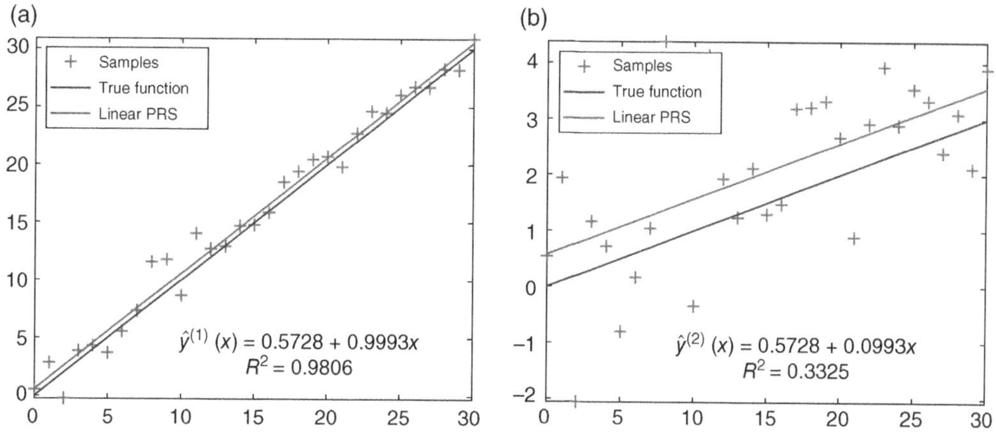


**Figure 2.8** Fitted linear polynomial response surface from (a) $y = x$ and (b) $y = 0.1x$.

```
SSy1= sum((y1-meany1).^2);
SSy2=sum((y2-meany2).^2);
SSr1=sum((y1p-meany1).^2);
SSr2=sum((y2p-meany2).^2);
R1=SSr1/SSy1;
R2=SSr2/SSy2;
figure(1)
plot(x,y1,'+',x,y1true,'b',x,y1p,'r');
legend('Samples','True function','Linear PRS');
figure(2)
plot(x,y2,'+',x,y2true,'b',x,y2p,'r');
legend('Samples','True function','Linear PRS');
```

The Matlab function `regress` can calculate the coefficient of multiple determination as a part of its model statistics. For example, the following code can be used to calculate $R^2$:

```
[b,~,~,~,stats]=regress(y, X);
R2=stats(1)
```

### 2.4.3 Cross-validation

The only true test of the predictive capabilities of the response surface is evaluating it at points not used in its construction. In that aspect, the coefficient of multiple determination cannot be a good measure to evaluate the goodness-of-fit. In order to evaluate the prediction capability of a surrogate, it would be necessary to have a separate set of samples that are not used in the fitting process. Therefore, it would be necessary to divide the entire set of samples into a training set and a validation set. The samples in the training set are used for fitting a surrogate, and the samples in the validation set are used for evaluating the accuracy or prediction capability of the surrogate. This may be considered wasteful because we do not use all the samples for fitting the best possible surrogate. Because additional tests or numerical evaluations of $y$ for validation are often expensive, it is worthwhile looking for a way of checking predictive capability without performing additional evaluations at new points.

Cross-validation is a resampling method that uses different portions of the samples to train and validate a model on different iterations. It is mainly used in settings where the goal is prediction, and one wants to estimate how accurately a predictive model will perform in practice. The goal of cross-validation is to test the model's ability to predict new samples that were not used in estimating it, in order to flag problems like overfitting or selection bias [18] and to give an insight into how the model will generalize to a new dataset. Cross-validation combines (averages) measures of fitness in prediction to derive a more accurate estimate of model prediction performance [19].

$k$-fold cross-validation: In $k$-fold cross-validation, the original samples are randomly partitioned into $k$ equal-sized subsets. Of the $k$ subsets, a single subset is retained as the validation sample set for testing the model, and the remaining $k-1$ subsets are used as training samples. The cross-validation process is then repeated $k$ times, with each of the $k$ subsets used exactly once as the validation set. The $k$ results can then be averaged to produce a single estimation. In this method, all samples are used for both training and validation, and each sample is used for validation exactly once. Figure 2.9 shows an example when $k = 5$. The entire samples are divided into five equal-sized subsets. Although the division looks sequential in the figure, in practice, subsets are generated randomly. The first four subsets are used to fit a surrogate, and the last subset is used to validate the surrogate. The outcome of the validation is in terms of prediction error $e_{p1}$. This process is repeated for other subsets and yields five prediction errors: $\mathbf{e}_p = \{e_{p1}, e_{p2}, e_{p3}, e_{p4}, e_{p5}\}^T$.

Leave-one-out cross-validation: Leave-one-out cross-validation involves using one sample as the validation set and the remaining $n_y - 1$ samples as the training set. That is, $n_y - 1$ samples are used to fit a surrogate, and the error between the surrogate prediction and the validation sample is calculated as a prediction error $e_{pi}$. This process is repeated in all $n_y$ samples to yield $n_y$ prediction errors. This is equivalent to k-fold cross-validation with $k = n_y$. The process looks similar to a jackknife; however, with cross-validation, one computes a statistic on the left-out sample(s), while with jackknife, one computes a statistic from the kept samples only. In this section, we will only explain leave-one-out cross-validation.

If the number of samples used for the fit is substantially larger than the number of coefficients, $n_y \gg n_\beta$, leaving out one sample will not change the quality of the fit significantly. We can therefore leave out one sample, fit the PRS to the remaining samples, and check the error at the point that was left out. Let us assume that we leave out the $i$th sample. The surrogate that is built without $i$th sample is denoted by $\hat{y}^{(i)}$. Then, the prediction error at the $i$th sample is

$$e_{pi} = y_i - \hat{y}^{(i)}(\mathbf{x}_i) \tag{2.28}$$

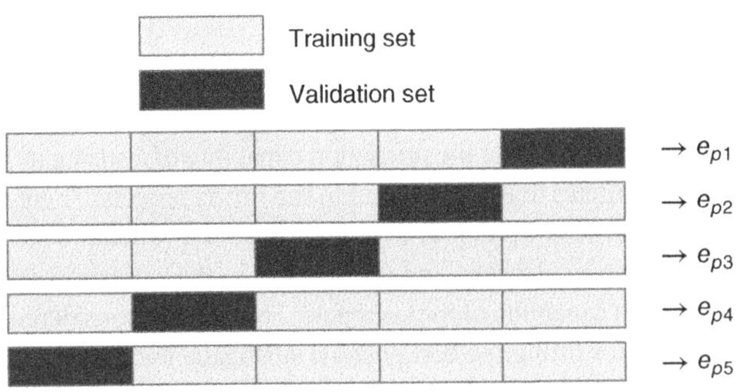


**Figure 2.9** Fivefold cross-validation.

Note that the prediction error $e_{pi}$ is different from the regression error: $e_i = y_i - \hat{y}(\mathbf{x}_i)$. We can then repeat the procedure at each sample to obtain the vector of prediction errors $\mathbf{e}_p^T = \{e_{p1}, e_{p2}, ..., e_{pn_y}\}$. Since we need a single scalar measure to evaluate the prediction accuracy of a surrogate, the square sum of the prediction errors can be used, which is known as PRESS (predicted residual error sum of squares). The PRESS can be defined as

$$\text{PRESS} = \sum_{i=1}^{n_y} e_{pi}^2 = \sum_{i=1}^{n_y} \left[y_i - \hat{y}^{(i)}(\mathbf{x}_i)\right]^2 \tag{2.29}$$

The PRESS statistic provides a summary measure of the fit of a model to a sample of observations that were not themselves used to estimate the model [20]. The PRESS statistic can be calculated for a number of surrogate models for the same set of samples, with the lowest values of PRESS indicating the best surrogate. Surrogate models that are overfit to samples would tend to give small residuals for samples included in the model-fitting, but large residuals for samples that are excluded.

Another advantage of the PRESS statistics is that it is independent of surrogate models. That is, it can be used in any surrogate model. Therefore, it is versatile. The only bottleneck of the PRESS statistic is that it can be expensive to fit surrogate models $n_y$ times. As $n_y$ increases, this can be expensive. However, the baseline assumption of surrogate modeling is that the major cost comes from obtaining samples, which means performing experiments or running expensive computer simulations. Therefore, the fitting cost of the surrogate can be ignorable.

In the case of linear regression, however, the requirement of fitting surrogates $n_y$ times can be relieved. The PRESS statistics can be evaluated with one surrogate fitting. First, we derive the linear relationship between the surrogate predictions and samples. By evaluating the surrogate predictions at all the sample points, we have

$$\hat{\mathbf{y}} = \mathbf{X} \cdot \mathbf{b} = \mathbf{X}\left(\mathbf{X}^T\mathbf{X}\right)^{-1}\mathbf{X}^T\mathbf{y} = \mathbf{E} \cdot \mathbf{y} \tag{2.30}$$

where $n_y \times n_y$ symmetric matrix $\mathbf{E}$ is an idempotent matrix. An important property of the idempotent matrix is that $\mathbf{E}^k = \mathbf{E}$. Using this relationship, the regression errors at sample locations in Eq. (2.11) can be written as

$$\mathbf{e} = \mathbf{y} - \hat{\mathbf{y}} = \left[\mathbf{I} - \mathbf{X}\left(\mathbf{X}^T\mathbf{X}\right)^{-1}\mathbf{X}^T\right]\mathbf{y} = [\mathbf{I} - \mathbf{E}]\mathbf{y} \tag{2.31}$$

where $[\mathbf{I} - \mathbf{E}]$ is also an idempotent matrix. In particular, the diagonal term $1 - E_{ii}$ represents the importance of sample $y_i$ on error $e_i$. It would be difficult to derive, but it can be shown that the relationship between the regression error $e_i$ and the prediction error $e_{pi}$ can be written as

$$e_{pi} = \frac{e_i}{1 - E_{ii}} \tag{2.32}$$

Basically, the prediction error at the $i$th sample is the regression error scaled by the diagonal term $1 - E_{ii}$. Therefore, the PRESS statistics can be calculated from a single surrogate fit as

$$\text{PRESS} = \sum_{i=1}^{n_y} \left(\frac{e_i}{1 - E_{ii}}\right)^2 \tag{2.33}$$

The RMS error of the PRESS statistic can be defined as

$$e_{\text{PRESS}} = \sqrt{\frac{1}{n_y - 1} \sum_{i=1}^{n_y} \left(\frac{e_i}{1 - E_{ii}}\right)^2} \tag{2.34}$$

The $e_{\text{PRESS}}$ is preferred over PRESS because the former has the same unit as QoI.

It should be noted that the matrix $\mathbf{X}^T\mathbf{X}$ is often ill-conditioned, especially for large problems, and then the calculation of the matrix $\mathbf{E}$ from Eq. (2.30) is not very reliable. However, even the direct calculation of $\mathbf{e}_p$, by performing the $n_y$ times of surrogate fits, is usually less expensive than carrying out additional experiments in order to test the accuracy of the surrogate.

**Example 2.7 Standard error versus PRESS error**

Generate 30 equal-interval samples from the true function $y = x$, $x \in [1, 30]$ with randomly distributed noise $\sim N(0, 1^2)$. Fit a linear PRS and compare (i) the constant term $b_1$ with the mean of noise, (ii) the standard error with the standard deviation of noise samples, and (iii) the standard error and the PRESS statistic.

**Solution:**

We used `rng default` command to keep the random number sequence the same. The mean of the noise is $\mu_{\text{noise}} = 0.5519$ and the standard deviation of the noise is $\sigma_{\text{noise}} = 1.3$. Fitting the 30 samples with a linear PRS yields the following expression:

$$\hat{y}(x) = 0.5981 + 0.997x$$

It is noted that the constant term $b_1 = 0.5981$ is the same as the mean of noise. This is because the linear regression is unbiased. That is, even if samples were generated from the true function since the noise samples were biased by 0.5519, the PRS surrogate is also shifted by a similar level of bias. This is not identical to the mean of noise because some errors are shifted to the linear coefficient $b_2$.

The standard error is calculated based on the square root of Eq. (2.20): $\hat{\sigma} = 1.3228$, which is close to the standard deviation of noise, 1.30. In addition, the RMS error of the PRESS statistic, $e_{\text{PRESS}}$, is calculated using Eq. (2.34): $e_{\text{PRESS}} = 1.3907$, which is also close to the standard error.

The actual RMS error between $\hat{y}(x)$ and the true function is 0.5619, which is much smaller than the standard error. This is because a large number of samples filtered the noise. As we discussed in Section 2.2, the surrogate is more accurate than the samples.

The following Matlab code is used for the example:

```
rng default;
x=[1:30]'; noise=randn(30,1);y=x+noise;
X=[ones(30,1) x];
b=regress(y,X)
yfit=b(1)+b(2)*x;
error=y-yfit;
sigma=sqrt(error'*error/28)
mean(noise)
std(noise)
M=X'*X; E=X*inv(M)*X';
d=diag(E)
ep=error./(1-d);
PRESS=ep'*ep
ePRESS=sqrt(PRESS/29)
```

**Example 2.8 Cross-validation error**

Consider the quadratic PRS in Example 2.5, calculate the PRESS statistic by repeating the fitting process by five times, and compare it with the PRESS statistic from a single fit in Eq. (2.33).

**Solution:**

First, we calculate the PRESS statistics from a single fit. From Example 2.5, the idempotent matrix can be obtained as

$$\mathbf{E}=\mathbf{X}\left(\mathbf{X}^T\mathbf{X}\right)^{-1}\mathbf{X}^T=\begin{bmatrix} 0.8857 & 0.2671 & -0.0857 & -0.1429 & 0.0857 \\ 0.2571 & 0.3714 & 0.3429 & 0.1714 & -0.1429 \\ -0.0857 & 0.3429 & 0.4857 & 0.3429 & -0.0857 \\ -0.1429 & 0.1714 & 0.3429 & 0.3714 & 0.2571 \\ 0.0857 & -0.1429 & -0.0857 & 0.2571 & 0.8857 \end{bmatrix}$$

Therefore, the regression errors and the prediction errors from a single fit can be obtained from Eq. (2.32) as

$$\mathbf{e}=\begin{Bmatrix} 0.243 \\ -0.521 \\ 0.107 \\ 0.379 \\ -0.207 \end{Bmatrix}, \quad \mathbf{e}_p=\left\{\frac{e_i}{1-E_{ii}}\right\}=\begin{Bmatrix} 2.125 \\ -0.8295 \\ 0.2083 \\ 0.6023 \\ -1.8125 \end{Bmatrix}$$

It is noted that the prediction error significantly increased in the first and last samples. This is because $E_{11}$ and $E_{55}$ are close to one. Physically, it means that when the first and the last sample is dropped, they belong to the extrapolation region after fitting the surrogate.

Second, we demonstrate the case when the third sample (0,0) is dropped. In this case, the following matrices and vectors can be defined:

$$\mathbf{X}=\begin{bmatrix} 1 & -2 & 4 \\ 1 & -1 & 1 \\ 1 & 1 & 1 \\ 1 & 2 & 4 \end{bmatrix}, \quad \mathbf{y}=\begin{Bmatrix} -1.5 \\ -1.5 \\ 1.25 \\ 1.75 \end{Bmatrix}, \quad \mathbf{X}^T\mathbf{X}=\begin{bmatrix} 4 & 0 & 10 \\ 0 & 10 & 0 \\ 10 & 0 & 34 \end{bmatrix}, \quad \mathbf{X}^T\mathbf{y}=\begin{Bmatrix} 0 \\ 9.25 \\ 0.75 \end{Bmatrix}$$

By solving the normal equation in Eq. (2.14), the unknown regression coefficients are identified as $b_1 = -0.2083$, $b_2 = 0.925$, and $b_3 = 0.0833$. Therefore, the following quadratic PRS can be obtained:

$$\hat{y}^{(3)}(x) = -0.2083 + 0.925x + 0.0833x^2$$

Now the prediction error is calculated at the dropped sample location (0, 0), which is

$$e_{p3} = y_3 - \hat{y}^{(3)}(0) = 0 + 0.2083 = 0.2083$$

This process is repeated for every sample position to obtain the following prediction errors:

$$\mathbf{e}_p = \left\{y_i - \hat{y}^{(i)}(x_i)\right\} = \begin{Bmatrix} 2.125 \\ -0.8295 \\ 0.2083 \\ 0.6023 \\ -1.8125 \end{Bmatrix}$$

Therefore, the prediction error from Eq. (2.32) is identical to the prediction error from the definition of cross-validation. The following Matlab code is used to calculate the prediction errors:

```
X=[1 -2 4;1 -1 1;1 0 0;1 1 1;1 2 4];
y=[-1.5 -1.5 0 1.25 1.75]';
for i=1:5
 Xt=X; Xt(i,:)=[];
 yt=y;yt(i)=[];
 bt=inv(Xt'*Xt)*Xt'*yt;
 ept(i)=y(i)-X(i,:)*bt;
end
```

Since the PRESS statistic measures the prediction accuracy of a surrogate, a surrogate is considered to be accurate when the PRESS statistic is small. When multiple surrogates are compared, the surrogate whose PRESS statistic is the smallest can be considered the best surrogate to fit the samples. In the case of PRS surrogates, however, the PRESS static will not converge to zero. This is expected because the PRS starts from the assumption that the samples have random noise, and a good surrogate is supposed to fit the trend of the samples, not the noise. Therefore, if a surrogate is identical to the true function, the diagonal term of the idempotent matrix $E_{ii} = 0$ and the PRESS statistic in Eq. (2.33) becomes the square-sum-error $SS_e$.

## 2.5 Confidence of Coefficients and Backward Elimination

In Example 2.5, we compared a linear PRS surrogate to a quadratic PRS surrogate. However, we do not have to limit ourselves to PRS surrogates that have all the terms up to a particular order. For example, we can consider a quadratic model without the constant term, $\hat{y}(x) = b_1x + b_2x^2$. Similarly, with two variables, it is common for people to consider a model of the form $\hat{y}(\mathbf{x}) = b_1 + b_2x_1 + b_3x_2 + b_4x_1x_2$. This model does not include quadratic terms in $x_1$ and $x_2$, but it includes the "interaction" term $x_1x_2$. We can stipulate such a partial model based on some knowledge of the behavior of the true function. However, most often, such partial models are created by discarding terms with coefficients that cannot be accurately estimated based on the samples. Such coefficients do not have much effect on the accuracy of the fit of the given samples and leaving them in the model can reduce its predictive quality for design points where these coefficients have a large effect on the prediction.

The first straightforward idea is to remove those basis functions whose coefficients are small compared to other coefficients. For example, let us consider the following quadratic PRS surrogate:

$$\hat{y}(x_1, x_2) = 10.5 + 0.01x_1 + 9.87x_2 - 5x_1^2 + 7.6x_1x_2 + 0.02x_2^2 \tag{2.35}$$

If we assume that both input variables $x_1$ and $x_2$ are normalized $x_1, x_2 \in [0, 1]$, then it is clear that the linear $x_1$ term and quadratic $x_2^2$ term are not important as their coefficients are small compared to other coefficients. That is, even if $x_1$ and $x_2^2$ vary, the surrogate prediction will not change much because of their small coefficients. Therefore, these two basis functions can be removed to yield

$$\hat{y}(x_1, x_2) = 10.5 + 9.87x_2 - 5x_1^2 + 7.6x_1x_2 \tag{2.36}$$

However, such a heuristic approach may cause a problem if other basis functions with a large coefficient vary significantly with a small change in samples. Therefore, it would be necessary to develop a systematic approach to determine which basis functions or which coefficients should be removed from the surrogate model.

As we discussed several times, PRS surrogates start from the assumption that there is a true function, and samples include random noise to the true function. We try to build a surrogate that can filter out random noise and approximate the true function as closely as possible. Because of this randomness in noise, however, different samples can be obtained by different realizations of the noise, which end up identifying different coefficients. Therefore, it is fair to say that the identified coefficients of the PRS are random as well, whose distribution we want to characterize. A straightforward way of identifying the distribution of coefficients is to generate many sets of samples by adding random noise to the true function. By fitting these multiple sets of samples, we can obtain multiple sets of coefficients, from which we can estimate the distribution. However, this approach is only possible when we know the true function and we can generate multiple sets of samples. In reality, it is already expensive to generate one set of samples. Therefore, it would be necessary to come up with a method of identifying the distribution of coefficients with one set of samples.

We can identify the distribution of these coefficients without changing the samples by estimating the standard deviation of the coefficients. That is, we assume that the coefficients are normally distributed with the values obtained from linear regression as a mean. Therefore, the standard deviation is the only unknown information needed to be estimated. Since a change in a sample can affect multiple coefficients simultaneously, it would be necessary to estimate the covariance of the coefficients. Let us first introduce the covariance matrix $\boldsymbol{\Sigma}_{\mathbf{b}}$ of the vector of random coefficients **b**, which is defined as

$$\boldsymbol{\Sigma}_{\mathbf{b}} = [\mathbf{b} - E(\mathbf{b})][\mathbf{b} - E(\mathbf{b})]^T \tag{2.37}$$

where $E(\cdot)$ is the expected value of a random variable. That is, due to random noise, we get different vectors **b** by fitting samples from different realizations. $E(\mathbf{b})$ is the expected value (or average over a very large number of experiments) of **b**, and $\mathbf{b} - E(\mathbf{b})$ is the deviation of **b** from its expected value. Then the covariance matrix is the expected value of the products of various components of the deviations. In particular, the diagonal terms of the matrix are by definition the variance of the components of **b**, while the off-diagonal terms are a measure of the correlation between components. It is possible to show that

$$\boldsymbol{\Sigma}_{\mathbf{b}} = \sigma^2 \left[\mathbf{X}^T\mathbf{X}\right]^{-1} \tag{2.38}$$

That is, due to the randomness in noise, the estimated coefficients from the regression are also random, and their randomness follows a normal distribution $\sim N(\mathbf{b}, \boldsymbol{\Sigma}_{\mathbf{b}})$.

Note that $\sigma$ is the standard deviation of random noise, which is unknown. Instead, we can use its estimate $\hat{\sigma}$ derived in Eq. (2.20). Now, we can estimate the standard deviation of individual coefficients as

$$\sigma_{b_i} = \hat{\sigma}\sqrt{\left[(\mathbf{X}^T\mathbf{X})^{-1}\right]_{ii}} \tag{2.39}$$

which is referred to as the standard error of the coefficient. A useful measure of the accuracy of a component of **b** is the estimate of the CoV, **c**, which is the ratio of the standard deviation to the absolute value of the component. The CoV of the $i$th component $c_i$ is defined as

$$c_i = \frac{\sigma_{b_i}}{|b_i|} \tag{2.40}$$

If $c_i > 1$, it means that the uncertainty (i.e., the standard deviation) is larger than the coefficient itself, which means that the identified coefficient has very low confidence.

The coefficient $|b_i|$ and its CoV $c_i$ can be used to identify insignificant basis functions of the PRS. An important basis function should have a large magnitude of the coefficient such that the surrogate prediction strongly depends on the term. Also, if the CoV is small, it means that the coefficient is well-identified. On the other hand, when the coefficient cannot be identified accurately for given samples, it means that the coefficient does not have much effect on the accuracy of the PRS.

In many PRS procedures, the quantity that is used to assess the need for a coefficient is the inverse of $c_i$, which is called the test statistic, or the t-statistic. The CoV is used in a strategy called backward elimination (Myers and Montgomery [4], p. 650). In this strategy, we eliminate the coefficient with the largest CoV. After removing the basis function, the surrogate is performed with the reduced basis function with the same samples. This process is repeated until the CoV of the remaining terms is small enough, or we can use a measure such as $R_a^2$ or $e_{\text{PRESS}}$ to indicate when eliminating additional terms reduces the accuracy of the PRS. This is illustrated by applying this strategy to the quadratic PRS in Example 2.5.

**Example 2.9 Backward elimination**

Use backward elimination, starting with the quadratic model of Example 2.5, to find the model with the highest value of $R_a^2$. Confirm your conclusion by recalculating the coefficients for a slight perturbation of the sample at $x = 1$ to $y(1) = 1.35$.

**Solution:**

The quadratic PRS that we obtained from Example 2.5 was $\hat{y}(x) = -0.1071 + 0.925x + 0.0536x^2$, which has the adjusted coefficient of multiple determination $R_a^2 = 0.8841$ and the estimated noise standard deviation $\hat{\sigma} = 0.5141$. In order to determine which coefficient should be removed, the inverse of the moment matrix and the covariance matrix of the coefficients are calculated as

$$(\mathbf{X}^T\mathbf{X})^{-1} = \begin{bmatrix} \frac{17}{35} & 0 & -\frac{1}{7} \\ 0 & \frac{1}{10} & 0 \\ -\frac{1}{7} & 0 & \frac{1}{14} \end{bmatrix}, \quad \boldsymbol{\Sigma}_b = \begin{bmatrix} 0.1284 & 0 & -0.03776 \\ 0 & 0.02643 & 0 \\ -0.03776 & 0 & 0.01888 \end{bmatrix}$$

Using these matrices, the coefficients of variation of all the coefficients can be calculated as

$$c_1 = \frac{\sqrt{0.1284}}{0.1071} = 3.35, \quad c_2 = \frac{\sqrt{0.02643}}{0.925} = 0.176, \quad c_3 = \frac{\sqrt{0.01888}}{0.0536} = 2.56$$

It turns out that the first coefficient $b_1$ has the largest CoV, and thus, we can remove the constant term. Therefore, the reduced PRS form becomes $\hat{y}(x) = b_2x + b_3x^2$. Note that we cannot use $b_2$ and $b_3$ from the initial surrogate because the reduced surrogate has different basis functions. Therefore, we need to fit the reduced surrogate again with the same samples, with the following matrices and vectors:

$$\mathbf{X} = \begin{bmatrix} -2 & 4 \\ -1 & 1 \\ 0 & 0 \\ 1 & 1 \\ 2 & 4 \end{bmatrix}, \quad \mathbf{y} = \begin{Bmatrix} -1.5 \\ -1.5 \\ 0 \\ 1.25 \\ 1.75 \end{Bmatrix}, \quad \mathbf{X}^T\mathbf{X} = \begin{bmatrix} 10 & 0 \\ 0 & 34 \end{bmatrix}, \quad \mathbf{X}^T\mathbf{y} = \begin{Bmatrix} 9.25 \\ 0.75 \end{Bmatrix}$$

The normal equation yields $b_2 = 0.925$ and $b_3 = 0.0221$. That is, $\hat{y}(x) = 0.925x + 0.0221x^2$. Therefore, the linear coefficient does not change, but the quadratic coefficient changes significantly from 0.0536 to 0.0221. The reduced surrogate has the following regression errors:

$$\mathbf{e} = \{-0.262, 0.597, 0.0, -0.3038, 0.188\}^T$$

And the various error metrics are $SS_y = 9.125$, $SS_e = 0.5522$, $R^2 = 0.9394$, $R_a^2 = 0.9193$, and $\hat{\sigma} = 0.429$. Compared to $R_a^2$ of the initial quadratic PRS, $R_a^2$ of the reduced PRS is increased from 0.8841 to 0.9193. Therefore, we can say that eliminating the constant term improves the prediction capability.

In order to check if a further elimination improves the surrogate, the coefficients of variation are calculated again. First, the following matrices are defined:

$$(\mathbf{X}^T\mathbf{X})^{-1} = \begin{bmatrix} \frac{1}{10} & 0 \\ 0 & \frac{1}{34} \end{bmatrix}, \quad \mathbf{\Sigma}_b = \begin{bmatrix} 0.0184 & 0 \\ 0 & 0.00541 \end{bmatrix}$$

Using these matrices, the coefficients of variation of all the coefficients can be calculated as

$$c_2 = \frac{\sqrt{0.0184}}{0.925} = 0.147, \quad c_3 = \frac{\sqrt{0.00541}}{0.0221} = 3.33$$

Based on the three coefficients of variation, the last coefficient $b_3$ has the largest CoV, and thus, we can remove the quadratic term. Therefore, the reduced PRS form becomes $\hat{y}(x) = b_2x$. Fitting this linear PRS yields $\hat{y}(x) = 0.925x$ with error metrics $R_a^2 = 0.9169$ and $\hat{\sigma} = 0.4354$. Therefore, $R_a^2$ is slightly decreased from the previous PRS surrogate, which means the previous PRS surrogate is better than this one. Therefore, we stop the elimination process and conclude that the best PRS surrogate to fit the samples is $\hat{y}(x) = 0.925x + 0.0221x^2$.

Now let the sample at $x = 1$ is perturbed from $y(1) = 1.25$ to $y(1) = 1.35$, which corresponds to 8% perturbation. Then, the coefficients of the original quadratic PRS are perturbed by

$$\hat{y}_{\text{original}}(x) = -0.1071 + 0.925x + 0.0536x^2$$

$$\hat{y}_{\text{perturbed}}(x) = -0.0729 + 0.935x + 0.0465x^2$$

That is, the constant coefficient is changed by 30%, the linear coefficient by 1%, and the quadratic coefficient by 13%. Therefore, it makes sense that the constant coefficient is the most uncertain and needs to be eliminated first. In the case of the reduced PRS, the coefficients are perturbed by

$$\hat{y}_{\text{original}}(x) = 0.925x + 0.02205x^2$$

$$\hat{y}_{\text{perturbed}}(x) = 0.935x + 0.025x^2$$

The linear coefficient is changed by 1%, while the quadratic coefficient is by 13%. However, the elimination of the quadratic coefficient may decrease the prediction capability of the surrogate.

## 2.6 Prediction Variance

In Section 2.4, we discussed cross-validation, which is a good metric to measure the prediction capability at unsampled locations. The good characteristic of cross-validation is that it is model-independent, which means that it can be applied to any kind of surrogate, not limited to linear regression PRS. However, cross-validation can provide the prediction accuracy of a surrogate in the form of prediction errors at sample locations and the RMS of prediction errors. The latter is a good estimate of the standard deviation of the noise. Therefore, cross-validation is a good metric to assess the overall accuracy of a surrogate. However, the ultimate goal of the accuracy study is to estimate the prediction accuracy of the surrogate at unsampled locations.

### 2.6.1 Prediction Uncertainty

If we limit ourselves to the PRS surrogate with linear regression, it is possible to estimate the prediction accuracy at unsampled locations. As mentioned before, the PRS surrogate that we obtained from regression in Eq. (2.10) is actually the mean of the prediction. This is because the PRS surrogate starts from the assumption of random noise. Because of the random distribution of noise $\sim N(0, \sigma^2)$, the regression coefficients are also random $\sim N(\mathbf{b}, \boldsymbol{\Sigma}_\mathbf{b})$, and thus, the prediction at unsampled locations as well. Since the relationship between the prediction and coefficients is linear and since the basis functions are not random, the prediction at an unsampled point $\mathbf{x}_p$will also follow a normal distribution $\sim N\left(\hat{y}(\mathbf{x}_p), \sigma_y^2(\mathbf{x}_p)\right)$. That is, the obtained PRS surrogate is a mean prediction and $\sigma_y^2(\mathbf{x}_p)$ is the prediction variance. Since this uncertainty in prediction is caused by the randomness in noise, it makes sense that the prediction uncertainty is proportional to the standard deviation $\sigma$ of the noise.

Also, the prediction accuracy is usually good close to the sample locations because even if the samples include random noise, the regression process filters out the noise, and the samples represent the trend of the true function. Therefore, as the prediction point moves away from sample locations, the prediction becomes inaccurate; that is, the prediction uncertainty increases. It would be better to explain the meaning of prediction uncertainty using a normal distribution. Figure 2.10 shows the probability density function of a normal distribution. The mean location corresponds to the surrogate prediction $\hat{y}(\mathbf{x}_p)$. Then the probability that the true function is located within the range $[\hat{y}(\mathbf{x}_p) - \sigma_y(\mathbf{x}_p), \hat{y}(\mathbf{x}_p) + \sigma_y(\mathbf{x}_p)]$ is 68%, $[\hat{y} - 2\sigma_y, \hat{y} + 2\sigma_y]$ is 95%, and

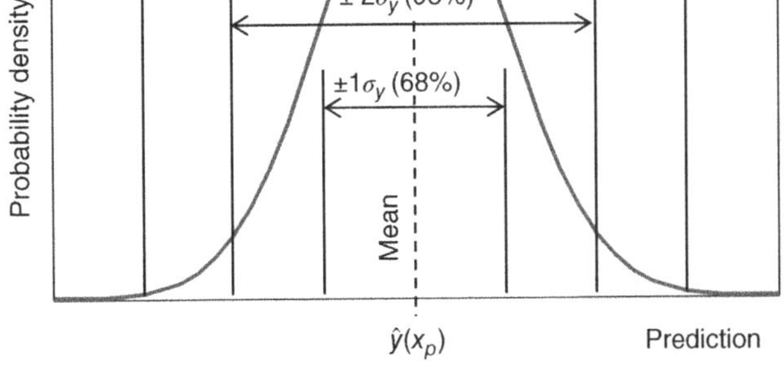


**Figure 2.10** Normal distribution of surrogate prediction.

$[\hat{y}-3\sigma_y, \hat{y}+3\sigma_y]$ is 99.7%. If the prediction uncertainty $\sigma_y(\mathbf{x}_p)$ is large, then the chance that the surrogate prediction is close to the true function is low. Therefore, it is important to have a low prediction uncertainty in order to have a good prediction capability. If $\sigma_y$ cannot be low everywhere, it should be low at the point of interest, such as the optimum design point.

An important thing to note in prediction uncertainty is that it tends to be low in the interpolation region, while it tends to increase fast in the extrapolation region. As shown in Figure 1.5, the interpolation region is within the convex hull of the sample points, while the extrapolation region is outside of the convex hull. The prediction in the extrapolation region is associated with large errors. Especially for a high-dimensional domain, extrapolation usually cannot be avoided.

The PRS expression in Eq. (2.10) can be written as $\hat{y}(\mathbf{x}_p) = \boldsymbol{\xi}(\mathbf{x}_p)^T\mathbf{b}$. From the viewpoint of uncertainty, the vector of basis functions $\boldsymbol{\xi}(\mathbf{x}_p)$ is deterministic, while the vector of coefficients $\mathbf{b}$ is uncertain. In probability theory, the variance of a random variable $y = ax$ with variance $V[x] = \sigma^2$ can be obtained by $V[y] = a^2V[x] = a^2\sigma^2$. When the random variable is a vector with multiple components, the covariance matrix is used instead. Therefore, the variance of the prediction $\hat{y}(\mathbf{x}_p)$ can be written as $V[\hat{y}(\mathbf{x}_p)] = \boldsymbol{\xi}(\mathbf{x}_p)^T\boldsymbol{\Sigma}_\mathbf{b}\boldsymbol{\xi}(\mathbf{x}_p) = \hat{\sigma}^2\boldsymbol{\xi}(\mathbf{x}_p)^T(\mathbf{X}^T\mathbf{X})^{-1}\boldsymbol{\xi}(\mathbf{x}_p)$. The square root of the prediction variance is called the standard error of prediction, defined as

$$\sigma_y = \hat{\sigma}\sqrt{\boldsymbol{\xi}(\mathbf{x}_p)^T(\mathbf{X}^T\mathbf{X})^{-1}\boldsymbol{\xi}(\mathbf{x}_p)} \tag{2.41}$$

where we used $\hat{\sigma}$ as an estimate of $\sigma$. The standard error of prediction increases proportionally to the standard deviation of the noise. Also, the term $\boldsymbol{\xi}(\mathbf{x}_p)^T(\mathbf{X}^T\mathbf{X})^{-1}\boldsymbol{\xi}(\mathbf{x}_p)$ increases fast as the prediction point $\mathbf{x}_p$ moves away from sample points.

### 2.6.2 Sample Sensitivity

Another interesting piece of information related to prediction accuracy is the sensitivity of prediction with respect to samples. That is, how much the prediction $\hat{y}(\mathbf{x}_p)$ will vary due to a change in one sample $y_i$? This information is particularly important if we want to find out the most important samples. Of course, the locations of samples significantly affect the prediction performance. However, this topic will be discussed separately in Chapter 3. In this section, it is assumed that the locations of all samples are fixed, and only the value of QoI $y_i$ varies.

This information is readily available from the definition of PRS surrogate with identified coefficients

$$\hat{y}(\mathbf{x}_p) = \boldsymbol{\xi}(\mathbf{x}_p)^T\mathbf{b} = \boldsymbol{\xi}(\mathbf{x}_p)^T(\mathbf{X}^T\mathbf{X})^{-1}\mathbf{X}^T\mathbf{y} \tag{2.42}$$

Since sample locations are fixed, the moment matrix $\mathbf{X}^T\mathbf{X}$ is fixed. Differentiating Eq. (2.42) with respect to $i$th component of $\mathbf{y}$ yields

$$\frac{\partial\hat{y}(\mathbf{x}_p)}{\partial y_i} = \left\{\boldsymbol{\xi}(\mathbf{x}_p)^T(\mathbf{X}^T\mathbf{X})^{-1}\mathbf{X}^T\right\}_i \tag{2.43}$$

It is noted that the sample sensitivity in Eq. (2.43) only shows the effect of the sample value on the mean prediction. Its effect on prediction uncertainty is more complicated through the standard deviation of the noise.

**Example 2.10 2-dimensional PRS with three samples**
The design space of a linear PRS $\hat{y}(x_1, x_2) = b_1 + b_2x_1 + b_3x_2$ is given by $-1 \leq x_1, x_2 \leq 1$. (a) When three sample locations are given as $(x_1, x_2) = (-1, -1), (-1, 1), (1, -1)$, calculate the standard error of prediction in terms of the noise standard deviation $\hat{\sigma}$ at all corner points and the center of the design space. (b) Find the location and value or the minimum standard error of prediction. (c) If an additional sample is given at (1, 1), repeat part (a).

**Solution:**

a) With the three samples, Figure 2.11(a) shows the interpolation and extrapolation regions along with the samples. For linear PRS, the vector of basis functions and design matrix are defined as

$$\boldsymbol{\xi} = \begin{Bmatrix} 1 \\ x_1 \\ x_2 \end{Bmatrix}, \quad \mathbf{X} = \begin{bmatrix} 1 & -1 & -1 \\ 1 & -1 & 1 \\ 1 & 1 & -1 \end{bmatrix}, \quad \mathbf{X}^T\mathbf{X} = \begin{bmatrix} 3 & -1 & -1 \\ -1 & 3 & -1 \\ -1 & -1 & 3 \end{bmatrix}, \quad (\mathbf{X}^T\mathbf{X})^{-1} = \frac{1}{4}\begin{bmatrix} 2 & 1 & 1 \\ 1 & 2 & 1 \\ 1 & 1 & 2 \end{bmatrix}$$

The standard error of prediction in Eq. (2.41) is given as

$$\sigma_y = \hat{\sigma}\sqrt{\boldsymbol{\xi}^T(\mathbf{X}^T\mathbf{X})^{-1}\boldsymbol{\xi}} = \hat{\sigma}\sqrt{0.5(1 + x_1 + x_2 + x_1^2 + x_2^2 + x_1x_2)}$$

The standard error at the three sample locations is $\sigma_y = \hat{\sigma}$. That is, the prediction uncertainty at the sample location is the same as the uncertainty of the sample itself. At the center of the design space $(x_1, x_2) = (0, 0)$, the standard error is $\sigma_y = \hat{\sigma}/\sqrt{2}$, which is lower than that of the sample locations. On the other hand, the unsampled corner $(x_1, x_2) = (1, 1)$ has the highest prediction uncertainty of $\sigma_y = \sqrt{3}\hat{\sigma}$. This is because this corner is the farthest extrapolation point.

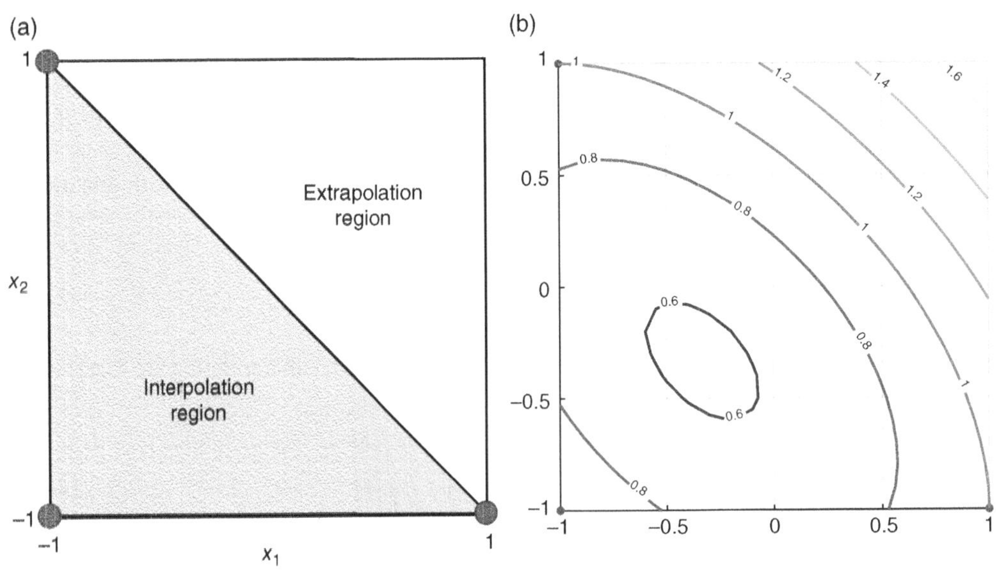


**Figure 2.11** Interpolation and extrapolation regions defined by samples in Example 2.10. (a) Interpolation and extrapolation regions and (b) prediction uncertainty.

b) It is interesting to note that the prediction uncertainty is not minimum at sample locations. Figure 2.11(b) shows the contour plot of the prediction uncertainty. It seems that the prediction uncertainty is minimum inside the interpolation region. The minimum point can be found by differentiating the standard error of prediction in Eq. (2.41), as

$$\frac{\partial \sigma_y^2}{\partial x_1} = \frac{\hat{\sigma}^2}{2}(1 + 2x_1 + x_2) = 0$$

$$\frac{\partial \sigma_y^2}{\partial x_2} = \frac{\hat{\sigma}^2}{2}(1 + 2x_2 + x_1) = 0$$

Solving the above two equations solves for $x_1 = x_2 = -1/3$. The prediction uncertainty at this point is $\sigma_y = \hat{\sigma}/\sqrt{3}$. The minimum point is, in fact, the centroid of the interpolation region. The following Matlab code is used to plot Figure 2.11(b).

```
x=[-1 -1 1]; y=[-1 1 -1];
[X,Y]=meshgrid(-1:.1:1, -1:.1:1);
Z=sqrt(.5*(1+X+Y+X.^2+Y.^2+X.*Y));
v=linspace(0.6,1.8,7)
scatter(x,y,'filled');
grid on; hold on
[C,h]=contour(X,Y,Z,v);
clabel(C,h)
```

c) When an additional sample is added at (1, 1), the vector of basis functions and design matrix are defined as

$$\xi = \begin{Bmatrix} 1 \\ x_1 \\ x_2 \end{Bmatrix}, \quad \mathbf{X} = \begin{bmatrix} 1 & -1 & -1 \\ 1 & -1 & 1 \\ 1 & 1 & -1 \\ 1 & 1 & 1 \end{bmatrix}, \quad \mathbf{X}^T\mathbf{X} = \begin{bmatrix} 4 & 0 & 0 \\ 0 & 4 & 0 \\ 0 & 0 & 4 \end{bmatrix}, \quad (\mathbf{X}^T\mathbf{X})^{-1} = \frac{1}{4}\begin{bmatrix} 1 & 0 & 0 \\ 0 & 1 & 0 \\ 0 & 0 & 1 \end{bmatrix}$$

The standard error of prediction in Eq. (2.41) is given as

$$\sigma_y = \hat{\sigma}\sqrt{\xi^T(\mathbf{X}^T\mathbf{X})^{-1}\xi} = \hat{\sigma}\sqrt{0.25(1 + x_1^2 + x_2^2)}$$

The standard error of prediction at the four corner points is $\sigma_y = \sqrt{3}\hat{\sigma}/2$, which is lower than the standard error of noise. This means that the surrogate is more accurate than the samples. The minimum prediction uncertainty occurs at the origin of the design space (0, 0), whose value is $\sigma_y = \hat{\sigma}/2$. Figure 2.12 shows the contour plot of the prediction uncertainty. It is noted that the additional sample does not improve the low prediction uncertainty but significantly reduces the large prediction uncertainty in the extrapolation region.

### 2.6.3 Prediction Variance with Variable Noise

So far, we consider the case when the noises of all samples are from an identical distribution $\sim N(0, \sigma^2)$. From this assumption, the prediction variance is determined as $V[\hat{y}(\mathbf{x}_p)] = \sigma^2\xi(\mathbf{x}_p)^T$

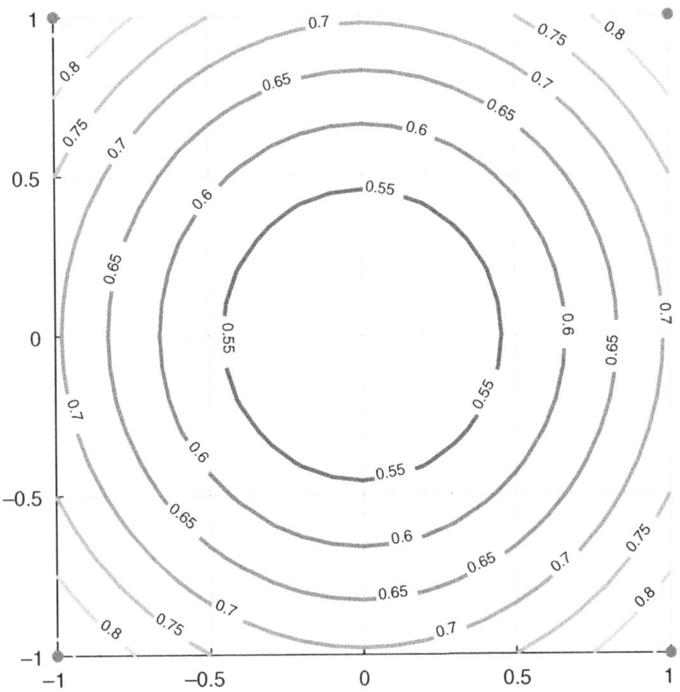


**Figure 2.12** Contour of the standard error of prediction with four samples.

$(\mathbf{X}^T\mathbf{X})^{-1}\xi(\mathbf{x}_p)$. In some cases, however, noises at different locations may have different magnitudes. This is particularly true when the magnitude of the noise is proportional to the magnitude of QoI. In such a case, the sample sensitivity in the previous subsection can be utilized to estimate the prediction variance. Let the noise variance of each sample point be $\sigma_i^2$. Then, the prediction variance can be defined as

$$V[\hat{y}(\mathbf{x}_p)] = \sum_{i=1}^{n_y} \left(\frac{\partial \hat{y}(\mathbf{x}_p)}{\partial y_i}\right)^2 \sigma_i^2 \tag{2.44}$$

**Example 2.11 Variable noise**

Consider the linear PRS in Example 2.5. (a) When all samples have the noise standard deviation $\sigma$, calculate the prediction variance at $x = 3$. (b) When all samples have different noise standard deviations, which sample has the most significant influence on the prediction variance at $x = 3$. Explain why.

**Solution:**

a) In the case of linear PRS, the following matrices were defined:

$$\mathbf{X} = \begin{bmatrix} 1 & -2 \\ 1 & -1 \\ 1 & 0 \\ 1 & 1 \\ 1 & 2 \end{bmatrix}, \quad \mathbf{y} = \begin{Bmatrix} -1.5 \\ -1.5 \\ 0 \\ 1.25 \\ 1.75 \end{Bmatrix}, \quad \mathbf{X}^T\mathbf{X} = \begin{bmatrix} 5 & 0 \\ 0 & 10 \end{bmatrix}, \quad \xi = \begin{Bmatrix} 1 \\ x \end{Bmatrix}$$

The linear PRS becomes

$$\hat{y}(x) = \xi^T(\mathbf{X}^T\mathbf{X})^{-1}\mathbf{X}^T\mathbf{y} = 0.925x$$

Prediction variance at $x = 3$ becomes

$$V[\hat{y}(x)] = \sigma^2\xi^T(\mathbf{X}^T\mathbf{X})^{-1}\xi = \sigma^2\{1, x\}\begin{bmatrix} 0.2 & 0 \\ 0 & 0.1 \end{bmatrix}\begin{Bmatrix} 1 \\ x \end{Bmatrix} = (0.2 + 0.1x^2)\sigma^2 = 1.1\sigma^2$$

Therefore, if all samples have the same noise standard deviation, the prediction variance at $x = 3$ is $1.1\sigma^2$.

b) When all samples have different noise standard deviations, it would be necessary to use sample sensitivity in Eq. (2.44). First sample sensitivity at $x = 3$ can be calculated as

$$\frac{\partial \hat{y}}{\partial y_i} = \xi^T(\mathbf{X}^T\mathbf{X})^{-1}\mathbf{X}^T = 0.1[\,2-2x \quad 2-x \quad 2 \quad 2+x \quad 2+2x\,]$$

$$= 0.1[\,-4 \quad -1 \quad 2 \quad 5 \quad 8\,]$$

When all samples have the same noise standard deviation, Eq. (2.44) yields $V[\hat{y}(x)] = 1.1\sigma^2$, which is the same outcome with Part (a). If all samples have different noise standard deviation, the most influential sample to the prediction variance at $x = 3$ is $y_5$, whose sample sensitivity is highest. This is because the sample location $x = 2$ is closest to the prediction point $x = 3$.

## 2.7 Outliers

In Section 2.5, we discussed the stepwise elimination of insignificant coefficients based on their CoV. That is, if a coefficient has a large uncertainty, it means that the coefficient is difficult to identify and insignificant in prediction. The true function may not have the basis function, and it justifies the removal of the coefficient. If a sample is suspicious, we may need systematic reasoning to eliminate the sample from the fitting process. The same concept can be applied to identify wrong/erroneous samples.

Both in physical experiments and computer simulations, we have samples with large errors occasionally. In computer simulations, these may reflect failures of the solution algorithm, software implementation, or mistakes by the user of the software. Sample points with large errors are called outliers. It is important to detect them and either remove or repair them because they can have a large detrimental effect on the prediction accuracy of the surrogate.

A standard tool for detecting outliers is the Iteratively Reweighted Least-Squares (IRLS) procedure. In order to understand their basis, let us first consider the weighted least-squares (WLS) procedure. Weighted least-squares procedures minimize a weighted sum of the squares of the residuals. Linear regression in Section 2.3 corresponds to using the same weights for all the samples. There are many possible reasons for using WLS rather than standard least squares. We may have more confidence in some samples than in others. We may want to weigh more heavily points that are close to the region where we will need to predict the surrogate than points far away. The noise at some points may be known to be higher than at other points.

In these cases, instead of using Eq. (2.12), we use the weighted-RMS error:

$$e_{w\text{RMS}} = \sqrt{\frac{1}{n_y}\sum_{i=1}^{n_y} w_i e_i^2} = \sqrt{\frac{1}{n_y}\mathbf{e}^T\mathbf{W}\mathbf{e}} \tag{2.45}$$

where $w_i$ is the weight associated with the $i$th sample, and $\mathbf{W}$ is a diagonal matrix with the weights on the diagonal. Minimizing the weighted RMS error, $e_{w\text{RMS}}$ yields a modified set of normal equations

$$\mathbf{X}^T\mathbf{W}\mathbf{X}\mathbf{b} = \mathbf{X}^T\mathbf{W}\mathbf{y} \tag{2.46}$$

IRLS procedures weigh points with large residuals with small weights, with the weight decreasing with increasing magnitude of the residual. Then the WLS procedure is performed. If the point is an outlier, the surrogate model will move away from it, so its residual will increase. We will assign a smaller weight to the point and repeat the procedure. Eventually, outliers are likely to end up with zero or very low weight. There are several weighting schemes, see e.g., Myers and Montgomery [4], p. 671. One of the simplest is Huber's scheme, defined as

$$w_i = \begin{cases} 1 & \text{if } |e_i|/\hat{\sigma} \le 1 \\ \hat{\sigma}/|e_i| & \text{otherwise} \end{cases} \tag{2.47}$$

In practice, the procedures can be repeated until the weight of the outlier decreases enough, or once a sample is identified as an outlier, the procedures can be stopped, and the samples can be removed in the regular linear regression process.

**Example 2.12 Identifying Young's modulus with four samples**

Stress and strain are linearly related by stress $= E \cdot$ strain, where Young's modulus $E$ needs to be estimated based on four stress–strain measurements. For values of strains of 1, 2, 3, and 4 millistrains, stresses are measured by 9, 22, 36, and 39 ksi. Using a linear PRS model stress $= E \cdot$ strain, identify Young's modulus after removing an outlier.

**Solution:**

Since we do not know which samples are outliers, we perform regular linear regression with a constant weight $w_i = 1$. Denoting the stress by $y$ as a QoI and the strain by $x$, the four samples are used to define the following normal equation:

$$\mathbf{X} = \begin{bmatrix} 1 \\ 2 \\ 3 \\ 4 \end{bmatrix}, \quad \mathbf{y} = \begin{Bmatrix} 9 \\ 22 \\ 36 \\ 39 \end{Bmatrix}, \quad \mathbf{X}^T\mathbf{X} = [30], \quad \mathbf{X}^T\mathbf{y} = \{317\}, \quad E = \left(\mathbf{X}^T\mathbf{X}\right)^{-1}\mathbf{X}^T\mathbf{y} = 10.569\,\text{Msi}$$

Note that there is no constant term in this PRS surrogate. The vector of residual errors is given as $\mathbf{e} = \mathbf{y} - \hat{\mathbf{y}} = \{1.567, -0.867, 4.3, -3.267\}^T$. Therefore, the standard errors of noise and coefficient become

$$\hat{\sigma} = \sqrt{\frac{\mathbf{e}^T\mathbf{e}}{4-1}} = 3.2846, \quad \hat{\sigma}_E = 3.2846\sqrt{\frac{1}{30}} = 0.6$$

The only residual error that is greater than $\hat{\sigma}$ is the third one. Therefore, its weight is calculated as $w_3 = 3.2846/4.3 = 0.764$.

Now, since samples have different weights, we perform weighted least-squares fitting with $\mathbf{W} = \text{diag}[1.0, 1.0, 0.764, 1.0]$ to yield

$$\mathbf{X}^T\mathbf{W}\mathbf{X} = [27.876], \quad \mathbf{X}^T\mathbf{W}\mathbf{y} = \{295.512\}, \quad E = \left(\mathbf{X}^T\mathbf{W}\mathbf{X}\right)^{-1}\mathbf{X}^T\mathbf{W}\mathbf{y} = 10.46\,\text{Msi}$$

Note that Young's modulus is slightly reduced. In the iterative reweighted least-squares process, the new vector of residual errors is calculated as $\mathbf{e} = \{1.457, -1.085, 4.628, 2.830\}^T$. The new linear PRS yields

$$\hat{\sigma} = \sqrt{\frac{\mathbf{e}^T\mathbf{W}\mathbf{e}}{4-1}} = 3.037, \quad \hat{\sigma}_E = 3.037\sqrt{\frac{1}{30}} = 0.56$$

Overall, the standard error of the coefficient is reduced slightly. Among the new residual errors, only the third one is larger than the standard error of noise. Therefore, its weight is recalculated as $w_3 = 3.037/4.628 = 0.656$. It is noted that the weight of the third sample decreases further. One more iteration yields

$$\mathbf{X}^T\mathbf{W}\mathbf{X} = [26.904], \quad \mathbf{X}^T\mathbf{W}\mathbf{y} = \{279.848\}, \quad E = \left(\mathbf{X}^T\mathbf{W}\mathbf{X}\right)^{-1}\mathbf{X}^T\mathbf{W}\mathbf{y} = 10.40\,\text{Msi}$$

We can continue iteration until the residual errors become smaller than the standard error of noise. Or, we can be satisfied that the process has identified the third sample as an outlier and eliminated it. If we remove the third sample, we can identify $E = 9.95$ Msi, with the standard error of coefficient $\hat{\sigma}_E = 0.376$.

## 2.8 Statistical View of Linear Regression

So far, we have developed the linear regression process as a minimization of the error between the samples and model predictions. This is indeed deterministic optimization, and due to the quadratic nature of the sum-of-square errors, optimization has a unique solution. At the same time, however, we also presented uncertainty associated with samples, regression coefficients, and predictions at unsampled points. These uncertainties stem from the fundamental assumption that samples include random noise. In this section, we will derive the uncertainty information from the statistical view of the linear regression process.

Although the purpose of surrogate modeling is to develop an approximate function using given samples, it would be better to express the samples in terms of the true model in order to explain their relationship from a statistical viewpoint. First, we assume that there is a true function $f(\mathbf{x}; \boldsymbol{\beta})$, which describes the behavior of QoI as a deterministic function of input variables $\mathbf{x}$ and model parameters $\boldsymbol{\beta}$. In particular, in the case of PRS surrogates, it is assumed that the true function is in the form of polynomials, and the model parameters $\boldsymbol{\beta}$ become the coefficients of the polynomials. Second, it is assumed that the PRS surrogate has an exact model form, but the coefficients are approximate. Therefore, the PRS surrogate can be represented by $f(\mathbf{x}; \mathbf{b})$, where $\mathbf{b}$ is the approximate coefficients. Lastly, the samples are assumed to be generated from the true function $f(\mathbf{x}; \boldsymbol{\beta})$ by adding random noise or some error defined by $\epsilon \sim N(0, \sigma^2)$. From this viewpoint, we can think of a sample $y$ at point $\mathbf{x}$ as a model given in the following form:

$$y = f(\mathbf{x}; \boldsymbol{\beta}) + \epsilon \tag{2.48}$$

That is, a sample has a deterministic part (true function) and a probabilistic part (noise or error).

Since the noise is assumed to be normally distributed with a zero mean and variance $\sigma^2$, the sample in Eq. (2.48) follows a normal distribution $y|\mathbf{x} \sim N(f(\mathbf{x}; \boldsymbol{\beta}), \sigma^2)$. Here, the notation, $y|\mathbf{x}$, means a conditional probability of $y$ given $\mathbf{x}$. Let $p(y|\mathbf{x})$ be the conditional probability density function (PDF), then it can be written as

$$p(y|\mathbf{x}) = N(f(\mathbf{x}; \boldsymbol{\beta}), \sigma^2) \tag{2.49}$$

In this viewpoint, each sample is a random variable whose mean is the true function and variance $\sigma^2$.

Now, the main task of surrogate modeling is to estimate the unknown coefficient $\boldsymbol{\beta}$. In the case of linear regression, the unknown coefficients are determined by minimizing the sum-of-square errors. From the statistical viewpoint, instead of minimizing a measure of error, the following question is asked: how likely is it that the samples can be obtained given the input variable $\mathbf{x}$ and parameter $\boldsymbol{\beta}$? Such a likelihood of obtaining the sample is represented in the form of conditional probability $p(y|\mathbf{x})$ in Eq. (2.49). Since there are $n_y$ numbers of samples, the likelihood of obtaining all samples can be written as

$$p\left(y_1, y_2, \ldots, y_{n_y} \middle| \mathbf{x}_1, \mathbf{x}_2, \ldots, \mathbf{x}_{n_y}\right) \tag{2.50}$$

This is indeed the condition joint PDF of $n_y$ numbers of random variables.

In PRS surrogate modeling, it is assumed that all samples are uncorrelated with each other. More specifically, they are independent and identically distributed (iid). That means, the random noises in $y_i$ and $y_j$ come from the same distribution$\sim N(0, \sigma^2)$, but their realizations $\epsilon_i$ and $\epsilon_j$ are independent. In such a case, the conditional joint PDF in Eq. (2.50) can be obtained by the product of all conditional PDFs of individual samples. That is, the likelihood function of all samples can be written as

$$\begin{aligned} p(\mathbf{y}|\mathbf{X}, \boldsymbol{\beta}, \sigma) &= \prod_{i=1}^{n_y} p(y_i|\mathbf{x}_i) = \prod_{i=1}^{n_y} N(f(\mathbf{x}_i; \boldsymbol{\beta}), \sigma^2) \\ &= \left(\sigma\sqrt{2\pi}\right)^{-n_y} \exp\left[-\frac{1}{2\sigma^2}\sum_{i=1}^{n_y}(y_i - f(\mathbf{x}_i; \boldsymbol{\beta}))^2\right] \end{aligned} \tag{2.51}$$

It is important to understand the difference between the conditional joint PDF and the likelihood function. Although both have the same expression, the interpretations are different. In the case of conditional joint PDF, $p(\mathbf{y}|\mathbf{X}, \boldsymbol{\beta}, \sigma)$ means the PDF value of $\mathbf{y}$ for given parameters $\mathbf{X}$, $\boldsymbol{\beta}$, $\sigma$. In this case, $\mathbf{y}$ is varied to obtain different probability density values. On the other hand, in the case of the likelihood function, $p(\mathbf{y}|\mathbf{X}, \boldsymbol{\beta}, \sigma)$ means the probability density value to obtain $\mathbf{y}$ for given $\mathbf{X}$, $\boldsymbol{\beta}$, $\sigma$. In this case, $\mathbf{y}$ is fixed, and $\mathbf{X}$, $\boldsymbol{\beta}$, $\sigma$ are varied to obtain different likelihood values at different parameters.

Figure 2.13 shows the relationship between the PDF and likelihood function in the case of one sample with one parameter. Figure 2.13(a) shows three PDFs of sample $y$ with three different parameter values $\beta_1$, $\beta_2$, $\beta_3$. The three PDFs have the same shape but with different mean values. This happens because we assume that sample $y$ has a mean $f(\mathbf{x}; \boldsymbol{\beta})$, which depends on the parameter $\beta$. If the variance $\sigma^2$ is also included as a parameter, the shape of the PDF will also change.

Now, when an actual sample turns out to be $y_1$ from the experiment, the PDFs have different values depending on parameter values (the three circle markers). This is basically likelihood. When $\beta = \beta_1$, the probability density of obtaining sample $y_1$ is $A$, when $\beta = \beta_2$ it is $B$, etc. Therefore, the

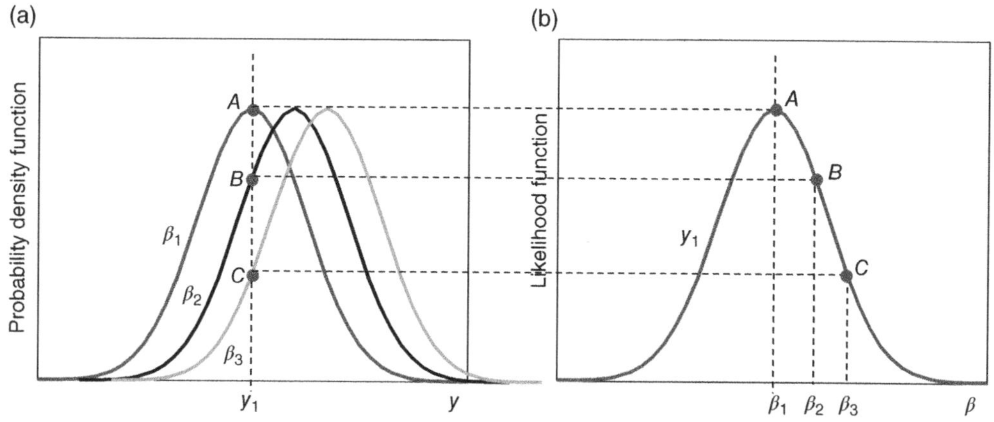


**Figure 2.13** Probability density function versus likelihood function. (a) Probabilities and (b) likelihoods.

likelihood of obtaining a sample $y_1$ can be plotted by varying the parameter $\beta$, as shown in Figure 2.13(b), which is the likelihood function. It is noted that sample $y$ is a variable in PDF, while parameter $\beta$ is a variable in the likelihood function. In general, the shape of the PDF and that of the likelihood are different.

Different from the least-squares method in linear regression, the idea behind the likelihood function is that the likelihood function in Figure 2.13(b) is our knowledge of the unknown parameter. That is, we cannot estimate the unknown parameter deterministically but only in the form of a probabilistic distribution. Because sample $y$ has uncertainty due to random noise, we cannot estimate the parameter exactly. Instead, the probability that the parameter is $\beta_1$ is high, while the probability that the parameter is $\beta_3$ is low. If we know sample $y_1$ is exact and there is no randomness, i.e., $\sigma = 0$, we can determine the unknown parameter $\beta_1$ exactly because the PDF is infinity at $y = y_1$ and zero everywhere else.

It is also interesting to note that the PDF in Figure 2.13(a) is a probability distribution of a random variable $y$, which is often known as aleatory uncertainty. That means different samples will have different values every time the sample is drawn. On the other hand, the distribution in Figure 2.13(b) does not represent any randomness. The model parameter $\beta$ is not random, but uncertain. This type of uncertainty is called epistemic uncertainty. It means the parameter is deterministic, but we do not know its value exactly. The top point $A$ in Figure 2.13(b) means that the probability that the actual parameter value being $\beta_1$ is highest, while the point $C$ means that the probability that the actual parameter value being $\beta_3$ is low based on the proof that sample has a value of $y_1$. Therefore, the distribution in Figure 2.13(b) should be understood as the probability that the parameter value is an accurate one, given the sample $y_1$. If the likelihood has a narrow distribution, it means the information is accurate and the uncertainty in the parameter is low. On the other hand, if the likelihood is widely distributed, it means that the information is vague, and it is difficult to identify accurate parameters.

When multiple samples are present, the individual likelihood functions are multiplied together. For example, when two samples, $y_1$ and $y_2$, are present, the PDF in Figure 2.13(a) becomes a two-dimensional joint PDF (the plot will be a three-dimensional bell shape). However, the likelihood function in Figure 2.13(b) will still be a one-dimensional function of $\beta$. If two samples $y_1$ and $y_2$ are close together, the distribution in Figure 2.13(b) will be narrowed, which means the information is accurate. If two samples are significantly different, the likelihood plot will be wide, and the parameter cannot be estimated accurately. On the other hand, when multiple parameters are present in the PRS model, the likelihood function in Figure 2.13(b) will be a multi-dimensional bell-shaped plot. Therefore, the likelihood function given in Eq. (2.51) can be viewed as a mapping from $n_y$-dimensional joint PDF to $n_\beta$-dimensional likelihood function.

Even if the likelihood function in Figure 2.13(b) is given in the form of a distribution, it is still valuable to obtain a single value of the parameter that can represent the best estimate. In fact, this estimate corresponds to the mode of the distribution in Figure 2.13(b). That is, the peak of the distribution is the best deterministic estimate of the parameter. Since the peak value corresponds to the maximum likelihood function, it is referred to as the maximum likelihood estimate. It is possible that the likelihood function in Eq. (2.51) is differentiated to obtain the maximum point, it would be better to work with a logarithm of it, which is called the log-likelihood function. This is because the product of exponential likelihood functions increases very fast and becomes a highly nonlinear function. The log-likelihood function can be defined as

$$L = \log p(\mathbf{y}|\mathbf{X}, \boldsymbol{\beta}, \sigma) = -\frac{1}{2\sigma^2}\sum_{i=1}^{n_y} |y_i - f(\mathbf{x}_i; \boldsymbol{\beta})|^2 - n_y \log \sigma\sqrt{2\pi} \tag{2.52}$$

Since logarithm is a monotonic function, even if the likelihood function in Eq. (2.51) and the log-likelihood function in Eq. (2.52) may have different values, the parameters, $\boldsymbol{\beta}$ and $\sigma$, that maximize both functions are the same.

In order to have the maximum likelihood estimate, the true function is expressed as a linear combination of basis functions as $f(\mathbf{x}_i; \boldsymbol{\beta}) = \boldsymbol{\xi}(\mathbf{x}_i)^T\boldsymbol{\beta}$. Then, the sum in the first term on the right-hand side of Eq. (2.52) becomes $(\mathbf{y} - \mathbf{X}\boldsymbol{\beta})^T(\mathbf{y} - \mathbf{X}\boldsymbol{\beta})$. Then, the maximum likelihood estimate determines the model parameter by differentiating the log-likelihood function as

$$\frac{\partial L}{\partial \boldsymbol{\beta}} = \frac{1}{\sigma^2}(\mathbf{X}^T\mathbf{y} - \mathbf{X}^T\mathbf{X}\boldsymbol{\beta}) = 0 \tag{2.53}$$

It is interesting to note that even if $\sigma$ is also considered a model parameter, the determination of the parameter $\boldsymbol{\beta}$ is independent of $\sigma$. This is because $\sigma$ is the same for all samples. Also, it can be seen from Figure 2.13(b) that the large $\sigma$ may cause a wide distribution, but the peak value will be the same.

An important conclusion from Eq. (2.53) is that the parameters obtained from the maximum likelihood estimate are identical to the parameters obtained from linear regression in Eq. (2.14), as

$$\tilde{\mathbf{b}} = (\mathbf{X}^T\mathbf{X})^{-1}\mathbf{X}^T\mathbf{y} \tag{2.54}$$

Here, $\tilde{\mathbf{b}}$ is used instead of $\boldsymbol{\beta}$ because the right-hand side includes the random variable $\mathbf{y}$. Therefore, $\tilde{\mathbf{b}}$ is an uncertain variable with a joint PDF given as $\sim N(\mathbf{b}, \boldsymbol{\Sigma}_\mathbf{b})$. Equation (2.54) shows how aleatory uncertainty (randomness) in samples is converted into epistemic uncertainty in the regression coefficients. In addition, the variance of noise can be estimated by differentiating the log-likelihood function with respect to $\sigma$, as

$$\hat{\sigma}^2 = \frac{1}{n_y}(\mathbf{y} - \mathbf{X}\tilde{\mathbf{b}})^T(\mathbf{y} - \mathbf{X}\tilde{\mathbf{b}}) = \frac{SS_e}{n_y} \tag{2.55}$$

which is the same as Eq. (2.20) except that the free degrees-of-freedom $n_y - n_\beta$ is used for the unbiased estimate.

The second-order derivative of the log-likelihood function with respect to parameter $\boldsymbol{\beta}$ is called the Hessian matrix or information matrix. By differentiating Eq. (2.53), the information matrix can be obtained as

$$\frac{\partial^2 L}{\partial \boldsymbol{\beta} \partial \boldsymbol{\beta}} = -\frac{1}{\sigma^2}\mathbf{X}^T\mathbf{X} \tag{2.56}$$

The uncertainty in the regression coefficients can be defined as a covariance matrix. The definition of a covariance matrix is $\text{cov}\left[\tilde{\mathbf{b}}\right] = \boldsymbol{\Sigma}_{\mathbf{b}} = E\left[\tilde{\mathbf{b}}\tilde{\mathbf{b}}^T\right] - E\left[\tilde{\mathbf{b}}\right]E\left[\tilde{\mathbf{b}}^T\right]$. The first term on the right-hand side can be expanded using the estimated coefficient in Eq. (2.54): $E\left[\tilde{\mathbf{b}}\tilde{\mathbf{b}}^T\right] = \left(\mathbf{X}^T\mathbf{X}\right)^{-1}$ $\mathbf{X}^T E[\mathbf{y}\mathbf{y}^T]\mathbf{X}\left(\mathbf{X}^T\mathbf{X}\right)^{-1}$. This is because only $\mathbf{y}$ is random in Eq. (2.54), all other terms can come out of the expectation operator. The mean of the sample matrix term can be simplified as $E[\mathbf{y}\mathbf{y}^T] = E\left[\left(\mathbf{X}\tilde{\mathbf{b}} - \boldsymbol{\epsilon}\right)\left(\mathbf{X}\tilde{\mathbf{b}} - \boldsymbol{\epsilon}\right)^T\right] = \mathbf{X}\mathbf{b}\mathbf{b}^T\mathbf{X}^T + \sigma^2\mathbf{I}$. The second term is straightforward as $E\left[\tilde{\mathbf{b}}\right]E\left[\tilde{\mathbf{b}}^T\right] = \mathbf{b}\mathbf{b}^T$. Therefore, combining these two terms, the covariance matrix of the coefficients becomes

$$\boldsymbol{\Sigma}_{\mathbf{b}} = \sigma^2\left(\mathbf{X}^T\mathbf{X}\right)^{-1} \tag{2.57}$$

which is identical to the one given in Eq. (2.38). It is interesting to note that the covariance matrix is the negative of the inverse of the information matrix in Eq. (2.56).

The last uncertainty that we used is the prediction variance. From the statistical viewpoint, the surrogate prediction can be written as $\hat{y}(\mathbf{x}) = \boldsymbol{\xi}(\mathbf{x})^T\tilde{\mathbf{b}}$. In this expression, the prediction is also an uncertain variable because of $\tilde{\mathbf{b}}$. However, since $\boldsymbol{\xi}(\mathbf{x})^T$ is a deterministic function, the variance of $\hat{y}(\mathbf{x})$ can be written as

$$V[\hat{y}(\mathbf{x})] = \boldsymbol{\xi}(\mathbf{x})^T\boldsymbol{\Sigma}_{\mathbf{b}}\boldsymbol{\xi}(\mathbf{x}) = \boldsymbol{\xi}(\mathbf{x})^T\sigma^2\left(\mathbf{X}^T\mathbf{X}\right)^{-1}\boldsymbol{\xi}(\mathbf{x}) \tag{2.58}$$

Therefore, the standard error of prediction can be written as

$$\sigma_y(\mathbf{x}) = \hat{\sigma}\sqrt{\boldsymbol{\xi}(\mathbf{x})^T\left(\mathbf{X}^T\mathbf{X}\right)^{-1}\boldsymbol{\xi}(\mathbf{x})} \tag{2.59}$$

which is identical to the one given in Eq. (2.41). In general, the standard error of prediction is used to find the confidence intervals of the prediction. For example, 95% confidence intervals can be written as

$$\text{CI}_{95} = \hat{y}(\mathbf{x}) \pm 2\sigma_y(\mathbf{x}) \tag{2.60}$$

**Example 2.13 Confidence intervals using different orders of PRS**

Generate 50 equally spaced samples in the design space $x \in [-5, 5]$ from the true function $y(x) = 5x^3 - x^2 + x$ and add random noise from a normal distribution $\sim N(0,100^2)$. Fit the samples using linear, cubic, 6th-order, and 10th-order PRS. Plot the mean prediction with 95% confidence intervals.

**Solution:**

The following MATLAB code is used to fit the PRS surrogate model and plot the prediction and confidence intervals. The code requires removing the comment symbol `%` in front of the design matrix $\mathbf{X}$ for the corresponding order of PRS. First, random noises are generated from a normal distribution. They are shifted such that their mean is zero. Even if the standard deviation of 100 is used to generate noise samples, the sample standard deviation turns out to be 126.3. The

MATLAB function `regress` is used to calculate the regression coefficients **b** and residuals **r**. The standard error of noise is calculated using the residuals, and the standard error of prediction is calculated at all sample points. Figure 2.14 shows the PRS predictions and 95% confidence intervals for linear, cubic, 6th-order, and 10th-order PRS. It is obvious that the linear PRS failed to follow the trend of the true function, and the estimated uncertainty is not consistent (30 samples were out of 95% confidence intervals). The cubic and 6th-order PRS seem to follow the trend of true function well, while the 10th-order PRS seems to be overfitting samples. In order to compare these surrogates $R^2$ and $R_a^2$ are calculated in Table 2.2. When $R^2$ is considered, it looks that the 10th-order PRS fits the

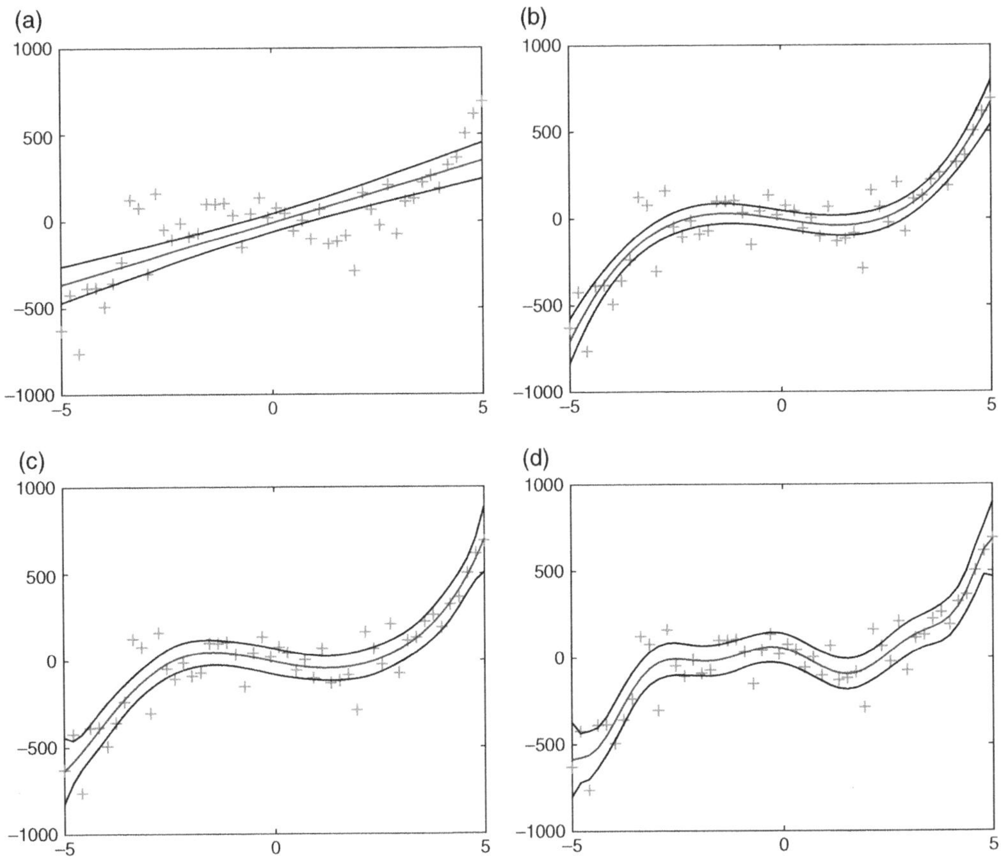


**Figure 2.14** Confidence intervals of different surrogate models. (a) 1st-order PRS, (b) 3rd-order PRS, (c) 6th-order PRS, and (d) 10th-order PRS.

**Table 2.2** Comparison of $R^2$ and $R_a^2$ for four PRS surrogates.

| | **1st-order PRS** | **3rd-order PRS** | **6th-order PRS** | **10th-order PRS** |
|---|---|---|---|---|
| $\hat{\sigma}$ | 186.3 | 120.8 | 123.5 | 122.3 |
| $R^2$ | 0.57 | 0.8267 | 0.8306 | 0.8494 |
| $R_a^2$ | 0.561 | 0.8154 | 0.8070 | 0.8108 |

best. But this is due to the fact that more coefficients are used. Based on $R_a^2$, the cubic PRS turns out to be the best, which has the same order of polynomials as the true function.

```
rng default;                                 % Control random number sequence
x=linspace(-5,5,50)';                        % Equally spaces sample locations
ytrue=5*x.^3 - x.^2 + x;                                     % True function
noise=100*randn(50,1); noise=noise-mean(noise); % Unbiased random noise
y=ytrue+noise;                                            % Generate samples
%X=[ones(50,1) x];
X=[ones(50,1) x x.^2 x.^3];                                  % Design matrix
%X=[ones(50,1) x x.^2 x.^3 x.^4 x.^5 x.^6];
%X=[ones(50,1) x x.^2 x.^3 x.^4 x.^5 x.^6 x.^7 x.^8 x.^9 x.^10];
[b,bint,r,rint,stats]=regress(y,X);                            % Fitting PRS
yfit=X*b;                                  % Prediction at sample points
s=sqrt(r'*r/(50-size(X,2)));                    % Standard error of noise
sy=s.*diag(sqrt(X*inv(X'*X)*X'));           %Standard error of prediction
plot(x,y,'+',x,yfit,'b',x,yfit-2*sy,'k',x,yfit+2*sy,'k');   % Plotting
```

## Exercise

**2.1** Fit a linear function in Figure 2.1 with 10, 30, 50, and 100 equal-interval samples. Show that $e_{\mathrm{RMS}}$ of the fitted curve converges to zero as the number of samples increases. Explain why that happens using the mean of noise samples.

**2.2** Generate 21 equal-interval samples for $x \in [0, 2]$ from the true function $y(x) = x^2$ and add random noise from the standard normal distribution $\sim N(0, 0.1^2)$. Shift the noise samples such that the samples have a zero mean. Fit these samples using a linear polynomial $\hat{y}(x) = b_1 + b_2 x$ and estimate $e_{\mathrm{RMS}}$ of the fitted model. Discuss the reason for the difference between the standard deviation of noise and $e_{\mathrm{RMS}}$.

**2.3** In a curve-fitting example with a true function $y = x$, noisy data are fitted to a linear polynomial $\hat{y} = 1.06x$. In addition, the data at $x = 10$ was $y_{10} = 11$. What are (i) $\epsilon$, (ii) $\epsilon_{10}$, and (iii) the surrogate error at $x = 10$?

**2.4** Generate 30 equal-interval samples for $x \in [1, 30]$ using a true function $y(x) = x$ and add random noise from $\sim N(0, 1^2)$. Use the three error metrics in Eqs. (2.7)–(2.9) to fit a linear PRS. Compare the three PRSs using the three error metrics as in Table 2.1. Hint: use the `fminsearch` function in Matlab.

**2.5** Repeat Problem 2.4 with the surrogate $\hat{y}(x) = bx$.

**2.6** Repeat Problem 2.4 with 10 points and compare the accuracy of the fit with respect to the true function.

**2.7** Find other error metrics for a fit besides the three discussed in Eqs. (2.7)–(2.9).

**2.8** Check the accuracy of the quadratic PRS in Example 2.2 in the region $1.5 \le D \le 3.5$, $3 \le H \le 7$. Find the maximum error, average error, and RMS error (i) for the nine sample points, and (ii) for the entire region (using the analytical expression from Example 2.2). You may cover the domain with a grid of (20 × 20) points and calculate the error in each one of the 400 points. (iii) Calculate the error using the PRESS procedure and compare it to the result of part (ii).

**2.9** Using the samples in Example 2.2, use backward elimination to find an incomplete quadratic PRS with the highest $R_a^2$. Then, check for accuracy of fit compared to the analytical profit per can over the entire region, and compare it to the results obtained in Problem 2.8.

**2.10** Consider the following experimental samples $(x, y) = (1, 1), (2, 2), (3, 5), (4, 9)$. (i) Construct a linear PRS $\hat{y}(x) = b_1 + b_2x$. (ii) Estimate $R_a^2$ and $e_{\text{PRESS}}$. (iii) If the quadratic PRS is $\hat{y}(x) = 1.25 - 1.05x + 0.75x^2$, compare the quality of quadratic PRS with linear PRS using $e_{\text{PRESS}}$. For quadratic PRS, use the following idempotent matrix:

$$E = \mathbf{X}\left(\mathbf{X}^T\mathbf{X}\right)^{-1}\mathbf{X}^T = \begin{bmatrix} 0.95 & 0.15 & -0.15 & 0.05 \\ 0.15 & 0.55 & 0.45 & -0.15 \\ -0.15 & 0.45 & 0.55 & 0.15 \\ 0.05 & -0.15 & 0.15 & 0.95 \end{bmatrix}$$

**2.11** A function is given at three data points: $y(-1) = 0$, $y(0) = 0$, and $y(1) = 3$. Fit a least-square constant and linear polynomials to the data. Calculate the RMS error of the two curve fits. The linear curve fit has a smaller error but check whether it could be expected to be a better predictor of the true function.

**2.12** A function $y = bx$ is fitted using two sample points, $y_1 = y(1) = 10$, $y_2 = y(2) = 25$. (i) Calculate the cross-validation error at the first point from the definition. (ii) If the true value of the coefficient $b$ is 11 ($y_{\text{true}} = 11x$) and the fitted value is 12, $\hat{y} = 12x$, calculate the value of $\epsilon$, $e$, and the surrogate error at $x = 1$. (iii) Calculate the coefficient $b$ to check that its value is indeed 12.

**2.13** A function $y = b/(1 + x)$ is fitted using two sample points: $y_1 = y(0) = 10$ and $y_2 = y(1) = 4$. (i) Estimate the coefficient $b$ using linear regression. (ii) Calculate the cross-validation error at the first point from the definition. (iii) The noise in the samples has been determined to be normally distributed with a standard deviation of 1.1. Use that to estimate the standard error in $b$. (iv) Provide a better choice of $b$ if the objective is to minimize the maximum error rather than the regression fit.

**2.14** A function $y = b(1 - x)/(1 + x)$ is fitted using two sample points: $y_1 = y(0) = 0$ and $y_2 = y(2) = -20$. (i) Estimate the coefficient $b$ using linear regression. (ii) Calculate the cross-validation error at $x = 2$ from definition.

**2.15** In Example 2.3, calculate the RMS error between the true function and fitted function with 50 equally spaced points (i) in $x \in [0, 10]$ and (ii) in $x \in [-2, 12]$. Discuss why the RMS error in (ii) is larger than (i).

**2.16** Generate 20 equally spaced samples of the true function $y(x) = x^3 - x$ in the design space $x \in [-1.5, 1.5]$ with random noise $\sim N(0, 0.5^2)$. Fit the samples using (i) linear PRS and (ii) cubic PRS. Compare the two surrogates in terms of the RMS error between the predictions and samples, as well as the RMS error between the predictions and the true function.

**2.17** Prove the relationship in Eq. (2.24).

**2.18** The sample pairs (0, 0), (1, 1), and (2, 1) represent strain (milli-strains) and stress (ksi) measurements. (i) Estimate Young's modulus using regression, and (ii) calculate the error in Young's modulus using cross-validation both from the repeated fitting and from the formula with one fitting.

**2.19** The pairs (1, 1), (2, 2), (4, 3) represent strain (millistrains) and stress (ksi) measurements. The relationship is given as stress $= E \times$ strain, where $E$ is Young's modulus. Estimate Young's modulus using the three error measures: $e_{RMS}$, $e_{av}$, and $e_{max}$. Estimate the error in Young's modulus using cross-validation of the three error measures. Use the range of Young's modulus $0.75 \leq E \leq 1$.

**2.20** Two samples (1, 1) and (4, 3) represent strain (milli-strains) and stress (ksi) measurements. For the material tested $\sigma = k\sqrt{\epsilon}$, (i) estimate $k$ using regression, (ii) estimate noise in samples, (iii) estimate the standard error in the estimate of $k$, and (iv) compare the standard error to the estimate $k$ by cross-validation.

**2.21** Two samples $y(0) = 10.0$, $y(1) = 4.0$ are fitted by $\hat{y}(x) = b/(1 + x)$. (i) Estimate the coefficient $b$ via regression. (ii) The noise in the samples has been determined to be normally distributed with a standard deviation of 1.1. Use that to estimate the standard error of the coefficient. (iii) If the maximum error is used for the fitting process, estimate a better coefficient than (i). Explain how you obtain the coefficient.

**2.22** Two samples $y(0) = 0.0$, $y(2) = -10.0$ are fitted by $\hat{y}(x) = b(1 - x)/(1 + x)$. (i) Estimate the coefficient $b$ via regression. (ii) Calculate the cross-validation error at $x = 2$ from the definition (not using E).

**2.23** Repeat Example 2.7 using only data at $x = 3, 6, 9, \ldots, 30$.

**2.24** We are trying to estimate the average cost of a gadget based on three sample points \$110, \$120, and \$130. You can think of it as a simple regression problem of fitting a polynomial $\hat{y} = b$, to three sample points. (i) Perform the linear regression fit to the samples, which is to estimate $b$ from the regression equations. (ii) Calculate the cross-validation error for the third sample from the definition (not the formula). (iii) If the true average (from a very large number of samples) is \$125, calculate the values of noise $\epsilon$, residual $e$, and the surrogate error for the third sample in Part (i).

**2.25** A PRS is given in the following model form: $\hat{y}(x) = bx$. Two samples are given as $y(1) = 10$ and $y(2) = 25$. (i) Calculate the cross-validation error at the first point from the definition, not the formula. (ii) If the true value of $b$ is 11 ($y_{true}(x) = 11x$) and the fitted value is 12, $\hat{y}(x) = 12x$, calculate the values of $\epsilon$, $e$, and the surrogate error at $x = 1$. (iii) Calculate $b$ to check that its value is indeed 12.

**2.26** The pairs (1,1) and (3,2) represent strain (milli-strains) and stress (ksi) measurements. Denoting the strain by $x$, the stress by $y$, and Young's modulus by $b$, we assume that the true relationship between stress and strain is $y = bx$. (i) Calculate the linear regression fit to the samples; that is, what the estimate of $b$ is. (ii) Calculate the cross-validation errors, from the definition (not the formula). (iii) If the true value of Young's modulus is 1, calculate the values of $\epsilon$, $e$, and the surrogate error at $x = 3$.

**2.27** The following table shows 15 test results with two variables. Since the magnitudes of the two variables are significantly different, it is a good idea to scale them such that $-1 \le \tilde{x}_1, \tilde{x}_2 \le 1$ where $\tilde{x}_i = \left(x_i - \left[x_i^{\max} + x_i^{\min}\right]/2\right)/\left(\left[x_i^{\max} - x_i^{\min}\right]/2\right)$.

(a) Construct a linear response surface using scaled variables: $\hat{y}(\tilde{x}_1, \tilde{x}_2) = b_1 + b_2\tilde{x}_1 + b_3\tilde{x}_2$.
(b) Calculate RMS, average, and maximum errors.
(c) Estimate the variance of random noise, coefficient of multiple determination, adjusted multiple determination, and PRESS.
(d) Construct a quadratic response surface: $\hat{y}(\tilde{x}_1,\tilde{x}_2) = b_1 + b_2\tilde{x}_1 + b_3\tilde{x}_2 + b_3\tilde{x}_1^2 + b_4\tilde{x}_2\tilde{x}_1 + b_5\tilde{x}_2^2$.
(e) Evaluate the quality of the quadratic PRS compared with the linear PRS.
(f) Perform backward elimination once for the least confident coefficient.

| Observation | $x_1$ | $x_2$ | $y$ |
|---|---|---|---|
| 1 | 195 | 4.0 | 1,004 |
| 2 | 255 | 4.0 | 1,636 |
| 3 | 195 | 4.6 | 852 |
| 4 | 255 | 4.6 | 1,506 |
| 5 | 225 | 4.2 | 1,272 |
| 6 | 225 | 4.1 | 1,270 |
| 7 | 225 | 4.6 | 1,269 |
| 8 | 195 | 4.3 | 903 |
| 9 | 255 | 4.3 | 1,555 |
| 10 | 225 | 4.0 | 1,260 |
| 11 | 225 | 4.7 | 1,146 |
| 12 | 225 | 4.3 | 1,276 |
| 13 | 225 | 4.72 | 1,225 |
| 14 | 230 | 4.3 | 1,321 |

**2.28** Repeat Example 2.10 when the four sample locations are not at the corners but at $(\pm 0.7, \pm 0.7)$. Discuss the difference from the results in Example 2.10.

**2.29** For a grid of $3 \times 3$ sample points in a two-dimensional design space $x_1, x_2 \in [-1, 1]$, compare the contours of the standard error of prediction for linear and quadratic PRS.

**2.30** The Branin-Hoo function is a popular analytical function that is used for testing surrogate models [21]. The functional form is defined in $x_1 \in [-5, 10]$ and $x_2 \in [0, 15]$.

$$f(x_1, x_2) = \left(x_2 - \frac{5.1}{4\pi^2}x_1^2 + \frac{5}{\pi}x_1 - 6\right)^2 + 10\left(1 - \frac{1}{8\pi}\right)\cos(x_1) + 10$$

(a) Generate an equally spaced $6 \times 6$ grid in the design space, among which use only internal $4 \times 4 = 16$ points as samples. Use the analytical function above to evaluate the samples. No need to add random noise.

(b) Fit the samples using cubic PRS with 10 coefficients. Evaluate the accuracy of the surrogate using $e_{\max}, e_{\text{av}}, e_{\text{RMS}}, R^2, R_a^2$, and $e_{\text{PRESS}}$.

**2.31** Tensile tests of composite plates with a hole are performed at different configurations, as shown in the table. Multiple nominally identical specimens were tested for different ratios of plate width $w$ to the hole diameter $D$ and for different fractions $R_{45}$ of the thickness of the plate occupied by $\pm 45°$ plies. The results were collected for 12 combinations of the two variables, and the failure stress (based on the cross-section without the hole) is given in the table. The standard deviation between nominally identical specimens is given for information only. For fitting a surrogate, use only the mean. (i) Fit the samples using a linear and quadratic PRS. (ii) Compare the two surrogates on the basis of global measures ($R_a^2, \hat{\sigma}$, and $e_{\text{PRESS}}$) and the maximum prediction variance in the domain.

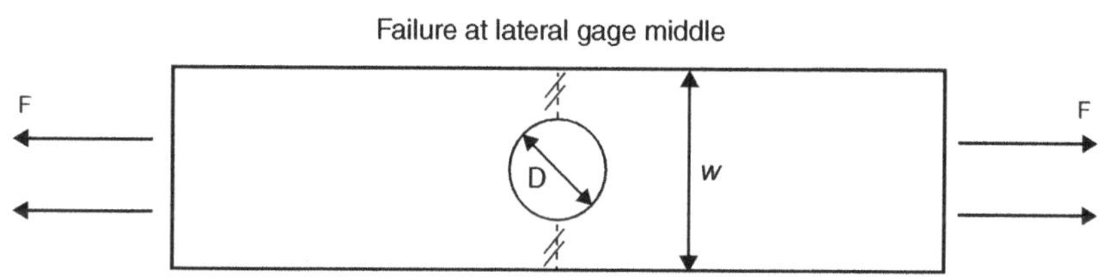


| No. | 1 | 2 | 3 | 4 | 5 | 6 | 7 | 8 | 9 | 10 | 11 | 12 |
|---|---|---|---|---|---|---|---|---|---|---|---|---|
| $R_{45}$ | 0.2 | 0.2 | 0.2 | 0.2 | 0.5 | 0.8 | 0.5 | 0.5 | 0.5 | 0.8 | 0.8 | 0.8 |
| $w/D$ | 3 | 4 | 6 | 8 | 6 | 6 | 3 | 4 | 8 | 3 | 4 | 8 |
| Mean (ksi) | 87.35 | 91.46 | 97.44 | 100.12 | 71.07 | 50.30 | 58.11 | 67.62 | 72.39 | 37.68 | 42.68 | 51.84 |
| Std (ksi) | 3.55 | 5.42 | 5.20 | 4.90 | 2.84 | 1.68 | 2.48 | 2.70 | 3.28 | 1.34 | 0.85 | 1.33 |

**2.32** Assuming a linear model, i.e., $y = \hat{y} + \epsilon = b_1 + b_2x + \epsilon$, where $\hat{y}$ is the regression model and $\epsilon$ is the error with independent normal distribution, i.e., $\epsilon \sim N(0, \sigma^2)$. (i) Compute 95% confidence interval of $\mathbf{b}$, i.e., computer 2.5% and 97.5% percentiles of $\mathbf{b}$. (ii) Compute a 95% confidence interval of $\hat{y}$ at $x = 0.2$. (iii) Plot $\hat{y}$ and confidence interval of $\hat{y}$ in the same graph $\in[0, 1]$. Samples are given as

```
x=[0 0.2 0.4 0.6 0.8 1.0]';
y=[0.4662, 0.5844, 0.7845, 0.8007, 0.9028, 0.8995]';
```

**2.33** In order to fit a linear polynomial $y(x) = b_1 + b_2x$, six equally spaced samples are generated in the design space of $x \in [0, 1]$. From the true model, the samples have normally distributed noise with a mean of zero and a standard deviation of 0.1. The random number generator in Matlab generates the following noise samples $(x, \epsilon) = (0, 0.119), (0.2, -0.04), (0.4, 0.033), (0.6, 0.017), (0.8, 0.019), (1.0, -0.072)$. Add the noise samples to the function $y = f(x) = x$, and fit it to a linear polynomial. Calculate the error measures $e_{\text{RMS}}, R^2, R_a^2, e_{\text{PRESS}}$ and compare $e_{\text{RMS}}$ and $e_{\text{PRESS}}$ to the RMS error calculated analytically between the line $y = x$ and the fitted line.

**2.34** Repeat Problem 2.33 when the true function is $y = x^2$. (i) Use a linear PRS. (ii) Use a quadratic PRS.

# 3

# Design of Experiments

## 3.1 Introduction

In Chapter 2, we assumed that the sample locations are fixed and studied how to fit a good polynomial response surface (PRS) through the given samples. We showed that different basis functions can affect the accuracy of the prediction, which was evaluated using prediction variance. We also studied a method of eliminating unimportant basis functions to improve the accuracy of the prediction. However, the choice of sample locations where experiments (whether numerical or physical) are performed has a very large effect on the quality of the surrogate. A partial result was shown in Example 2.10 and Exercise Problem 2.28 of Chapter 2, where the prediction variance changes significantly depending on the locations of the samples. As the purpose of surrogate modeling is to approximate a quantity of interest (QoI) using a set of samples, the goal is to make the surrogate model as accurate as possible using as few samples as possible. To achieve this goal, the first important task is to determine how many samples to use and where to locate the samples.

In this chapter, we will explore methods for selecting a good set of sample locations for carrying out experiments. The selection of these sample locations is known as the design of experiments (DoE) or experimental designs. This terminology was developed because PRS was initially designed to approximate QoI from experiments. It is generally accepted that DoE is as important as the surrogate itself because the accuracy of the surrogate depends on the number and location of samples. In addition, there is no single DoE scheme that is the best for all surrogates. Various DoE techniques cater to different sources of errors, in particular, errors due to noise in samples or errors due to an improper surrogate model.

Although DoE is used to find sample locations to build a surrogate in this text, DoE was originally developed in agricultural applications to identify the effect of input variables (factors) on the output QoI [22]. DoE is defined as a branch of applied statistics that deals with planning, conducting, analyzing, and interpreting controlled tests to evaluate the factors (i.e., input variables) that control the value of a parameter or group of parameters. It is used (i) to determine if input variables have an effect on QoIs, (ii) to determine if multiple inputs interact in their effect on the QoIs, (iii) to model the behavior of the QoI as a function of input variables, and (iv) to optimize the QoI. It is noted that DoE is used for defining sample locations for surrogate modeling in this text.

It would be beneficial to understand common DoE terms and concepts before we study detailed DoE methods. The most commonly used terms in DoE include input variables, uncontrollable parameters, and output QoI. Figure 3.1 illustrates the relationship between these terms. Input variables are those variables that can be varied in experiments or numerical simulations (the process in the

*Surrogate Modeling and Optimization: Theories, Applications, and Limitations*, First Edition. Nam-Ho Kim.

Published 2026 by John Wiley & Sons, Inc.
Companion website: www.wiley.com/go/kim/surrogate_modeling

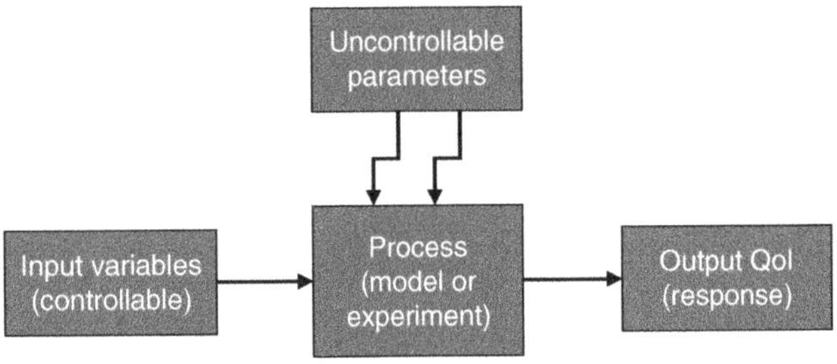


**Figure 3.1** The effect of input variables on the quantity of interest in a numerical model or experiment.

figure). In DoE literature, input variables are referred to as factors. It is assumed that input variables are deterministic, and the users have full control of them. Input variables include dimensions of test specimens, applied loads, mass, time, etc. Output QoI is the response of the model or measurement from the experiment. It is assumed that the output QoI varies depending on input variables. Since the QoI is the output from an experiment or simulation, it may include measurement error or simulation error. It is also assumed that the output QoI is a continuous and smooth function of input variables. Uncontrollable parameters are those parameters that cannot be controlled by the user, such as ambient temperature during the experiment. In practice, variation of these parameters during the process can cause random noise in the output QoI. Even if these parameters affect the output QoI, their contribution is not modeled as a functional relationship in the surrogate. The process represents physical experiments or numerical simulations that can produce output QoI for given input variables and uncontrollable parameters. In some sense, the process is considered a black box to the user. That means the user does not need to know the complex process within the experiment or simulation. As long as the user can provide different input variables and obtain output QoI, the surrogate model can be generated. In this sense, the surrogate model is considered a data-driven approach as it can be applied to any physical model or process. The main purpose of DoE is to determine input variables such that their effects on the output QoI can be identified in the best way.

As mentioned before, in this textbook, DoE is used for the purpose of generating sample locations to build a surrogate. In that sense, DoE is inherently a multi-objective optimization problem. We would like to select points so that we maximize the accuracy of the information that we get from the experiments. As shown in Chapter 2, however, the accuracy can be improved by increasing the number of samples as well. On the other hand, we also would like to minimize the number of experiments, because they are expensive. In some cases, the objective of the experiments is to estimate some physical characteristics, and in these cases, we would like to maximize the accuracy of these characteristics. However, in the design applications, which are of primary interest in this textbook, we would like to construct a surrogate that could be used to predict performance at unsampled locations. In this case, our primary goal is to choose the points for the experiments to maximize the predictive capability of the model. Therefore, we focus on how to reduce/minimize the prediction variance at unsampled locations.

Samples do not need to be generated simultaneously. Initially with a small number of samples, additional samples can be added sequentially to improve the accuracy of surrogate prediction. The strategy of adding additional samples is called adaptive sampling, sequential design, or active learning. Usually, the adaptive sampling strategies require a criterion to determine the next sample locations. Many criteria have been proposed, such as D-optimal design and minimum bias design. These criteria are often based on a trade-off between minimizing bias (i.e., error) and/or minimizing variance (i.e., uncertainty).

When a surrogate model is flexible enough, a better surrogate can be obtained with more samples. With a fixed size of design space, the distance between samples decreases as the number of samples increases. Therefore, the meaning of the number of samples is equivalent to the distance between the samples. It is easy to imagine that if the distance between samples is short, the surrogate approximate can be accurate because the samples can capture the trend of functional behavior well. At the same time, for a fixed number of samples, it is better to locate the samples such that they are more or less uniformly distributed in the entire sampling space, which is called "space-filling design." In order to have a space-filling design, some sampling strategies minimize the maximum distance between samples. Or some strategies uniformly divide the sampling space and put a sample in each segment. Section 3.4 introduces two important space-filling DoEs: Latin hypercube sampling (LHS) and orthogonal arrays (OAs).

A lot of work has been done on experimental designs in regular design domains. Such domains occur when each design variable is bounded by simple lower and upper limits so that the design domain is box-like. Occasionally, spherical domains are also considered. Sometimes, each design variable can take only two or three values, often called levels. These levels are termed low, nominal, and high. In other cases, the design space is approximately box-like, but it is possible to carry out experiments with the design variables taking values outside the box for the purpose of improving the properties of the surrogate. In Section 3.2, we will summarize briefly some of the properties of experimental design in box-like domains and present some of the more popular experimental designs in such domains. The readers are referred to Myers and Montgomery [3] for the in-depth analysis of this subject.

For design optimization, however, it is common for us to try and create surrogates in irregularly shaped domains. In that case, we have to create our own experimental design. Section 3.3 will discuss several techniques available for finding good designs in a generally shaped domain. These strategies mostly use an optimization technique based on different criteria. Since additional sample locations can be found based on pre-existing sample locations, these strategies are good for an adaptive sampling scheme.

## 3.2 Design of Experiments in Box-like Domains

### 3.2.1 Scaling of Input Variables

When the design space of input variables is given in a box-like shape, the design space is defined by simple lower and upper limits on each of the input variables

$$x_i^l \leq x_i' \leq x_i^u, \quad i = 1, \ldots, n \tag{3.1}$$

where $x_i^l$ and $x_i^u$ are the lower and upper bounds, respectively, of the input variable $x_i'$. The prime indicates that the design variable has not been normalized. For convenience, we scale the design variable as

$$x_i = \frac{2x_i' - x_i^l - x_i^u}{x_i^u - x_i^l} \tag{3.2}$$

The normalized variables are then all bound in the cube

$$-1 \leq x_i \leq 1, \quad i = 1, \ldots, n \tag{3.3}$$

Such normalization is useful when different input variables have different dimensions with different orders of magnitude. For example, the thickness of a plate may vary in $x'_1 \in [0.5, 1.5] \times 10^{-3}$ m, while Young's modulus of the material may vary in $x'_2 \in [190, 210] \times 10^9$ Pa. Therefore, using the original variables can cause numerically ill-conditioning of the moment matrix during the regression process. The normalization in Eq. (3.2) should not affect the PRS with linear regression because of the linear relationship between the original and normalized variables.

Although the main purpose of surrogate models in this text is optimization, they are also frequently used for uncertainty quantification, where input variables follow a specific probability distribution, and the probability distribution of output QoI needs to be evaluated. In such a case, it would be better to scale the input variables based on their distribution. For example, when the user wants the design space to cover six times the standard deviation, input variables can be scaled by

$$x_i = \frac{x'_i - \mu_{x_i}}{\sigma_{x_i}} \tag{3.4}$$

where $\mu_{x_i}$ and $\sigma_{x_i}$ are, respectively, the mean and standard deviation of the input variable $x'_i$. Then, the scaled variables are all bound in the cube

$$-6 \le x_i \le 6, \quad i = 1, \dots, n \tag{3.5}$$

When the mean and standard deviation of input variables are not available, the mean and standard deviation of samples can also be used as an approximation.

It is also possible that the output QoI can be normalized in a similar way to the input variables. Unfortunately, the output QoI is not what the user can control; it is not possible to establish the lower and upper bounds of the output QoI. Instead, the minimum and maximum of samples can be used to normalize the output QoI, as

$$y_i = \frac{y'_i - y_{\min}}{y_{\max} - y_{\min}} \tag{3.6}$$

where $y_{\min}$ and $y_{\max}$ are, respectively, the minimum and maximum samples.

**Example 3.1 Normalization of input variables**

Consider the quadratic PRS in Example 2.5 with five samples. Normalize the input variable based on Eq. (3.2) and the QoI based on Eq. (3.6) and show that the resulting surrogate is identical to the one given in Example 2.5 after conversion.

**Solution:**

With the original samples before scaling, the quadratic PRS was

$$\hat{y}'(x') = -0.1071 + 0.925x' + 0.0536x'^2$$

Note that the notation $'$ is used for the variables in the original domain. Since the lower and upper bounds of input variables are not given in the example, the minimum and maximum values of the samples are used as the two bounds: $x^l = -2$, $x^u = 2$, $y_{\min} = -1.5$, and $y_{\max} = 1.75$. Then the input variables and output QoI are converted to

$$x = \frac{1}{2}x', \quad y_i = \frac{y'_i + 1.5}{3.25}$$

Then, the five samples are scaled by (−1, 0), (−0.5, 0), (0, 6/13), (0.5, 11/13), and (1, 1). The PRS is of the form $\hat{y}(x) = b_1 + b_2x + b_3x^2$, and we can define the following matrices and vectors for regression:

$$\mathbf{X} = \begin{bmatrix} 1 & -1 & 1 \\ 1 & -0.5 & 0.25 \\ 1 & 0 & 0 \\ 1 & 0.5 & 0.25 \\ 1 & 1 & 1 \end{bmatrix}, \quad \mathbf{y} = \begin{Bmatrix} 0 \\ 0 \\ 6/13 \\ 11/13 \\ 1 \end{Bmatrix}, \quad \mathbf{X}^T\mathbf{X} = \begin{bmatrix} 5 & 0 & 2.5 \\ 0 & 2.5 & 0 \\ 2.5 & 0 & 2.125 \end{bmatrix}, \quad \mathbf{X}^T\mathbf{y} = \begin{Bmatrix} 2.3077 \\ 1.4213 \\ 1.2115 \end{Bmatrix}$$

By solving the regression equation in Eq. (2.14), the unknown regression coefficients are identified as $b_1 = 0.4286$, $b_2 = 0.5692$, and $b_3 = 0.0659$. Therefore, with the normalized variables, the quadratic PRS can be written as

$$\hat{y}(x) = 0.4286 + 0.5692x + 0.0659x^2$$

Now, after applying the conversion relationship, the above quadratic PRS can be converted into

$$\hat{y}(x) = \frac{\hat{y}'(x) + 1.5}{3.25} = 0.4286 + 0.5692\left(\frac{1}{2}x'\right) + 0.0659\left(\frac{1}{2}x'\right)^2$$

$$\hat{y}'(x') = -0.1071 + 0.925x' + 0.0536x'^2$$

Therefore, the same surrogate model is obtained after converting it to the original variables.

### 3.2.2 Interpolation, Extrapolation, and Prediction Variance

The simplest experimental design for the cube of design space is one experiment at each of the $2^n$ vertices. This design is called a two-level full-factorial design, where the word "factorial" refers to "factor," a synonym for design variable, rather than the factorial function. This means that two samples are generated in each design variable at their minimum and maximum. Such a two-level DoE is good when the functional relationship between the design variables and the QoI is linear with possible interactions (i.e., including $x_ix_j$, $i \neq j$ terms). If three samples are used in each design variable, it is called a three-level design. A three-level full-factorial design would require $3^n$ numbers of samples. A three-level DoE is good when the functional relationship is quadratic.

For a small number of design variables (a low-dimensional problem), the number of samples $2^n$ or $3^n$ can be a manageable number of experiments. However, for larger values of $n$, we usually cannot afford the full-factorial design. For example, for $n = 10$, the number of samples becomes $2^n = 1{,}024$ for two-level, and $3^n = 59{,}049$ for a three-level DoE. Therefore, the number of samples increases exponentially proportional to the number of dimensions, which is called the "curse of dimensionality." This is considered a major bottleneck of surrogate modeling for applications with many input variables. Therefore, full-factorial design is limited only to low-dimensional problems ($n \leq 4$). For high values of $n$, it may be necessary to consider fractional factorial designs, which do not include all the vertices. Different DoE methods are available depending on how to choose sample locations in fractional factorial designs.

If we want to fit a linear PRS to samples, it certainly appears that we will not need anywhere near $2^n$ samples for a good fit. For example, for $n = 10$, the linear PRS has 11 unknown coefficients to fit, and using 1,024 experiments to fit these 11 coefficients may appear excessive even if we could afford that many experiments. Theoretically, 11 samples should be good enough if we know that the functional relationship is linear, and the samples do not include measurement noise or error. In practice, however, we do not know the exact functional relationship, and therefore, we need more samples than the number of unknown coefficients. In general, it is acceptable that the number of samples is about two or three times more than that of the unknown coefficients.

However, with fewer samples, we lose an important property of using a surrogate as an interpolation tool rather than as a tool for extrapolation. To understand this aspect, we will first define what we mean by interpolation and extrapolation. Intuitively, we say that a surrogate will interpolate samples at a point if that point is "completely surrounded" by sample points. In one-dimensional space, as shown in Figure 3.2(a), this means that there is a sample to the right of the interpolated point, as well as a sample to the left of it. In two-dimensional space, as shown in Figure 3.2(b), we would like the point to be surrounded by at least three samples so that it falls inside the triangle defined by these three samples. Similarly, in three-dimensional space, we would like the point to be surrounded by at least four samples; that is, the point lies inside the tetrahedron defined by these four samples. In $n$-dimensional space, we would like the point to be surrounded by $n+1$ sample points, or in other words, fit inside the simplex defined by the $n+1$ samples. A simplex is the generalization of a triangle and a tetrahedron; a shape in $n$-dimensional space defined by linearly $n+1$ independent points.

Given a set of $n+1$ points in $n$-dimensional space, $x_1, x_2, \ldots, x_{n+1}$, the simplex defined by these points includes all the points that can be obtained by a convex sum of these points. That is, it includes any point $\mathbf{x}$, which may be written as

$$\mathbf{x} = \sum_{i=1}^{n+1} \alpha_i \mathbf{x}_i \tag{3.7}$$

with

$$\sum_{i=1}^{n+1} \alpha_i = 1 \tag{3.8}$$

and $\alpha_i \geq 0$, $i = 1, \ldots, n+1$. Given a set of $n_y$ sample points, the set of points where the surrogate performs interpolation is the union of these simplices, which is called the convex hull of the sample points.

In general, surrogate prediction is more accurate in the interpolation region than in the extrapolation region. As shown in Example 2.10, when three samples were used, the standard error of prediction at the extrapolated point (1, 1) was highest $\sigma_y = \sqrt{3}\hat{\sigma}$. On the other hand, the centroid of the three samples has the lowest standard error $\sigma_y = \hat{\sigma}/\sqrt{3}$. Therefore, it is best to locate samples such that the convex hull of the samples can cover the design space as much as possible. However,

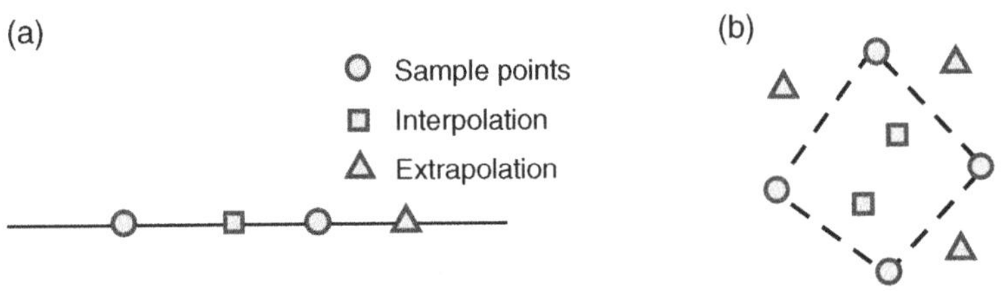


**Figure 3.2** Interpolation versus extrapolation in one- and two-dimensional design space. (a) One dimension and (b) two dimensions.

as shown in Figure 1.5, the volume of the convex hull of samples becomes very small compared to that of the design space as the dimension of the design space increases. We showed that in 10-dimensional space, if two samples were generated at the center of each orthant ($x_i = -0.5$, $0.5$ in the normalized space), 1,024 samples from the two-level full-factorial design can only cover 0.1% volume of the design space. Therefore, most prediction points belong to the extrapolation region, where the prediction accuracy of the surrogate deteriorates quickly.

One may argue that what if we generate samples at all vertices ($x_i = -1.0, 1.0$ in the normalized space) instead of the center of each orthant. In such a case, the entire design space becomes the interpolation region. However, in such a choice of sample locations, the distance between samples increases rapidly. In two-dimensional space, if all four samples are located at the vertices, the maximum distance between samples is $2\sqrt{2}$. In three-dimensional space, it becomes $2\sqrt{3}$. Therefore, the distance between samples increases along with the dimension of the design space. Even if surrogate prediction in the interpolation is better than in the extrapolation region, the prediction deteriorates quickly as the distance between samples increases. Instead of discussing the accuracy of surrogate prediction qualitatively, it would be necessary to define a quantitative measure to assess the quality of prediction accuracy of a surrogate.

One measure that we can use to estimate the loss of prediction accuracy incurred when we use extrapolation is the prediction variance. In the case of PRS, we developed the prediction variance in Chapter 2. In summary, the PRS assumes that (i) the true function is described by a linear combination of monomials with unknown coefficients, (ii) samples are obtained by adding random noise to the true function, where the noise follows a Gaussian distribution $\sim N(0, \sigma^2)$, and (iii) noises at different sample points have the same standard deviation and are not correlated. Under these assumptions, the noise standard deviation (standard error of noise) was estimated by

$$\hat{\sigma} = \sqrt{\frac{\mathbf{e}^T\mathbf{e}}{n_y - n_\beta}} \tag{3.9}$$

where $n_y$ and $n_\beta$ are, respectively, the number of samples and the number of model parameters, and $e_i = y_i - \hat{y}_i$ is the residual at $i$th sample location.

Recall that the linear regression PRS model that we used in Eq. (2.10) can be written as $\hat{y}(\mathbf{x}) = \boldsymbol{\xi}(\mathbf{x})^T\mathbf{b}$, where $\boldsymbol{\xi}(\mathbf{x})$ is the vector of monomial basis functions and $\mathbf{b}$ is the vector of regression coefficients. With noise in the samples, $\mathbf{b}$ has some uncertainty, while $\boldsymbol{\xi}(\mathbf{x})$ is deterministic. At the prediction point $\mathbf{x}$, the prediction variance is given as $V[\hat{y}(\mathbf{x})] = \hat{\sigma}^2\boldsymbol{\xi}(\mathbf{x})^T\left(\mathbf{X}^T\mathbf{X}\right)^{-1}\boldsymbol{\xi}(\mathbf{x})$. The square root of the prediction variance is called the standard error of prediction, defined as

$$\sigma_y(\mathbf{x}) = \hat{\sigma}\sqrt{\boldsymbol{\xi}(\mathbf{x})^T\left(\mathbf{X}^T\mathbf{X}\right)^{-1}\boldsymbol{\xi}(\mathbf{x})} \tag{3.10}$$

The standard error gives us an estimate of the sensitivity of the surrogate prediction at different points. From a statistical perspective, when the surrogate prediction at an unsampled point $\mathbf{x}$ is $\hat{y}(\mathbf{x})$, the true function at that point is unknown (i.e., uncertain), and the probability of the true function can be given as $\sim N\left(\hat{y}, \sigma_y^2\right)$. In the extreme case, if $\sigma_y(\mathbf{x}) = 0$ at a point, it means the surrogate prediction $\hat{y}(\mathbf{x})$ is exact and there is no uncertainty. Therefore, a surrogate prediction is considered accurate when $\sigma_y(\mathbf{x})$ is small. Note that $\sigma_y(\mathbf{x})$ varies at different locations.

The main goal of DoE is that we would like to select sample locations to make the prediction variance as small as possible anywhere in the domain where we would like to estimate the QoI.

Intuitively, it appears that this would be helped if the standard error did not vary much from one point to another. This property is called stability, which is defined as the ratio, $\sigma_y^{max}/\sigma_y^{min}$. The following example demonstrates the effect of using extrapolation on the stability of the standard error.

**Example 3.2 Prediction variances from different designs**

Consider the problem of fitting a linear PRS $\hat{y}(\mathbf{x}) = b_1 + b_2x_1 + b_3x_2$ to samples in the square domain $-1 \leq x_1, x_2 \leq 1$. Compare the maximum value of the prediction variance for two cases: (i) a full-factorial design (samples at all four vertices), and (ii) a fractional factorial design including three vertices, obtained by omitting the vertex (1, 1).

**Solution:**

*Full-factorial design*: We number the four sample locations as $\mathbf{x}_1 = [-1, -1]^T$, $\mathbf{x}_2 = [-1, 1]^T$, $\mathbf{x}_3 = [1, -1]^T$, and $\mathbf{x}_4 = [1, 1]^T$. For this case, the moment matrix can be obtained as

$$\xi(\mathbf{x}) = \begin{Bmatrix} 1 \\ x_1 \\ x_2 \end{Bmatrix}, \quad \mathbf{X} = \begin{bmatrix} 1 & -1 & -1 \\ 1 & -1 & 1 \\ 1 & 1 & -1 \\ 1 & 1 & 1 \end{bmatrix}, \quad \mathbf{X}^T\mathbf{X} = \begin{bmatrix} 4 & 0 & 0 \\ 0 & 4 & 0 \\ 0 & 0 & 4 \end{bmatrix}, \quad (\mathbf{X}^T\mathbf{X})^{-1} = \frac{1}{4}\begin{bmatrix} 1 & 0 & 0 \\ 0 & 1 & 0 \\ 0 & 0 & 1 \end{bmatrix}$$

The standard error of prediction in Eq. (3.10) is given as

$$\sigma_y = \hat{\sigma}\sqrt{\xi^T(\mathbf{X}^T\mathbf{X})^{-1}\xi} = \hat{\sigma}\sqrt{0.25(1 + x_1^2 + x_2^2)}$$

Therefore, the minimum standard error occurs at the origin ($x_1 = x_2 = 0$), $\sigma_y = \hat{\sigma}/2$, while the maximum occurs at the vertices, $\sigma_y = \sqrt{3}\hat{\sigma}/2$. This case represents a fairly stable variation of the standard error of only $\sqrt{3}$ between the smallest and highest value in the domain of interest.

*Fractional factorial design*: Without a sample at (1,1), we have

$$\mathbf{X} = \begin{bmatrix} 1 & -1 & -1 \\ 1 & -1 & 1 \\ 1 & 1 & -1 \end{bmatrix}, \quad \mathbf{X}^T\mathbf{X} = \begin{bmatrix} 3 & -1 & -1 \\ -1 & 3 & -1 \\ -1 & -1 & 3 \end{bmatrix}, \quad (\mathbf{X}^T\mathbf{X})^{-1} = \frac{1}{4}\begin{bmatrix} 2 & 1 & 1 \\ 1 & 2 & 1 \\ 1 & 1 & 2 \end{bmatrix}$$

Therefore, the standard error of prediction is given as

$$\sigma_y = \hat{\sigma}\sqrt{0.5(1 + x_1 + x_2 + x_1^2 + x_1x_2 + x_2^2)}$$

At the origin, the standard error is $\sigma_y = \hat{\sigma}/\sqrt{2}$, which is increased from the full-factorial design. At the three vertices with samples, the standard error of prediction is the same as the standard error of noise: $\sigma_y = \hat{\sigma}$. This is expected, with only three sample points, the PRS passes through the samples so that the error at the sample points should be the same as the measurement error. Therefore, the standard error increases slightly at the origin and the sample locations. However, at the fourth vertex (1, 1), where the PRS represents extrapolation, the standard error becomes $\sigma_y = \sqrt{3}\hat{\sigma}$, much increased from the case of full-factorial design. This is because this is the farthest extrapolation point from the convex hull of the three samples. By setting the derivatives of the prediction error to zero, we easily find that the minimum error is at the centroid of the three sample points, at (−1/3, − 1/3). At the centroid $\sigma_y = \hat{\sigma}/\sqrt{3}$. Now the ratio between the smallest and highest standard errors is 3, with the highest errors in the region of extrapolation. Therefore, the full-factorial design is more stable than the fractional factorial design.

It can be checked that when we use a full-factorial design for a linear PRS with $n$ variables, the moment matrix becomes diagonal, $\mathbf{X}^T\mathbf{X} = 2^n\mathbf{I}$, where $\mathbf{I}$ is a unit matrix of order $n + 1$. Therefore, the standard error of prediction becomes

$$\sigma_y = \frac{\hat{\sigma}}{2^{n/2}}\sqrt{1 + x_1^2 + x_2^2 + \dots + x_n^2} \tag{3.11}$$

In such a case, the maximum prediction error (achieved at any vertex) is $\hat{\sigma}\sqrt{(n+1)/2^n}$. That is, the quality of the fit becomes very good with increasing $n$. This reflects the fact that we use $2^n$ points to calculate $n + 1$ coefficients so that we filter out the effect of noise. This estimate is misleading in actual situations, however, because rarely do we have a true linear model. When the response we measure is not linear, we will have modeling errors, also called bias errors, which are not averaged out.

On the other hand, if the number of samples becomes the same as $n + 1$ unknown coefficients, it is called a saturated design. In that case, the standard error progressively increases along with the dimension $n$, because the portion of design space covered by the simplex containing the sample points becomes progressively smaller. For example, in Example 3.2, the three points used for the saturated design form a triangle covering half of the design domain. For a three-dimensional cube, four vertices will span a tetrahedron with a volume of one-sixth of the volume of the enclosing cube. For the $n$-dimensional case, the full-factorial design is the vertices of a cube of volume $2^n$, while $n + 1$ vertices obtained by perturbing one variable at a time from one vertex span a simplex of volume $2^2/n!$. Therefore, the fraction of the extrapolation region rapidly increases with the dimension. As the extrapolation region increases, so does the prediction variance. For example, for a three-dimensional design space, the maximum prediction error with a full-factorial design is $\sqrt{0.5}\sigma$, while the maximum prediction error for the saturated four-point fractional factorial design is $\sqrt{7}\sigma$ (see Exercise Problem 3.3).

### 3.2.3 Designs for Linear Polynomial Response Surfaces

For fitting linear PRS, we typically use designs with only two levels for each design variable, and the most popular fractional designs are the so-called orthogonal designs. An orthogonal design means that different coefficients in linear regression are uncorrelated; that is, $\mathrm{CoV}(b_i, b_j) = 0$. It means that the uncertainty in one coefficient is not related to the uncertainty in other coefficients, which is a desirable property as the effect of different basis functions can be determined individually. An orthogonal design is one where the moment matrix $\mathbf{X}^T\mathbf{X}$ is diagonal, which is equivalent to the inner product of two different columns of the design matrix being zero, which means they are orthogonal. Since each column of the design matrix represents the basis function, the orthogonality means that the coefficient of each basis function can be estimated independently from other coefficients. The popularity of orthogonal designs is partly based on the following theorem (see Myers and Montgomery [3], p. 284):

> For the first-order model (linear PRS) and fixed sample size, if all variables lie between $-1$ and 1, then the variance of the coefficients is minimized if the design is orthogonal, and all the variables are at their outer positive or negative limits (i.e., $-1$ or $+1$).

It is easy to check that the full-factorial design is orthogonal, but it is not trivial to produce orthogonal designs with a smaller number of samples. Various orthogonal designs can be found in books on the DoE (see Myers and Montgomery [3]). Important orthogonal designs will be presented as part of the OA in Section 3.4. To demonstrate the beneficial properties of orthogonal designs, we will consider the two-dimensional case that we have studied in Example 3.2. In that example, we fitted a two-variable linear PRS first with a full-factorial design (four samples at corners) and then with only three samples (simplex in two dimensions). Fitting a linear PRS in $n$ variables on the basis of $n + 1$ points requires the points to be linearly independent, so that they form a simplex. There is no redundancy in the design, in that the number of points is equal to the number of coefficients, and this is called a saturated design. In order to get an orthogonal design with three samples in two-dimensional space, we have to give up on having the variables only in the corners. Instead, a perfect simplex is used, where the distances between all points are the same.

**Example 3.3 Orthogonal design**

Consider the equilateral triangle, which results in a scalar matrix (a scalar matrix is a scalar multiple of the unit matrix) $\mathbf{X}^T\mathbf{X}$. It includes the points $\left(\sqrt{3/2}, -1/\sqrt{2}\right), \left(-\sqrt{3/2}, -1/\sqrt{2}\right), (0, \sqrt{2})$. Check for the stability of the prediction variance and its maximum value in the unit square for the linear model $\hat{y}(\mathbf{x}) = b_1 + b_2x_1 + b_3x_2$.

**Solution:**

First, it is noted that the three samples are the corners of an equilateral triangle, whose centroid is the same as that of the design space, i.e., the origin. In order to calculate the standard error of prediction in Eq. (3.10), the inverse of the moment matrix needs to be calculated first using the three samples:

$$\mathbf{X} = \begin{bmatrix} 1 & \sqrt{3/2} & -1/\sqrt{2} \\ 1 & -\sqrt{3/2} & -1/\sqrt{2} \\ 1 & 0 & \sqrt{2} \end{bmatrix}, \quad \mathbf{X}^T\mathbf{X} = \begin{bmatrix} 3 & 0 & 0 \\ 0 & 3 & 0 \\ 0 & 0 & 3 \end{bmatrix}, \quad (\mathbf{X}^T\mathbf{X})^{-1} = \frac{1}{3}\begin{bmatrix} 1 & 0 & 0 \\ 0 & 1 & 0 \\ 0 & 0 & 1 \end{bmatrix}$$

Therefore, the three samples proved to be an orthogonal design. Then, the standard error of prediction in Eq. (3.10) becomes

$$\sigma_y(\mathbf{x}) = \hat{\sigma}\sqrt{(1 + x_1^2 + x_2^2)/3}$$

Therefore, the standard error of prediction becomes $\sigma_y = \hat{\sigma}/\sqrt{3}$ at the origin and $\sigma_y = \hat{\sigma}$ at the vertices. That is, the stability ratio is $\sigma_y^{\max}/\sigma_y^{\min} = \sqrt{3}$, which is an improvement from 3 in Example 3.2. Also, $\sigma_y^{\max}$ is reduced from $\sqrt{3}\hat{\sigma}$ to $\hat{\sigma}$. However, this has come at the price of obtaining the samples outside the unit square, as shown in Figure 3.3.

As will be shown later, this reduces the variance error but increases the so-called bias error. A bias error is the error introduced when the model that we try to fit is different from the true function. For example, if the model is linear and the true function is quadratic, the simplex model that we have used in this example is likely to increase the error rather than decrease it.

### 3.2.4 Designs for Quadratic Polynomial Response Surfaces

Quadratic PRS with $n$ variables have $n_\beta = (n + 1)(n + 2)/2$ coefficients, as shown in Eq. (2.16). To fit quadratic PRS, we need at least that many samples, and at least three levels for each design variable in order to capture a quadratic functional change. For $n > 3$, it is possible to have the required number of samples with only two levels. For example, a quadratic PRS with four variables has 15 coefficients, and a full-factorial design in two levels has $2^4 = 16$ points. Therefore, theoretically, it is

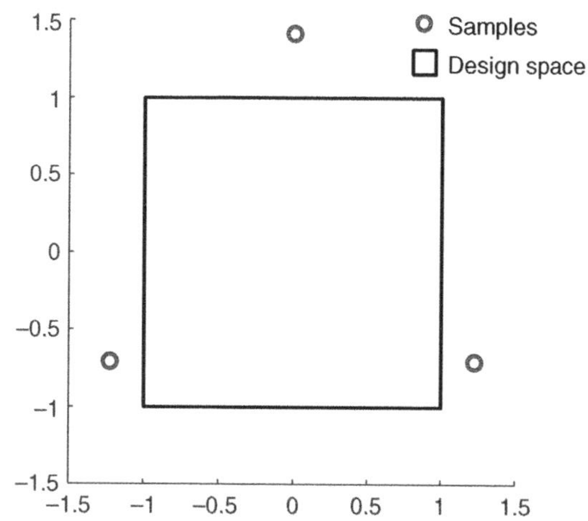


**Figure 3.3** Design space and sample locations for Example 3.3.

possible to identify unknown coefficients with a two-level full-factorial design. However, if we let only one design variable vary at a time, we can easily check that we are left with three coefficients, and we need three different levels of that design variable. That is, a quadratic PRS with $n = 2$ can be defined as $\hat{y}(x_1, x_2) = b_1 + b_2x_1 + b_3x_2 + b_4x_1^2 + b_5x_1x_2 + b_6x_2^2$. When $x_1$ is fixed, the PRS becomes one-dimensional and can be simplified as $\hat{y}(x_2) = \tilde{b}_1 + \tilde{b}_2x_2 + \tilde{b}_3x_2^2$ with three coefficients. Therefore, it would make sense to have three levels in each design variable.

We can use a full-factorial three-level design for a quadratic PRS, which will have $3^n$ samples, as shown in Figure 3.4. In addition to samples in all vertices (either −1 or +1 location in the normalized design space), the full-factorial three-level design has a sample at the center of each design variable (i.e., 0 location). In most cases, however, we cannot afford such many samples even for fairly small values of $n$. For example, for $n = 6$, we require $3^6 = 729$ samples, which are too many to identify $n_\beta = 28$ coefficients. In general, in order to have a stable estimation, the number of samples needs to be two or three times more than that of the unknown coefficients. Therefore, 80 samples should be good enough instead of 729. Many alternative DoEs are proposed that can still capture a quadratic change of the function with a much smaller number of samples than the full-factorial design.

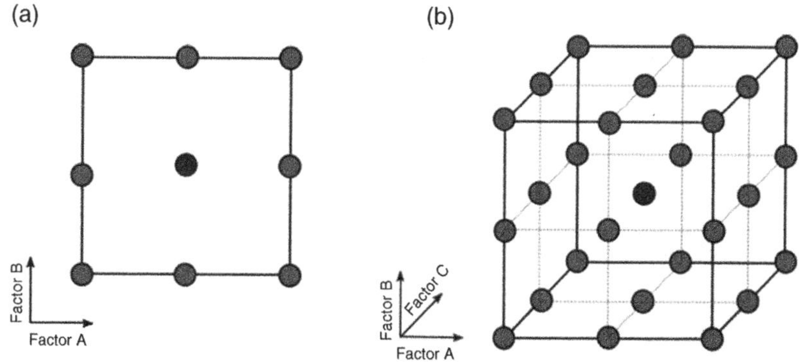


**Figure 3.4** Sample locations for full-factorial designs. (a) Two dimensions and (b) three dimensions.

A popular compromise that reduces the number of samples to close to the two-level full-factorial design is the central composite design (CCD). The CCD is a fractional factorial design with center points, augmented with a group of axial points that allows for estimating curvature. The CCD is composed of the $2^n$ points of the full-factorial two-level design, with all the variables at their extremes, plus a number of repetitions $n_c$ at the center of the design space, and the $2n$ points obtained by changing one design variable at a time by a distance $\alpha \geq 1$, which are referred to as axial points. Therefore, the total number of samples for CCD is $n_y = 2^n + 2n + n_c$. In determining the range of design space, the center point is often close to or near the local optimum. The replicate center points are used to test if the curvature near the design point is significant, which is important for design exploration. For this reason, the two-level factorial design is augmented with the design center to capture the second-order behavior of QoI. In addition to the center points, it is necessary to have some axial points in order to construct the second-order model and to estimate the parameters related to each second-order term. Without the axial points, the variation due to the second-order terms cannot be orthogonally decomposed into the variations of each second-order term. If quadratic coefficients $b_{ii}$ turn out to be all negative or positive, then we can ensure that the center point is in the vicinity of the local optimum. If the center point is the current operating condition, the estimates of the second-order term give information on whether the region of exploration is close to the local optimum (maximum or minimum). The unique feature of the CCD is that the $2n$ axial samples at the distance $\alpha \geq 1$ are out of the design space. This can be a limitation of the CCD because, in some situations, we may not allow generating samples out of the design space.

Figure 3.5 shows the CCD for $n = 2$ and $n = 3$. The value of $\alpha$ chosen in the figures are such that all the points outside the origin are of the same distance from the origin, so that we have a spherical design. This placement of the points is at the higher end of the typical choice for $\alpha = \sqrt{n}$. A more popular choice is based on the concept of rotatability. The property of rotatability requires that the prediction variance is dependent only on the distance from the origin and not on the orientation with respect to the coordinate axes [23].

It can be shown that for the CCD, the rotatability requirement will be satisfied for

$$\alpha = 2^{n/4} \tag{3.12}$$

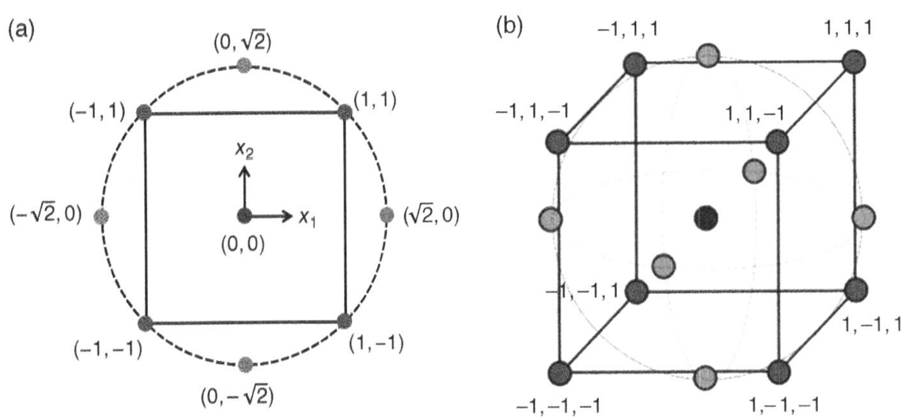


**Figure 3.5** Central composite designs for two- and three-dimensional design spaces. (a) Two dimensions and (b) three dimensions.

This equation gives $\alpha = \sqrt{2}$ for $n = 2$, which is the same as the spherical design; however, for $n = 3$, we get $\alpha = 1.682$, which is slightly smaller than the equal distance of $\sqrt{3}$. According to Myers and Montgomery [3], it is not necessary to have exact rotatability in a second-order design in the practical sense.

**Example 3.4 Central composite design**

When input variables are temperature and pressure in the injection-molding process of plastic material, generate a table of input variables using the CCD. The temperature varies ∈[190°, 210°], while the pressure varies ∈[50 MPa, 100 MPa]. Use $n_c = 5$.

**Solution:**

Including four vertices, four faces, and five repetitions at the center, the total number of samples is 13. The following table shows sample locations. First, 13 sample locations are determined in the normalized design space, and then, they are transformed to the original design space using the inverse relationship of Eq. (3.2), as

$$x_i' = \frac{1}{2}(1 + x_i)x_i^u + \frac{1}{2}(1 - x_i)x_i^l$$

| Sample ID | Temperature (°) | Pressure (MPa) |
|---|---|---|
| 1 | 190 | 50 |
| 2 | 210 | 50 |
| 3 | 210 | 100 |
| 4 | 190 | 100 |
| 5 | 185.9 | 75 |
| 6 | 214.1 | 75 |
| 7 | 200 | 39.6 |
| 8 | 200 | 110.4 |
| 9, 10, 11, 12, 13 | 200 | 75 |

It may sound like we waste many samples by using $n_c$ repeated samples at the origin. The reason that repeated samples are used at the origin is related to reducing the prediction variance. When we choose sample locations, the objective is to maximize the prediction accuracy; that is, to minimize prediction variance everywhere in the design space. If a single center point is used in the CCD, the prediction variance at the origin is highest. This is unfortunate because normally we choose the center of the design space as the best candidate for optimization, where we need the most accuracy. We want the prediction to be reliable throughout the region, especially near the center, since we hope the optimum is near the center of the design space. By picking five to six center points, the prediction variance at the center is the smallest. At the same time, the prediction variance on the edge is also reduced compared to a single sample at the center. If one or two center points are used in CCD, then the prediction variance at the origin is higher than that of the edges. The prediction variance increases again beyond the edge of the design space. Therefore, the main reason for repeated samples at the origin is to balance the precision at the edge of the design relative to the center. With either a spherical design or a rotatable one, we find that we need a number of replicate center points (points at the origin) to obtain good prediction variance stability.

**Example 3.5 Standard error of prediction of CCD**

In a two-dimensional quadratic PRS with CCD, calculate the minimum and maximum standard errors of prediction in Eq. (3.10) by gradually increasing the number of samples at the center from $n_c = 1$ to $n_c = 5$. Plot the contour plots of the standard error of prediction and discuss the difference between $n_c = 1$ and $n_c = 5$.

**Solution:**

With two input variables, the CCD uses four corner points, (−1, − 1), (1, − 1), (1, 1), (−1, 1), and four-axial points, $(-\sqrt{2}, 0)$, $(\sqrt{2}, 0)$, $(0, -\sqrt{2})$, $(0, \sqrt{2})$, and $n_c$ repetitions of the center point (0, 0). For a quadratic PRS, the moment matrix with these sample locations becomes

$$\mathbf{X}^T\mathbf{X} = \begin{bmatrix} 8+n_c & 0 & 0 & 8 & 0 & 0 \\ 0 & 8 & 0 & 0 & 0 & 0 \\ 0 & 0 & 8 & 0 & 0 & 0 \\ 8 & 0 & 0 & 12 & 0 & 4 \\ 0 & 0 & 0 & 0 & 4 & 0 \\ 8 & 0 & 0 & 4 & 0 & 12 \end{bmatrix}$$

Therefore, the effect of repeated samples at the center occurs at the first diagonal element of the moment matrix. The following Matlab code is used to calculate the standard error of prediction at a $21 \times 21$ grid and plot the contours. The users can remove the Matlab comment % in the design matrix in order to change the number of repetitions. The code also calculates the maximum and minimum standard errors of prediction along with their ratios (stability).

```
a=sqrt(2); b=2;
X=[1 -1 -1  1  1  1;
   1  1 -1  1 -1  1;
   1  1  1  1  1  1;
   1 -1  1  1 -1  1;
   1  a  0  b  0  0;
   1 -a  0  b  0  0;
   1  0  a  0  0  b;
   1  0 -a  0  0  b;
%   1  0  0  0  0  0;
%   1  0  0  0  0  0;
%   1  0  0  0  0  0;
%   1  0  0  0  0  0;
   1  0  0  0  0  0];
XTX=X'*X;
XTXi=inv(XTX);
[X, Y]=meshgrid(-1:.1:1, -1:.1:1);
Z=zeros(size(X));
[n, m]=size(X);
for i=1:n
    for j=1:m
        xi=[1 X(i,j) Y(i,j) X(i,j).^2 X(i,j).*Y(i,j) Y(i,j).^2];
        Z(i,j)=sqrt(xi*XTXi*xi');
    end
```

```
end
v=linspace(min(min(Z)),max(max(Z)),10);
[C,h]=contour(X,Y,Z,v);
clabel(C,h)
Zmax=max(max(Z))
Zmin=min(min(Z))
stability=Zmax/Zmin
```

The following table shows the minimum and maximum standard errors of prediction along with their ratios.

| $n_c$ | $\sigma_y^{min}$ | $\sigma_y^{max}$ | $\sigma_y^{max}/\sigma_y^{min}$ |
|---|---|---|---|
| 1 | 0.6657 | 1.0000 | 1.5021 |
| 2 | 0.5825 | 0.7906 | 1.3572 |
| 3 | 0.5216 | 0.7906 | 1.5157 |
| 4 | 0.4743 | 0.7906 | 1.6667 |
| 5 | 0.4361 | 0.7906 | 1.8127 |

It is interesting to note that the maximum standard error $\sigma_y^{\max} = 1.0$ occurs at the center of the design space when $n_c = 1$. When $n_c \geq 2$, the maximum standard error occurs at the corners $\sigma_y^{\max} = 0.7906$, and a further increase of $n_c$ does not change the maximum standard error. On the other hand, the increase in $n_c$ reduces the minimum standard error. Therefore, even if the stability ratio $\sigma_y^{\max}/\sigma_y^{\min}$ increases, the prediction accuracy is improved as the maximum is fixed, while the minimum is reduced. Figure 3.6 shows contour plots of prediction variance with one central point and five central points. It is obvious that more center samples can increase the area of the region that has a low standard error of prediction.

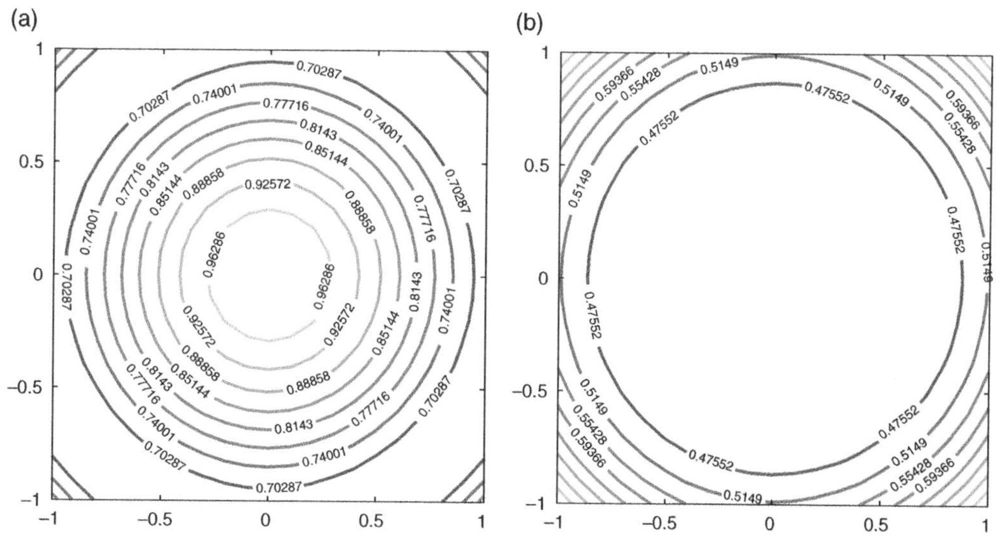


**Figure 3.6** Contour plots of $\sigma_y/\hat{\sigma}$ for $n = 2$, $\alpha = \sqrt{2}$. (a) $n_c = 1$ and (b) $n_c = 5$.

The sampling strategy of replicating central points with rotatable CCD can cause a problem with numerical simulations that give exactly the same answer when the simulation is repeated at the same point. This is because the source of errors in numerical simulation is modeling error, not random noise. Fortunately, however, with $\alpha = 1$, there is no need for repeated central points. The case of $\alpha = 1$, which is called the face-center CCD, is very attractive in many applications because it does not require using any other levels except (−1, 0, 1). Therefore, all samples are within the design space. Figure 3.7 shows the contours of the standard error of prediction for the face-center CCD in a two-dimensional design space. In the case of a two-dimensional space, the face-center CCD becomes identical to a three-level full-factorial design. As shown in the figure, the standard error of prediction at the center of the design space is $\sigma_y(0, 0) = 0.7454$, while at the four corners $\sigma_y(\pm 1, \pm 1) = 0.8975$, which is the maximum value (see Exercise Problem 3.11). The minimum standard error of prediction occurs at $\sigma_y\left(\pm\sqrt{0.4}, \pm\sqrt{0.4}\right) = 0.5980$. The stability of face-center CCD is $\sigma_y^{\max}/\sigma_y^{\min} = 1.50$. Therefore, while the design is not rotatable, the stability is quite good. This performance is better than the CCD with $n_c = 1$ in Example 3.5 in terms of the maximum standard error of prediction.

Even if CCD requires a smaller number of samples than the full-factorial design, it is still no longer practical for high-dimensional design spaces because the number of samples increases too fast. CCD is popular for $3 \le n \le 6$, where it gives reasonable ratios between the number of points and the number of coefficients of a quadratic polynomial. For example, when $n = 10$, the number of samples from the CCD is $n_y = 2^{10} + 2 \times 10 + n_c = 1{,}044 + n_c$, while the number of unknown coefficients for a quadratic PRS is $n_\beta = 66$. Therefore, $n_y \approx 200$ should be enough. One possible solution is to keep the $2n$ axial samples that perturb a single variable but have a fractional factorial design to replace the $2^n$ vertices. However, while the number of vertices increases as $2^n$, the number of

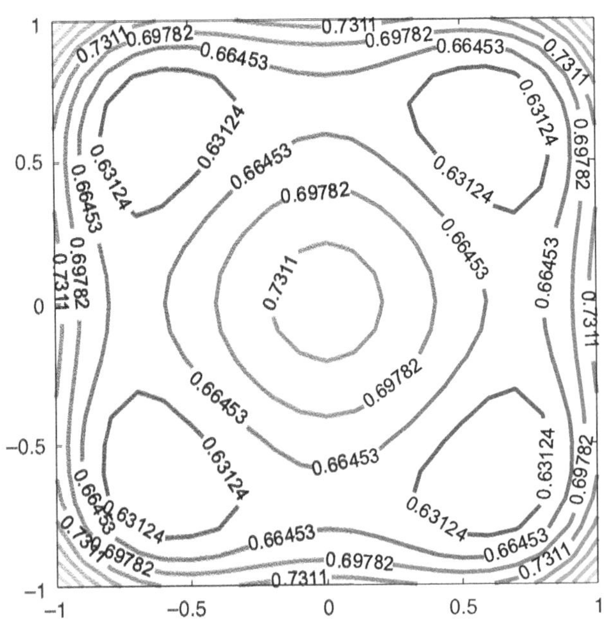


**Figure 3.7** Contour plots of $\sigma_y/\hat{\sigma}$ for face-center central composite design for a single center point.

polynomial coefficients increases as $n_\beta = (n+1)(n+2)/2$. Consequently, the type of fractional design used has to be modified as $n$ increases.

There is a block design, first introduced by Box and Behnken [24], where the number of samples increases at the same rate as the number of polynomial coefficients. The two-variable block design is based on perturbing only two variables from the nominal value. That is, at each point, we have a pair $(i; j)$, such that $|x_i| = 1$, $|x_j| = 1$, and $x_k = 0$ for all $k \neq i, j$. The two variables are perturbed in all four combinations of $\pm 1$. For example, for $n = 3$, we will have the following sample locations:

| $x_1$ | $x_2$ | $x_3$ |
|---|---|---|
| −1 | −1 | 0 |
| −1 | 1 | 0 |
| 1 | −1 | 0 |
| 1 | 1 | 0 |
| −1 | 0 | −1 |
| −1 | 0 | 1 |
| 1 | 0 | −1 |
| 1 | 0 | 1 |
| 0 | −1 | −1 |
| 0 | −1 | 1 |
| 0 | 1 | −1 |
| 0 | 1 | 1 |
| 0 | 0 | 0 |

where the last point is the central point, which may be repeated. Figure 3.8 shows the sample locations of two-variable block designs for a three-dimensional design space. Unfortunately, the block design cannot work for a two-dimensional design space because a two-dimensional block design yields a total of five samples (four edge points and one center point), while there are six unknown coefficients in a quadratic PRS.

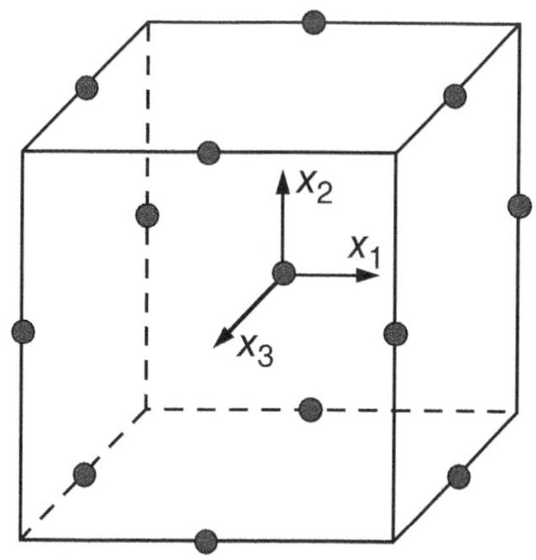


**Figure 3.8** Sample locations of a two-variable block design for three-dimensional space.

For the general case, we can select two variables out of $n$ in $n(n-1)/2$ ways. For each such combination of two variables, we have four design points, with each one of the variables taking the values of $\pm1$. Therefore, the total number of points in this block design is $n_y = n_c + 2n(n-1)$. For large values of $n$, this tends asymptotically to be four times larger than the number of coefficients that we need to fit. However, this happens for very large values of $n$. For example, for $n = 10$, the number of coefficients is 66, while the number of points in the block design with one center point is 181 (for comparison, the number of points in the CCD is 1,045). Block designs are spherical, in that all the points are at the same distance from the origin. For example, all the two-variable block design points are at a distance of $\sqrt{2}$ from the origin. For large values of $n$, this distance can be much smaller than the distance of the vertices (which is $\sqrt{n}$). Therefore, extrapolating to the vertices based on the block design may be risky.

**Example 3.6 Block design**

Although the box design does not work for a quadratic PRS in two dimensions, it is still possible for a linear PRS. Calculate the minimum and maximum standard errors of prediction for the two-variable block design in a two-dimensional design space and compare it with that of the full-factorial design in Example 3.2.

**Solution:**

In a two-dimensional design space, the sample locations for the block design are $\mathbf{x}_1 = (-1, 0)$, $\mathbf{x}_2 = (1, 0)$, $\mathbf{x}_3 = (0, -1)$, $\mathbf{x}_4 = (0, 1)$, and $\mathbf{x}_5 = (0, 0)$. Therefore, the design matrix and the moment matrix can be defined as

$$\mathbf{X} = \begin{bmatrix} 1 & 1 & 0 \\ 1 & -1 & 0 \\ 1 & 0 & 1 \\ 1 & 0 & -1 \\ 1 & 0 & 0 \end{bmatrix}, \quad \mathbf{X}^T\mathbf{X} = \begin{bmatrix} 5 & 0 & 0 \\ 0 & 2 & 0 \\ 0 & 0 & 2 \end{bmatrix}, \quad (\mathbf{X}^T\mathbf{X})^{-1} = \begin{bmatrix} 1/5 & 0 & 0 \\ 0 & 1/2 & 0 \\ 0 & 0 & 1/2 \end{bmatrix}$$

Since the moment matrix is diagonal, the box design yields an orthogonal design. The standard error in Eq. (3.10) can be calculated as

$$\sigma_y(\mathbf{x}) = \hat{\sigma}\sqrt{\frac{1}{5} + \frac{1}{2}(x_1^2 + x_2^2)}$$

The minimum standard error of prediction occurs at the center of the design space $\sigma_y(0,0) = 1/\sqrt{5} = 0.4472$, while the maximum standard error or prediction occurs at the four corner points $\sigma_y(\pm1, \pm1) = \sqrt{6/5} = 1.0954$. Therefore, stability becomes $\sigma_y^{\max}/\sigma_y^{\min} = 2.4495$. Compared to the full-factorial design, the block design has a lower minimum standard error, but the maximum standard error increases significantly. This happens because the corner points become an extrapolation region for the block design. Figure 3.9 shows the contour plot of the standard error of prediction for the block design. Note that the block design satisfies the property of rotatability, where the prediction variance is dependent only on the distance from the origin and not on the orientation with respect to the coordinate axes.

Matlab software provides some useful functions for full and fractional factorial designs. Matlab function `ff2n(n)` Generates samples from a two-level full-factorial design of $n$ input variables. The output is $n_y \times n$ matrix, where each row represents the sample location in normalized space. The total number of samples will be $n_y = 2 \times 2 \times \ldots \times 2 = 2^n$. Since the two-level full-factorial design

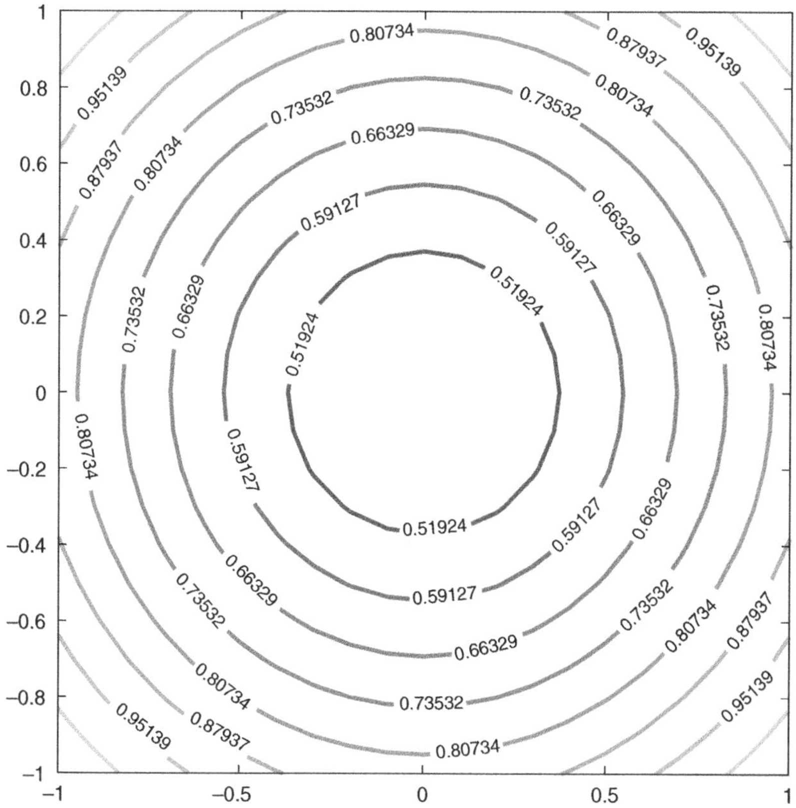

**Figure 3.9** Contour plot of the standard error of prediction for block design.

has sample location only at the lower and upper bounds, the matrix is composed of zero (lower bound) and one (upper bound). For a more general full-factorial design, the Matlab function `fullfact([l_1 ... l_n])` can be used, where each variable can have different levels. That is, variable $x_1$ has $l_1$ levels, $x_2$ has $l_2$ levels, etc. The total number of samples will be $n_y = l_1 \times l_2 \times ... \times l_n$. The output is $n_y \times n$ matrix, where each row represents the level of samples.

For CCD, Matlab function `ccdesign(n, 'Name', value)` can be used for $n$ design variables with options in `'Name'` and `value`. The number of repeated center points can be specified as `'center'`, $n_c$. In addition, three different types of CCD can be specified: `'circumscribed'`, `'inscribed'`, and `'faced'`. Figure 3.10 shows these three types of CCD. It is noted that the last option `'faced'` yields face-centered CCD. The following Matlab code can generate samples using CCD and plot them:

```
dCC = ccdesign(2,'type','circumscribed');
plot(dCC(:,1),dCC(:,2),'ro','MarkerFaceColor','b')
X = [1 -1 -1 -1; 1 1 1 -1];
Y = [-1 -1 1 -1; 1 -1 1 1];
line(X,Y,'Color','b')
axis square equal off
```

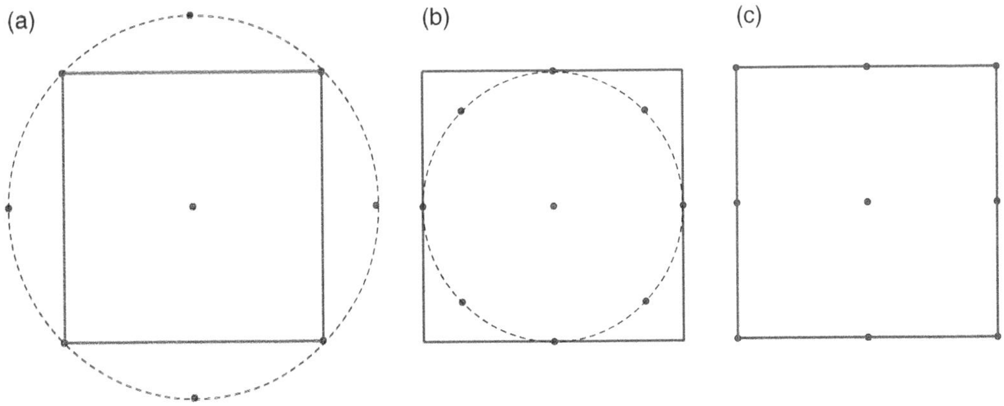


**Figure 3.10** Three different central composite designs from Matlab. (a) Circumscribed, (b) inscribed, and (c) faced.

For block design, Box-Behnken design [24] is implemented in the Matlab function `bbdesign (n, 'Name', value)`. The number of repeated center points can be specified as `'center'`, $n_c$. Matlab command `dBB = bbdesign(3)` will produce sample locations using a block design, as shown in Figure 3.8.

## 3.3 Optimal Design of Experiments

The various DoE methods that we considered in the previous section, as well as other DoE methods available in the literature, may perform satisfactorily if the design space is in a box-like domain. After normalizing design variables, the design space will be a square in two dimensions, a cube in three dimensions, etc. When the shape of the design space is fixed, it is possible to pre-determine the number and location of samples that can balance the number of samples and prediction variance. For example, the full-factorial design is good for lowering prediction variance but requires a large number of samples for high dimensions. On the other hand, the box design can provide a reasonable number of samples for high dimensions but has a high prediction variance because of a large extrapolation region.

For design optimization, however, an optimization problem normally comes with various constraints, which limit the design variables in a certain region, which is referred to as a feasible design. For a given number of samples, the accuracy of PRS is improved as the volume of design space becomes smaller. Therefore, it is desirable to invoke as many constraints as we can to reduce the volume of the design domain. The difficulty is that the standard DoE methods in the previous section cannot be applied to irregular design spaces. Therefore, it would be necessary to develop non-standard DoE methods.

Not only an irregular design space but also a box-like design space may take advantage of using a non-standard DoE method. This is because, in the standard DoE methods, the users cannot choose an arbitrary number of samples. Different DoE methods have their own required number of samples. Therefore, if the users want to make a specific number of samples, there may not be any standard DoE methods that allow the same number of samples. In practice, the users may want to choose the best DoE method for a given amount of resources, which is equivalent to a given number of samples. In these cases, the users may have to create their own DoE; that is, to select the best set of sample locations for the given design space and the number of samples.

For a fixed number of samples $n_y$, non-standard DoE methods basically try to find the locations of samples to satisfy a certain criterion, which is often posed as an optimization problem. That is, the best locations of samples are sought by minimizing a criterion, which represents the quality of the surrogate prediction. Because these types of DoE require solving an optimization problem, they are often referred to as optimal DoE.

Since the optimality criterion of most optimal DoE is based on some function of the moment matrix, the "optimality" of a given design is model-dependent. That is, the users must specify a model for the design (linear or quadratic PRS, etc.) and the number of samples $n_y$ desired before the "optimal design" can be generated. Therefore, the generated DoE is "optimal" only for the given model, the given criterion, and the given number of samples.

The optimization problem of optimal DoE has a considerable challenge in practice. In the case of determining the location of $n_y$ samples in $n$-dimensional space, there are $n \times n_y$ number of variables. Finding an optimum $n \times n_y$ variables can easily become impractical as the number of variables increases quickly. For example, in the case of a six-dimensional design space with 20 samples, there will be 120 variables to optimize. Therefore, traditional optimization algorithms are not a good choice for optimal DoE. Instead, optimal DoE problems are commonly treated as combinatorial problems. A pool of candidate points is defined first, where the QoI can possibly be evaluated. Then, out of the pool, the "best" $n_y$ points are selected. For example, $3^n$ samples from the full-factorial design can be defined as the pool of candidate points, and then, $n_y = 3 \times n_\beta$ best samples can be selected from them.

As mentioned before, since optimal DoE is an optimization problem, different optimal DoE can be obtained when different criteria are used. In the case of PRS, most of them are related to the moment matrix $\mathbf{X}^T\mathbf{X}$, because linear regression relies on it. The D-optimal design seeks to maximize $|\mathbf{X}^T\mathbf{X}|$, the determinant of the moment matrix. This criterion results in minimizing the variance of the PRS coefficients. The A-optimal design seeks to minimize the trace of the inverse of the moment matrix. This criterion results in minimizing the average variance of the parameter estimates. The G-optimal design seeks to minimize the maximum prediction variance, i.e., minimize the maximum of $\xi(\mathbf{x})^T(\mathbf{X}^T\mathbf{X})^{-1}\xi(\mathbf{x})$, over a specified set of design points. There are other optimal designs that will not be covered in this text. Interested readers are referred to Atkinson et al. [25].

### 3.3.1 D-Optimal Design

In Chapter 2, we discussed different types of uncertainty in linear regression. From the assumption that the model form of PRS is correct but samples have random noise, different realizations of noise can cause different fitting results, which is the source of uncertainty. This uncertainty first causes the uncertainty of the vector of regression coefficients $\mathbf{b}$ in the form of a covariance matrix, as shown in Eq. (2.38): $\boldsymbol{\Sigma}_\mathbf{b} = \hat{\sigma}^2\left[\mathbf{X}^T\mathbf{X}\right]^{-1}$. Then, the uncertainty in the coefficients can cause the uncertainty in the prediction, which is given in Eq. (3.10): $\sigma_y(\mathbf{x}) = \hat{\sigma}\sqrt{\xi(\mathbf{x})^T\left(\mathbf{X}^T\mathbf{X}\right)^{-1}\xi(\mathbf{x})}$. Most optimal design methods try to find sample locations to reduce these two types of uncertainty. Since the standard deviation of noise $\hat{\sigma}$ depends on the samples, the users cannot control it. It is aleatory uncertainty, and we have to live with it. The vector of basis functions $\xi(\mathbf{x})$ is fixed when the model form of PRS is given. Therefore, the only term that can vary is the inverse of the moment matrix $(\mathbf{X}^T\mathbf{X})^{-1}$, which depends on the number and locations of samples. Therefore, most optimal DoE methods try to reduce $(\mathbf{X}^T\mathbf{X})^{-1}$ so that the standard error of prediction can be reduced.

The idea behind the D-optimal design is that since the model form is accurate, if there is no uncertainty in the regression coefficients, the prediction will also be accurate. Therefore, a good design makes the covariance matrix $\boldsymbol{\Sigma}_{\mathbf{b}}$ small, which is equivalent to making $(\mathbf{X}^T\mathbf{X})^{-1}$ small or making the moment matrix $\mathbf{X}^T\mathbf{X}$ large. Since $\mathbf{X}^T\mathbf{X}$ is a matrix, the magnitude of the moment matrix is represented by the determinant of the matrix. Therefore, D-optimal DoE is to find $\left(\mathbf{x}_1, \mathbf{x}_2, \ldots, \mathbf{x}_{n_y}\right)$, such that

$$\text{maximize } D = \left|\mathbf{X}^T\mathbf{X}\right| \tag{3.13}$$

The determinant of the moment matrix can be shown to be directly related to the confidence region of the coefficients of the PRS (a confidence region for a coefficient is the region where the coefficient will lie with a given probability). In fact, it is inversely proportional to the square of the volume of the confidence region of the coefficients. Therefore, maximizing the determinant increases our confidence in the coefficients. Unlike traditional DoEs, D-optimal designs do not require orthogonal design matrices, and as a result, parameter estimates may be correlated. Parameter estimates may also be locally, but not globally, D-optimal.

The details of the D-optimal algorithm can be found in Atkinson et al. [25]. First, since the optimal designs depend on the surrogate model, it is necessary to specify an approximate mathematical model that defines the functional form of the relationship between the QoI and the independent variables (the design variables). Next, generate a set of possible candidate points based on this model. Finally, from these candidates, select the subset that maximizes the determinant of the moment matrix $|\mathbf{X}^T\mathbf{X}|$. The number of possible designs grows rapidly as the complexity of the model increases. This number is usually so large that an exhaustive search of all possible designs for a given sample size is not feasible. The D-optimal algorithm begins with a randomly selected set of points. Points in and out of the current design are exchanged until no exchange can be found that increases the determinant of the moment matrix. To cut down on the running time, the number of points considered during any iteration may be limited.

Unfortunately, this method does not guarantee that the global maximum can be found. To overcome this, the algorithm is repeated several times in the hope that at least one iteration leads to the global maximum. For example, the algorithm can repeat 50 or 100 times by starting with a random initial $n_y$ set of samples. However, since the criterion is not convex and since the dimension is high, there is no guarantee to obtain the global maximum of the determinant.

Finding the D-optimal set of points from a given set of points is often a difficult combinatorial problem. For example, if we need to find 10 D-optimal points out of 50 sample locations, we can have ${}_{10}C_{50} \approx 10^{10}$ possible combinations. Therefore, the number of combinations becomes huge even for moderate-sized problems. Solution algorithms can rarely find the D-optimal set and usually settle on a suboptimal but good set. Some solution algorithms are based on replacing one point at a time and taking advantage of inexpensive expressions for updating the determinant when one point is changed. Genetic algorithms have also been used to find a good design based on D-optimality.

**Example 3.7 D-optimal design**

In order to fit a linear PRS $\hat{y} = b_1x_1 + b_2x_2$, two sample locations were given as (0, 0) and (1, 0). Find the third sample location using a D-optimal design in the design space $x_1, x_2 \in [0, 1]$.

**Solution:**

Let the third sample location be $(p, q)$. Then the design matrix and the moment matrix are defined as

$$\mathbf{X} = \begin{bmatrix} 0 & 0 \\ 1 & 0 \\ p & q \end{bmatrix}, \quad \mathbf{X}^T\mathbf{X} = \begin{bmatrix} 1+p^2 & pq \\ pq & q^2 \end{bmatrix}, \quad |\mathbf{X}^T\mathbf{X}| = q^2$$

For a given design space, the maximum of $|\mathbf{X}^T\mathbf{X}|$ occurs at $q = 1$, while $p$ is arbitrary. Therefore, based on the D-optimal design, the third sample point is selected at $(p, 1)$ with $p$ being arbitrary.

It would be interesting to check how the D-optimal design actually reduces the uncertainty in the regression coefficients. Based on Eq. (2.38), the covariance matrix of regression coefficients can be calculated as

$$\mathbf{\Sigma_b} = \hat{\sigma}\left[\mathbf{X}^T\mathbf{X}\right]^{-1} = \hat{\sigma}\begin{bmatrix} 1 & -\dfrac{p}{q} \\ -\dfrac{p}{q} & \dfrac{1}{q^2} + \dfrac{p^2}{q^2} \end{bmatrix}$$

Therefore, it is obvious that $q = 1$ gives lower variances (diagonal terms) and lower correlations (off-diagonal terms). Although the D-optimal design concludes that $p$ can be arbitrary, the covariance matrix shows that when $p = 0$, the uncertainty in the coefficients is minimum. For example, when $p = 0$, the correlation between the two coefficients is zero, and the variance of $b_1$ is minimum. Therefore, the D-optimal design may not capture the true minimum of the covariance matrix. This is because a single scalar measure may not fully represent the complex behavior of the entire matrix.

Matlab provides various functions for D-optimal designs. Since D-optimal designs depend on the PRS model, either the users specify the PRS model or provide the design matrix. Also, as mentioned before, it is possible that the users generate candidate samples first and the Matlab function can choose the best sample locations out of all candidate locations. Otherwise, it is also possible that $n_y$ sample locations are randomly generated first, and then, they are moved to different locations to improve the D-optimality criterion. In order to specify the PRS model, the following options are available: `'linear'`, `'interaction'`, `'quadratic'`, and `'purequadratic'`. `'linear'` and `'quadratic'` are the standard linear and quadratic PRS surrogates, respectively. `'interaction'` only includes interaction terms, such as $x_1x_2$, $x_1x_3$, etc., without squared terms, while `'purequadratic'` only includes squared terms, such as $x_1^2$, $x_2^2$, etc., without interaction terms.

Matlab function `rlist = candexch(C,nrows)` uses a row-exchange algorithm to select `nrows` number of D-optimal designs from the candidate set. The matrix `C` is the candidate design matrix, and the output `rlist` is the list of selected rows. Possible options are how to choose initial `nrows` samples, the maximum number of iterations, and the number of times to try to generate a design from new starting locations. In order to find a better local optimum, the algorithm can repeat multiple times with random starting locations. This function has the advantage of using non-standard basis functions (e.g., without a constant term), while the disadvantage is that the users need to generate the design matrix at all candidate points.

Matlab function `[dRE,X] = rowexch(n,ny,model)` uses a row-exchange algorithm to generate a D-optimal design with `ny` samples for a linear additive model with `n` design variables. The output `dRE` is a $n_y \times n$ matrix of sample locations, and `X` is the design matrix. The `rowexch` function first generates candidate samples automatically and then utilizes `candexch` function to find the D-optimal design. The only difference is that `rowexch` can only use specific polynomial forms that are specified in the `model`. Both `candexch` and `rowexch` can only find D-optimal design out of candidate sample locations.

Matlab function `[dCE,X] = cordexch(n,ny,model)` is different from `rowexch` in the sense that it does not choose a D-optimal design from candidate locations. Instead, it starts with `ny` initial sample locations that are generated randomly, and it moves the previous locations to new locations to increase the determinant of the moment matrix. At each step, the coordinate-exchange algorithm exchanges a single element of the design matrix **X** with a new element evaluated at a neighboring point in the design space. Since the algorithm searches a small neighborhood of the current sample location, the algorithm is more likely to become trapped in a local minimum.

**Example 3.8 D-optimal design using Matlab**

Use Matlab's `cordexch` function to generate 6 and 12 samples for two-dimensional quadratic PRS using D-optimal design. Discuss the sample locations and the ratio of uncertainty between the two sample sets.

**Solution:**

Since two-dimensional quadratic PRS has six regression coefficients, a minimum of six samples is required. The following Matlab code can calculate D-optimal designs for 6 and 12 samples.

```
ny=6;                          %With 6 samples:
[dce,X]=cordexch(2,ny,'quadratic');
scatter(dce(:,1),dce(:,2),200,'filled')
D6=det(X'*X)

ny=12;                         %With 12 samples:
[dce,X]=cordexch(2,ny,'quadratic');
scatter(dce(:,1),dce(:,2),200,'filled')
D12=det(X'*X)
```

With six samples, the determinant of the moment matrix was $|\mathbf{X}^T\mathbf{X}| = 256$, while that of 12 samples was $|\mathbf{X}^T\mathbf{X}| = 30{,}320$. Therefore, it is expected that the uncertainty in the regression coefficient is much lower for the 12 samples. In terms of the ratio between the two determinants, the 12 samples may have less than 1% uncertainty than the 6 samples. The sample locations are shown in Figure 3.11, and the individual locations are shown in the following tables.

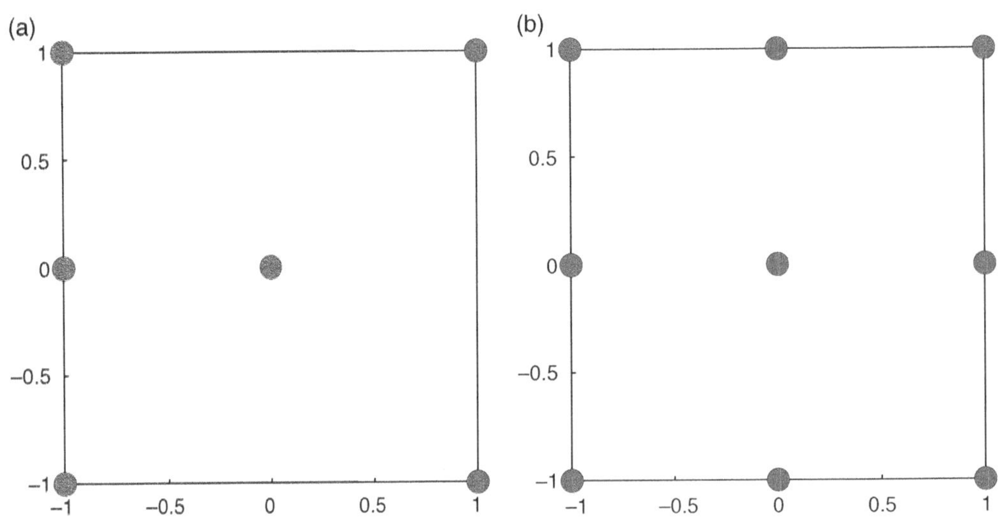


**Figure 3.11** D-optimal DoE. (a) 6 samples and (b) 12 samples.

Six sample locations:

| Sample no. | 1 | 2 | 3 | 4 | 5 | 6 |
|---|---|---|---|---|---|---|
| $x_1$ | 1 | −1 | −1 | 1 | 0 | 0 |
| $x_2$ | −1 | 1 | −1 | 1 | 0 | −1 |

12 sample locations:

| Sample no. | 1 | 2 | 3 | 4 | 5 | 6 | 7 | 8 | 9 | 10 | 11 | 12 |
|---|---|---|---|---|---|---|---|---|---|---|---|---|
| $x_1$ | 1 | −1 | 0 | −1 | 1 | −1 | 1 | 1 | 1 | −1 | 0 | 0 |
| $x_2$ | −1 | −1 | −1 | 1 | 1 | 0 | 1 | 0 | −1 | 1 | 1 | 0 |

It is noticed that all samples are on the edge of the design space, except for the center locations. In the case of six samples, four samples were located in the corners, and the fifth sample is at the center. The last sample was located at (−1, 0), but it could have been positioned in either (1, 0), (0, −1), or (0, 1). In the case of 12 samples, 9 samples were located in two-level full-factorial designs. The remaining three samples actually overlapped with some of these nine samples (samples 7, 9, and 10).

### 3.3.2 A-Optimal Design

A-optimal design is similar to D-optimal design in the sense that both focus on the variance of regression coefficients. In order to reduce the covariance matrix of coefficients, the D-optimal design maximizes the determinant of the moment matrix. The A-optimal design tries to reduce the variance of individual coefficients. The variance of each coefficient relies on the diagonal terms of the inverse of the moment matrix $(\mathbf{X}^T\mathbf{X})^{-1}$. The A-optimal design seeks to minimize the sum of these elements, that is, the trace of $(\mathbf{X}^T\mathbf{X})^{-1}$. Therefore, the A-optimal design is to find $(\mathbf{x}_1, \mathbf{x}_2, \ldots, \mathbf{x}_{n_y})$, such that

$$\text{minimize} \sum_{i=1}^{n_\beta} (\mathbf{X}^T\mathbf{X})_{ii}^{-1} \tag{3.14}$$

Jones et al. [26] showed that the A-optimal design is more consistent with the screening objective than the D-optimal design. An A-optimal design minimizes the average variance of the parameter estimates, which is directly related to that goal. While there are many cases where A- and D-optimal designs coincide, the A-optimal designs tend to have better statistical properties when the A- and D-optimal designs differ. In such cases, A-optimal designs generally have more uncorrelated columns in their model matrices than D-optimal designs. Also, even though A-optimal designs minimize the average variance of the parameter estimates, various cases exist where they outperform D-optimal designs in terms of the variances of all individual parameter estimates. A-optimal designs can also substantially reduce the worst prediction variance compared with D-optimal designs.

**Example 3.9 A-optimal design**

Repeat Example 3.7 with an A-optimal design.

**Solution:**

From the covariance matrix given in Example 3.7, the sum of the diagonal terms can be written as

$$\sum_{i=1}^{2} (\mathbf{X}^T\mathbf{X})_{ii}^{-1} = \left(1 + \frac{1+p^2}{q^2}\right)$$

The above term has a minimum when $p^2 = 0$ and $q^2 = 1$. Therefore, the third sample location using the A-optimal design becomes $(p, q) = (0, 1)$. Note that in the D-optimal design $p$ was arbitrary, while the A-optimal design found the best location for minimizing the variance of the two coefficients.

### 3.3.3 G-Optimal Design

D-optimal and A-optimal designs try to reduce uncertainty in the covariance matrix of coefficients, which can indirectly reduce the prediction variance. However, it makes more sense if an optimal design tries to reduce prediction variance directly, which is called the G-optimal design. Since prediction variance is a function of the prediction point, the G-optimal design minimizes the maximum prediction variance as

$$\text{minimize } \max v(\mathbf{x}) = \xi(\mathbf{x})^T \left(\mathbf{X}^T\mathbf{X}\right)^{-1} \xi(\mathbf{x}) \tag{3.15}$$

It can be shown that under the standard statistical assumptions about the error, the maximum prediction variance in the domain defined by the sample points is always larger than or equal to the number of terms in the response surface, $n_\beta$ (see Myers and Montgomery [3], p. 367). Therefore, with a G-optimal design, we have a target to shoot for. We would like to seek a set of points that will bring the maximum prediction variance close to $n_\beta/n_y$. This value is achieved by a two-level full-factorial design for a linear model, see Example 3.2, where the prediction variance was $3\hat{\sigma}^2/4$ (four samples with three coefficients).

If all sample locations are sought, the total number of design variables would be $n_y \times n$, which becomes impractical for a large number of samples or a high-dimensional design space. In adaptive sampling, some sample locations are fixed, and additional $k$ samples can be sought. In such a case, the number of design variables would be $k \times n$.

In general, finding $n_y$ sample locations such that the maximum prediction variance is minimized, as in Eq. (3.15), is a two-level optimization problem. In the inner level, for a given set of sample locations, we need to find the location where the prediction variance is maximum and the maximum prediction variance value. This optimization problem is formulated as follows: find $\mathbf{x} = \{x_1, \ldots, x_n\}^T$ to satisfy

$$\begin{aligned} &\min f(\mathbf{x}) = -\xi(\mathbf{x})^T \left(\mathbf{X}^T\mathbf{X}\right)^{-1} \xi(\mathbf{x}) \\ &\text{such that } \; -1 \le \mathbf{x} \le 1 \end{aligned} \tag{3.16}$$

In this equation, it is assumed that the design matrix is known (or the sample locations are known). The maximum prediction variance is then given as $-f$. The outer-level optimization is to find the sample locations, i.e., $\left\{\mathbf{x}^{(1)}, \mathbf{x}^{(2)}, \ldots, \mathbf{x}^{(n_y)}\right\}$ such that the maximum prediction variance is minimized. The optimization problem is formulated as

$$\begin{aligned} &\min g\left(\mathbf{x}^{(1)}, \mathbf{x}^{(2)}, \ldots, \mathbf{x}^{(n_y)}\right) = \max\left(-\xi^T \left(\mathbf{X}^T\mathbf{X}\right)^{-1} \xi\right) \\ &\text{such that } \; -1 \le \mathbf{x}^{(1)}, \mathbf{x}^{(2)}, \ldots, \mathbf{x}^{(n_y)} \le 1 \\ &\left|\mathbf{X}^T\mathbf{X}\right| > 0 \end{aligned} \tag{3.17}$$

The last condition on the determinant of the function $|\mathbf{X}^T\mathbf{X}|$ is imposed to avoid the repetition of the points, which makes the matrix singular.

The optimization problem is difficult to solve because the lower-level optimum is not a smooth function of the upper-level design variables (the position of the sample points). This is because, as the data points change, the position of the maximum variance can jump. Therefore, most gradient-based optimization ends up with a local optimum. Gradient-free global search algorithms, such as a genetic algorithm, can be computationally expensive. In practice, the inner-level optimization problem, finding the maximum prediction variance, is often replaced by grid evaluation, where the design space is discretized by a grid and the maximum prediction variance is sought among the grids.

**Example 3.10 G-optimal design**

Repeat Example 3.7 with a G-optimal design. Compare the maximum standard error of prediction for D-optimal, A-optimal, and G-optimal designs.

**Solution:**

Since Eq. (3.15) is a min-max problem, it would be difficult to solve it analytically. We will solve the problem heuristically with some reasoning. Using the inverse of the moment matrix in Example 3.7, the prediction variance term in Eq. (3.15) can be written as

$$v(\mathbf{x}) = \xi(\mathbf{x})^T(\mathbf{X}^T\mathbf{X})^{-1}\xi(\mathbf{x}) = \left(x_1 - 2\frac{p}{q}x_2\right)^2 + \frac{1+p^2}{q^2}x_2^2$$

First, since $q$ only occurs in the denominator, if $q = 0$, the variance term goes to infinity. Therefore, it is obvious that $q$ must be in its upper bound to reduce the variance. Therefore, we heuristically determine $q = 1$. Then, the variance can be simplified as

$$v(\mathbf{x}) = (x_1 - 2px_2)^2 + (1 + p^2)x_2^2$$

Since the maximum variance normally occurs at corner points, we evaluate the variance at four corners:

$$v(0, 0) = 0$$
$$v(1, 0) = 1$$
$$v(0, 1) = 1 + p^2$$
$$v(1, 1) = 2 - 2p + p^2$$

It is obvious that the maximum may occur either at (0, 1) or (1, 1), depending on the value of $p$. When $p \geq 0.5$, the maximum occurs at (0, 1), and the minimum of the maximum occurs at $p = 0.5$. When $p \leq 0.5$, the maximum occurs at (1, 1), and the minimum of the maximum occurs at $p = 0.5$. Therefore, the G-optimal design yields $(p, q) = (0.5,1)$, which is the middle of the upper bound.

Note that the sample location of the D-optimal design in Example 3.7 is not fixed. For the purpose of comparison, however, we choose the third sample location of the D-optimal design at (1, 1), just to make it different from the other two designs. With the third sample at (0.5, 1), the maximum standard error of prediction of the G-optimal design becomes $\sigma_y^{\max} = \sqrt{5}\hat{\sigma}/2$. It is interesting to note that the maximum standard error of prediction from D-optimal and A-optimal designs is $\sigma_y^{\max} = \sqrt{2}\hat{\sigma}$. Therefore, G-optimal design yields the sample location that has the smallest maximum standard error of prediction.

Figure 3.12 compares the location of the third sample and the standard error of prediction for D-optimal, A-optimal, and G-optimal designs. When the third sample is located in one corner, the

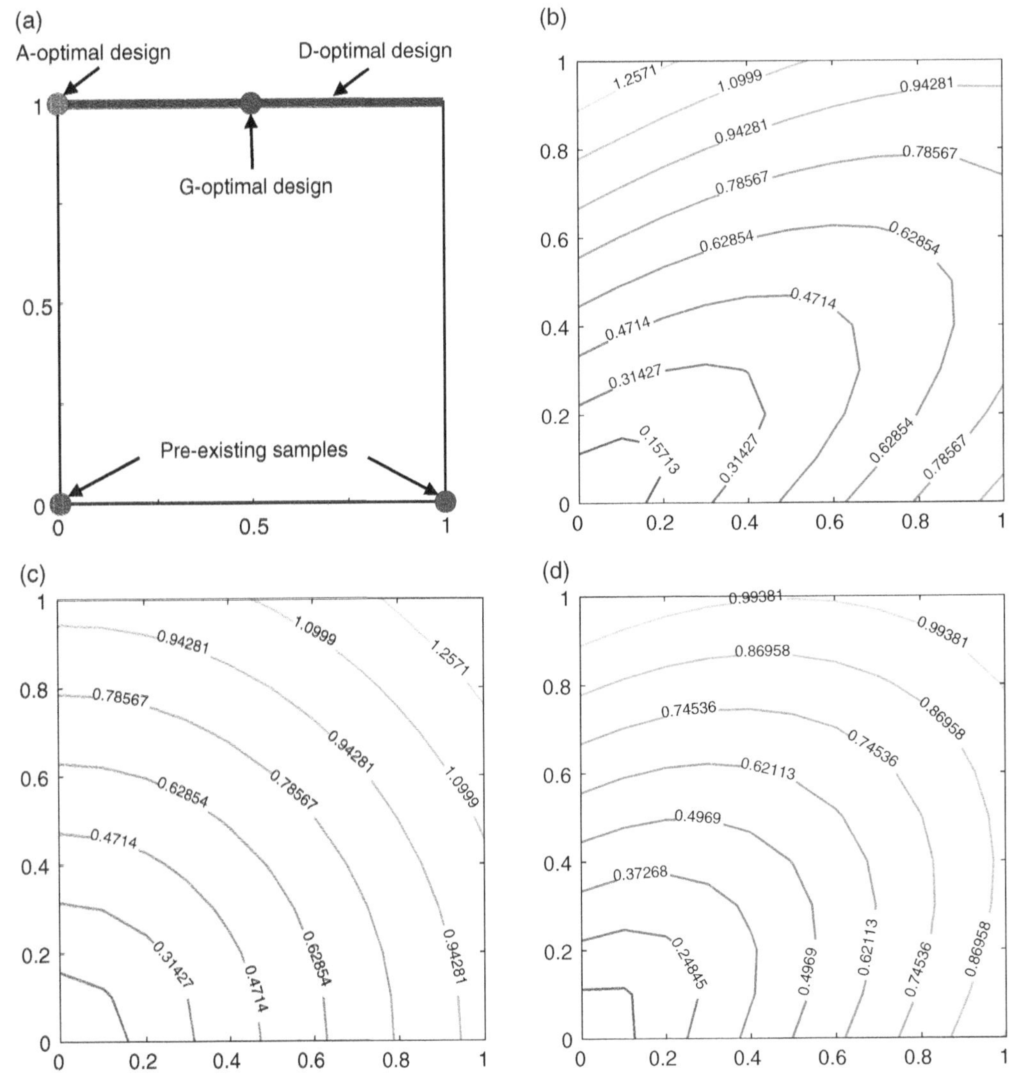


**Figure 3.12** Comparison of sample locations and standard error of prediction for D-optimal, A-optimal, and G-optimal designs. (a) Sample locations, (b) $\sigma_y$ of D-optimal design, (c) $\sigma_y$ of A-optimal design, and (d) $\sigma_y$ of G-optimal design.

maximum standard error occurs in the other corner as this is the farthest extrapolation point. On the other hand, since the G-optimal design locates the third sample in the middle of the edge, the distance to the farthest extrapolation point is shorter than the other two designs. It is noted that the minimum standard error of all three designs is zero at the origin. This happens because the linear PRS $\hat{y} = b_1x_1 + b_2x_2$ does not have a constant term. The following Matlab code is used to plot the contour of the standard error of prediction:

```
X=[0 0;1 0;1 1];       % D-optimal design
%X=[0 0;1 0;0 1];       % A-optimal design
%X=[0 0;1 0;0.5 1];     % G-optimal design
```

```
%
XTX=X'*X;
XTXi=inv(XTX);
[X, Y]=meshgrid(0:.1:1, 0:.1:1);
Z=zeros(size(X));
[n, m]=size(X);
for i=1:n
    for j=1:m
        xi=[X(i,j) Y(i,j)];
        Z(i,j)=sqrt(xi*XTXi*xi');
    end
end
v=linspace(min(min(Z)),max(max(Z)),10);
[C,h]=contour(X,Y,Z,v);
clabel(C,h)
Zmax=max(max(Z))
Zmin=min(min(Z))
```

It would be interesting to compare the covariance matrix of the three designs. Remember that D-optimal and A-optimal designs are for minimizing the covariance matrix. The covariance matrices of the three designs are summarized in the following table:

| Designs | D-optimal | A-optimal | G-optimal |
|---|---|---|---|
| Covariance matrix | $\Sigma_b = \hat{\sigma}\begin{bmatrix} 1 & -1 \\ -1 & 2 \end{bmatrix}$ | $\Sigma_b = \hat{\sigma}\begin{bmatrix} 1 & 0 \\ 0 & 1 \end{bmatrix}$ | $\Sigma_b = \hat{\sigma}\begin{bmatrix} 1 & -0.5 \\ -0.5 & 1.25 \end{bmatrix}$ |

It turns out that the variance of regression coefficients is the smallest for the A-optimal design. In addition, the A-optimal design shows uncorrelated coefficients. Even if the D-optimal design is for minimizing the determinant of the moment matrix, it shows the largest variance among the three and also a significant correlation. G-optimal design is good for maximum prediction variance, but it has a relatively large variance and correlation between coefficients.

### 3.3.4 Minimum Bias Design

The three optimal designs that we discussed in the previous subsections only focus on reducing uncertainty in regression coefficients or prediction. Unfortunately, reducing uncertainty does not always guarantee prediction accuracy. It has an assumption that first, the model form is accurate, and second, the mean of regression coefficients is identical to the true coefficients. When the model form is not accurate or the regression process cannot estimate the coefficients accurately, there exists an error in surrogate prediction. This error is called a modeling error by engineers and a bias error by statisticians.

Bias measures the difference between the surrogate prediction $\hat{y}(\mathbf{x})$ of QoI and the true QoI $y(\mathbf{x})$. Since the true QoI is often unknown and measurements have random noise, the average of measurements is commonly considered the true QoI. On the other hand, variance measures how surrogate $\hat{y}(\mathbf{x})$ is affected by a particular dataset. Samples are assumed to be generated from the true QoI by adding random noise. If the surrogate fit is perfect, it filters out the random noises in the samples, and different datasets should yield the same surrogate prediction. However, this can be achieved when the number of samples approaches infinity. With a finite number of samples, different datasets will yield different surrogate predictions. This uncertainty in prediction is measured in terms of variance.

Unfortunately, variance and bias are competing objectives in DoE. The bias error can be reduced by fitting the surrogate close to samples (e.g., by adding more basis functions), but this causes a large variance. On the other hand, the variance error can be reduced by smoothing the surrogate model (e.g., by adding a penalty term as in Eq. (1.6)), but this smoothing penalty inevitably increases the bias error. Therefore, it is necessary to balance the bias and variance errors through a trade-off. In theory, it is possible to improve the bias and variance error simultaneously by adding more samples and more basis functions.

For PRS surrogate models, using the assumption that the true model is also in a polynomial form, Myers and Montgomery [3] proposed minimum bias designs. To consider the model errors, it would be necessary to introduce the concepts of design moments. Let $R$ be the region of interest in terms of predicting QoI, and let $\mu_i$ be the first moments of the domain, which is defined as

$$\mu_i = \frac{1}{V}\int_R x_i \, dR, \quad i = 1, \ldots, n \tag{3.18}$$

where $V$ is the volume of the domain $R$. Similarly, the second moment $\mu_{ij}$ can be defined as

$$\mu_{ij} = \frac{1}{V}\int_R x_i x_j \, dR, \quad i,j = 1, \ldots, n \tag{3.19}$$

Higher-order moments can also be defined in a similar way.

Since samples are given at discrete points, the above definition of moments needs to be extended to the case with discrete samples. First, let there are $n_y$ samples in $n$-dimensional design space. Then, $x_{ik}$ ($i = 1, \ldots, n$, $k = 1, \ldots, n_y$) represents the $i$th component of the $k$th sample. Therefore, the discrete counterpart of the moment is defined as

$$m_i = \frac{1}{n_y}\sum_{k=1}^{n_y} x_{ik}, \quad i = 1, \ldots, n \tag{3.20}$$

Higher-order moments can also be found in a similar way.

The concept of minimum bias design is based on the equivalence of the discrete moments with the continuous moments. Let us consider a PRS surrogate model given in the form of $\hat{y}(\mathbf{x}) = \xi(\mathbf{x})^{(1)^T}\mathbf{b}^{(1)}$, and it does not has a bias term. Assuming that the bias term is given in a polynomial form with higher orders, the true function can be defined as $y(\mathbf{x}) = \xi(\mathbf{x})^{(1)T}\mathbf{b}^{(1)} + \xi(\mathbf{x})^{(2)T}\mathbf{b}^{(2)}$, where the second term corresponds to the bias. With given $n_y$ samples, the relationship between the samples and regression coefficients can be written as

$$\mathbf{y} = \mathbf{X}^{(1)}\mathbf{b}^{(1)} + \mathbf{X}^{(2)}\mathbf{b}^{(2)} \tag{3.21}$$

where $\mathbf{X}^{(1)}$ and $\mathbf{X}^{(2)}$ are the design matrices using $\xi(\mathbf{x})^{(1)}$ and $\xi(\mathbf{x})^{(2)}$, respectively. The average moment matrix of the combined design matrix can be partitioned as

$$\mathbf{M} = \left[\begin{array}{c|c} \mathbf{M}_{11} & \mathbf{M}_{12} \\ -- & -- \\ \mathbf{M}_{21} & \mathbf{M}_{22} \end{array}\right] = \frac{1}{n_y}\left[\begin{array}{c|c} \mathbf{X}^{(1)^T}\mathbf{X}^{(1)} & \mathbf{X}^{(1)^T}\mathbf{X}^{(2)} \\ ------ & ------ \\ \mathbf{X}^{(2)^T}\mathbf{X}^{(1)} & \mathbf{X}^{(2)^T}\mathbf{X}^{(2)} \end{array}\right] \tag{3.22}$$

On the other hand, the moment in the continuous domain can be defined using the basis functions as

$$\boldsymbol{\mu} = \left[\begin{array}{c|c} \boldsymbol{\mu}_{11} & \boldsymbol{\mu}_{12} \\ \hline \boldsymbol{\mu}_{21} & \boldsymbol{\mu}_{22} \end{array}\right] = \frac{1}{V}\left[\begin{array}{c|c} \int_R \xi(\mathbf{x})^{(1)}\xi(\mathbf{x})^{(1)^T} dR & \int_R \xi(\mathbf{x})^{(1)}\xi(\mathbf{x})^{(2)^T} dR \\ \hline \int_R \xi(\mathbf{x})^{(2)}\xi(\mathbf{x})^{(1)^T} dR & \int_R \xi(\mathbf{x})^{(2)}\xi(\mathbf{x})^{(2)^T} dR \end{array}\right] \tag{3.23}$$

Comparing the discrete moment in Eq. (3.22) with the continuous moment in Eq. (3.23), the regression process is nothing but approximating the integrals of basis functions by a sum over the sample points. Therefore, good samples make this approximation accurate. The idea of minimum bias design is to choose sample locations such that the mean-squared-bias error of PRS is minimized, which can be achieved by making the discrete moment equal to the continuous moment. If there is enough freedom in selecting sample locations, it is possible to make the difference disappear. If that is not possible, optimization methods can be used to seek the best points to minimize the difference.

Box and Draper [27] proved that minimum bias designs satisfy the following relationship:

$$\mathbf{M}_{11}^{-1}\mathbf{M}_{12} = \boldsymbol{\mu}_{11}^{-1}\boldsymbol{\mu}_{12} \tag{3.24}$$

Box and Draper concluded that the following conditions:

$$\mathbf{M}_{11} = \boldsymbol{\mu}_{11}, \quad \mathbf{M}_{12} = \boldsymbol{\mu}_{12} \tag{3.25}$$

are sufficient conditions for a minimum bias design (Myers and Montgomery [3], p. 411). When the true functional form is known, the vector of basis functions, $\xi(\mathbf{x})^{(2)}$, is available. Without having the true functional form, however, the second condition in Eq. (3.25) cannot be imposed. Therefore, in most cases, the first condition is used to define sample locations with additional constraints. For a given vector of basis functions, the continuous moment matrix, $\boldsymbol{\mu}_{11}$, is calculated first. Then, the unknown sample locations are found by matching the discrete moment matrix, $\mathbf{M}_{11}$, with $\boldsymbol{\mu}_{11}$.

In contrast to the minimum variance designs that tend to put points in the periphery of the design domain, minimum bias designs tend to bring them closer to the centroid. In addition, minimum bias designs often have low prediction variance, but the reverse is not true. That is, minimum variance designs tend to have large bias errors. Compromise designs, in which $\mathbf{M}_{11}$ and $\mathbf{M}_{12}$ are slightly larger than $\boldsymbol{\mu}_{11}$ and $\boldsymbol{\mu}_{12}$, respectively, are occasionally used.

**Example 3.11 Symmetric minimum bias design**

The true function is a quadratic polynomial, given in the form $y(\mathbf{x}) = x_1^2 + x_2^2$, which is only used to generate samples. A linear PRS, $\hat{y}(\mathbf{x}) = b_1 + b_2x_1 + b_3x_2$, will be constructed in the design space $x_1, x_2 \in [-1, 1]$. Find four symmetric sample locations using a minimum bias design. Compare $e_{\text{RMS}}$ of minimum bias design with that of a two-level full-factorial design.

**Solution:**

Since the true functional form is unknown, we will only use the condition, $\mathbf{M}_{11} = \boldsymbol{\mu}_{11}$. For the linear model with the vector of basis functions, $\xi(\mathbf{x})^{(1)} = \{1, x_1, x_2\}^T$, the continuous moment matrix is

$$\boldsymbol{\mu}_{11} = \frac{1}{V}\int_{-1}^{1}\int_{-1}^{1}\begin{bmatrix} 1 & x_1 & x_2 \\ x_1 & x_1^2 & x_1x_2 \\ x_2 & x_1x_2 & x_2^2 \end{bmatrix} dx_1dx_2 = \begin{bmatrix} 1 & 0 & 0 \\ 0 & \frac{1}{3} & 0 \\ 0 & 0 & \frac{1}{3} \end{bmatrix}$$

Note that because of the symmetry of the domain, all the integrals with odd powers of either $x_1$ or $x_2$ are zero.

The discrete moment matrix $\mathbf{M}_{11}$ can be calculated from its definition in Eq. (3.22)

$$\mathbf{M}_{11} = \frac{1}{n_y}\mathbf{X}^{(1)T}\mathbf{X}^{(1)} = \frac{1}{n_y}\begin{bmatrix} \sum_{k=1}^{n_y} 1 & \sum_{k=1}^{n_y} x_{1k} & \sum_{k=1}^{n_y} x_{2k} \\ \sum_{k=1}^{n_y} x_{1k} & \sum_{k=1}^{n_y} x_{1k}^2 & \sum_{k=1}^{n_y} x_{1k}x_{2k} \\ \sum_{k=1}^{n_y} x_{2k} & \sum_{k=1}^{n_y} x_{1k}x_{2k} & \sum_{k=1}^{n_y} x_{2k}^2 \end{bmatrix}$$

If we pick four points that are symmetric with respect to both the $x_1$- and $x_2$-axis, there are two possibilities. One is the set of samples along the axis of each variable: $(\pm\alpha, 0)$, $(0, \pm\alpha)$, where $0 \le \alpha \le 1$ is a constant. The second is the set of samples along $\pm 45°$ lines: $(\pm\beta, \pm\beta)$, where $0 \le \beta \le 1$ is a constant. Because of the symmetry, the sums involving odd powers will vanish, so that all the zeros in $\mathbf{M}_{11}$ will match the zeroes in $\boldsymbol{\mu}_{11}$. Only the diagonal terms are non-zero. The value of $\alpha$ and $\beta$ is calculated by setting the diagonal terms in $\mathbf{M}_{11}$ matrix equal to the corresponding component in $\boldsymbol{\mu}_{11}$. That is,

$$\frac{1}{4}\sum_{k=1}^{4} 1 = 1, \quad \frac{1}{4}\sum_{k=1}^{4} x_{1k}^2 = \frac{1}{3}, \quad \frac{1}{4}\sum_{k=1}^{4} x_{2k}^2 = \frac{1}{3}$$

The first term is trivial. The remaining two terms are used to calculate $\alpha$ or $\beta$.

For the first set $(\pm\alpha, 0)$ and $(0, \pm\alpha)$, these equations yield

$$\frac{1}{4}\left(\alpha^2 + \alpha^2\right) = \frac{1}{3}, \quad \rightarrow \alpha = \sqrt{\frac{2}{3}} = 0.8165$$

Therefore, the four sample locations are along each axis: $(\pm 0.8165, 0)$ and $(0, \pm 0.8165)$. With these sample locations, the regression coefficients can be calculated as

$$\mathbf{X} = \begin{bmatrix} 1 & -\alpha & 0 \\ 1 & \alpha & 0 \\ 1 & 0 & -\alpha \\ 1 & 0 & \alpha \end{bmatrix}, \quad \mathbf{y} = \begin{Bmatrix} \alpha^2 \\ \alpha^2 \\ \alpha^2 \\ \alpha^2 \end{Bmatrix}, \quad \mathbf{b} = \left(\mathbf{X}^T\mathbf{X}\right)^{-1}\mathbf{X}^T\mathbf{y} = \begin{Bmatrix} \alpha^2 \\ 0 \\ 0 \end{Bmatrix}$$

Therefore, the linear PRS becomes $\hat{y}(\mathbf{x}) = b_1 + b_2x_1 + b_3x_2 = \alpha^2 = 2/3$, which is a constant.

For the second set $(\pm\beta, \pm\beta)$, these equations yield

$$\frac{1}{4}\left(\beta^2 + \beta^2 + \beta^2 + \beta^2\right) = \frac{1}{3}, \quad \rightarrow \beta = \sqrt{\frac{1}{3}} = 0.5774$$

Therefore, the four sample locations are along each axis: $(\pm 0.5774, \pm 0.5774)$. With these sample locations, the regression coefficients can be calculated as

$$\mathbf{X} = \begin{bmatrix} 1 & -\beta & -\beta \\ 1 & -\beta & \beta \\ 1 & \beta & -\beta \\ 1 & \beta & \beta \end{bmatrix}, \quad \mathbf{y} = \begin{Bmatrix} 2\beta^2 \\ 2\beta^2 \\ 2\beta^2 \\ 2\beta^2 \end{Bmatrix}, \quad \mathbf{b} = \left(\mathbf{X}^T\mathbf{X}\right)^{-1}\mathbf{X}^T\mathbf{y} = \begin{Bmatrix} 2\beta^2 \\ 0 \\ 0 \end{Bmatrix}$$

Therefore, the linear PRS becomes $\hat{y}(\mathbf{x}) = b_1 + b_2x_1 + b_3x_2 = 2\beta^2 = 2/3$, which is a constant. Note that the two surrogates are identical.

Now, let us compare these two sets to the full-factorial design $(\pm1, \pm1)$ for fitting the function $y(\mathbf{x}) = x_1^2 + x_2^2$. With these sample locations, the regression coefficients can be calculated as

$$\mathbf{X} = \begin{bmatrix} 1 & -1 & -1 \\ 1 & -1 & 1 \\ 1 & 1 & -1 \\ 1 & 1 & 1 \end{bmatrix}, \quad \mathbf{y} = \begin{Bmatrix} 2 \\ 2 \\ 2 \\ 2 \end{Bmatrix}, \quad \mathbf{b} = (\mathbf{X}^T\mathbf{X})^{-1}\mathbf{X}^T\mathbf{y} = \begin{Bmatrix} 2 \\ 0 \\ 0 \end{Bmatrix}$$

Therefore, for the full-factorial design, the surrogate model is also a constant function $\hat{y}(\mathbf{x}) = 2$. Obviously, the surrogate $\hat{y}(\mathbf{x}) = 2/3$ of the minimum bias design is better than that of the full-factorial design. Figure 3.13 shows the sample locations of the minimum bias design. The circular markers are for the first design, while the square markers are for the second design.

To appreciate that the fit is optimal, consider fitting the function by a general linear polynomial. Because of the double symmetry of the function, all the linear terms should vanish, so that the response surface should indeed be of the form $\hat{y}(\mathbf{x}) = b_1$. The mean square error over the region is then

$$e_{\text{RMS}}^2 = \frac{1}{4}\int_{-1}^{1}\int_{-1}^{1} (x_1^2 + x_2^2 - b_1)^2 dx_1 dx_2 = \frac{28}{45} - \frac{4}{3}b_1 + b_1^2$$

Differentiating the error with respect to $b_1$ and setting it to zero confirms the fact that $b_1 = 2/3$ gives the minimum error, $e_{\text{RMS}}^2 = 8/45$. In contrast, $b_1 = 2$ gives $e_{\text{RMS}}^2 = 88/45$.

If this example appears impressive, note that we did not have any noise at all in the function, and the example was selected to make the minimum bias design look good. In other cases, the results may be less dramatic, and a compromise between minimum bias and minimum variance may be called for.

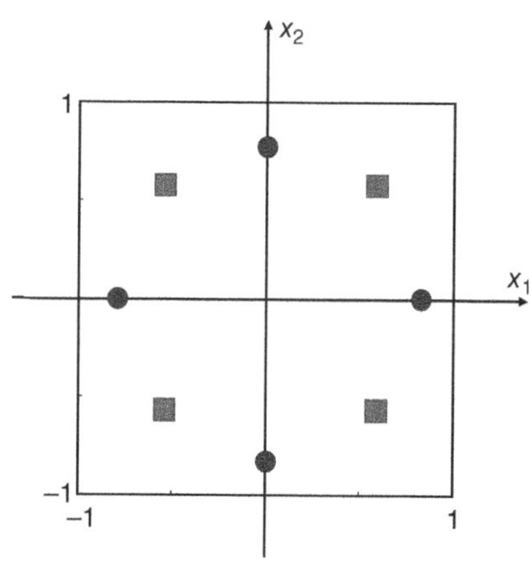


**Figure 3.13** Sample locations of minimum bias designs.

## 3.4 Space-Filling Design of Experiments

The optimal designs in the previous section are good for finding sample locations that minimize prediction uncertainty and/or bias error of surrogate prediction. The optimal designs are especially useful for adaptive sampling, where the locations of additional samples are sought in addition to pre-existing samples. The limitations of optimal designs are as follows: (i) it can be computationally challenging if the dimension of the design space is high and the number of samples is large, and (ii) the optimal designs depend on the specific surrogate model and the number of samples. Especially, most optimal designs in Section 3.3 are based on PRS surrogate, which uses linear regression with polynomial basis functions. Therefore, optimal designs are difficult to generalize for a general surrogate model with an arbitrary number of samples.

DoE methods that are independent of surrogate models are based on the simple fact that surrogate predictions are accurate when the prediction point is close to the sample location. Therefore, it is always a good idea to distribute samples regularly in the design space. But as shown in Section 3.2, it would require too many samples to construct dense full-factorial designs with many levels. Instead, fractional factorial designs are sought in order to reduce the required number of samples, and at the same time, satisfy the uniformity as much as possible. This can be achieved by either maximizing the minimum distance [28] or minimizing correlation measures among samples [29]. Clusters of samples can improve prediction accuracy but can be considered a waste of resources. A large empty space in the interpolation region or a large extrapolation region can cause poor surrogate predictions. Therefore, a good DoE method means filling the design space as much as possible for a given number of samples. This strategy of sampling is referred to as space-filling DoE in this text.

In addition to the space-filling property, it would be beneficial if samples were randomly distributed. The traditional DoE in Section 3.2 has specific locations for samples. In addition, traditional DoE tends to locate samples on the boundary of the design space, either at corners or on the edges. Therefore, there is a chance that a surrogate might have a good performance at these sample locations but a bad performance at unsampled locations. Considering the fact that the optimal design normally presents at the center of the domain, these boundary samples can cause a bias error in the central region. If a space-filling DoE has randomness in selecting locations, it can provide a robust estimate of prediction accuracy.

The term "space-filling" might mislead reality. It is appropriate only for low-dimensional spaces. For high-dimensional spaces, we cannot afford to "fill" the design space. Therefore, it is inevitable to anticipate a large extrapolation region in a high-dimensional design space. In addition, space-filling DoE tends to put samples inside the design space instead of at boundaries. Therefore, in a low-dimensional design space, it would be useful to add additional samples at the corners of the design space. Adding samples at all corners is impractical for a high-dimensional design space.

There are many different methods to generate space-filling DoEs. Among them, OAs [30] and LHS [31] are considered practical options. OA produces uniform designs but can generate particular forms of sample replications, while LHS does not produce replicates but can lack uniformity. As a result, OA-based LHS [32] and other optimal LHS schemes [33, 34] have been proposed. Among many space-filling DoEs, we will only consider Monte Carlo simulation (MCS), LHS, and OA designs in this section.

### 3.4.1 Monte Carlo Simulation

MCS is a popular method for generating random samples based on probability distribution. Originally, this is a method to generate random samples for the purpose of uncertainty quantification, but it can be extended for the purpose of DoE. Normally, MCS assumes that a random variable $X$ has a probability density function $f_X(x)$ and generates samples that follow this distribution. The idea to extend MCS for the purpose of DoE is to use a uniform distribution for $f_X(x)$ such that samples are generated uniformly in the design space. Although the idea seems reasonable, in practice, uniformity can be achieved only with a large number of samples. It is likely that some regions will be poorly sampled, and some regions will be overly sampled. In five-dimensional space, for example, it will require $2^5 = 32$ samples if one sample is to be located at each orthant. However, the probability of having one sample at each orthant is

$$\frac{31}{32} \times \frac{30}{32} \times \ldots \times \frac{1}{32} = 1.8 \times 10^{-13}$$

Therefore, it would be very unlikely to have evenly distributed samples.

**Example 3.12 Monte Carlo sampling**

Generate 20 samples of $(x_1, x_2) \in [0, 1]$ using MCS and plot the marginal histogram. Check if samples are uniformly distributed in the design space.

**Solution:**

The following Matlab code is used to generate 20 samples from uniform distribution and plot sample locations and marginal histogram of each variable:

```
rng default;
x=rand(20,2);
subplot(2,2,1); plot(x(:,1), x(:,2), 'o');
subplot(2,2,2); hist(x(:,2),20);
subplot(2,2,3); hist(x(:,1),20);
```

Matlab command `rand` generates random samples from a uniform distribution in the interval of [0, 1]. Therefore, if the range of input variables is different, it would be necessary to scale the samples appropriately. Figure 3.14 shows the sample locations and marginal histograms. The distribution of points shows both clustering in the lower-right and upper-left corners and scattering in the lower-left corner. Obviously, the distribution of samples does not look like a uniform distribution. If the samples are uniformly distributed, it is expected that the marginal histograms show a uniform height. The histograms were divided into 20 bins, and therefore, if samples were uniformly distributed, it is expected to have one sample in each bin. However, as shown in the figure, with 20 samples, there is evidence of both clustering and scarcity of samples.

### 3.4.2 Latin Hypercube Sampling

LHS is a semi-random sampling scheme, where the design space is gridded, and samples are randomly located within the grid. The design range of each variable is divided by $n_y$ intervals depending on its probability distribution. In the case of space-filling designs (i.e., the design variable is assumed to be uniformly distributed), the design space is divided by uniform intervals of size $1/n_y$. For example, Figure 3.15(a) shows seven LHS samples from a uniform distribution. First, the range of design space is divided into seven intervals, and one sample is placed at each interval. However, the sample location is random in the sense that the sample location within an interval is randomly selected.

Although uniform distribution is mostly used here for space-filling DoE, LHS can generate samples from any distribution. For example, Figure 3.15(b) shows 10 LHS samples from a normal distribution. In order to do that, first, the range of probability [0, 1] is equally divided into 10 intervals

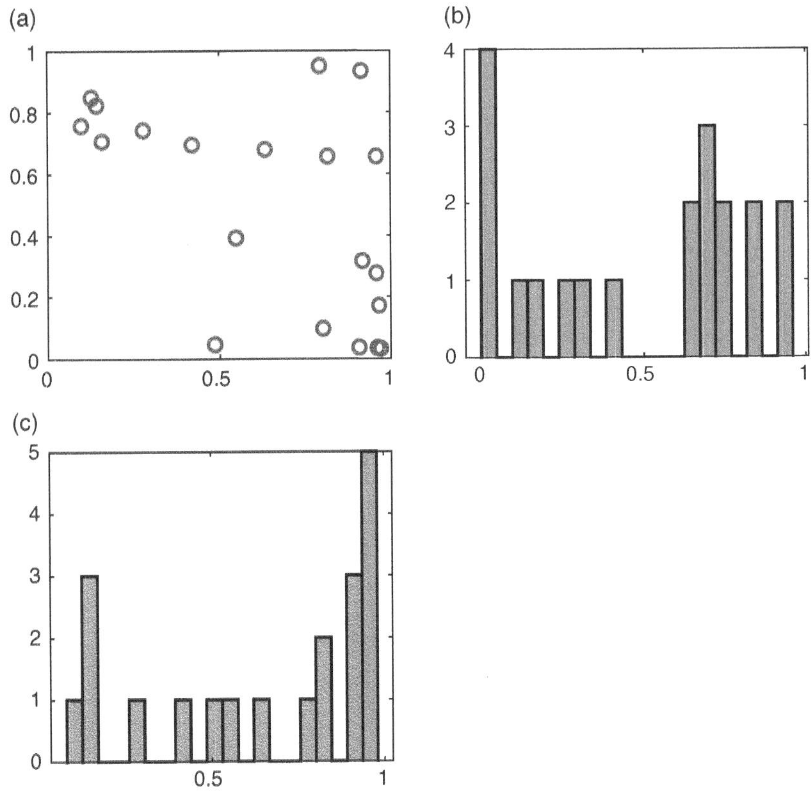


**Figure 3.14** Marginal histogram of 20 samples in two-dimensional design space. (a) Sample location, (b) histogram of $x_2$, and (c) histogram of $x_1$.

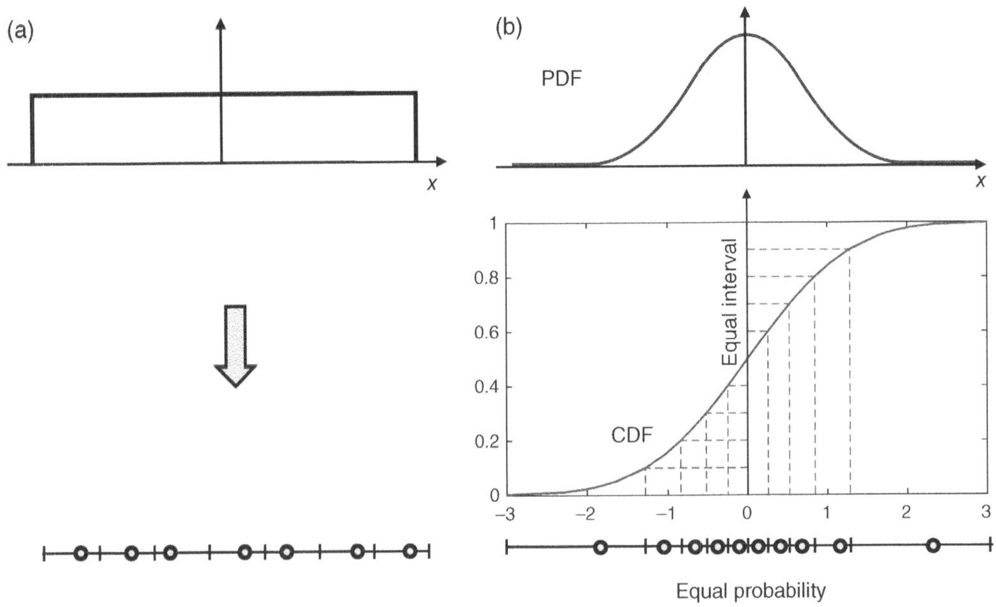


**Figure 3.15** Latin hypercube sampling scheme in one dimension. (a) LHS of uniform distribution and (b) LHS of normal distribution.

(vertical axis). Second, as shown in the figure, the inverse transformation is performed to find the corresponding variable's intervals (horizontal axis). Therefore, all intervals may have different sizes, but they have the same probability. Lastly, one sample is randomly generated at each interval. If the purpose of the surrogate is to propagate uncertainty from input to output, it makes sense to sample according to the distribution of the input variables. However, for the purpose of optimization, it would be better to fill the design space evenly, and therefore, a uniform distribution makes sense.

In the case of a one-dimensional variable, the randomness in the LHS is only from the fact that the sample location within an interval is random. However, in the case of multi-dimensional variables, there is another source of randomness in the LHS, which is more significant than the first. In the case of two designs with three samples (i.e., $n = 2$, $n_y = 3$), as shown in Figure 3.16, each variable has three intervals. Because of two variables, the grid has a total of nine cells. Therefore, there is randomness associated with selecting three sample locations out of nine cells. However, this is not a pure combination problem of ${}_3C_9$ because LHS requires one sample at each column or row. More specifically, in the first column of the figure, LHS can select one row out of three. In the second column, LHS can select one row out of two because the row that was selected in the first column cannot be selected. Then, for the last column, the row is fixed because two rows are already selected by the previous two columns. Therefore, the possibility of locating samples is $3! = 3 \times 2 \times 1$. This is the major source of uncertainty in LHS.

Often, the distribution of sample locations in LHS is defined using a definition table. As shown in Figure 3.16, the first column defines the intervals for $x_1$, and the second column defines the intervals for $x_2$ where the sample is located. It is noted that the first column is in a sequence, but the second column is a random permutation of sequences 1, 2, 3. If there is a third variable, the definition matrix would have one more column with another random permutation of the three numbers. In general, for $n_y$ samples in $n$ variables, the location of the samples will be defined by an $n_y \times n$ definition matrix, with each column a random permutation of 1, 2, ..., $n_y$.

In general, the samples from LHS will be better distributed for each individual variable than that of MCS, but still, there could be a large portion of the design space that is not sampled. An extreme example would be if all the permutations happen to be in a sequence, then all the samples will be aligned on one of the diagonals in the box. Due to the randomness in selecting intervals, there are many possible LHS designs that still satisfy the basic requirement of one sample in each interval of each design variable. Therefore, it makes sense to generate many LHS designs and pick the best one. In order to do that, it would require having criteria to choose the best LHS design. The desirable criteria would be maximizing the minimum size of the empty hypersphere (i.e., unsampled region) or minimizing correlation among samples. The first criterion tries to remove clustered samples. Unfortunately, since calculating the minimum size of an empty hypersphere is tricky, the

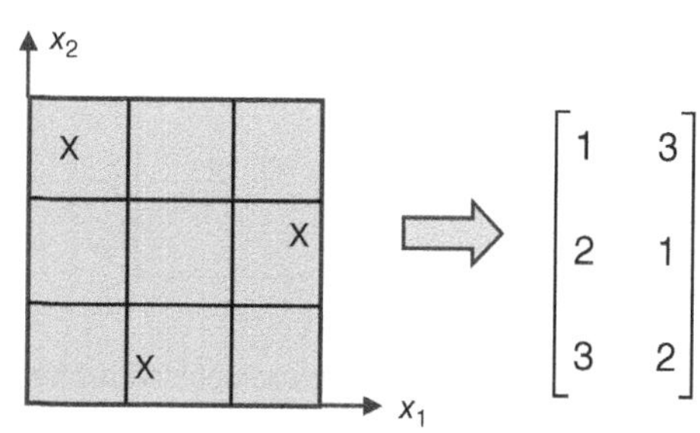


**Figure 3.16** Latin hypercube samples and definition matrix ($n = 2$, $n_y = 3$).

minimum distance between samples is often used as an LHS design criterion that is easy and cheap to calculate.

Matlab command `lhsdesign` can either maximize the minimum distance (default option) or minimize correlation. The command iteratively generates samples to improve the criterion. Figure 3.17(a) shows 10 samples using LHS with two variables. The following Matlab code is used to generate samples and plot them. The variable `x` includes samples from the minimum distance option, while the variable `xr` includes samples from the minimum correlation option.

```
x=lhsdesign(10,2);  plot(x(:,1), x(:,2), 'o');
xr=lhsdesign(10,2,'criterion','correlation');
hold on; plot(xr(:,1), xr(:,2), 'r+');
r=corrcoef(x)
%r = 1.0000    -0.6999
%    -0.6999    1.0000
r=corrcoef(xr)
%r = 1.0000    -0.0545
%    -0.0545    1.0000
```

The circle markers in the figure are generated with the default option (maximizing the minimum distance) and have a correlation coefficient of −0.7. The cross markers are obtained by minimizing the correlation option, and it is −0.0545. Note that even though the minimum distance between the

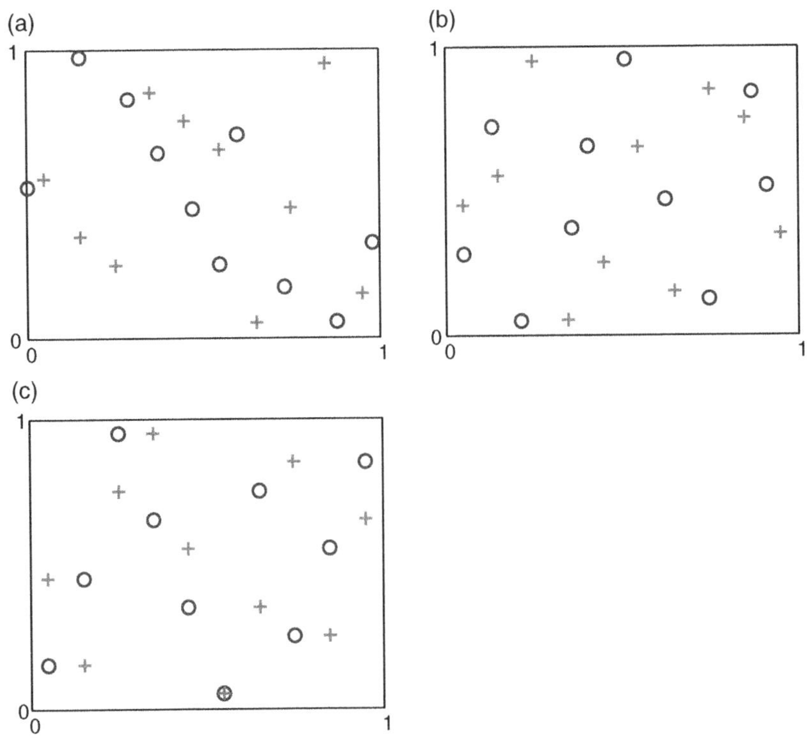


**Figure 3.17** Latin hypercube samples with different options ($n = 2, n_y = 10$). (a) LHS with five iterations, (b) LHS with 5,000 iterations, and (c) LHS with fixed center points.

circles is larger than between the crosses, the circles have a much larger empty space in the lower-left corner. This is because Matlab uses only five iterations as the default for optimizing the design.

Since the five default iterations in Matlab often yield a poor DoE, it makes sense to use more iterations. Figure 3.17(b) shows the results with 5,000 iterations. Maximizing the minimum distance (circle markers) actually reduces the correlation as well, where the correlation is dropped to 0.236. Of course, minimizing the correlation leads to a lower correlation of 0.042. With more iterations, maximizing the minimum distance eliminates the large unsampled region that was present in Figure 3.17(a). On the other hand, even with more iterations, minimizing the correlation did not change the minimum distance or eliminate the unsampled region. Therefore, for the cross markers, there are still large empty spaces near (0.45,0.75), (0, 0), and (0, 1). This explains why the minimum distance is the default criterion in Matlab.

If the sample location within an interval does not have to be random, the randomness of LHS can be reduced by putting the samples at the center of the intervals, using `'smooth'` parameter in `lhsdesign`. Samples in Figure 3.17(c) are generated by the following command: `x=lhsdesign(10, 2, 'iterations',5000,'smooth','off')`. With 10 samples, each variable is divided into 10 intervals, and the samples are located at the center of these intervals, such as 0.05, 0.15, 0.25, and so on. Therefore, in the figure, the circles (max minimum distance) and crosses (minimum correlation) are aligned, and one point even overlaps.

### 3.4.3 Orthogonal Arrays

OAs are a type of general fractional factorial design. It is a highly fractional orthogonal design that is based on a design matrix proposed by Dr. Genichi Taguchi [35]. DoE from OAs allows an arbitrary number of variables with an arbitrary number of levels. Consider the matrix of samples in Figure 3.18(a), which has four samples ($n_y = 4$) of three variables ($n = 3$) with two levels. In the normalized design variables, the two levels mean that a design variable has a value of $-1$ or $+1$; i.e., the lower and upper bounds of design space. The notation of an OA is $\mathrm{OA}_{n_y}(s^n)$, where $n_y$ is the number of samples, $n$ the number of design variables, and $s$ the number of levels. Therefore, the OA in Figure 3.18(a) can be denoted as $\mathrm{OA}_4(2^3)$. An OA is written in the form of $n_y \times n$ matrix.

(a)

| Experiment no. | Design variable 1 | 2 | 3 |
|---|---|---|---|
| 1 | −1 | −1 | −1 |
| 2 | −1 | 1 | 1 |
| 3 | 1 | −1 | 1 |
| 4 | 1 | 1 | −1 |

(b)

| Experiment no. | Design variable 1 | 2 | 3 |
|---|---|---|---|
| 1 | −1 | −1 | −1 |
| 2 | −1 | 0 | 0 |
| 3 | −1 | 1 | 1 |
| 4 | 0 | −1 | 1 |
| 5 | 0 | 0 | −1 |
| 6 | 0 | 1 | 0 |
| 7 | 1 | −1 | 0 |
| 8 | 1 | 0 | 1 |
| 9 | 1 | 1 | −1 |

**Figure 3.18** Examples of OAs. (a) $\mathrm{OA}_4(2^3)$ and (b) $\mathrm{OA}_9(3^3)$.

For linear PRS, DoE from OAs yields an orthogonal design; that is, the moment matrix becomes diagonal.

OAs are balanced to ensure that all levels of all variables are considered equally. For this reason, the variables can be evaluated independently of each other despite the fractionality of the design. An important property of an OA is that in every pair of columns, each of the possible ordered pairs of elements appears the same number of times. In Figure 3.18(a), for example, in every pair of columns, each of the four ordered pairs $(-1, -1)$, $(-1, 1)$, $(1, -1)$, and $(1, 1)$ appears exactly once. If we choose columns 1 and 2, they are $(-1, -1)$, $(-1, 1)$, $(1, -1)$, and $(1, 1)$. If we choose columns 1 and 3, they are $(-1, -1)$, $(-1, 1)$, $(1, 1)$, and $(1, -1)$. Therefore, all possible pairs appear once.

Figure 3.18(b) shows an OA of $OA_9(3^3)$. In this case, since each variable can have three levels: $-1$, $0$, $+1$, all possible pairs of levels are $(-1, -1)$, $(-1, 0)$, $(-1, 1)$, $(0, -1)$, $(0, 0)$, $(0, 1)$, $(1, -1)$, $(1, 0)$, $(1, 1)$. In every pair of columns, each of the nine ordered pairs appears exactly twice. Therefore, it is an OA $OA_9(3^3)$.

It is interesting to note that $s^n$ is in fact the number of samples in the $s$-level full-factorial design, while $n_y$ is the actual number of samples in the OA. Therefore, the ratio $(n_y/s^n)$ is the fraction of a full-factorial design. For $OA_4(2^3)$ in Figure 3.18(a), the ratio is $4/2^3 = 1/2$ fraction of a full-factorial design, and for $OA_9(3^3)$ in Figure 3.18(b), the ratio becomes $9/3^3 = 1/3$.

A sub-matrix formed by deleting some columns of an OA is also an OA. Thus, by deleting certain columns of a given OA, it is possible to generate many different plans for multifactor experiments. If the third column in Figure 3.18(a) is deleted, it becomes $OA_4(2^2)$ and it is nothing but the two-level full-factorial design. The same is true for Figure 3.18(b), when the third column is removed, it becomes the three-level full-factorial design. In practice, since each column represents a design variable, the effect of the OA will be the same if any single column is removed. In Figure 3.18(a), for example, the resulting OA by deleting the third column will be identical to that by deleting the first column. They are only different by row indices.

A fractional factorial design that enables uncorrelated estimation of every factorial effect included in the underlying linear model, assuming that all other effects are zero, is called an orthogonal plan. Fractional factorial designs based on OAs, irrespective of the degree of fractionation, are necessarily orthogonal plans. This is the primary reason for the popularity of fractional factorials based on OAs.

Since each row of an OA represents a sample, exchanging rows would not affect the property of DoE. Also, exchanging columns simply means relabeling variables. It would not affect the orthogonality of DoE either. Lastly, the levels of factors, $-1, 0, +1$, can be changed to different orders, such as $+1, -1, 0$ because all levels will appear the same number of times. Therefore, exchanging levels would not change the property either. In summary, two OAs are defined to be equivalent if one can be obtained from the other via the following operations: (i) exchanging rows, (ii) exchanging columns, and (iii) exchanging labels of levels.

Although OAs are a powerful tool to generate effective samples, it has limitations to using them for the purpose of DoE. First, it is not general enough to generate an arbitrary number of samples, variables, and levels. Only a limited number of combinations are available. For a list of available OAs, theory, and applications, see, for example, Owen [36], Hedayat et al. [37], and references therein. Second, samples in some OAs can be repeated. This is undesirable for numerical experiments where the QoI is also repeated. In such a case, it would be necessary to perturb the repeated sample locations slightly.

**Example 3.13 Orthogonal array**

Calculate the standard error of prediction for the OA shown in Figure 3.18(a) when a linear PRS $\hat{y}(x) = b_1 + b_2x_1 + b_3x_2 + b_4x_3$ is fit with the four samples.

**Solution:**

For the linear PRS, the design matrix and the moment matrix at the four sample locations become

$$\mathbf{X} = \begin{bmatrix} 1 & -1 & -1 & -1 \\ 1 & -1 & 1 & 1 \\ 1 & 1 & -1 & 1 \\ 1 & 1 & 1 & -1 \end{bmatrix}, \quad \mathbf{X}^T\mathbf{X} = \begin{bmatrix} 4 & 0 & 0 & 0 \\ 0 & 4 & 0 & 0 \\ 0 & 0 & 4 & 0 \\ 0 & 0 & 0 & 4 \end{bmatrix}$$

As expected, since the OA yields an orthogonal design, the moment matrix becomes diagonal. Using the inverse of the moment matrix, the standard error of prediction can be calculated as

$$\sigma_y(\mathbf{x}) = \hat{\sigma}\sqrt{\xi(\mathbf{x})^T\left(\mathbf{X}^T\mathbf{X}\right)^{-1}\xi(\mathbf{x})} = \frac{\hat{\sigma}}{2}\sqrt{1 + x_1^2 + x_2^2 + x_3^3}$$

The standard error of prediction has its minimum $\sigma_y^{\min} = \hat{\sigma}/2$ at the origin, while its maximum $\sigma_y^{\max} = \hat{\sigma}$ at the corners of the design space. It is noted that the standard error of prediction is the same for both sampled or unsampled corners. This happened because the OA captures the behavior of an individual variable while fixing other variables.

## 3.5 Review of Various Designs of Experiments

In this chapter, we discussed several different DoE methods. Some DoEs are well structured, while others have randomness embedded in the process. Also, different DoEs have different characteristics, and it is difficult to tell whether one DoE is better than the others for all possible configurations. From the perspective of users, it might be necessary to provide a guideline to choose an appropriate DoE for their applications. In this section, different application conditions are considered, and DoEs that are useful for such conditions are recommended.

### 3.5.1 Guideline for Selecting Designs of Experiments

Since the conditions of using DoE can be different in terms of the number of samples, variables, and levels, there would not be a single DoE that works best in all possible conditions. In addition, the samples obtained from DoE can have different levels of noise. Therefore, it would be a good idea to find some recommendations for finding a good DoE for the given conditions. The conditions that we need to consider for choosing a good DoE are (i) the level of noise, (ii) the number of variables, (iii) the number of regression coefficients, and (iv) the number of samples. Also, it is possible to consider if all the experiments are conducted simultaneously or sequentially. The latter does not require all sample locations in advance. Sample locations are sequentially determined based on previous sample results.

Low dimension with high noise: The first condition that we want to consider is the case of low dimension with a high level of noise. This corresponds to the case where the QoI depends on two or three input variables, but the experiments include a large level of noise. When the design space is box-like, the full-factorial design or CCD is recommended. This is because the number of required samples is relatively small in low dimensions. Both full-factorial design and CCD are good for reducing the prediction variance. When the domain is irregular, either D-optimal or A-optimal design is recommended as these designs are optimized to reduce the uncertainty in coefficients. Since the domain is irregular, an adaptive sampling scheme can be a good strategy to add samples until the uncertainty reaches an acceptable level.

Low dimension with low noise: The second condition is the case of low dimension with a low level of noise. This corresponds to the case when the QoI depends on two or three input variables, but the level of noise in experiments is low. In such a case, the focus is on the accuracy of the surrogate rather than reducing prediction variance. When the design space is box-like, the minimum bias design, LHS design, and OAs would be good choices for DoE. The LHS design can fill the low-dimensional space with a small number of samples. The OAs can capture the trend of all design variables with a reasonably small number of samples. When the design space is irregular, MCS might be a good choice by removing samples that belong outside of the design space.

High dimension with high noise: This would be considered the most challenging case because it is difficult for samples to fill the design space, and the experimental result at each sample location is not accurate. When the design space is box-like, it would be good to use CCD, block design, and fractional factorial design. CCD would be a good choice if the number of design variables is not too many (i.e., $n = 5{\sim}7$) because the number of samples would be too many compared to the regression coefficients for large $n$. For a large number of variables, block design or fractional factorial design are the only practical methods, but both may come with a large extrapolation region. For irregular domains, D-optimal or A-optimal designs are recommended using the adaptive sampling option.

High dimension with low noise: When the number of input variables is large, but simulations and/or experiments are relatively accurate, it is better to use space-filling DoEs, such as LHS design. Especially, it would be good to use the maximizing the minimum distance option in the LHS design. However, due to the curse of dimensionality, it is inevitable to accept a large portion of the extrapolation region in DoE, which can lead to a large prediction variance.

### 3.5.2 Good Practice for Design of Experiments

Although it is possible to choose different DOE methods as we discussed in this section, there are several good practices to follow in general.

1) It is always good to normalize the design space to $-1 \leq x_k \leq 1$ such that the design space is a hypercube. Theoretically, it is possible to have an arbitrary range of design variables. However, when the ranges of design variables are significantly different, a numerical difficulty can occur. For example, let us consider the case where both Young's modulus and Poisson's ratio are design variables. In the MKS unit system, Young's modulus of metal is in the order of $10^{11}$ Pa, while the Poisson's ratio is $0 < \nu < 0.5$. Such a huge difference makes it difficult to measure the distance between sample locations. Therefore, it would be better to normalize all design variables with the same range using Eq. (3.2). When a design variable does not have lower and/or upper bounds, it is still required to select low or high values for the bounds. If a design variable varies in different orders of magnitude, it is possible to use a logarithmic scale.
2) Similar to the normalization of design variables, it would be better to normalize QoI using Eq. (3.6), where the range of QoI becomes $0 \leq y \leq 1$. If the QoI that is approximated varies over many orders of magnitude, it would be better to use the logarithm of the function or a similar transformation so that the region of small values of QoI is not ignored due to large values of QoI.
3) If the design space is a box-like domain, it would be better to use a well-known pattern of sampling methods, such as the CCD or fractional factorial design. These methods are well established, along with in-depth error analysis.
4) The most difficult part of the DoE is the curse of dimensionality. Working with five design variables is much easier than six variables. Therefore, it is recommended to reduce the design domain and the number of design variables as much as possible. Not all variables significantly affect the QoI. Therefore, it would be better to fix those variables that do not change the QoI

significantly. Sensitivity analysis or parameter study can be used to identify important/significant variables.

5) As discussed in Chapter 2, the number of samples should be large enough so that the surrogate fitting process satisfies the regression property. In the case of polynomial response surfaces in Chapter 2, it is recommended that the number of samples should be at least two or three times more than that of unknown model parameters. If the number of samples is not large enough, it is possible that the surrogate fits noise, not the trend. This phenomenon is called over-fitting.
6) The basic assumption of polynomial response surfaces is that the functional form is correct, while samples have noise. Therefore, it would be best to start with an accurate functional form. Any domain knowledge or prior experience would be useful to find an appropriate functional form. Also, it would be a good strategy to start with a high-order polynomial and remove unimportant basis functions progressively. If the functional form of the surrogate model is significantly different from the true function, the estimated noise standard deviation includes not only the noise in the data but also the bias error of the model.
7) When the backward elimination process is applied, it would be better to remove those coefficients that have low t-statistics. Also, during the backward elimination process, it would not be a good idea to eliminate more than one coefficient at a time. This is because the other coefficient might be significant after the previous coefficient is eliminated.

## Exercise

**3.1** Consider a surrogate model $\hat{y}(x) = b_1 + b_2(x+1) + b_3(x+2)^2$ with design space $x \in [-2, 2]$. Five samples are given to fit the surrogate: $(-2, -1)$, $(-1, -0.05)$, $(0, 1)$, $(1, 2)$, and $(2, 3.5)$. (i) Fit the surrogate to the samples as they are. (ii) Scale the input variables and output QoIs and then fit the surrogate to the samples. Check if the two surrogates are identical or not.

**3.2** Consider the problem of fitting a linear PRS $\hat{y}(\mathbf{x}) = b_1 + b_2x_1 + b_3x_2 + b_4x_3$ to samples in the square domain $-1 \le x_1, x_2, x_3 \le 1$. Compare the maximum value of the prediction variance of the full-factorial design (samples at all eight vertices) with that of two-dimensional linear PRS in Example 3.2.

**3.3** Find the maximum prediction variance in the unit cube for a linear PRS, when the samples are given in the four points $(-1, -1, -1)$, $(-1, -1, 1)$, $(-1, 1, -1)$, and $(1, -1, -1)$. Compare this result with that of Exercise Problem 3.2.

**3.4** In Example 3.2, it was shown that the two-level full-factorial design in a two-dimensional problem is an orthogonal design, and the prediction variance is minimum at the origin. Show the same conclusion holds for a three-dimensional problem.

**3.5** In a three-dimensional design space, four non-planar samples are the minimum number of samples to define a simplex. Show that the following four samples yield an orthogonal design for a linear PRS and calculate the minimum and maximum standard error of prediction: $\mathbf{x}_1 = (-1, -1, 1)$, $\mathbf{x}_2 = (1, -1, -1)$, $\mathbf{x}_3 = (-1, 1, -1)$, $\mathbf{x}_4 = (1, 1, 1)$.

**3.6** In one-dimensional linear PRS, show that the two samples $x_1 = -\alpha$ and $x_2 = \alpha$, $\alpha > 0$, produce an orthogonal design. Discuss the value of $\alpha$ when the standard error of prediction can be minimized.

**3.7** In Example 3.3, the three samples are given at $\left(\alpha\sqrt{3/2}, -\alpha/\sqrt{2}\right)$, $\left(-\alpha\sqrt{3/2}, -\alpha/\sqrt{2}\right)$, $\left(0, \alpha\sqrt{2}\right)$ with $\alpha \in [0.1, 1.5]$. Show that these samples yield an orthogonal design. Plot the minimum and maximum standard error of prediction as a function of $\alpha$.

**3.8** For two-dimensional quadratic PRS, use the three-level full-factorial design to calculate the minimum and maximum standard error of prediction and stability ratio.

**3.9** In Example 3.5, vary the location of the axial point $\alpha \in [1.0, 1.7]$ in the increment of 0.1 and plot $\sigma_y^{\min}$ and $\sigma_y^{\max}$ as a function of $\alpha$ when $n_c = 1$.

**3.10** Repeat Problem 3.9 when $n_c = 5$.

**3.11** For the face-center CCD design shown in Figure 3.7, calculate the minimum and maximum standard error of prediction along with its stability.

**3.12** For the three-dimensional CCD with $n_c = 1$, calculate the minimum and maximum of the standard error of prediction along with its stability.

**3.13** For the three-dimensional block design shown in Figure 3.8, calculate the minimum and maximum standard error of prediction along with its stability. Compare the results with that of the CCD in Exercise Problem 3.12.

**3.14** Find the maximum prediction variance in the unit cube for a linear polynomial, when the data is given in the four points $(1, 1, -1)$, $(1, 1, 1)$, $(1, -1, -1)$, $(-1, 1, -1)$.

**3.15** Repeat Problem 3.14 when samples are given in the four points $(-1, -1, -1)$, $(-1, -1, 1)$, $(-1, 1, -1)$, $(1, -1, -1)$.

**3.16** When the first data point is given by $(-1, -1)$, find additional two points in the unit square that will minimize the maximum prediction variance in the unit square for a linear response surface.

**3.17** Find the three points in the unit square that will minimize the maximum prediction variance in the unit square for a linear response surface.

**3.18** For Example 3.11, find the maximum prediction variance for the minimum bias designs and compare it to that of the full-factorial design.

**3.19** Construct a minimum-bias CCD by finding $\alpha$ that minimizes the maximum standard error of prediction. Use $n_c = 1$.

**3.20** A linear response surface $\hat{y}(\mathbf{x}) = b_1x_1 + b_2x_2$ is fit using the following 4 DoEs: $(-1, -1)$, $(1, -1)$, $(-1, 1)$ and $(1, p)$. (i) Determine unknown $-1 \le p \le 1$ using the D-optimality criterion. (ii) Calculate the prediction variance at the center location $(0, 0)$.

**3.21** Three LHS samples are given in the form of a definition matrix for two uniformly distributed variables $x_1$ and $x_2$. The design space is given as $x_1, x_2 \in [0, 1]$. (i) Draw the intervals of the variables and one realization of this design, using the maximum minimum distance criterion to place them well in the cells. (ii) Give the coordinates of these points and the value or the criterion.

$$\begin{bmatrix} 1 & 2 \\ 2 & 3 \\ 3 & 1 \end{bmatrix}$$

**3.22** Find the minimum and maximum standard error of prediction for the orthogonal design given in Figure 3.18(b) when a linear PRS is used.

**3.23** Compare prediction variance for an orthogonal design based on equilateral triangle with three vertices $\left(\sqrt{3/2}, -1/\sqrt{2}\right), \left(-\sqrt{3/2}, -1/\sqrt{2}\right), \left(0, \sqrt{2}\right)$ to the right triangle $(-1, -1)$, $(1, -1)$, $(-1, 1)$. Use linear polynomial $y = b_1 + b_2x_1 + b_3x_2$.

**3.24** Repeat Problem 3.23 when the equilateral triangle lies inside of the box. That is, three vertices are located at $(1, -0.5)$, $(-1, -0.5)$, $(0, 1)$ to the right triangle $(-1, -1)$, $(1, -1)$, $(-1, 1)$. Use linear polynomial $y = b_1 + b_2x_1 + b_3x_2$.

**3.25** Show that the four points: $(\pm 0.1876, \pm 0.7947)$ constitute a minimum bias design for the one-dimensional response surface $y = b_1 + b_2x + b_3x^2$ on the interval $(-1, 1)$. Assume that the true model is a cubic polynomial.

**3.26** Find the D-optimal design for the one-dimensional response surface $y = b_1 + b_2x$ on the interval $(-1, 1)$ using three points. You can assume that two of the three points are $x = 1$ and $x = -1$.

**3.27** For a linear model, with tests conducted at three vertices of the unit square $(-1, 1)$, $(1, -1)$, $(-1, -1)$, verify that the prediction variance is given as $\text{Var}[y] = \sigma^2\left(1 + x_1 + x_2 + x_1^2 + x_1x_2 + x_2^2\right)/2$. Hint: You do not need to invert the matrix $\mathbf{X}^T\mathbf{X}$. You can extract that matrix from the expression for the variance so that you will just need to check that it is the correct inverse.

**3.28** If the region of interest includes only the triangle defined by the three points in Problem 3.27, formulate the optimization problem of finding the maximum variance in the triangle. Find where the variance is maximal in the triangle and check your answer from the Kuhn–Tucker conditions of the optimization problem you formulated. If you can add one more experiment, where would you add it to reduce the maximum prediction variance in the triangle.

**3.29** You need to fit a linear function $y = b_1 + b_2x$ to data. You can run experiments in the interval $(-1, 1)$, but you are interested only in results in the interval $(-1, 0.5)$. For three experiments, find the D-optimal design, and find the equations that the coordinates of the points need to satisfy for a minimum bias design. Find the minimum bias design, and compare the two for the function $y = e^x$ in the interval $(-1, 0.5)$.

**3.30** A response surface can be constructed for the purpose of predicting the results of future experiments or for the purpose of evaluating some physical constants (that is the coefficients in the response surface could be these constants). In addition, a response surface could be constructed under conditions of high experimental noise or conditions of large modeling errors (the assumed model being different from the true model). Please indicate which optimality criterion you will use for selecting experimental points for each one of the four possible combinations of these conditions: (i) Physical constants, high noise. (ii) Physical constants, large modeling error. (iii) Predicting future experiments, high noise. (iv) Predicting future experiments, large modeling error.

**3.31** Create a 13-point D-optimal design in two-dimensional space using Matlab cordexch function and compare its prediction variance to that of the CCD design shown in Example 3.5.

# Part II

# Design Optimization

# 4

# Optimization Definition and Formulation

## 4.1 Introduction

In Part 1 of this book, we discussed the basics of surrogate modeling using polynomial response surfaces. In most engineering applications, surrogate modeling is not a goal but a tool. The purpose of surrogate modeling is to construct an approximate model such that the cost of experiments or simulations can be reduced in subsequent analysis. Therefore, the effectiveness of a surrogate model is increased when an application requires numerous experiments or simulations with different values of input variables. Common applications include design optimization and uncertainty quantification, where hundreds or thousands of simulations are required. This book focuses on design optimization, where input variables represent design variables, and the quantity of interest (QoI) is either the objective or constraint function. Since an approximate model is used, it is referred to as surrogate-based analysis and optimization (SBAO) in this text. In addition to saving computational time, surrogates also allow for the optimization of applications with nonsmooth or noisy responses and can provide insights into the nature of the design space [38].

In order to understand the basic procedure and concept of SBAO, it would be necessary to discuss design optimization first. Part 2 of this book will discuss the fundamentals of design optimization. The contents of this part are by far brief and incomplete. We will define basic terminologies that are used in design optimization and basic algorithms to solve optimization problems. For a comprehensive understanding of design optimization, users are referred to the books of Arora [39] and Haftka and Gürdal [40].

As a first step toward design optimization, this chapter will present the definition of design optimization, optimization formulation, and optimality criteria. First, we will define the design optimization definition in Section 4.2, where various terminologies in optimization are defined. The definition of an optimization problem boils down to defining design parameters, objective functions, and constraint functions. When the number of design variables is less than two, graphical optimization is also presented. In Section 4.3, the standard form of an optimization problem is introduced, which makes it possible for engineers to focus on solving optimization problems in a unified fashion. The concept of convexity is briefly introduced as well, which can guarantee the global optimum design. In Section 4.4, an important topic of optimality criteria is presented, which leads engineers to find an optimum design. In practice, optimality criteria are used to determine if a given design is an optimum design or not.

*Surrogate Modeling and Optimization: Theories, Applications, and Limitations*, First Edition. Nam-Ho Kim.

Published 2026 by John Wiley & Sons, Inc.
Companion website: www.wiley.com/go/kim/surrogate_modeling

## 4.2 Design Optimization Definition

Design is a procedure to improve or enhance the performance of a system by changing its parameters. In the previous chapters, the performance of a system is referred to as a quantity of interest (QoI). A QoI can be quite general in engineering fields and can include: the weight, stiffness, and compliance of a structure; the fatigue life of a mechanical component; the noise in the passenger compartment; the vibration level; the output power of a generator; the safety of a vehicle in a crash, etc. However, aesthetic measures such as whether a design is attractive to customers are not considered. All QoIs are presumed to be measurable quantities. System parameters are variables that a design engineer can change during the design process. For example, the thickness of a vehicle body panel can be changed to improve vehicle performance. The cross-section of a beam can be changed in designing bridge structures. System parameters that can be changed during the design process are called design variables.

### 4.2.1 Design Optimization Process

The engineering design of a system in the simulation-based design process consists of design problem formulation, mathematical modeling, parameterization, simulation, and optimization. Figure 4.1 is a flowchart of the design optimization process in which computational simulation and mathematical programming (optimization algorithm) play essential roles. The success of the system-level, simulation-based design process strongly depends on consistent design parameterization, accurate simulation, and an efficient mathematical programming algorithm.

A design engineer simplifies a physical engineering problem into a mathematical model that can represent the physical problem up to the desired level of accuracy. A mathematical model has

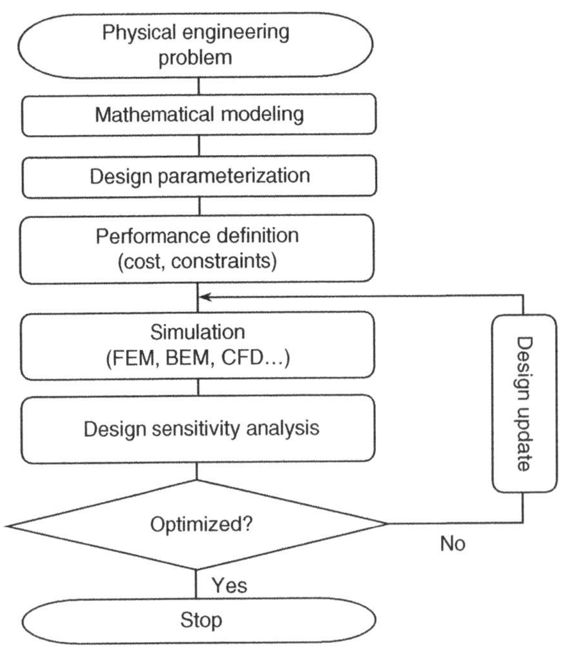


**Figure 4.1** Flowchart of the design optimization process.

parameters that are related to the parameters of the physical problem. Among them, the design engineer identifies those design variables to be used during the design process. Design parameterization, which allows the design engineer to define the properties of each design component of the system being designed, is one of the most important steps in the design process. The principal role of design parameterization is to define the system parameters that characterize the system model and to collect a subset of the parameters as design variables. Design parameterization forces engineering teams in design, analysis, and manufacturing to interact at an early design stage, and supports a unified design variable set to be used as the common ground to carry out all simulations, design, and manufacturing processes. Only proper design parameterization will yield a good optimum design since the optimization algorithm will search within a design space that is defined for the design problem. The design space is defined by the type, number, and range of design variables.

Simulation can be carried out using experiments on an actual or reduced scale, which is a straightforward and still prevalent method for industrial applications. However, the expense and the inefficiency involved in fabricating prototypes make this approach difficult to apply. An analytical model may resolve these difficulties since it approximates the physical problem as a mathematical model and solves it in a simplified form. However, the analytical method has limitations even for very simple problems. With the emergence of various computational capabilities, most analytical approaches to mathematical problems have been converted to numerical approaches, which can solve very complicated, real engineering applications. Finite element analysis (FEA), boundary element analysis, and computational fluid dynamics (CFD) are short lists of numerical tools used in engineering analysis. However, the numerical models themselves can be computationally expensive and may include numerical errors. The main purpose of surrogate modeling is to replace the numerical simulation with a surrogate model such that the computational cost can be reduced significantly.

The selection of a design space and an analysis method must be carefully determined since the analysis, both in terms of accuracy and efficiency, must be able to handle all possible designs in the chosen design space. That is, the larger the design space, the more sophisticated the analysis capability must be. For example, if larger shape design changes are expected during design optimization, mesh distortion in FEA could be a serious problem, and a finite element model that can handle large shape design changes must be used.

A performance measure in a simulation-based design is the result of numerical simulations. Based on the evaluation of simulation results, such engineering concerns as high stress, clearance, natural frequency, output torque, or mass can be identified as performance measures for design improvement. Typical examples of performance measures are mass, volume, displacement, stress, compliance, buckling, natural frequency, noise, fatigue life, and crashworthiness. It is expected that the users should define all relevant performances that are of concern during the design process.

Cost and constraints can be defined by combining certain performance measures with appropriate constraint bounds for interactive design optimization. The cost function, sometimes called the objective function, is minimized (or maximized) during optimization. The selection of a proper cost function is an important decision in the design process. A valid cost function has to be influenced by the design variables of the problem; otherwise, it is not possible to reduce the cost by changing the design. In many situations, an obvious cost function can be identified. In other situations, the cost function is a combination of different performance measures. This is called a multiobjective cost function. The contribution of different cost functions must be either weighted or traded off.

Constraint functions are the criteria that the system has to satisfy for each feasible design. Among all design ranges, those that satisfy the constraint functions are candidates for the optimum design. For example, a design engineer may want to design a bridge whose weight is minimized and whose maximum stress is less than the yield strength. In this case, the cost function, or weight, is the most important criterion to be minimized. However, as long as stress, or constraint, is less than the yield strength, the stress level is not important.

Once the design variables, cost, and constraint functions are defined, an optimization engine starts changing the design variables to find the best cost function that satisfies all constraint functions. This is an iterative process, which requires numerous evaluations of functions. Surrogate models can play a critical role in this stage to approximate the expensive function evaluations with cheap surrogate models. In general, optimization algorithms are categorized into gradient-based and gradient-free algorithms. Most gradient-based optimization algorithms are based on the mathematical programming method, which requires the function values and gradient (i.e., sensitivity) information at given design variables. For a given design variable that defines the numerical model, simulation provides the values of the cost and constraint functions for the algorithm. The sensitivities of the cost and constraint functions must also be supplied to the optimization algorithm. Then, the optimization algorithms calculate the best possible design for the problem. Chapter 5 will introduce several numerical optimization algorithms. Each algorithm has its own advantages and disadvantages. The performance of an optimization algorithm critically depends on the characteristics of the design problem and the types of cost and constraint functions. Since most gradient-based algorithms try to improve the design by searching nearby designs starting from the current design using gradient information, the search is limited to finding local minima. In addition, the mathematical programming algorithms assume that both the cost and constraint functions are smooth and continuous functions of design variables.

The main difference between gradient-based and gradient-free algorithms is the requirement of gradient information, which tends to be expensive and sometimes difficult to calculate. The gradient-free algorithms explore the design space to find a better design. Therefore, they have a better chance to find the global minimum. However, there is no guarantee that gradient-free algorithms can find a global minimum. The main bottleneck of gradient-free algorithms is that the number of function evaluations is often too high, which is the main advantage of surrogate modeling. Therefore, gradient-free algorithms are popular in conjunction with surrogate models. In Chapter 6, several popular gradient-free optimization algorithms will be introduced, such as the genetic algorithm, particle swarm optimization algorithm, and direct search algorithm.

### 4.2.2 Design Variables and Feasible Domain

Design variables are those system parameters that design engineers want to change during the optimization process. Theoretically, any system parameter can be a design variable, but the process and performance of optimization strongly depend on the types of design variables. The optimization engine generates a set of design variables that require performance values from simulation and sensitivity information to find an optimum design. Thus, the numerical model has to be updated for a different set of design variables supplied by the optimization engine. If the cost function reaches a minimum with all constraint requirements satisfied, then an optimum design is obtained.

Common design variables include material properties, boundary and loading conditions, the thickness of a plate, cross-sectional dimensions of a beam, and the shape of a domain. Many design variables are related to the geometry of a system, but there are also mathematical quantities, such as the stiffness of a spring and the damping coefficient of a dashpot, which can serve as design variables. From the perspective of mathematical programming, the most important characteristic of design variables is continuous or discrete. A continuous design variable can take any value between the lower and upper bounds in real space. However, discrete variables can only take a specific set of candidates. For example, if the number of students is a design variable, then it is inherently a discrete variable.

Another important characteristic of design variables is independence; that is, design engineers can change all design variables independently. A common example is shown in Figure 4.2, where the cross-sectional dimensions of a tube are defined as design variables. In this case, three variables, $r_o$, $r_i$, and $t$, are not independent because there exists a relationship $t = r_o - r_i$. Therefore, design engineers cannot change all three variables independently.

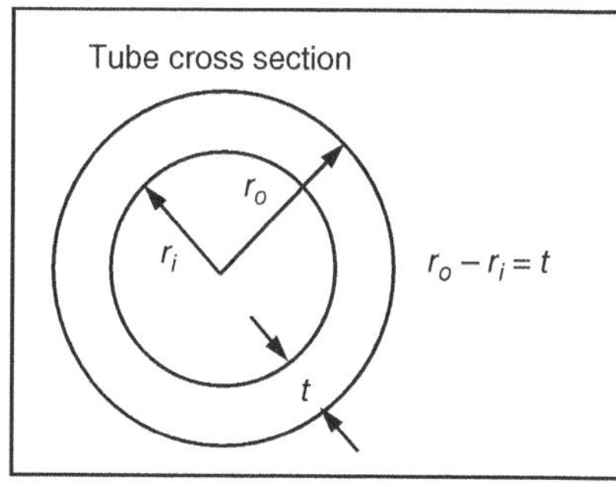


**Figure 4.2** Independence of design variables.

The purpose of an optimization problem is to find the best design among many possible candidates. For this reason, design engineers have to specify the best possible design and the best possible candidates. Although the design space is given, not all designs in the design space can be a candidate because some of them may not satisfy constraints. The subset of design space that satisfies all constraints is called a feasible set. Mathematically, the feasible set $S$ is defined as

$$S = \{\mathbf{x} \mid \mathbf{x} \text{ satisfies all constraints}\} \tag{4.1}$$

A possible candidate must exist within a feasible design region to satisfy problem constraints. Every design in the feasible region is an acceptable design, even if it is not the best one. The best design is usually the one that minimizes (or maximizes) the cost function within the feasible set. Thus, the goal of the design optimization problem is to find a design that minimizes the cost function among all feasible designs.

**Example 4.1 Fitting the stress-strain relationship**

In Chapter 2, we discussed the least-squares method, where the error between the samples and surrogate prediction is minimized. This is indeed an optimization problem. Consider fitting the stress–strain relationship $\sigma = E\epsilon$ using the following three measured samples:

| | | | |
|---|---|---|---|
| Strain (με) | 1 | 2 | 4 |
| Stress (ksi) | 1 | 2 | 3 |

Find Young's modulus $E$ which minimizes the difference between the samples and the model using (i) the maximum difference and (ii) the root-mean-square (RMS) difference.

**Solution:**

The goal is to find Young's modulus $E$ which minimizes the differences between the samples and the model. In this case, the design variable is Young's modulus $E$, and the objective function is the goodness of fit. For a given Young's modulus $E$, it is possible to define three errors between the samples and the curve. It is clear when we say the best fit, we want all three errors to be small. However, if we want a single measure of goodness, we need to obtain a single measure of the smallness of the three errors. As we discussed in Chapter 2, different measures can be used for the goodness of fit. Therefore, the following two different optimization problems can be formulated:

$$\text{Maximum difference} \quad d_{\max}(E) = \max_{i=1,2,3} |\sigma_i - E\epsilon_i|$$

$$\text{RMS} \quad d_{\text{RMS}}(E) = \sqrt{\frac{1}{3}\sum_{i=1}^{3}(\sigma_i - E\epsilon_i)^2}$$

a) In the case of maximum difference, the minimum $d_{\max}(E)$ occurs when the maximum positive difference is in equal magnitude with the maximum negative difference. For the given samples, this occurs between the second and the third samples (the difference in the first sample is always less than that of the second sample). Therefore, the minimum $d_{\max}(E)$ occurs when $2-2E=-(3-4E)$, which yields $E_{\max}=5/6$ ksi/με.
b) In the case of RMS difference, the square of $d_{\mathrm{RMS}}$ can be written as

$$d_{\mathrm{RMS}}^2(E)=\frac{1}{3}\left[(1-E)^2+(2-2E)^2+(3-4E)^2\right]=\frac{1}{3}\left(14-34E+21E^2\right)$$

The $d_{\mathrm{RMS}}^2(E)$ has its minimum when its derivative with respect to $E$ becomes zero; that is, $E_{\mathrm{RMS}}=17/21$ ksi/με.

It is noted that the two formulations yield different optimum designs. Therefore, it is important to define the cost function carefully as the optimum design depends on how the cost function is defined. In practice, the RMS difference is popular because the cost function is a smooth (nonlinear) function of design variables such that the maximum/minimum can be found using gradient information. Figure 4.3 shows the optimization results using the two different cost functions. $E_{\mathrm{RMS}}$ is slightly lower than $E_{\max}$ because the first data do not have any role in the maximum difference cost function. Also, the figure shows that $d_{\max}$ is not continuous at the optimum design, while $d_{\mathrm{RMS}}$ is continuous. The following Matlab code is used to generate the plot in Figure 4.3:

```
eps=[1 2 4]; sig=[1 2 3];
e=linspace(0.5,1,101);
sigmodel=e'*eps;
sigr=ones(101,1)*sig;
diff=abs(sigr-sigmodel);  maxdiff=max(diff');
sumsquares=diag(diff*diff');  rms=sqrt(sumsquares/3);
plot(e,maxdiff); xlabel('E'); ylabel('cost function')
hold on;  plot(e,rms,'r-'); legend('max error','RMS
error','Location','North')
```

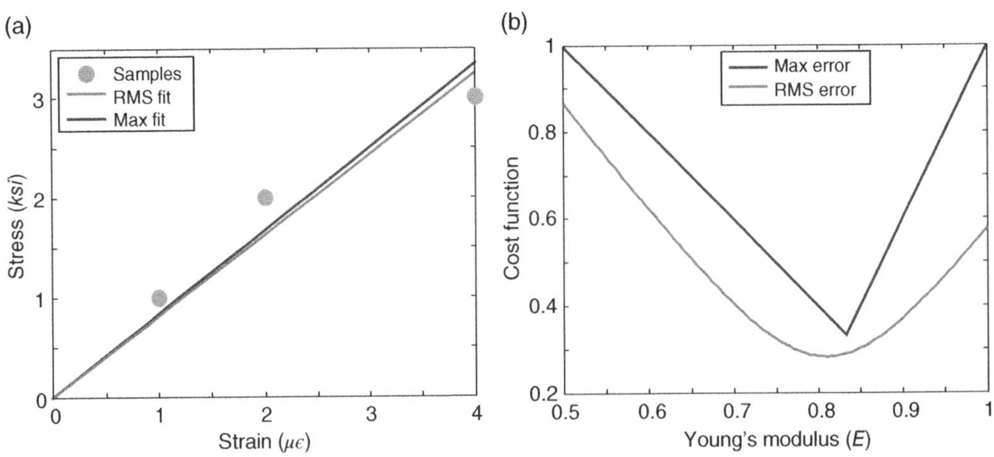


**Figure 4.3** Stress–strain curve and the variation of the cost function.

**Example 4.2 Optimization problem definition**

The weight and value of the five items are shown in the table. We want to put items in the knapsack to maximize the total value, while the total weight is less than 20 lbs. Define the optimization problem, i.e., define design variables and cost function.

| Item | 1 | 2 | 3 | 4 | 5 |
|---|---|---|---|---|---|
| Weight (lb) | 4 | 6 | 7 | 10 | 3 |
| Value ($) | 12 | 12 | 12 | 27 | 5 |

**Solution:**

In this case, design variables would be if an item is chosen or not. Therefore, the design variables are not continuous, but discrete. Indeed, design variables are binary. If $i$th item is chosen, the design variable $x_i = 1$; otherwise, $x_i = 0$. The cost function is the total value of the chosen items. However, since the total weight must be less than 20 lbs, the cost is reduced by $10 if the total weight is over 20 lbs. This obviously may not prevent the total weight of over 20 lbs, but we will use this approach for the moment. Therefore, the objective function can be written as

$$\begin{aligned} \underset{x_i \in (0,1)}{\text{maximize}}\ \text{value} = &\ 12(x_1 + x_2 + x_3) + 27x_4 + 5x_5 \\ &- 10 \times \text{sgn}(4x_1 + 6x_2 + 7x_3 + 10x_4 + 3x_5 - 20) \end{aligned}$$

where sgn() is a signum function, whose value is one if the argument is greater than zero, otherwise zero.

The previous two examples are referred to as unconstrained optimization because the optimization problem has only an objective function, but no constraints. Of course, there is a constraint for a design variable, such as binary design or lower and upper bounds of design variables. However, these types of constraints can be handled directly when the design variable is changed. Therefore, they are referred to as side constraints.

The condition of the total weight being less than 20 lbs in Example 4.2 can be written in the form of a constraint. In this case, the constrained optimization problem can be written as

$$\underset{x_i \in (0,1)}{\text{maximize}}\ \text{value} = 12(x_1 + x_2 + x_3) + 27x_4 + 5x_5$$

$$\text{subject to}\ 4x_1 + 6x_2 + 7x_3 + 10x_4 + 3x_5 \le 20$$

In practice, there are many different ways of formulating an optimization problem. For example, the optimization problem in Example 4.1 yields a smooth objective function in the case of RMS difference. However, in the case of maximum difference, the objective function becomes nonsmooth. To avoid a nonsmooth objective function, we can add a bound design variable $\alpha$, as well as error bound constraints, such that the following constrained optimization problem is equivalent to the one that was defined in Example 4.1:

$$\underset{\alpha, E}{\text{minimize}}\ \alpha$$

$$\text{subject to} - \alpha \le \sigma_i - E\epsilon_i \le \alpha, \quad i = 1, 2, 3$$

The optimization problem has three constraints. It is noted that the objective function is equal to one design variable $\alpha$, and the other variable $E$ appears only in the constraints. It is also noted that both the objective and constraint functions are smooth and linear. Since we know the sign of the differences, we can rewrite the optimization problem as

$$\underset{\alpha, E}{\text{minimize}} \ \alpha \tag{4.2}$$

$$\text{subject to } \sigma_1 - E\epsilon_1 \leq \alpha, \quad \sigma_2 - E\epsilon_2 \leq \alpha, \quad \sigma_3 - E\epsilon_3 \geq -\alpha$$

Therefore, by introducing an additional variable $\alpha$, it is possible to make the objective and constraint functions smooth. We will discuss how to define an optimization problem in Section 4.3 in detail.

### 4.2.3 Graphical Optimization

In general optimization problems, it is difficult to see the relationship between objective and constraint functions with respect to design variables because the number of design variables is usually large. Therefore, visualizing objective and constraint functions in high dimensions is a challenging task. However, when the number of design variables is one or two, it is possible to plot the objective and constraint functions and find the optimum design graphically. Since graphical optimization plots all functions in the entire design space, it is expensive but helps to visualize the design space and to understand the nature of the design problem. The procedures of graphical optimization are as follows:

- Draw the design space (lower and upper bounds of design variables).
- Plot constraints on the graph and find the feasible set.
- Plot contour lines of the objective function.
- Find the optimum point (the objective function has the lowest value within the feasible set).

The following example illustrates the graphical optimization for finding Young's modulus in Example 4.1 using the constrained optimization problem in Eq. (4.2).

**Example 4.3 Graphical optimization**
Find the optimum Young's modulus of Example 4.1 using the constrained optimization problem in Eq. (4.2).

**Solution:**
The optimization problem in Eq. (4.2) is linear in the sense that both the objective function and constraint functions are linear functions of design variables. The three inequality constraints are shown in Figure 4.4, where three lines are constraint boundaries with hatching marking the region where the constraint is violated. This is a standard way of marking an inequality constraint. The feasible set $S$ is defined as the area that satisfies all constraints. Since the objective function is equal to $\alpha$, there is no need for objective function contours. Within the feasible set, the optimum design that minimizes $\alpha$ is the lowest point in the vertical axis, which is point A in the figure. At optimum design $E = 5/6$, and the maximum error $d_{\max} = 1/3$.

It is seen that there are two active constraints at the optimum design at point A; they are the second (solid line) and third (dashed line) constraints. It is easy to check that the error at the first constraint is half of that of the second constraint for any value of $E$, so that constraint is not binding. The optimum is found, therefore, at the intersection, where the differences between the model and the second and third constraints have the same magnitude and opposite signs.

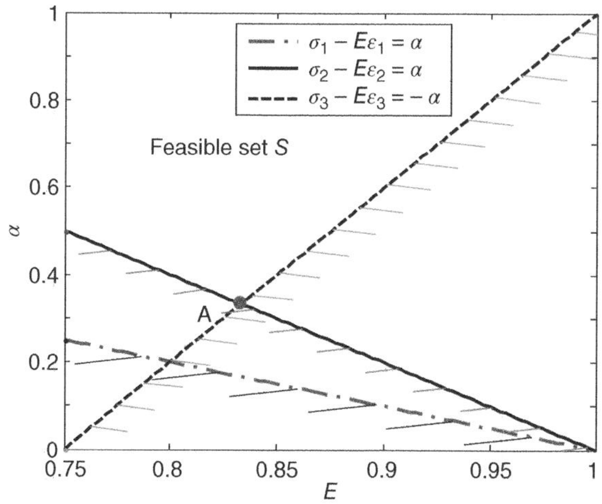


**Figure 4.4** Graphical optimization of finding Young's modulus.

Graphical optimization can provide additional benefits in understanding the optimization problem at hand. When the feasible domain is not well-defined, the optimization problem can have a problem. These problems are often difficult to identify in a high-dimensional design space. However, using graphical optimization, the design space along with all constraints can be visualized, from which possible problems in the feasible domain can be identified first. The possible problems that are related to the feasible domain are (i) problems with unbounded feasible regions, (ii) problems with multiple optima, (iii) problems with empty feasible regions, and (iv) problems with no active constraints. Figure 4.5 shows some examples of problems related to feasible regions. The blue lines are the contour of the objective function, whose values decrease in the direction of $\Delta f < 0$. The lines are the constraint bounds, where the design becomes infeasible on the hatched side. In Figure 4.5(a), the objective function is reduced as the design moves away from the constraint bounds. Therefore, in this problem, the optimum solution is unbounded, and the constraints do not play any role in the optimization problem. In Figure 4.5(b), two constraints

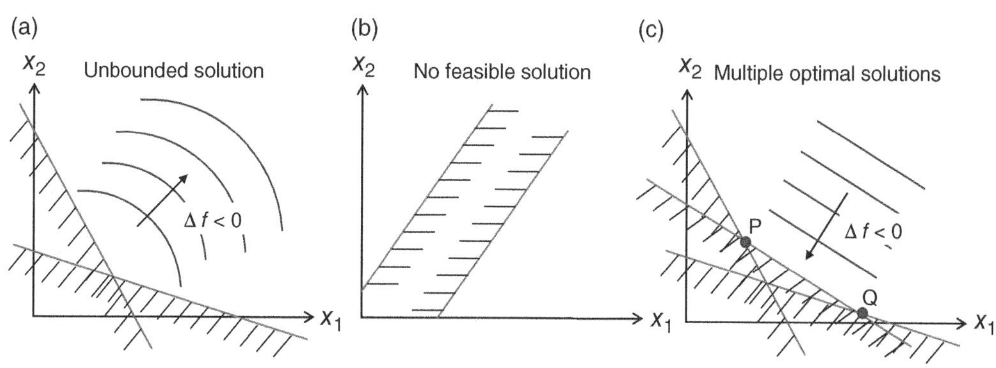


**Figure 4.5** Examples of problems with feasible domains. (a) Unbounded solution, (b) no feasible solution, and (c) multiple optimal solutions.

conflict with each other, and there is no feasible domain at all. In this case, there is no optimal solution exists. The last difficulty in Figure 4.5(c) is when the contour line of the objective function is parallel to the constraint bound. In this case, all designs along the line PQ can be optimum designs as the objective function has the same value along the line.

## 4.3 Optimization Problem Formulation

Optimization exists in almost all engineering applications. Therefore, instead of developing optimization procedures for different applications, it makes sense to develop a unified optimization procedure that can be used for any application. Such unification can be possible by defining the standard form of optimization problem formulation. We first present the three-step optimization problem definition, which defines the three ingredients of optimization problems. Next, the standard form of an optimization problem is presented. As long as an optimization problem can be written in the standard form, the subsequent optimization procedures can be applied independently of applications. Lastly, the convex optimization problem is presented, which is an important property of an optimization problem.

### 4.3.1 Three-step Problem Definition

Most engineering optimization problem formulation can be achieved in three steps: design parameterization, defining an objective function, and defining constraint functions. Design parameterization is to define/select design variables from system parameters. Although any parameter can be selected as a design variable, it is important to identify design variables clearly. In addition, as mentioned in the previous section, it is important that all design variables can maintain independence. In this text, the vector of design variables is defined as $\mathbf{x} = \{x_1, x_2, \ldots, x_n\}^T$, where $n$ is the dimension of design variables. In general, design variables can be discrete or continuous. However, in the context of surrogate modeling, since a surrogate model assumes that input variables are continuous and the QoI is a smooth function of input variables, design variables are considered continuous as well. In order to define the design space, it would be necessary to specify the lower and upper bounds of each design variable. These bounds are specified by side constraints $\mathbf{x}_L \leq \mathbf{x} \leq \mathbf{x}_U$, where $\mathbf{x}_L$ and $\mathbf{x}_U$ are, respectively, the lower and upper bounds of design variables.

An objective function must be a function of design variables. In the context of surrogate modeling, the objective function is a continuous and smooth function of design variables. In general, the objective function can be minimized or maximized. However, in the standard form of optimization, it is assumed that the objective function is minimized during the optimization process. When an objective function is supposed to be maximized, the negative objective function is minimized instead. That is,

$$\underset{\mathbf{x}}{\text{maximize}}\ F(\mathbf{x}) \quad \rightarrow \quad \underset{\mathbf{x}}{\text{minimize}}\ f(\mathbf{x}) = -F(\mathbf{x})$$

Constraint functions can be given in the form of equality or inequality constraints. The functional requirements of constraint functions are identical to those of the objective function: dependence on design, smoothness, and continuity. The standard form of inequality constraint is less than or equal to zero. That is, $g(\mathbf{x}) \leq 0$. However, many engineering applications have constraints in the form that the constraint function is less than (or larger than) a threshold. In such a case, the constraint can be converted into

$$g(x) \le g_{\max} \rightarrow g(x) - g_{\max} \le 0$$
$$g(x) \ge g_{\min} \rightarrow g_{\min} - g(x) \le 0$$

The standard form of equality constraint is $h(\mathbf{x}) = 0$. If the right-hand side is not zero, a similar conversion as in the inequality constraint can be done. Although there is no limit to the number of inequality constraints, the number of equality constraints must be less than the number of design variables.

### 4.3.2 Standard Form

From the viewpoint of algorithm development, it would be convenient if all optimization problems were written in a single form. Then, the developers do not need to consider solving different optimization problems. To facilitate the description of algorithms for solving optimization problems, there is a fairly standard notation for writing them. The letter $f$ is typically used for objective functions, $g$ for inequality constraints, and $h$ for equality constraints. Lower and upper bounds on design variables are often called side constraints and are written separately in different forms. The standard form of an optimization problem in this text is defined as

$$\begin{aligned} &\text{minimize } f(\mathbf{x}) \\ &\text{subject to } g_i(\mathbf{x}) \le 0, \quad i = 1, \dots, K \\ &\qquad\qquad h_j(\mathbf{x}) = 0, \quad j = 1, \dots, M \\ &x_l^L \le x_l \le x_l^U, \quad l = 1, \dots, n \end{aligned} \tag{4.3}$$

where $\mathbf{x} = \{x_1, x_2, \dots, x_n\}^T$ is the vector of design variables, $f(\mathbf{x})$ is the objective function, $g_1(\mathbf{x}), \dots, g_K(\mathbf{x})$ are inequality constraints, $h_1(\mathbf{x}), \dots, h_M(\mathbf{x})$ are equality constraints, and $\mathbf{x}^L$ and $\mathbf{x}^U$ are lower and upper bounds of design variables, respectively. Of course, not all optimization problems can be written in standard form. For a more general form of optimization problems, the readers are referred to the paper published by Svanberg [41]. Once the inequality and equality constraints are all defined, the feasible set in Eq. (4.1) can be written as

$$S = \{\mathbf{x} \mid g_i(\mathbf{x}) \le 0, \quad i = 1, \dots, K; \quad h_j(\mathbf{x}) = 0, \quad j = 1, \dots, M\} \tag{4.4}$$

Therefore, the optimization problem is to find the optimum design variable $\mathbf{x}^* \in S$ that minimizes the objective function $f(\mathbf{x}^*)$.

When an inequality constraint is strictly less than zero, i.e., $g_i(\mathbf{x}) < 0$, the constraint is inactive. When the constraint is on the bound, $g_i(\mathbf{x}) = 0$, it is active. Lastly, the constraint is violated when $g_i(\mathbf{x}) > 0$. On the other hand, an equality constraint can be either active, $h_j(\mathbf{x}) = 0$, or violated, $h_j(\mathbf{x}) \neq 0$. In general, when all constraints are linearly independent, the total number of active constraints should be less than or equal to the number of design variables. This can be understood from the basic law of linear algebra, where the number of independent equations cannot be more than the number of unknown variables.

Optimization problems can be categorized depending on the functional form of the objective and constraint functions. First, when both the objective function and all constraint functions are linear functions of design variables, it is called a linear programming (LP) problem. In such a case, many specialized numerical algorithms are available. A quadratic programming (QP) problem occurs when the objective function is a quadratic function of design variables, while all constraints are linear. As we will discuss in the next subsection, the global optimum design can be found when the optimization problem is LP or QP. When both objective function and constraint functions are nonlinear functions of design variables, it is called a nonlinear programming (NLP) problem, which is the most general but at the same time most challenging to find the optimum design.

### 4.3.3 Normalization

Even if an optimization problem is well defined in the standard form, it may cause some numerical difficulties to solve. This is because engineering optimization may cover a broad range of quantities of interest. Different objectives and constraints may have different orders of magnitude. For example, the allowable strength of steel is about 500 MPa, which is $5 \times 10^8$ Pa. On the other hand, the allowable displacement can be in the order of $10^{-3}$ m. When the objective and constraint functions have a huge difference in magnitude, the standard form has no problem from the mathematical viewpoint, but it is numerically difficult to handle such a huge difference in the orders of magnitude. Therefore, it makes sense to normalize the objective and constraints such that their magnitude is approximately in the order of one. For example, stress constraint is often given in the form of $\sigma_{\max} \leq \sigma_{\text{allowable}}$. Then, it is possible to normalize the constraint using $\sigma_{\text{allowable}}$ and make it in the standard form as

$$\frac{\sigma_{\max}}{\sigma_{\text{allowable}}} - 1 \leq 0 \tag{4.5}$$

Since all constraints have their limits, they can be normalized using their limits. In the case of the objective function, it can be normalized using its initial value or using the target value.

**Example 4.4 Standard form**

Write the standard form of the optimization problem in Example 4.3.

**Solution:**

The optimization problem formulation given in Eq. (4.2) is not the standard form. Without normalizing the objective and constraints, the standard form of the optimization problem can be written as

$$\begin{aligned}
&\underset{\alpha, E}{\text{minimize}} \ \alpha \\
&\text{subject to } \sigma_1 - E\epsilon_1 - \alpha \leq 0 \\
&\qquad \sigma_2 - E\epsilon_2 - \alpha \leq 0 \\
&\qquad -\sigma_3 + E\epsilon_3 - \alpha \leq 0 \\
&\qquad 0.5 \leq E \leq 1
\end{aligned}$$

Note that the side constraint is added to limit Young's modulus. Note that the normalization of this optimization problem can be tricky. Since the limit of inequality constraints, $\alpha$, is not fixed but a design variable itself, therefore, in this case, it would be better not to normalize the optimization problem.

**Example 4.5 Optimization problem formulation of a tube under compressive load**

A tubular column is under the compressive load $P$ as shown in Figure 4.6. The height of the tube is fixed, and the mid radius $R$ and thickness $t$ are design variables. The goal is to minimize the weight of the tube, which is equivalent to the cross-sectional area of the tube. The following three failure modes are considered constraints. (i) Stress failure: the axial compressive stress should be less than the allowable strength $\sigma_a$. (ii) Global buckling: the tube should not buckle for the given applied load $P$. (iii) Local buckling: the tube should not wrinkle into a diamond pattern, as shown in Figure 4.6. The design variables have lower and upper bounds. Write the standard form of the optimization problem.

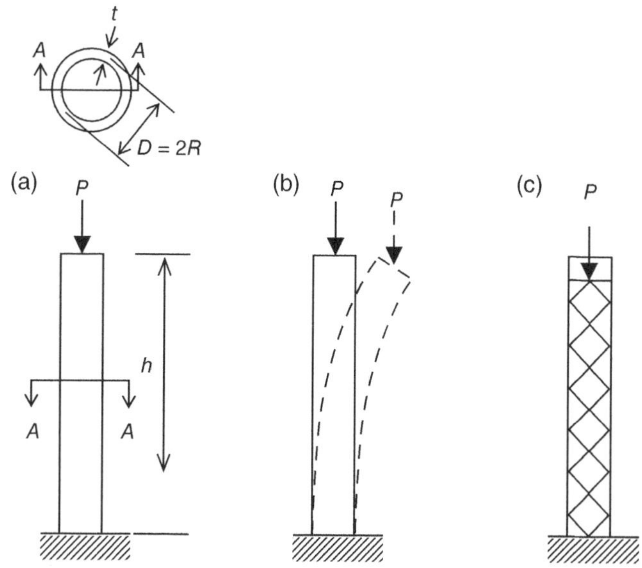


**Figure 4.6** A tube under compressive load. (a) Column, (b) global buckling, and (c) local buckling.

**Solution:**

Assuming that the radius is much larger than the thickness, $R \gg t$, the cross-sectional area and the moment of inertia can be approximated as $A = 2\pi Rt$ and $I = \pi R^3 t$. Then, the objective function can be defined using the density of the material and the volume of the tube as

$$\text{mass} = \rho(hA) = 2\pi\rho hRt$$

The equations for the three failure modes are given as

$$\text{Stress constraint: } \sigma = \frac{P}{2\pi Rt} \leq \sigma_a$$

$$\text{Buckling load: } P_{cr} = \frac{\pi^3 ER^3 t}{4h^2} \geq P$$

$$\text{Local buckling: } \sigma_s = \frac{2Et}{2R\sqrt{3(1-\nu^2)}} \leq \sigma_a$$

$$\text{Side constraints: } R_{\min} \leq R \leq R_{\max}, \quad t_{\min} \leq t \leq t_{\max}$$

Each failure mode provides for different importance of the diameter and the thickness. The stress depends only on the area, which is also the objective function, and the area depends only on the product of the thickness and the diameter. This means that with a stress constraint only, any combination of the diameter and thickness that is at the stress limit $\sigma_a$ will have the same weight.

The global buckling load depends on the moment of inertia $I$ that depends on $D$ more strongly than on $t$. Therefore, it will drive the design toward a thin tube. Local buckling pushes in the other direction, in that it benefits from large $t$ and small $D$. This indicates that if buckling is critical (happens if the column is slender enough, i.e., $h$ is large), both failure modes will happen simultaneously.

The standard formulation requires us to write the failure constraints so that all are of the form $g(D, t) \leq 0$. Therefore, the standard form of the optimization problem can be written as

$$\underset{R,t}{\text{minimize}}\ 2\pi\rho hRt$$

$$\text{subject to } g_1(R,t) = \frac{\sigma}{\sigma_a} - 1 \leq 0$$

$$g_2(R,t) = 1 - \frac{P}{P_{cr}} \leq 0$$

$$g_3(R,t) = \frac{\sigma_S}{\sigma_a} - 1 \leq 0$$

$$R_{\min} \leq R \leq R_{\max}, \quad t_{\min} \leq t \leq t_{\max}$$

Also, all constraints are normalized by their limits. This provides a handy measure of constraint satisfaction. For example, if the constraint is equal to −0.1, it tells us that we have a 10% margin between the response and the allowable response. If the constraint is equal to 0.05, it means we exceed the allowable by 5%. This normalized and nondimensional formulation of constraints is also typically better for the numerical performance of optimization algorithms. That is, using an optimization routine, it is likely to lead to faster convergence and increase the chance that we will find the true optimum.

**Example 4.6 Optimization of beer can design**

A beer company wants to design a new can so that the minimum amount of sheet metal can be used. This is equivalent to minimizing the manufacturing cost. The can is required to hold at least 400 ml of fluid. The diameter of the can should be no more than 8 cm. In addition, it should not be less than 3.5 cm for shipping and handling reasons. The height of the can should be no more than 18 cm and no less than 8 cm. Write the standard form of the optimization problem and solve it using the graphical optimization method.

**Solution:**

As shown in Figure 4.7, the design variables are the diameter $D$ and the height $H$. The amount of sheet metal is the surface area of the can, which is $\pi DH + \pi D^2/2$. The volume of the can is defined in terms of design variables as $\pi D^2 H/4$. Therefore, the standard form of the optimization problem can be written as

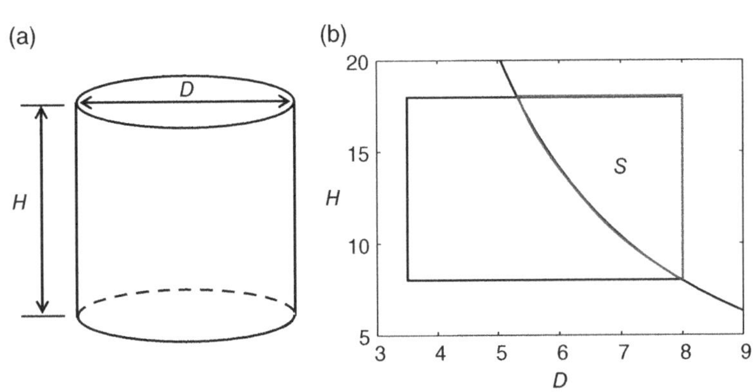


**Figure 4.7** Beer can design parametrization and feasible set. (a) Design parametrization and (b) feasible set.

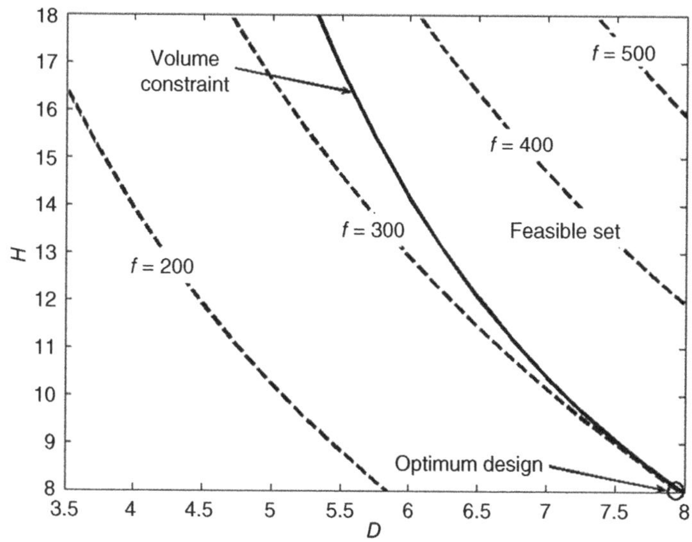


**Figure 4.8** Graphical optimization of a beer can design problem.

$$\underset{D,H}{\text{minimize}} \ \pi DH + \frac{\pi D^2}{2}$$

$$\text{subject to } 400 - \frac{\pi D^2 H}{4} \leq 0 \, \text{cm}^3$$

$$3.5 \leq D \leq 8 \, \text{cm}$$

$$8 \leq H \leq 18 \, \text{cm}$$

Note that the inequality constraint is not normalized. Figure 4.7(b) shows the feasible set that satisfies the inequality constraint.

The objective function is a quadratic function of $D$ and a linear function of $H$. As shown in Figure 4.8, the objective function gradually increases as both design variables increase. However, there is no feasible region for $f = 200$. When $f = 300$, the objective contour meets a corner of the feasible set for the first time. This point $(D, H) = (8, 8)$ is indeed the optimum design. Therefore, the diameter is at its upper bound, while the height is at its lower bound.

### 4.3.4 Convex Function and Convex Problem

The optimization problem formulation defined in the previous subsection is general enough that it can be applied to many engineering applications. However, it does not guarantee that there is a single optimum design, especially when the objective and constraints are a nonlinear function of design variables. In fact, an important challenge in numerical optimization is to find the global optimum design. As we will discuss in the next section, most optimization algorithms are designed to find a local optimum, not the global one. However, if a function is convex, it has a single minimum, and the local minimum is the global minimum. Therefore, the concept of a convex function and a convex problem is important, as it can provide a nice theoretical foundation to determine the uniqueness of the global optimum design.

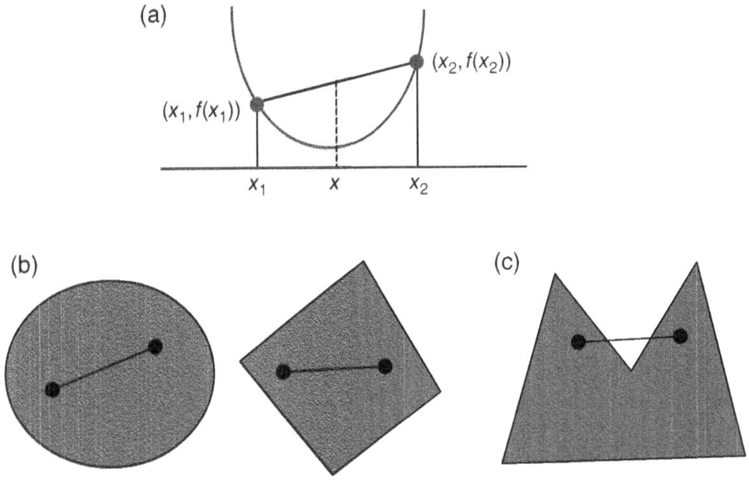


**Figure 4.9** Convex function and convex set. (a) Convex function, (b) convex set, and (c) nonconvex set.

A function is convex if a straight line connecting two points on its graph will not dip below the function values in between. That is, the straight line that connects two points on the function is always larger than the function itself. As shown in Figure 4.9(a), for $0 \leq \alpha \leq 1$, $x = \alpha x^{(2)} + (1 - \alpha)x^{(1)}$ is a point between two points $x^{(1)}$ and $x^{(2)}$. Then, a function is convex if it satisfies the following condition for all $x$:

$$f(x) \leq \alpha f\left(x^{(2)}\right) + (1 - \alpha) f\left(x^{(1)}\right) \tag{4.6}$$

The left-hand side is the function value at $x$, while the right-hand side is the straight line that connects $f(x^{(1)})$ and $f(x^{(2)})$. This can apply to $n$-dimensional function rather than just to a function of a single variable, as shown in the figure. The weighted sum of all the quantities, where the sum of the weights is equal to one and all the weights are positive, is called a convex combination. Therefore, the equation above says that the value of the function at a convex combination of two points cannot be larger than the convex combination with the same weights of the function values at these two points.

Although the convexity of a function is defined using Eq. (4.6), it is not a convenient form to show convexity because it is impractical to show it for all arbitrary combinations of $x^{(1)}$ and $x^{(2)}$. In practice, convexity is shown using Hessian information. In a multidimensional function, let the vector of input variables be defined as $\mathbf{x} = \{x_1, x_2, \ldots, x_n\}^T$. Then, the gradients (first-order derivatives) of the function $f(\mathbf{x})$ can be defined as

$$\nabla f(\mathbf{x}) = \left\{ \frac{\partial f}{\partial x_1}, \frac{\partial f}{\partial x_2}, \ldots, \frac{\partial f}{\partial x_n} \right\}^T \tag{4.7}$$

which is an $n$-dimensional vector. The gradients are the slope of the function in the direction of each variable. The second derivatives of function $f(\mathbf{x})$ are called the Hessian matrix, which is defined as

$$\mathbf{H} \equiv \nabla^2 f(\mathbf{x}) = \begin{bmatrix} \dfrac{\partial^2 f}{\partial x_1^2} & \cdots & \dfrac{\partial^2 f}{\partial x_1 \partial x_n} \\ \vdots & \ddots & \vdots \\ \dfrac{\partial^2 f}{\partial x_1 \partial x_n} & \cdots & \dfrac{\partial^2 f}{\partial x_n^2} \end{bmatrix} \tag{4.8}$$

The Hessian matrix is symmetric. The Hessian matrix is related to the curvature of the function.

A symmetric matrix $\mathbf{H}$ is positive definite if $\mathbf{x}^T\mathbf{H}\mathbf{x} > 0$ for all $\mathbf{x} \neq 0$. A symmetric matrix is positive semi-definite if $\mathbf{x}^T\mathbf{H}\mathbf{x} \geq 0$ for all $\mathbf{x} \neq 0$. This property is important for the Hessian matrix. If the Hessian matrix $\nabla^2 f$ is positive semi-definite, then the original function $f(\mathbf{x})$ is convex. In addition, if the Hessian matrix is positive definite, then the function $f(\mathbf{x})$ is strictly convex. That is, the function has a single minimum, which is the global minimum.

In order to apply the above concept of convex function to an optimization problem, it is necessary to adopt the concept of a convex set as well because an optimization problem is to minimize the objective function within the feasible set. A feasible set is convex if for all $\mathbf{x}^{(1)}$ and $\mathbf{x}^{(2)}$ in the feasible set $S$, the convex combination $\mathbf{x} = \alpha\mathbf{x}^{(2)} + (1 - \alpha)\mathbf{x}^{(1)}$ also belongs to $S$. This will happen if all the inequality constraints are convex, and all the equality constraints are linear. If an equality constraint is nonlinear, then it is curved, and if we connect two points on the line by a straight segment, the interior of the segment will not satisfy the equality constraint, and this will violate the convexity requirement. Figure 4.9(b) shows examples of a convex set, while Figure 4.9(c) shows an example of a nonconvex set. When the feasible set is convex and the objective function is convex, the optimization problem is called a convex problem. For a convex problem, a local minimum is also a global minimum.

## 4.4 Optimality Criteria

Once an optimization problem is formulated, it is often solved using numerical optimization techniques, which will be discussed in Chapters 5 and 6. If the graph of the objective function over the entire feasible set is known, it would be possible to find the optimum design graphically. However, this information is barely available for most applications, especially for high-dimensional design problems. Therefore, most optimization algorithms try to find the optimum design by moving from one design to another iteratively. In such an iterative process, it would be necessary to determine if a new design is optimum or not. Optimality criteria are conditions that an optimum design must satisfy. Optimality criteria are considered one of the most important concepts in design optimization. Therefore, it would be important to fully understand them.

### 4.4.1 Global Versus Local Optimum

Before discussing optimality criteria, it would be necessary to understand the local and global optima. The global optimum is the design that has the lowest value of the objective function within the feasible set. The goal of an optimization problem is to find the global optimum. The issue related to the global optimum is that it is difficult to find it. For most engineering applications, the functional form of the objective function is usually unknown. Optimization algorithms iteratively improve the current design by moving to a new design that has a lower objective function. If there is no way to lower the objective function from the current design, this design should satisfy optimality criteria. However, even if the current design is the optimum design in the immediate neighborhood, it does not mean that it is the best design in the entire feasible set. Figure 4.10 shows an illustration of an objective function as a function of a design variable. The circles with a light color represent local optima, which are the best designs in the immediate neighborhood. The circle with a dark color is indeed the global optimum design, which has the lowest value of the objective function in the feasible set. If an optimization algorithm starts from different designs, it is possible that it may find different local optima or a global optimum. The challenge is that it is difficult to determine if the optimum design that is found by an optimization algorithm is a local or global optimum.

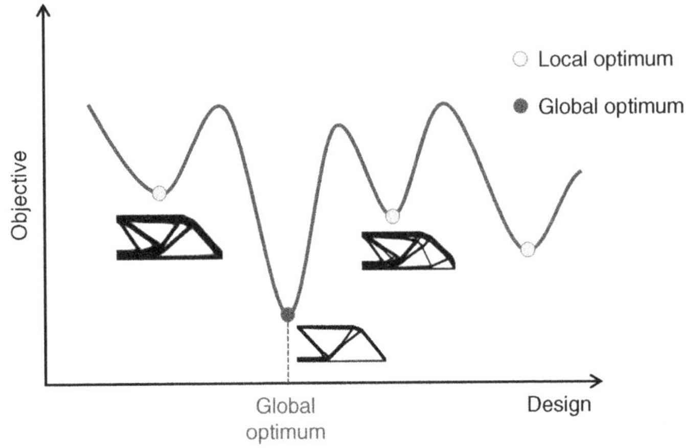


**Figure 4.10** Global versus local optima.

Mathematically, a design $\mathbf{x}^*$ is called a global optimum for $f(\mathbf{x})$ if

$$f(\mathbf{x}^*) \leq f(\mathbf{x}), \quad \text{for all } \mathbf{x} \in S \tag{4.9}$$

where $S$ is the feasible set that satisfies all constraints. The practical difficulty associated with the global optimum is that there is no mathematical method to find it. However, the existence of the global optimum can be provided by the Weierstrass theorem or the extreme value theorem [42]: If $f(\mathbf{x})$ is continuous and the feasible set $S$ is closed and bounded, then there is a global minimum. The unbounded feasible set was graphically illustrated in Figure 4.5(a). A bounded set means that the area (or volume in 3D or hyper-volume in high dimension) that is covered by the feasible set is finite. The closed set means that the inequality constraints are in the form of less-than-or-equal-to ($\leq$) so that the constraint boundary is included in the feasible set.

A design $\mathbf{x}^*$ is called a local optimum for $f(\mathbf{x})$ if

$$f(\mathbf{x}^*) \leq f(\mathbf{x}), \quad \text{for all } \mathbf{x} \in S \text{ in a small neighborhood of } \mathbf{x}^* \tag{4.10}$$

In Figure 4.10, all circles with a light color are local optima, where they are the lowest points in a small neighborhood. Since it is impractical to search the entire feasible set, almost all numerical optimization algorithms can only guarantee a local optimum, not the global one.

### 4.4.2 Unconstrained Optimization

In order to develop the optimality criteria, the local and global optimal shown in Figure 4.10 give a clue. Both the local and global optima have their extreme value, where the slope becomes zero. That is, at optimum, the function may satisfy $df(x)/dx = 0$; that is, the design is at a stationary point or an extreme value. However, this cannot be generalized for arbitrary functions. Figure 4.11(a) shows two functions $f_1(x) = |x - 5|$ and $f_2(x) = 0.2(x - 5)^2$. Both functions have the global optimum at $x = 5$. However, $df_1(x)/dx \neq 0$ at $x = 5$. In fact, $df_1(x)/dx$ cannot be defined at $x = 5$. This is because $f_1(x)$ is not a smooth function of $x$ and nondifferentiable. Therefore, optimality criteria assume that the function is smooth with respect to design variables.

Even if a function is smooth, the condition of $df(x)/dx = 0$ is not enough to determine if a design $x$ is optimum or not. Figure 4.11(b) shows three functions. All three functions have $df(x)/dx = 0$ at $x = 5$. Among the three, however, only $f(x) = (x - 5)^2$ has a minimum at $x = 5$. The other function

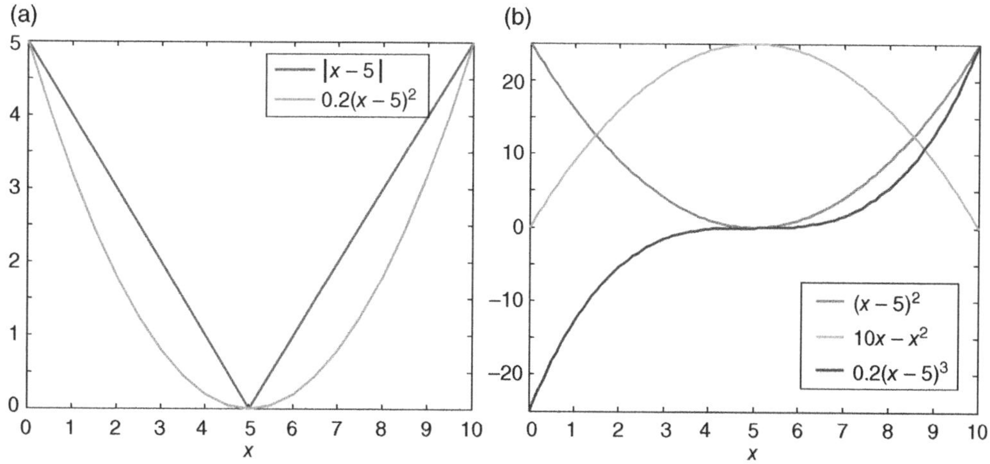


**Figure 4.11** Optimum design of a smooth and nonsmooth function. (a) Two functions have the same global minimum and (b) three functions have the same slope.

$f(x) = 10x - x^2$ has its maximum, and $f(x) = 0.2(x - 5)^3$ has no minimum or maximum at $x = 5$. In fact, this point is an inflection point. In optimization theory, the requirement of $df(x)/dx = 0$ is called a necessary condition. If a design is optimum, it must satisfy the necessary condition. However, not all designs that satisfy the necessary condition are optimum.

Consider the following unconstrained optimization problem with a single variable:

$$\underset{x}{\text{minimize}}\, f(x) \tag{4.11}$$

If $x^*$ is a local minimum, for an arbitrary neighboring point $x = x^* + \Delta x$, the function must satisfy $f(x) \geq f(x^*)$. Using the Taylor series expansion, the function $f(x)$ can be expanded with respect to $f(x^*)$ as

$$f(x) = f(x^* + \Delta x) = f(x^*) + f'(x^*)\Delta x + \frac{1}{2}f''(x^*)\Delta x^2 + \text{H.O.T.}$$

where H. O. T. represents high-order terms, and $f'$ and $f''$ are, respectively, the first- and second-order derivatives. If $f(x^*)$ is a local minimum, then the following functional difference must be non-negative:

$$\Delta f = f(x) - f(x^*) = f'(x^*)\Delta x + \frac{1}{2}f''(x^*)\Delta x^2 + \text{H.O.T.} \tag{4.12}$$

For $x^*$ to be a local minimum, it would be necessary that $\Delta f \geq 0$. Assuming $\Delta x$ is small; the term including $\Delta x^2$ and higher-order terms can be ignored. Therefore, $\Delta f \approx f'(x^*)\Delta x \geq 0$ for arbitrary $\Delta x$. The only possibility that this inequality satisfies for arbitrary $\Delta x$ is that its coefficient $f'(x^*)$ must vanish. Otherwise, $\Delta f$ can be negative if $-\Delta x$ is used instead. Therefore, from this argument, it is possible to obtain the following first-order necessary condition of optimum design:

$$f'(x^*) \equiv \frac{df(x^*)}{dx} = 0 \tag{4.13}$$

This seemingly obvious condition is called Karush–Kuhn–Tucker (KKT) condition [43, 44], which is the most popular condition in optimum design.

Once $f'(x^*) = 0$ is satisfied, Eq. (4.12) becomes

$$\Delta f = \frac{1}{2} f''(x^*)\Delta x^2 + \text{H.O.T.} \geq 0 \tag{4.14}$$

Since $\Delta x^2 > 0$ always, the second-order necessary condition would be

$$f''(x^*) \equiv \frac{d^2 f(x^*)}{dx^2} \geq 0 \tag{4.15}$$

Therefore, if a design $x^*$ is a local optimum, Eqs. (4.13) and (4.15) must satisfy. That is, they are first- and second-order necessary conditions. On the other hand, the sufficient condition is given as

$$f'(x^*) = 0 \text{ and } f''(x^*) > 0 \tag{4.16}$$

The positiveness of the second-order derivative is an important property of a function. This is indeed the Hessian information in Eq. (4.8). The positive second-order derivative is equivalent to the positive definite matrix for a multivariable function. Therefore, it is related to the convexity of a function. The curvature of a function is related to the second-order derivative. That is, the curvature $\kappa(x)$ can be defined by $\kappa(x) = f''(x)/(1 + f'^2)^{1.5}$. Geometrically, this makes sense as a function with a positive curvature has a minimum point.

**Example 4.7 Stationary points**

Consider a polynomial function $f(x) = x^3 - 3x$ in the design space $x \in [-2, 2]$. Classify the stationary points of the function from the optimality criteria and check by plotting them.

**Solution:**

Since optimum designs must satisfy the KKT condition, it makes sense to find $x$ where the gradient vanishes.

$$f'(x) = 3x^2 - 3 = 0 \quad \rightarrow \quad x = -1 \text{ or } x = 1$$

That means both $x = -1$ and $x = 1$ are candidate points for optima. At $x = -1$, the second-order necessary condition $f''(x) = 6x = -6 < 0$ does not satisfy. In fact, this is the maximum point A, as shown in Figure 4.12. The function has a maximum value $f(-1) = 2$ at point A. At $x = 1, f''(x) = 6x = 6 > 0$; that is, the second-order necessary condition and sufficient condition are satisfied. Indeed, this is the minimum point B, as shown in Figure 4.12.

In this particular problem, it is interesting to note that point C in Figure 4.12 is also a minimum point, while point D is a maximum point. However, these points do not satisfy both the necessary and sufficient conditions. This is because they are on the constraint bound. The optimality conditions with constraints will be discussed in the next subsection.

The optimality criteria for a single-variable optimization problem can be extended to multivariable optimization problems. Let the multidimensional design variables be defined as $\mathbf{x} = \{x_1, x_2, ..., x_n\}^T$. Let $\mathbf{x}^*$ be the optimum design and $\mathbf{x} = \mathbf{x}^* + \Delta\mathbf{x}$ be an arbitrary neighborhood. Then the Taylor series can be used to expand the objective function $f(\mathbf{x})$ with respect to $f(\mathbf{x}^*)$ as

$$\begin{aligned} f(\mathbf{x}) &= f(\mathbf{x}^*) + \sum_{i=1}^{n} \frac{\partial f(\mathbf{x}^*)}{\partial x_i}\Delta x_i + \frac{1}{2}\sum_{i=1}^{n}\sum_{j=1}^{n} \frac{\partial^2 f(\mathbf{x}^*)}{\partial x_i \partial x_j}\Delta x_i \Delta x_j + \text{H.O.T.} \\ &= f(\mathbf{x}^*) + \Delta\mathbf{x}^T \nabla f(\mathbf{x}^*) + \frac{1}{2}\Delta\mathbf{x}^T \mathbf{H}(\mathbf{x}^*)\Delta\mathbf{x} + \text{H.O.T.} \end{aligned} \tag{4.17}$$

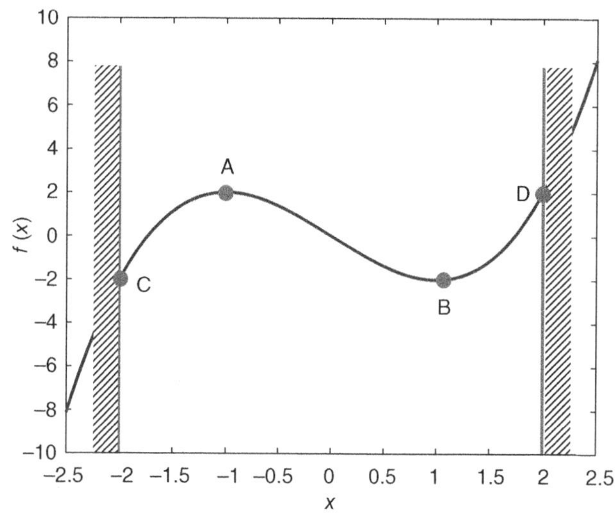


**Figure 4.12** Optimum points of a cubic function with design space $x \in [-2, 2]$.

where $\nabla f(\mathbf{x}^*)$ is the vector of gradients in Eq. (4.7) and $\mathbf{H}(\mathbf{x}) = \nabla^2 f(\mathbf{x})$ is the Hessian matrix defined in Eq. (4.8).

In order to satisfy $\Delta f = f(\mathbf{x}) - f(\mathbf{x}^*) \geq 0$ for arbitrary $\mathbf{x}$ in the neighborhood of $\mathbf{x}^*$, it would be necessary that $\nabla f(\mathbf{x}^*) = 0$. If $\partial f/\partial x_i \neq 0$, it is possible to choose $\left(x_i - x_i^*\right)$ in the opposite direction and other $\left(x_j - x_j^*\right) = 0$ to make $\Delta f < 0$. Therefore, in order to satisfy $f(\mathbf{x}^*)$ being the minimum, it would be necessary that $f(\mathbf{x}^*)$ needs to be stationary; that is, $\nabla f(\mathbf{x}^*) = 0$.

Now with $\nabla f(\mathbf{x}^*) = 0$, the change of the objective function becomes $\Delta f = \frac{1}{2}\Delta\mathbf{x}^T\mathbf{H}(\mathbf{x}^*)\Delta\mathbf{x} + \text{H.O.T.}$ Therefore, the sufficient condition for $f(\mathbf{x}^*)$ being the minimum is that

$$\Delta\mathbf{x}^T\mathbf{H}(\mathbf{x}^*)\Delta\mathbf{x} > 0 \text{ for all } \Delta\mathbf{x} \neq 0 \tag{4.18}$$

This means that the matrix of second-order derivatives (Hessian) is positive definite. In the same argument in Eq. (4.15), the second-order necessary condition would satisfy if the Hessian matrix is positive semi-definite. It would be impractical to check the positive definiteness for all $\Delta\mathbf{x} \neq 0$. Instead, the simplest way to check positive definiteness is to check if all eigenvalues of the Hessian matrix are positive. On the other hand, if the Hessian matrix is positive semi-definite, then it is the second-order necessary condition. The eigenvalues of the positive semi-definite matrix are non-negative.

In summary, the following optimality criteria can be summarized for multivariable optimization problems:

$$\text{KKT condition} \quad \nabla f(\mathbf{x}^*) = 0 \tag{4.19}$$

$$\text{Second} - \text{order necessary condition} \quad \Delta\mathbf{x}^T\mathbf{H}(\mathbf{x}^*)\Delta\mathbf{x} \geq 0 \text{ for all } \Delta\mathbf{x} \neq 0 \tag{4.20}$$

$$\text{Sufficient condition} \quad \Delta\mathbf{x}^T\mathbf{H}(\mathbf{x}^*)\Delta\mathbf{x} > 0 \text{ for all } \Delta\mathbf{x} \neq 0 \tag{4.21}$$

The positive definite matrix in Eq. (4.21) is indeed a quadratic form defined as $q = \Delta\mathbf{x}^T\mathbf{H}(\mathbf{x}^*)\Delta\mathbf{x}$. Assuming that the KKT condition is satisfied, the minimum or maximum of the function can be determined by the property of the quadratic form. The following table summarizes them:

| Quadratic form | Hessian matrix | Extreme |
|---|---|---|
| $q > 0$ | Positive definite | Minimum |
| $q \geq 0$ | Positive semi-definite | Possibly minimum |
| $q < 0$ | Negative definite | Maximum |
| $q \leq 0$ | Negative semi-definite | Possibly maximum |
| $q \geq 0$ or $q \leq 0$ | Indefinite | Saddle point |

**Example 4.8 Stationary points**

Find a stationary point of the following objective functions and determine if the stationary point is minimum, maximum, or saddle point: (a) $f_1(x_1,x_2) = x_1^2 + x_1x_2 + x_2^2$ and (b) $f_2(x_1,x_2) = x_1^2 + 3x_1x_2 + x_2^2$.

**Solution:**

a) For $f_1(x_1,x_2) = x_1^2 + x_1x_2 + x_2^2$, the stationary points can be found using the KKT condition as

$$\nabla f_1 = \begin{Bmatrix} 2x_1 + x_2 \\ x_1 + 2x_2 \end{Bmatrix} = 0$$

By solving the system of equations, the stationary point turns out to be $\{x_1, x_2\} = \{0, 0\}$. At this stationary point, the Hessian matrix becomes

$$\mathbf{H} = \begin{bmatrix} 2 & 1 \\ 1 & 2 \end{bmatrix}$$

The eigenvalue of the Hessian matrix can be obtained by solving $|\mathbf{H} - \lambda\mathbf{I}| = 0$, where $\mathbf{I}$ is the identity matrix. The eigenvalues that make the determinant zero are $\lambda_1 = 1$ and $\lambda_2 = 3$. Since both eigenvalues are positive, the Hessian matrix is positive definite, and the stationary point $\{x_1, x_2\} = \{0, 0\}$ satisfies sufficient condition, i.e., it is a minimum point. Figure 4.13(a) shows the plot of $f_1$, which confirms that it is the global minimum.

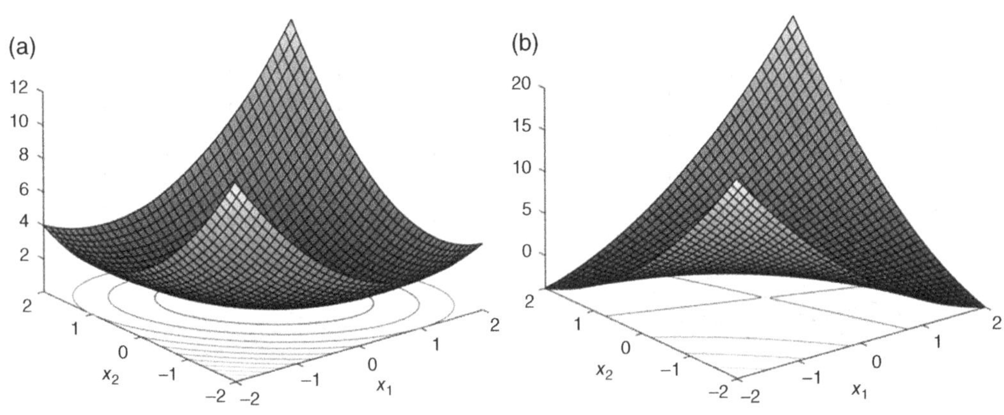


Figure 4.13 Surface plots of two functions in Example 4.8. (a) $f_1(x_1,x_2) = x_1^2 + x_1x_2 + x_2^2$ and (b) $f_2(x_1,x_2) = x_1^2 + 3x_1x_2 + x_2^2$.

b) For $f_2(x_1, x_2) = x_1^2 + 3x_1x_2 + x_2^2$, the stationary points can be found using the KKT condition as

$$\nabla f_2 = \begin{Bmatrix} 2x_1 + 3x_2 \\ 3x_1 + 2x_2 \end{Bmatrix} = 0$$

By solving the system of equations, the stationary point turns out to be $\{x_1, x_2\} = \{0, 0\}$. At this stationary point, the Hessian matrix becomes

$$\mathbf{H} = \begin{bmatrix} 2 & 3 \\ 3 & 2 \end{bmatrix}$$

The eigenvalues that make the determinant zero are $\lambda_1 = -1$ and $\lambda_2 = 5$. Since one eigenvalue is negative, while the other is positive, this stationary point is not a minimum or a maximum. In fact, this point is a saddle point as shown in Figure 4.13(b). It is interesting to note that the saddle point is a minimum along the line $x_1 = x_2$, while it is a maximum along the line $x_1 = -x_2$.

### 4.4.3 Constrained Optimization

The necessary and sufficient conditions in the previous section are purely determined by the objective function because there was no constraint. However, as shown in Example 4.7, the two points C and D do not satisfy the optimality criteria, but they were a minimum and a maximum point because of constraints. Therefore, it is obvious that constraints play an important role in optimality criteria. Before we formally develop optimality criteria for a constrained optimization problem, it would be beneficial to investigate the effect of constraints for optimum design. The following examples can provide useful insights into constrained optimization.

**Example 4.9 Effect of constraint on optimum design**
First, let us consider the following optimization problem with an inequality constraint:

$$\begin{aligned} &\text{minimize } f(\mathbf{x}) = (x_1 - 1)^2 + (x_2 - 1)^2 \\ &\text{subject to } g(\mathbf{x}) = x_1 + x_2 - 4 \le 0 \end{aligned}$$

The stationary point for the objective function can be obtained from

$$\nabla f = \begin{Bmatrix} 2(x_1 - 1) \\ 2(x_2 - 1) \end{Bmatrix} = 0$$

which can be solved for $\{x_1, x_2\} = \{1, 1\}$. The Hessian matrix at this point is $\mathbf{H} = 2\mathbf{I}$, which is positive definite. Also, at this point, the inequality constraint $g(\mathbf{x}) = x_1 + x_2 - 4 = -2 < 0$. Since this point satisfies the inequality constraint, this point is an optimum point. As shown in Figure 4.14(a), the objective function has its minimum $f^* = 0$ at $\{x_1, x_2\} = \{1, 1\}$, and it belongs to the feasible domain. In this case, the constraint does not play any role in optimum design. In general, when an optimum design is inside of the feasible domain, the constraint is inactive and does not play any role in optimum design.

As a second example, let us consider a different objective function but with the same inequality constraint:

$$\begin{aligned} &\text{minimize } f(\mathbf{x}) = (x_1 - 3)^2 + (x_2 - 3)^2 \\ &\text{subject to } g(\mathbf{x}) = x_1 + x_2 - 4 \le 0 \end{aligned}$$

In this case, the stationary point of the objective function is at $\{x_1, x_2\} = \{3, 3\}$, and the Hessian is also positive definite. However, the inequality constraint $g(\mathbf{x}) = x_1 + x_2 - 4 = 2 > 0$ is violated at this

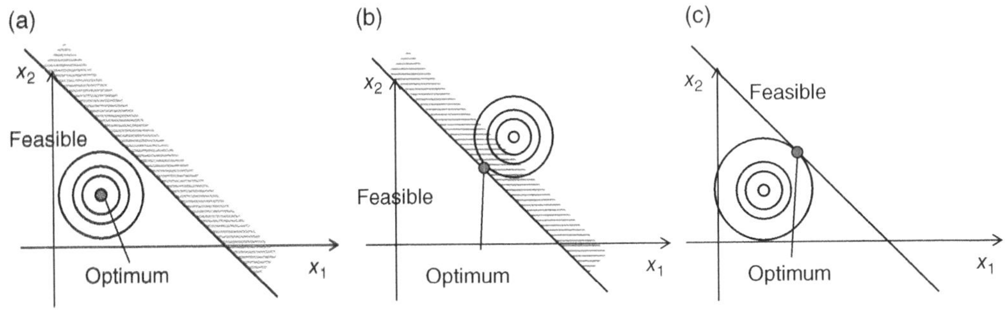


**Figure 4.14** The effect of constraints on optimum design. (a) Inequality constraint has no effect, (b) optimum is on the boundary of inequality constraint, and (c) optimum is always on the equality constraint.

point. Therefore, this point cannot be an optimum design. As shown in Figure 4.14(b), the stationary point is out of the feasible domain. The concentric circles in the figure are the contour of the objective function. The objective function has its minimum at $\{x_1, x_2\} = \{3, 3\}$ and gradually increases until it meets a point $\{x_1, x_2\} = \{2, 2\}$, where it is the lowest objective function while satisfying the inequality constraint. Therefore, this is an optimum point, although the gradient of the objective function $\nabla f \neq 0$. In this case, the inequality constraint is active and plays a critical role in determining the objective function. The optimum design is located on the boundary of the feasible domain. When an inequality constraint is active, it becomes equality; that is, $g(\mathbf{x}) = x_1 + x_2 - 4 = 0$.

Lastly, let us consider the same objective function with an equality constraint:

$$\begin{aligned} &\text{minimize } f(\mathbf{x}) = (x_1 - 1)^2 + (x_2 - 1)^2 \\ &\text{subject to } g(\mathbf{x}) = x_1 + x_2 - 4 = 0 \end{aligned} \tag{4.22}$$

Different from inequality constraints, an equality constraint is always active. As shown in Figure 4.14(c), the feasible domain is the line of constraint $g(\mathbf{x}) = x_1 + x_2 - 4 = 0$. The contour of the objective function meets with the feasible set at $\{x_1, x_2\} = \{2, 2\}$ first time, which is the optimum design. An equality constraint is always active, and the optimum design is located on the constraint boundary.

It is interesting to note that an equality constraint can reduce the number of design variables. In the previous example, the relationship between the two design variables can be obtained using the equality constraint; that is, $x_2 = 4 - x_1$. If this relationship is used to remove variable $x_2$ from the objective function, the original constrained optimization problem becomes unconstraint one $f(x_1) = (x_1 - 1)^2 + (3 - x_1)^2$ with a single design variable. In general, it may not be possible to reduce design variables explicitly because an equality constraint can be an implicit function of design variables. However, its effect should be the same as reducing design variables.

In the previous example, if an additional equality constraint exists, then the optimum design is nothing but the intersection of the two equality constraints because that is the only feasible point. This is basically the same as two equations in a two-dimensional domain, where a unique solution is expected. If there are three independent equality constraints, then the problem is overly constrained and does not have a solution. Therefore, the number of equality constraints should be less than the number of design variables. Considering the fact that an active inequality constraint plays the same role as an equality constraint, the total number of active inequality constraints and equality constraints should be less than the number of design variables.

Now, optimality criteria for a general constrained optimization problem are presented. First, only a single equality constraint is considered. Consider the following constrained optimization problem with an equality constraint:

$$\begin{aligned} &\text{minimize } f(\mathbf{x}) \\ &\text{subject to } h(\mathbf{x}) = 0 \end{aligned} \tag{4.23}$$

Instead of solving the constrained optimization problem, it is converted into an unconstrained optimization problem by using the Lagrange multiplier method. That is, the constrained optimization problem in Eq. (4.23) is equivalent to the following unconstrained optimization problem:

$$\text{minimize } L(\mathbf{x}, \lambda) = f(\mathbf{x}) + \lambda h(\mathbf{x}) \tag{4.24}$$

where $\lambda$ is the Lagrange multiplier and $L(\mathbf{x}, \lambda)$ is referred to as the Lagrangian function. That is, the objective and constraint functions are linearly combined using the Lagrange multiplier. The main difference is that an additional variable $\lambda$ is introduced in addition to design variables $\mathbf{x}$.

Then, the necessary conditions for the stationary Lagrangian function are that the derivatives with respect to the design variables and the Lagrange multiplier are zero. That is,

$$\nabla L(\mathbf{x}, \lambda) = 0 \rightarrow \begin{cases} \dfrac{\partial f(\mathbf{x})}{\partial \mathbf{x}} + \lambda \dfrac{\partial h(\mathbf{x})}{\partial \mathbf{x}} = 0 \\ h(\mathbf{x}) = 0 \end{cases} \tag{4.25}$$

Therefore, in the case of a constrained optimization problem, the necessary conditions do not require the stationary condition of the objective function: $\nabla f \neq 0$. Instead, the linear combination of the gradient of the objective function and that of the constraint, scaled by the Lagrange multiplier, vanishes. Also, the equality constraint must satisfy at the optimum point. Because of the relation $\nabla f(\mathbf{x}) = -\lambda \nabla h(\mathbf{x})$, the Lagrange multiplier is often called "shadow prices," which means the price of constraints. If $\nabla h(\mathbf{x})$ is small, a large value of $\lambda$ is required to equilibrate it with the gradient of the objective function.

The KKT condition in Eq. (4.25) is for a single equality constraint. There are $n + 1$ variables, and Eq. (4.25) provides $n + 1$ equations. For a general optimization problem with $M$ equality constraints, we need the same number of Lagrange multipliers. Therefore, the equivalent unconstrained optimization problem can be written as

$$\text{minimize } L(\mathbf{x}, \boldsymbol{\lambda}) = f(\mathbf{x}) + \sum_{j=1}^{M} \lambda_j h_j(\mathbf{x}) \tag{4.26}$$

where $\boldsymbol{\lambda} = \{\lambda_1, \lambda_2, \ldots, \lambda_M\}^T$ is the vector of Lagrange multipliers. Each constraint has a Lagrange multiplier. The KKT condition becomes

$$\nabla L(\mathbf{x}, \boldsymbol{\lambda}) = 0 \rightarrow \begin{cases} \dfrac{\partial f(\mathbf{x})}{\partial \mathbf{x}} + \displaystyle\sum_{j=1}^{M} \lambda_j \dfrac{\partial h_j(\mathbf{x})}{\partial \mathbf{x}} = 0 \\ h_j(\mathbf{x}) = 0, \quad j = 1, \ldots, M \end{cases} \tag{4.27}$$

The above KKT conditions have $n + M$ variables and $n + M$ equations.

**Example 4.10 KKT condition for an equality constraint**
Find an optimum point that satisfies the KKT condition for the equality constraint in Example 4.9.

**Solution:**
For the constrained optimization problem in Eq. (4.22), the Lagrangian function can be defined as

$$L(\mathbf{x}, \lambda) = (x_1 - 1)^2 + (x_2 - 1)^2 + \lambda(x_1 + x_2 - 4)$$

The KKT condition is the stationary condition of the Lagrangian function as

$$\frac{\partial L}{\partial x_1} = 2(x_1 - 1) + \lambda = 0$$

$$\frac{\partial L}{\partial x_2} = 2(x_2 - 1) + \lambda = 0$$

$$\frac{\partial L}{\partial \lambda} = x_1 + x_2 - 4 = 0$$

The above three equations can be solved for $\{x_1, x_2\} = \{2, 2\}$, and $\lambda = -2$.

**Example 4.11 KKT conditions**
Consider a quadratic objective function and a quadratic constraint that requires the design to be on a circle with a radius of 10 and a center at the origin. Find the points that satisfy the KKT condition and determine if they are minimum or maximum.

$$\begin{aligned} &\text{minimize } f(\mathbf{x}) = x_1^2 + 10x_2^2 \\ &\text{subject to } h(\mathbf{x}) = 100 - \left(x_1^2 + x_2^2\right) = 0 \end{aligned} \tag{4.28}$$

**Solution:**
Since the objective function prefers $x_1$ to $x_2$, the minimum may be expected for high $x_1$ and low $x_2$. The Lagrangian function is defined as

$$L(\mathbf{x}, \lambda) = x_1^2 + 10x_2^2 + \lambda\left(100 - x_1^2 - x_2^2\right)$$

Taking derivatives with respect to $x_1$, $x_2$, and $\lambda$, we get three equations, with the last being the constraint equations.

$$\frac{\partial L}{\partial x_1} = 2x_1 - 2\lambda x_1 = 0$$

$$\frac{\partial L}{\partial x_2} = 20x_2 - 2\lambda x_2 = 0$$

$$\frac{\partial L}{\partial \lambda} = 100 - x_1^2 - x_2^2 = 0$$

The first two equations will produce contradictory values for $\lambda$ if both $x_1$ and $x_2$ are nonzero. That indicates that a minimum will be obtained when $x_2 = 0$ and a maximum when $x_1 = 0$. Since all the terms are quadratic, we can change the sign of $x_1$ or $x_2$ without changing the results. These correspond to moving 180° around the circle. Then, we have four points that satisfy the KKT conditions (Figure 4.15):

$$x_1 = 0, x_2 = \pm 10 \; (f = 1{,}000, \text{maxima})$$

$$x_1 = \pm 10, x_2 = 0 \; (f = 100, \text{minima})$$

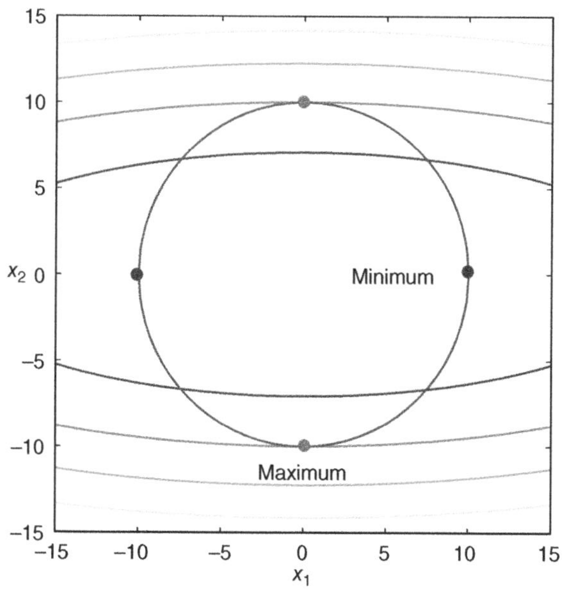


**Figure 4.15** Maxima and minima of a quadratic objective function with a circular constraint.

The optimality criteria for inequality constraint are based on the fact that when the inequality constraint is inactive, it does not play a role in the optimum design, and when the inequality constraint is active, it is basically the same as an equality constraint in the previous subsection. Consider the following optimization problem with inequality constraints:

$$\begin{aligned} &\text{minimize } f(\mathbf{x}) \\ &\text{subject to } g_i(\mathbf{x}) \le 0, i = 1, \ldots, K \end{aligned} \tag{4.29}$$

Since we have KKT conditions for equality constraints, we can convert the inequality constraints to equality ones by adding a slack variable. That is,

$$g_i(\mathbf{x}) \le 0 \rightarrow \quad g_i(\mathbf{x}) + s_i^2 = 0 \tag{4.30}$$

where $s_i^2 \ge 0$ is the slack variable. Therefore, an inequality constraint introduces an additional variable to the system. When the slack variable $s_i^2 = 0$, the inequality constraint becomes an equality constraint and active. When the slack variable is positive $s_i^2 > 0$, the constraint is inactive. When the slack variable is negative, $s_i^2 < 0$, it means that the constraint is violated.

After converting inequality constraints to equality ones, we use the same Lagrangian function to make an unconstrained optimization problem, as

$$\text{minimize } \; L(\mathbf{x}, \boldsymbol{\lambda}, \mathbf{s}) = f(\mathbf{x}) + \sum_{i=1}^{K} \lambda_i \left(g_i(\mathbf{x}) + s_i^2\right) \tag{4.31}$$

The Lagrangian is a function of $n$ design variables, $K$ Lagrange multiplier, and $K$ slack variables. Therefore, the stationary condition of the Lagrangian function will produce $n + 2K$ numbers of equations. The KKT condition becomes

$$\nabla L(\mathbf{x}, \boldsymbol{\lambda}, \mathbf{s}) = 0 \rightarrow \begin{cases} \dfrac{\partial f(\mathbf{x})}{\partial \mathbf{x}} + \displaystyle\sum_{i=1}^{K} \lambda_i \dfrac{\partial g_i(\mathbf{x})}{\partial \mathbf{x}} = 0 \\ g_i(\mathbf{x}) + s_i^2 = 0, \quad i = 1, \ldots, K \\ 2\lambda_i s_i = 0, \quad i = 1, \ldots, K \end{cases} \tag{4.32}$$

The last $K$ equations, $\lambda_i s_i = 0$, are called complementary slackness or switching conditions. This condition can be interpreted as $\lambda_i g_i = 0$ as $g_i = 0$ when $s_i = 0$. This condition implies that either the Lagrange multiplier is zero or the constraint is zero, which is why it is called complementary slackness. When the constraint is active, $g_i = 0$, the Lagrange multiplier is nonzero. If the constraint is inactive, it has a zero value. In particular, when the inequality constraint is less-than-or-equal-to type, the Lagrange multiplier of an active constraint becomes positive. Therefore, the complementary slackness can be summarized as

$$\begin{aligned} &\lambda_i = 0 \quad \rightarrow \quad g_i < 0 \ \text{ inactive constraint} \\ &\lambda_i > 0 \quad \rightarrow \quad g_i = 0 \ \text{ active constraint} \end{aligned} \tag{4.33}$$

This complementary slackness is consistent with the previous discussion that inactive constraints do not play any role in determining the optimum design. In Eq. (4.32), the summation in the first equation is only for active constraints because $\lambda_i$ for inactive constraints will be zero. It is noted that the Lagrange multiplier of an equality constraint can be either positive or negative.

**Example 4.12 Constrained optimization problem**

Find designs that satisfy the KKT conditions for the following constrained optimization problem:

$$\begin{aligned} &\text{minimize } f(\mathbf{x}) = (x_1 - 3)^2 + (x_2 - 3)^2 \\ &\text{subject to } h(x) = x_1 - 3x_2 - 1 = 0 \\ &\qquad\qquad g(\mathbf{x}) = x_1 + x_2 - 4 \le 0 \end{aligned} \tag{4.34}$$

**Solution:**

The inequality constraint needs to introduce a slack variable. The Lagrangian function can be defined as

$$L(\mathbf{x}, \boldsymbol{\lambda}, s) = (x_1 - 3)^2 + (x_2 - 3)^2 + \lambda_1(x_1 - 3x_2 - 1) + \lambda_2\left(x_1 + x_2 - 4 + s^2\right)$$

The KKT conditions become

$$\nabla L(\mathbf{x}, \boldsymbol{\lambda}, s) = 0 \rightarrow \begin{cases} 2(x_1 - 3) + \lambda_1 + \lambda_2 = 0 \\ 2(x_2 - 3) - 3\lambda_1 + \lambda_2 = 0 \\ x_1 - 3x_2 - 1 = 0 \\ x_1 + x_2 - 4 + s^2 = 0 \\ s\lambda_2 = 0 \end{cases}$$

It is noted that all equations are linear, except for the complementary slackness term. Therefore, we can consider two cases separately: $\lambda_2 = 0$ or $s = 0$. Each case needs to satisfy all constraints.

Case 1: $\lambda_2 = 0$. In this case, the KKT conditions become

$$\begin{cases} 2(x_1 - 3) + \lambda_1 = 0 \\ 2(x_2 - 3) - 3\lambda_1 = 0 \\ x_1 - 3x_2 - 1 = 0 \\ x_1 + x_2 - 4 + s^2 = 0 \end{cases}$$

The above equations can be solved for $x_1 = 3.7, x_2 = 0.9, \lambda_1 = -1.4$, and $s^2 = -0.6$. The objective at this point is $f^* = 4.9$. Since it is required to be $s^2 \geq 0$, this condition violates the inequality condition.

Case 2: $s = 0$. In this case, the KKT conditions become

$$\begin{cases} 2(x_1 - 3) + \lambda_1 + \lambda_2 = 0 \\ 2(x_2 - 3) - 3\lambda_1 + \lambda_2 = 0 \\ x_1 - 3x_2 - 1 = 0 \\ x_1 + x_2 - 4 = 0 \end{cases}$$

The above equations can be solved for $x_1 = 3.25, x_2 = 0.75, \lambda_1 = -1.875$, and $\lambda_2 = 1.125 > 0$. The objective at this point is $f^* = 5.125$, which is larger than that of Case 1. Since the Lagrange multiplier of the inequality constraint is positive, this case is valid. Figure 4.16 shows the contour of the objective function along with the two constraints. Due to the equality constraint, the feasible domain is the blue-colored line. Case 2 is the point when both constraints are active, while Case 1 is the point where the inequality constraint is violated.

As shown in Example 4.12, the Lagrange multipliers play an important role in determining optimization. As mentioned before, the Lagrange multipliers are called "shadow prices." This means how much constraint gradients need to change in order to compensate for the gradient of the objective function at the optimum design. From the first equation of the KKT condition in Eq. (4.32), the gradient of the objective function can be written as

$$\nabla_{\mathbf{x}} L(\mathbf{x}, \boldsymbol{\lambda}, \mathbf{s}) = 0 \rightarrow \frac{\partial f(\mathbf{x})}{\partial \mathbf{x}} = -\sum_{i=1}^{K} \lambda_i \frac{\partial g_i(\mathbf{x})}{\partial \mathbf{x}} \tag{4.35}$$

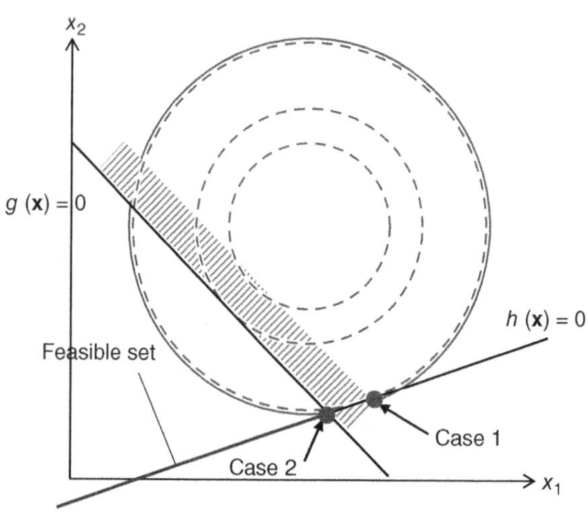


**Figure 4.16** Objective and constraints of the optimization problem in Example 4.12.

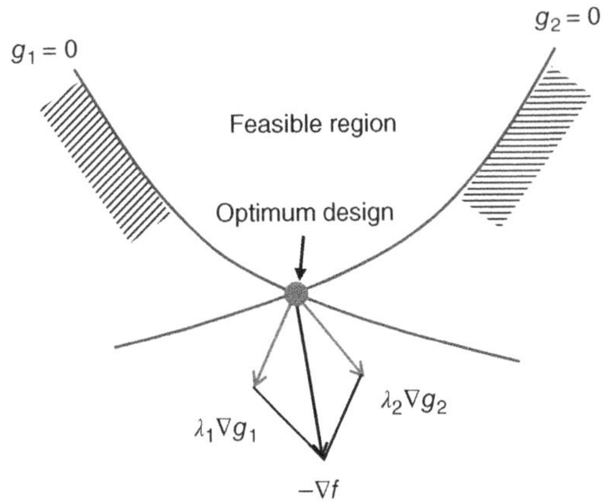


**Figure 4.17** The relationship between objective gradient and constraint gradients at the optimum design.

where $\nabla_{\mathbf{x}}$ means the derivative with respect to design variables $\mathbf{x}$. This relation means that the gradient of the objective function can be represented by a linear combination of active constraints. The Lagrange multipliers are the coefficients of each constraint gradient. As we discussed in the surrogate chapters, the coefficients represent the importance of the term. Therefore, the Lagrange multipliers represent how important the constraint is to compensate for the objective change. An inactive constraint would not contribute to the optimum design. An important constraint would contribute significantly to the objective change. At the optimum, as shown in Figure 4.17, the objective gradient can be represented by a linear combination of constraint gradients with Lagrange multipliers as a proportional constant.

The KKT conditions in Eq.(4.27) or Eq. (4.32) are necessary conditions; i.e., those points that satisfy the necessary conditions are the candidates for optimum design. Similar to the unconstrained optimization problem, it is possible to develop second-order necessary conditions and sufficient conditions. Consider the following general form of Lagrangian function:

$$\text{minimize } L(\mathbf{x}, \boldsymbol{\mu}, \boldsymbol{\lambda}, \mathbf{s}) = f(\mathbf{x}) + \sum_{j=1}^{M} \mu_j h_j(\mathbf{x}) + \sum_{i=1}^{K} \lambda_i \left(g_i(\mathbf{x}) + s_i^2\right) \tag{4.36}$$

where we used different notations for the Lagrange multipliers for the equality and inequality constraints. The first-order necessary condition in Eq. (4.27) or Eq. (4.32) can be represented by

$$\nabla L(\mathbf{x}, \boldsymbol{\mu}, \boldsymbol{\lambda}, \mathbf{s}) = 0 \tag{4.37}$$

where the gradient operator is the derivative of all variables: $\nabla = \left\{\nabla_{\mathbf{x}}^T, \nabla_{\boldsymbol{\mu}}^T, \nabla_{\boldsymbol{\lambda}}^T, \nabla_{\mathbf{s}}^T\right\}^T$.

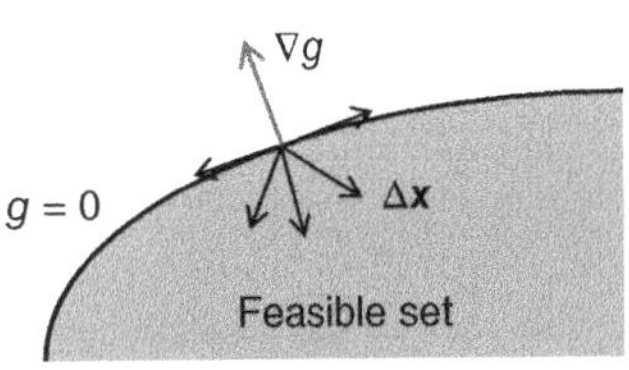


**Figure 4.18** Constraint boundary and feasible direction.

In the case of unconstrained optimization problems, the sufficient condition is that the Hessian matrix $\nabla^2 f$ is positive definite. In the case of constrained optimization problems, the sufficient condition is that the Hessian of the Lagrangian function $\nabla_{\mathbf{xx}}L$ is positive definite in all feasible directions. A feasible direction is a direction in which a small change in the design keeps the design feasible. As shown in Figure 4.18, the optimum design is located on the boundary of active constraints. At this point, a feasible direction keeps the design in the feasible domain after a small movement in that direction. In the case of equality constraint, the movement direction should be perpendicular to the gradient of the constraints. The gradient direction is the direction to change the constraint. Therefore, in order to keep the constraint satisfied after movement, the direction needs to satisfy $\nabla h_j \cdot \Delta\mathbf{x} = 0$. In the case of an inequality constraint, in order to make keep the constraint satisfied after movement, the direction needs to satisfy $\nabla g_i \cdot \Delta\mathbf{x} \le 0$. However, the direction that makes $\nabla g_i \cdot \Delta\mathbf{x} < 0$ will make the constraint inactive after the movement. Therefore, in order to remain an active constraint, the inequality constraint also satisfies $\nabla g_i \cdot \Delta\mathbf{x} = 0$.

Therefore, the second-order necessary condition can be stated as

$$q = \Delta\mathbf{x}^T[\nabla_{\mathbf{xx}}L]\Delta\mathbf{x} \ge 0 \text{ for all } \Delta\mathbf{x} \ne 0 \text{ satisfying} \tag{4.38}$$

$$\begin{cases} \nabla g_i^T \Delta\mathbf{x} = 0 \text{ for all active inequalities} \\ \nabla h_j^T \Delta\mathbf{x} = 0 \text{ for all equalities} \end{cases} \tag{4.39}$$

In a similar way, the second-order sufficient condition can be stated as

$$q = \Delta\mathbf{x}^T[\nabla_{\mathbf{xx}}L]\Delta\mathbf{x} > 0 \text{ for all } \Delta\mathbf{x} \ne 0 \text{ satisfying Eq.(4.39)} \tag{4.40}$$

The feasible direction in Eq. (4.39) relaxes the requirement of positive definiteness of the Hessian matrix. It is enough that the Hessian matrix is positive definite only for feasible directions.

### 4.4.4 Effect of Constraint Limit

In the previous subsection, it was shown that the Lagrange multipliers can be interpreted as "shadow prices" of constraint gradients in order to compensate for the gradient of the objective function at optimum design. There are different roles of the Lagrange multipliers in a practical sense. The first important property of the Lagrange multipliers is that it can linearly approximate how much the optimum objective function will change when constraint bounds change.

Although the standard form of a constraint has a zero on the right-hand side, let us consider that the constraint limit is on the right-hand side for this subsection. That is, the equality constraints are written as $h_i(\mathbf{x}) = a_i$, and the inequality constraints are written as $g_j(\mathbf{x}) \le b_j$. Therefore, $a_i$ and $b_j$ are constraint limits. Then, if $a_i$ and $b_j$ change, the optimum design will also change. Therefore, we can say that the optimum design is a function of these constraint bounds: $\mathbf{x}^* = \mathbf{x}^*(\mathbf{a}, \mathbf{b})$. If the optimum design changes, the objective function at the optimum design will also change: $f^* = f^*(\mathbf{a}, \mathbf{b})$. Then, the gradients of the objective function with respect to the constraint bounds are nothing but the negative of the Lagrange multipliers. In order to show this property, consider the following form of Lagrangian function:

$$L(\mathbf{x}, \boldsymbol{\mu}, \boldsymbol{\lambda}, \mathbf{s}) = f(\mathbf{x}, \mathbf{a}, \mathbf{b}) + \sum_{j=1}^{M} \mu_j\left(h_j(\mathbf{x}) - a_j\right) + \sum_{i=1}^{K} \lambda_i\left(g_i(\mathbf{x}) - b_i + s_i^2\right) \tag{4.41}$$

At the optimum design, the Lagrangian function needs to be stationary. Therefore, if the above equation is differentiated with respect to $a_j$ and $b_i$, we can obtain the following important results:

$$\frac{\partial f}{\partial a_j} = -\mu_j^*, \quad \frac{\partial f}{\partial b_i} = -\lambda_i^* \tag{4.42}$$

where $\mu_j^*$ and $\lambda_i^*$ are Lagrange multipliers at the optimum design.

This information can be used to linearly approximate the change of optimum objective function due to the change in constraint bounds. For that purpose, we use the Taylor series expansion again. For a change of $a_j$ and $b_i$, the objective function at the optimum design can be expanded as

$$f^*(a_j + \Delta a_j, b_i + \Delta b_i) = f^*(a_j, b_i) + \frac{\partial f^*}{\partial a_j}\Delta a_j + \frac{\partial f^*}{\partial b_i}\Delta b_i + \text{H.O.T.} \tag{4.43}$$

Then, the change in the objective function can be estimated by

$$\Delta f^* = -\sum_{j=1}^{M} \mu_j^* \Delta a_j - \sum_{i=1}^{K} \lambda_i^* \Delta b_i \tag{4.44}$$

Therefore, when the constraint bounds are changed in a small amount, instead of solving the optimization problem again, it is possible to estimate how much the optimum objective function will change using Eq. (4.44).

Figure 4.19 shows how the change in the optimum design can be approximated using the Lagrange multiplier. The original optimum design is located at point A where both constraints are active: $g_1 = b_1$ and $g_2 = b_2$. When the constraint bound of $g_2$ is changed to $b_2 + \Delta b_2$, the true modified optimum design should be located at point B if the optimization problem is solved again with the new constraint bound. Instead, Eq. (4.44) can be used to approximate the change of the optimum objective function at point C. It is noted that Eq. (4.44) cannot approximate the new optimum design. It only approximates the change of the optimum objective function.

**Example 4.13 Optimization of a container box**

Formulate an optimization problem to minimize the cost of building a top-open container box, while the volume of the box should be greater than 125 ft$^3$. The side panel cost is \$10/ft$^2$, while the floor and end panels are \$15/ft$^2$. The three design variables are the depth, length, and height of the box. (a) Formulate the optimization problem and solve it for the optimum design. (b) When the required volume increases to 130 ft$^3$, estimate the change in cost using the Lagrange multiplier. (c) Solve the optimization problem with the new constraint bound and compare the change in optimum objective function with that of (b).

**Solution:**

a) Let design variables are $\{x_1, x_2, x_3\} = \{\text{depth, length, height}\}$. The optimization problem can be formulated as

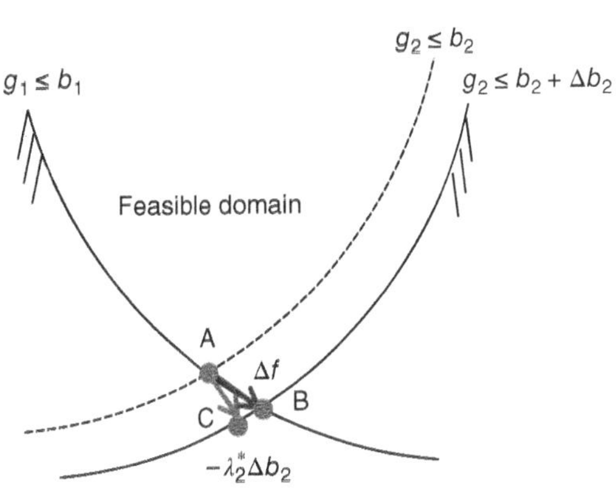


Figure 4.19 Change of the optimum objective function due to the change in constraint bound.

$$\text{minimize } f(\mathbf{x}) = 20x_2x_3 + 30x_1x_3 + 15x_1x_2$$
$$\text{subject to } g(\mathbf{x}) = 125 - x_1x_2x_3 \le 0$$

The Lagrangian function is defined as

$$L(\mathbf{x}, \lambda, s) = 20x_2x_3 + 30x_1x_3 + 15x_1x_2 + \lambda\left(125 - x_1x_2x_3 + s^2\right)$$

The stationary condition of the Lagrangian function becomes

$$\nabla L(\mathbf{x}, \lambda, s) = 0 \rightarrow \begin{cases} 15x_2 + 30x_3 - \lambda x_2x_3 = 0 \\ 15x_1 + 20x_3 - \lambda x_1x_3 = 0 \\ 30x_1 + 20x_2 - \lambda x_1x_2 = 0 \\ 125 - x_1x_2x_3 + s^2 = 0 \\ s\lambda = 0 \end{cases}$$

The optimum design that satisfies the above KKT conditions is $x_1^* = 4.8075$, $x_2^* = 7.2112$, and $x_3^* = 3.6056$. At the optimum design, the objective function and the Lagrange multiplier are $f^* = 1560$ and $\lambda^* = 8.320$.

b) When the constraint bound increases to 130 $\text{ft}^3$, its effect on the constraint bound changes from $-125$ to $-130$. This is because the constraint is written in standard form. Therefore, $\Delta b = -5$. The increase in the optimum objective function can be estimated by

$$\Delta f = -\lambda\Delta b = -8.320 \times -5 = 41.6$$

c) The modified optimization problem is

$$\text{minimize } f(\mathbf{x}) = 20x_2x_3 + 30x_1x_3 + 15x_1x_2$$
$$\text{subject to } g(\mathbf{x}) = 130 - x_1x_2x_3 \le 0$$

The solution to the above optimization problem is $f^* = 1601.3$, where $\Delta f = 41.3$. Therefore, the approximation using the Lagrange multiplier ($\Delta f = 41.6$) is close to that of the actual change ($\Delta f = 41.3$).

### 4.4.5 Sensitivity of Optimum Solution to Parameters

In optimization problems, design variables are the ones that can be changed by engineers to find an optimum design. However, most engineering applications have parameters, which are not design variables, but they can be changed, or the engineers want to see their effects on the optimum design. For example, when an engineering application includes frictional contact, the friction coefficient is uncontrollable, but its exact value is difficult to identify. Therefore, engineers might want to see the effect of the friction coefficient on the optimum design. If the optimum objective is very sensitive to the friction coefficient, it would be necessary to identify it more carefully or use a conservative value to compensate for inaccurate information. As another example, Young's modulus is an important parameter for structural simulation. However, Young's modulus of steel has about 15% variability. Therefore, engineers may want to try different values of Young's modulus to understand its effect on the optimum design. Other common parameters are constraint bounds, as we discussed in the previous subsection. In the case of a stress constraint, for example, $\sigma_{\text{max}} \le \sigma_{\text{limit}}$, where $\sigma_{\text{limit}}$ is an allowable strength of a material. In practice, since the allowable strength has uncertainty, we may want to get an estimate of how much the optimum objective will change if the allowable strength is increased by going to a better grade of material.

When an engineering application has uncontrollable parameters, it would be valuable to know the sensitivity (i.e., the gradient) of the optimum objective function with respect to the parameters. If the sensitivity is high, the parameter is important and needs to be identified accurately or used a conservative value to compensate for its effect. If the sensitivity is low, the parameter will not have much effect on the optimum design. The Lagrange multipliers play an important role in this parameter sensitivity. In order to show it, it is assumed that the objective and constraints depend on a parameter $p$. For simplicity of presentation, we only consider the following optimization problem with an inequality constraint:

$$\begin{aligned}&\text{minimize } f(\mathbf{x}, p)\\&\text{subject to } g(\mathbf{x}, p) \le 0\end{aligned} \tag{4.45}$$

More complicated optimization problems with equality and inequality constraints can be considered in a similar way. Since both objective and constraint depend on the parameter $p$, it is obvious that the optimum design also depends on the parameter $\mathbf{x}^*(p)$.

The objective function at the optimum design can be written as $f^*(p) = f(\mathbf{x}^*(p), p)$, as well as the constraint $g^*(p) = g(\mathbf{x}^*(p), p)$. Since the constraint is a function of both the parameter and the design variable, the sensitivity of the constraint with respect to the parameter can be written as

$$\frac{dg}{dp} = \frac{\partial g}{\partial p} + \frac{\partial g}{\partial \mathbf{x}}^T \frac{d\mathbf{x}}{\partial p} = 0 \tag{4.46}$$

where $dg/dp$ is the total derivative, and $\partial g/\partial p$ is the partial derivative. The second term on the right-hand side uses the chain rule of differentiation. The sensitivity of the objective function with respect to the parameter can be written in a similar form. By using Eqs. (4.35) and (4.46), the sensitivity of the objective function at the optimum design can be calculated as

$$\begin{aligned}\frac{df}{dp} &= \frac{\partial f}{\partial p} + \frac{\partial f}{\partial \mathbf{x}}^T \frac{d\mathbf{x}}{\partial p}\\&= \frac{\partial f}{\partial p} - \lambda \frac{\partial g}{\partial \mathbf{x}}^T \frac{d\mathbf{x}}{\partial p}\\&= \frac{\partial f}{\partial p} + \lambda \frac{\partial g}{\partial p}\end{aligned} \tag{4.47}$$

Equation (4.47) shows that the Lagrange multiplier is a measure of the effect of a change in the constraint on the objective function. Consider, for example, the constraint is given in the form of $g(\mathbf{x}) = G(\mathbf{x}) - p \le 0$. By increasing $p$, we make the constraint easier to satisfy (i.e., stress limit is increased for stress constraint). If a constraint is relaxed, the optimum objective function can be reduced further. Since $\partial g/\partial p = -1$, from Eq. (4.47), $df/dp = -\lambda$. That is, the optimum objective function can be reduced by increasing $p$. In the opposite case (i.e., increasing a parameter makes the constraint more difficult to satisfy), $\lambda$ is the marginal price that we pay in terms of an increase in the objective function for making the constraint more difficult to satisfy.

There are two special cases that are worth noting. (i) When only the objective function depends on the parameter, it is remarkable that the total derivative is equal to the partial derivative. That means that the effect of changing the optimum position $\mathbf{x}^*$ as a function of $p$ can be neglected. (ii) When $p$ is the bound on a single constraint; that is, when the constraint can be written as $g(\mathbf{x}) = G(\mathbf{x}) - p \le 0$, $df/dp = -\lambda$, which is why the Lagrange multipliers are called shadow prices.

**Example 4.14 Sensitivity of optimization solution**

Consider the following optimization problem with an inequality constraint:

$$\text{minimize } f(\mathbf{x}) = x_1^2 + 10x_2^2$$
$$\text{subject to } g(\mathbf{x}, p) = p - \left(x_1^2 + x_2^2\right) \le 0$$

where $p$ is the square of the radius of the circle. (i) When $p = 100$, calculate the optimum design and optimum objective function along with the Lagrange multiplier. (ii) Calculate the sensitivity of the optimum objective function with respect to the parameter $p$.

**Solution:**

The Lagrangian function and the KKT conditions can be written as

$$\text{minimize } L(\mathbf{x}, \boldsymbol{\lambda}, \mathbf{s}) = x_1^2 + 10x_2^2 + \lambda\left(p - x_1^2 - x_2^2 + s^2\right)$$

$$\nabla L(\mathbf{x}, \lambda, s) = 0 \rightarrow \begin{cases} 2x_1 - 2\lambda x_1 = 0 \\ 20x_2 - 2\lambda x_2 = 0 \\ p - x_1^2 - x_2^2 + s^2 = 0 \\ \lambda s = 0 \end{cases}$$

The above KKT conditions can be used to find the optimum design $x = \left\{ \pm\sqrt{p}, 0 \right\}$ where the objective function is $f^*(p) = p$, and its sensitivity is $df^*/dp = 1$. The sensitivity of the optimum objective function can also be calculated using Eq. (4.47) as

$$\frac{df}{dp} = \frac{\partial f}{\partial p} + \lambda \frac{\partial g}{\partial p} = 0 + 1 \times 1 = 1$$

This result can easily be verified since for any value of $p$ we get that the optimum is at $x_2^* = 0$ and $x_1^* = \pm\sqrt{p}$, so $f^*(p) = p$.

## Exercise

**4.1** Formulate the optimization problem for each of the following problems. Identify design variables and objective function.

(a) Find the aspect ratio of a rectangle with the highest ratio of area to the square of the perimeter.

(b) We need to fly to a city in Florida, rent a car, and visit Gainesville, Jacksonville, and Tampa, and fly back from the last city you visit. What should your itinerary be to minimize your driving distance?

(c) You need to perform a task once a month, on the same day of each month (e.g., the 13th). It is more inconvenient to do on a weekend. Select the day of the month to minimize the number of times it will fall on a weekend in one given year (not a leap year).

**4.2** A rectangular underground storage tank is to be constructed and installed. Specifications require that the volume of the tank be 1,000 $m^3$ and that the ratio of the lengths of any two sides is no greater than two. The top surface of the tank is to be 3 m below the ground surface when installed. The cost of construction of the tank is \$150/$m^2$ based on the surface area. Installation cost in the dollar is equal to 200 times the product of the area and the square of the depth of the hole to be excavated. For convenience, assume the tank and the hole's cross-sectional area are the same, i.e., no clearance is required. Formulate the optimization problem to minimize the total project cost and write the result in standard form.

**4.3** A coal mining company has three coal mines and four coal cellars. The company is trucking coal from three mines to four cellars before shipping out to the customer. To make efficient usage of trucks, the total delivery distance from mines to cellars by the trucks needs to be minimized. The following tables show the distance between three mines (Mine numbers 1, 2, and 3) and four cellars (Cellar numbers A, B, C, and D). Each truck can carry two tons of coal at one time. The amount of coal produced at each coal mine and the amount of coal that can be stored at each cellar are given in the following table. Each delivery of the coal by truck means a round trip of the distance between the mine and cellar, since the truck will return empty to get more coal. Assume that the company has enough trucks so each truck can move between a selected mine and cellar only. Formulate the problem in standard form. Do not normalize or attempt to solve. (Hint: There are 12 design variables).

| | Distance (meter) from mine to cellar | | | |
|---|---|---|---|---|
| **Mine number** | **Cellar A** | **Cellar B** | **Cellar C** | **Cellar D** |
| 1 | 3,500 | 2,900 | 3,450 | 1,290 |
| 2 | 1,670 | 4,500 | 2,390 | 4,230 |
| 3 | 2,340 | 1,250 | 2,880 | 3,770 |

| **Coal production at mines (per day)** | |
|---|---|
| 1 | 3,500 ton |
| 2 | 4,000 ton |
| 3 | 4,500 ton |

| **Cellar storage capacity (per day)** | |
|---|---|
| A | 2,000 ton |
| B | 3,000 ton |
| C | 4,500 ton |
| D | 2,500 ton |

**4.4** Find the optimum design of the following optimization problem using graphical optimization.

$$\underset{x_1, x_2}{\text{minimize}}\ f = -x_1 - x_2$$

$$\text{subject to } 2x_1 + 3x_2 \leq 12$$

$$2x_1 + x_2 \leq 8$$

$$x_1, x_2 \geq 0$$

**4.5** Solve Example 4.5 when design variables are the outer radius $R_o$ and inner radius $R_i$.

**4.6** A two-bar truss, as shown in the figure, is under a vertical load $W$. The goal of the optimization problem is to minimize the weight of the truss. The constraints are (i) each member's stress should be less than the yield stress $\sigma_Y$, (ii) the tip deflection should be less than the allowable displacement $u_{\text{allowable}}$, and (iii) the compressive stress should be less than the buckling stress. When design variables are the two cross-sections, $A_1$ and $A_2$, the height $h$, and the length $L$, write the standard form of the optimization problem.

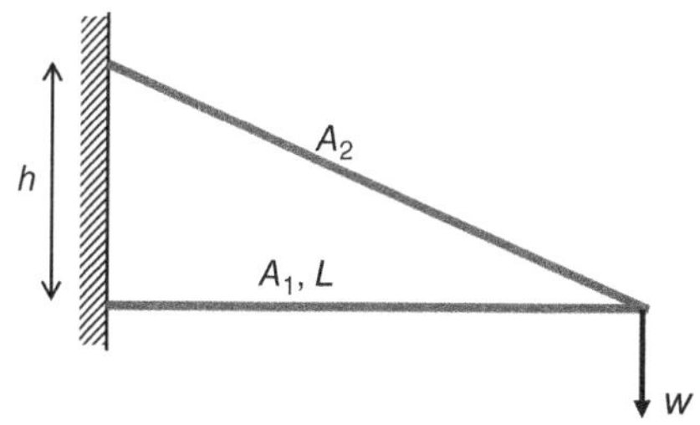


**4.7** Provide two formulations in standard form for minimizing the surface area of a cylinder of a given volume when the diameter and height are the design variables. One formulation should use the volume as equality constraint, and another use it to reduce the number of design variables.

**4.8** Formulate in standard form the problem of finding an open-top rectangle container with a volume of at least 50 and minimum surface area.

**4.9** You need to go from point A to point B in minimum time while maintaining a safe distance from point C. Formulate an optimization problem in standard normalized form to find the path with no more than three design variables when $A = (0, 0)$, $B = (10, 10)$, $C = (4, 4)$, and the minimum safe distance is 5.

**4.10** Check for the convexity of the following functions. If the function is not convex everywhere, check its domain of convexity.

(a) $f(x_1, x_2) = x_1^3 + 2x_2^2$

(b) $f(x_1, x_2) = 3x_1^2 + 2x_1x_2 + 2x_2^2 - 8$

(c) $f(x_1, x_2) = x_1^3 + 12x_1x_2^2 + 2x_2^2 + 5x_1^2$

**4.11** Consider the following optimization of a quadratic objective function with a ring constraint

$$\underset{x_1, x_2}{\text{minimize}}\ f(x_1, x_2) = x_1^2 + 10x_2^2$$

$$\text{subject to } 81 \leq x_1^2 + x_2^2 \leq 1{,}000$$

(a) Where is the minimum? What starting points (note the plural) will not find it when using `fmincon`? What is the problem preventing `fmincon` from reaching the minimum?

(b) For the problem above, we found that the inner ring constraint was active, and its Lagrange multiplier was equal to 1. Calculate an estimate based on the Lagrange multiplier of the effect of changing the constraint from 81 to 79?

**4.12** Classify the stationary points of the following functions from the optimality criteria, then check by plotting them. (i) $f(x) = 2x^3 + 3x^2$, (ii) $f(x) = 3x^4 + 4x^3 - 12x^2$, (iii) $f(x) = x^5$, and (iv) $f(x) = x^4 + 4x^3 + 6x^2 + 4x$.

**4.13** Find the stationary points of the following function and classify them: $f(x_1, x_2, x_3) = -x_1^2 + 2x_2x_3 + x_2^2 + 4x_3^2$.

**4.14** The objective function $f = x_1 + 2x_2$ is to be minimized subject to the constraint that $x_2$ is larger than $-5$, and that the point is inside the circle $x_1^2 + x_2^2 = R^2$. (i) Write the Kuhn–Tucker conditions for the problem. (ii) For $R^2 = 34$, check that the optimum is at $x_1 = -3, x_2 = -5$, and calculate the Lagrange multipliers. (iii) Calculate the derivative of the optimum objective with respect to $R$, by using the Lagrange multipliers. (iv) Use the equations for the derivatives of an optimum solution to obtain the derivatives of $x$ and the Lagrange multiplier with respect to $R$. Use the results to check on (iii). (v) Use the results to extrapolate the optimal solution for the smallest value of $R$ for which you think that extrapolation makes sense. Explain.

**4.15** The objective function $f = 3x_1 + 4x_2$ is to be minimized subject to the constraint that both $x_1$ and $x_2$ are less than 5 in magnitude, and that the point is inside the circle $x_1^2 + x_2^2 = R^2$. (i) Write the optimization problem in standard form. (ii) For $R = 5$, check that the optimum is at $x_1 = -3, x_2 = -4$, and the Lagrange multiplier associated with the circle constraint is equal to 0.5. (iii) Calculate the derivative of the optimum objective with respect to $R$, by using the Lagrange multiplier. (iv) Use the equations for the derivatives of an optimum solution to obtain the derivatives of $\mathbf{x}$ and the Lagrange multiplier with respect to $R$. Use the results to check on (iii). (v) Use the results to extrapolate the optimal solution for the largest value of $R$ for which you think that extrapolation makes sense. Explain.

**4.16** Given the objective function $f = (x_1 + x_2)^2 / x_1$, what is the error in the forward difference derivatives at that point if the increments in $x_1$ and $x_2$ are one percent of their value?

**4.17** Consider the problem of minimizing the function $f(x_1, x_2) = x_1 + R^2$ with respect to $x_1$ and $x_2$, subject to the constraint $x_1^2 + x_2^2 \leq R^2$. Obviously, the solution for this problem is $x_1 = -R$, $x_2 = 0$. (i) Write the problem in standard form. (ii) Check that the solution satisfies the Kuhn–Tucker conditions at the optimum for $R = 1$. (iii) Use the solution at $R = 1$ and the Lagrange multiplier, to predict what the solution will be for $R = 1.5$. Check how accurate the prediction is.

**4.18** Consider the function $f(x) = 3x^4 + 12x^3 + 18x^2 + 12x$. (i) Check that it has a stationary point at $x = -1$. (ii) Calculate what kind of stationary point it is. (iii) Specify a region where you are sure the function is convex.

**4.19** Consider the problem of minimizing the surface area of a cylinder with a required volume of at least $250\pi$ in$^3$. (i) Formulate the optimization problem in the standard form. (ii) Write the optimality conditions (KKT conditions). Use a Lagrange multiplier only for the volume constraint. (iii) At the optimum, the height and diameter are both 10 in. Use this to calculate the Lagrange multiplier. (iv) Identify the type of stationary point. (v) Calculate the sensitivity of the optimum surface area to the value of the volume.

**4.20** Consider a function $f(x) = (x - 2)^5$. (i) Check if $f$ has a stationary point at $x = 2$. (ii) Classify the definiteness of the Hessian matrix at the stationary point and classify the stationary point.

**4.21** Consider a function $f(x_1, x_2) = 4 - (x_1 - 2)^2 - (x_2 - 3)^2$. (i) Check if $f$ has a stationary point at $(x_1, x_2) = (2, 3)$. (ii) Classify the definiteness of the Hessian matrix at the stationary point and classify the stationary point.

**4.22** Calculate the derivative of the surface area $f = 2\pi r^2 + 2\pi rh$ with respect to change in volume $g = V = \pi r^2 h$ using the Lagrange multiplier and compare to the derivative obtained by differentiating the exact solution.

**4.23** Consider the problem of maximizing the volume of a cylinder $V = \pi r^2 h$ with a surface area $S = 2\pi r^2 + 2\pi rh$ of no more than $150\pi$ in$^2$. (i) Formulate the optimization problem in the standard form. (ii) Write the optimality conditions (KKT conditions). Use a Lagrange multiplier only for the surface constraint. (iii) At the optimum, the height and diameter are both 10 in. Use this to calculate the Lagrange multiplier. (iv) Identify the type of stationary point. (v) Calculate the sensitivity of the optimum volume to the value of the surface area constraint.

**4.24** An engineering design problem is formulated as

$$\text{minimize } f(\mathbf{x}) = x_1^2 + 2x_2^2 - 5x_1 - 2x_2 + 10$$

$$\text{subject to } h(\mathbf{x}) = x_1 + 2x_2 - 3 = 0$$

$$g(\mathbf{x}) = 3x_1 + 2x_2 - 6 \le 0$$

(i) Write KKT necessary condition. (ii) How many cases are there to be considered? Identify those cases. (iii) Find the solution for the case where $g(\mathbf{x})$ is active. Is this an acceptable case? (iv) Suppose the Lagrange multiplier for $h(\mathbf{x}) = 0$ is $\mu = -2$ and that for $g(\mathbf{x}) \le 0$ is $\lambda = 1$. If the constant in $h(\mathbf{x})$ is changed to 3.2 and the constant in $g(\mathbf{x})$ is changed to 6.2, approximate the change of the optimum objective.

**4.25** An engineering design problem is formulated as

$$\text{minimize } f(\mathbf{x}) = -px_1 - 4x_2$$

$$\text{subject to } x_1 - 3x_2 \ge -10p$$

$$x_1 + x_2 \le 6$$

$$x_1 - x_2 \le 2$$

$$x_1 + 3x_2 \ge 6$$

where $x_1$ and $x_2$ non-negative. Use the Kuhn–Tucker conditions to check for (not solve for!) the optimality of the solution $x_1 = 2, x_2 = 4$ for $p = 1$. (Hint: you do not need to worry about inactive constraints in the optimality conditions!). Then, estimate the optimal objective function for $p = 1.1$ without solving the problem again. Estimate how much we can change $p$ before a similarly obtained estimate of the optimum objective function will not work.

**4.26** Classify the stationary points of the following functions: (i) $f(x) = 2x^3 + 3x^2 - 6x + 7$, (ii) $f(x) = x$, (iii) $f(x) = x^4$, and (iv) $f(x_1, x_2) = 7 + 2(x_1^3 + x_2^3) - 6x_1x_2$.

**4.27** For the given rational function, determine (i) all stationary points, and (ii) check whether the stationary points are strictly local minima using the sufficient conditions

$$f = \frac{x_1 + x_2}{3 + x_1^2 + x_2^2 + x_1 x_2}$$

**4.28** A design problem is formulated as

$$\text{minimize } f(\mathbf{x}) = (x_1 - 8)^2 + (x_2 - 5)^2$$
$$\text{subject to } g_1(\mathbf{x}) = 2x_1 - 3x_2 - 4 \leq 0$$
$$g_2(\mathbf{x}) = 3x_1 + 3x_2 - 16 \leq 0$$

(i) Construct the Lagrangian and derive the equations for the KKT necessary conditions. (ii) How many cases are there to be considered? (iii) Consider the cases in which the second inequality constraint $g_2(\mathbf{x})$ is active. Find the point that satisfies the KKT necessary conditions.

**4.29** Consider the following design optimization problem:

$$\text{minimize } f(\mathbf{x}) = x_1^2 + x_2^2 - 4x_1 + 4$$
$$\text{subject to } g_1(x) = -x_1 \leq 0$$
$$g_2(x) = -x_2 \leq 0$$
$$g_3(x) = x_2 - (1 - x_1)^3 \leq 0$$

(i) Find the optimum point graphically. (ii) Show that the optimum point does not satisfy the KKT condition. Explain why.

**4.30** Consider the following optimization problem:

$$\text{minimize } f(\mathbf{x}) = (x_1 - 3)^2 + (x_2 - 3)^2$$
$$\text{subject to } g(x) = 2x_1 + x_2 - 2 \leq c$$
$$-x_1 \leq 0, \ -x_2 \leq 0$$

(i) Find an optimum point and objective function when $c = 0$. (ii) Estimate the objective function when $c$ changes to 0.005, 0.010, and 0.040. Compare these estimations with the true optimum objective functions.

**4.31** Answer true or false for the following statements:
(a) A function can have a negative value at its maximum point.
(b) If a constant is added to a function, the location of its minimum point can change.
(c) If the curvature of a function is negative at a stationary point, then the point is a maximum.

**4.32** Find the stationary points of the following functions and classify them: (i) $f(\mathbf{x}) = x_1^2 + 4x_1x_2 + 2x_1x_3 - 7x_2^2 - 6x_2x_3 + 5x_3^2$, (ii) $f(\mathbf{x}) = x_1^2 + 2x_2x_3 + x_2^2 + 4x_3^2$, and (iii) $f(\mathbf{x}) = 40x_1 + x_1^2x_2 + x_2^2/x_1$.

**4.33** Derive the first-order necessary conditions of a constrained optimization problem with equality constraints using the Lagrange multiplier method.

**4.34** For $f(x, p) = \sin x + px,\ 0 \leq x \leq 2\pi$, find the minimum for $p = 0$, estimate the derivative $df^*/dp$, and check by solving again for $p = 0.1$ and comparing to finite difference derivative.

**4.35** An optimization problem is to minimize the surface area of a cylinder, while the volume of the cylinder must be larger than $128\pi$. Calculate the derivative of the cylinder surface area with respect to change in volume using the Lagrange multiplier and compare to the derivative obtained by differentiating the exact solution.

# 5

# Numerical Optimization Algorithms

## 5.1 Introduction

Although the optimality criteria in Section 4.4 are powerful tools to find an optimum design, they are limited to apply for solving general engineering optimization problems. They can only be applicable when both the objective and constraints are available as an explicit function of design variables. They require to solve a system of equations, which are composed of the gradients of the objective and constraint functions. Therefore, solving optimality criteria is only possible when the analytical expressions of the objective and constraint functions are available. This is a significant limitation because most advanced engineering applications use numerical methods to solve for system responses, from which the quantity of interest (QoI) can be calculated. The finite element method, computational fluid dynamics, and rigid-body dynamics are a short list of examples of numerical methods. For a given design, it is possible to evaluate the values of the objective and constraint functions, and possibly their gradients, but their analytical expressions are mostly not available. Due to this critical limitation of optimality criteria, the optimality criteria are used only for limited applications, such as structural optimization [45, 46] and topology optimization [47, 48].

Most engineering optimization problems are solved using numerical methods. Still, the optimality criteria can be used to check the validity of the optimum design. In general, numerical optimization algorithms can be categorized into two groups. The first group of algorithms starts from a design and moves to a new design that can reduce the objective function while satisfying all constraints. This process is repeated until the algorithm cannot find a new design that is better than the previous one. These algorithms are the main focus of this chapter. In order to find a better design, not only the value of functions but also the gradients are used to find the direction to change the design. Because of this, the first group of algorithms is referred to as gradient-based algorithms.

The second group of optimization algorithms does not require gradient information, which is why they are referred to as gradient-free algorithms. These algorithms explore the design space until they cannot find a design that can improve the objective function while satisfying constraints. Sometimes these algorithms are referred to as a global optimization algorithm, but this name can be misleading, as if they can guarantee to find a global optimum design. As we discussed in Section 4.4, it is possible to guarantee the existence of a global optimum design, but it does not mean that we can always find it. Unfortunately, there is no numerical optimization algorithm that can guarantee to find a global optimum for general problems. A proper name would be global search algorithms, as they tend to explore the entire design space rather than moving from the current design to a better one. We will discuss global search algorithms in Chapter 6.

*Surrogate Modeling and Optimization: Theories, Applications, and Limitations*, First Edition. Nam-Ho Kim.

Published 2026 by John Wiley & Sons, Inc.
Companion website: www.wiley.com/go/kim/surrogate_modeling

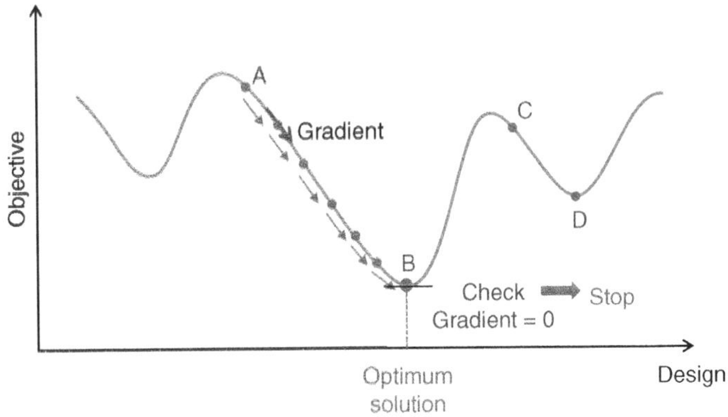


**Figure 5.1** Finding an optimum design from a starting point.

In numerical optimization algorithms, simulation provides the values of the objective and constraint functions for given design variables. Gradient information is also supplied to the optimization algorithms, which can be obtained using sensitivity analysis [49] or finite difference. Then, the optimization algorithms, discussed in this chapter, calculate the best possible design of the problem. Each algorithm has its own advantages and disadvantages. The performance of an optimization algorithm critically depends on the characteristics of the design problem and the types of the objective and constraint functions.

As mentioned earlier, since the entire graphs of the objective and constraint functions are not available, optimization algorithms can make a decision based on the current function value and its sensitivity. Referring to Figure 5.1, let us assume that the current design is located in Point A. Since the gradient of the function is negative at this point, the function will decrease if the design increases. Therefore, the numerical algorithm gradually increases the design until it reaches Point B, where the gradient becomes zero. At Point B, the objective function will increase if the design is either increased or decreased. Therefore, Point B is claimed as an optimum design.

It is interesting to note that the optimum design at Point B is obtained because the optimization algorithm starts from Point A. It was sheer luck that Point B happened to be the global optimum in this case. If the optimization algorithm starts from Point C, it would end up at Point D as an optimum design. Therefore, there is no guarantee of finding the global optimum design. A real challenge is that even if the algorithm finds the global optimum design at Point B, there is no way to tell if that is the global optimum. In order to increase the chance of finding the global optimum, it would be necessary to repeat the optimization algorithm with different starting points.

## 5.2 Overview of the Numerical Optimization Process

Table 5.1 shows the procedure of basic optimization algorithms. As mentioned before, gradient-based numerical optimization algorithms start from an initial design and move to a new design that is better than the previous one. Therefore, it is necessary to choose the initial design by the users (Step 1). The initial design may or may not belong to the feasible set, but it is always good to start with a design within the feasible set. For a given design, the optimization algorithm evaluates the

**Table 5.1** Procedure of basic optimization algorithms.

| Step | Procedure | Comment |
|---|---|---|
| 1 | Start with $\mathbf{x}^{(k)}$ with $k = 0$ | Initial design must be given |
| 2 | Evaluate function values and their gradients: | $f^{(k)}, g_i^{(k)}, h_j^{(k)}, \nabla f^{(k)}, \nabla g_i^{(k)}, \nabla h_j^{(k)}$ |
| 3 | Using information from Step 2, determine design change $\Delta\mathbf{x}^{(k)}$ | |
| 4 | Check for the termination condition | Stop if converged |
| 5 | Update design: $\mathbf{x}^{(k+1)} = \mathbf{x}^{(k)} + \Delta\mathbf{x}^{(k)}$ | |
| 6 | Increase $k = k + 1$ and go to Step 2 | Design iteration |

values of the objective and constraint functions (Step 2). In most optimization problems, it is considered that the computational cost of the optimization algorithm itself is ignorable compared to that of the evaluation of the objective and constraint functions. Therefore, this step takes a major part of the computational cost. The algorithm also requires calculating gradient information. Some software has the capability of calculating gradient information using an adjoint method [49], but most software calculates function values but not gradient information. In such a case, a finite difference method is used to approximate the gradient information, where each design variable is perturbed by a small amount and the simulation is repeated. Since each perturbation requires a full simulation, gradient calculation causes major computational costs. Once the function values and gradient information are available, optimization algorithms calculate the design change (Step 3). This is the key step for optimization algorithms. Different algorithms have different ways of calculating design change $\Delta\mathbf{x}^{(k)}$, which affects the convergence of the algorithm. Once the design is changed, optimization algorithms check if the design is converged (Step 4). It is natural to use optimality criteria to determine the convergence, but various other criteria can be used for this purpose. For example, if the objective and design variables do not change for several iterations, it means that the algorithm cannot improve the objective anymore and is converged. Also, if the algorithm failed to converge within the maximum number of function evaluations or the maximum number of iterations, it may stop without convergence. If the algorithm is not converged, the design is updated using $\Delta\mathbf{x}^{(k)}$ (Step 5) and moved to the next iteration (Step 6).

Among the steps shown in Table 5.1, the key step in optimization is the determination of design change $\Delta\mathbf{x}^{(k)}$ in Step 3. In most algorithms, the design change is further decomposed into (i) the direction and (ii) the magnitude of the design change. The former is often called the search direction, and the latter is referred to as the step size. Therefore, the design change can be written as

$$\Delta\mathbf{x}^{(k)} = \alpha_k \mathbf{d}^{(k)} \tag{5.1}$$

where $\alpha_k$ is the step size and $\mathbf{d}^{(k)}$ is the search direction. The search direction is calculated in such a way that the objective function initially decreases along with the direction. Due to nonlinearity, however, the function may increase if the design is changed too much. Therefore, the step size determines how much the design should change before the objective function increases again.

## 5.3 Determination of Step Size

### 5.3.1 Descent Direction

In general, the search direction is not unique, but not any direction can be a search direction. Therefore, it is necessary to determine the condition that a search direction must satisfy. It is obvious that the search direction must reduce the objective function. Let us consider an unconstrained optimization problem to minimize the objective function $f(\mathbf{x})$. The required condition is that the objective function should decrease after iteration: $f(\mathbf{x}^{(k+1)}) < f(\mathbf{x}^{(k)})$. This condition is used in the following Taylor series expansion of $f(\mathbf{x}^{(k+1)})$ at $\mathbf{x}^{(k)}$ as

$$\begin{aligned} f\left(\mathbf{x}^{(k+1)}\right) &= f\left(\mathbf{x}^{(k)}\right) + \nabla f\left(\mathbf{x}^{(k)}\right)^T \Delta\mathbf{x}^{(k)} + \text{H.O.T.} \\ &\cong f\left(\mathbf{x}^{(k)}\right) + \mathbf{c}^{(k)T} \Delta\mathbf{x}^{(k)} < f\left(\mathbf{x}^{(k)}\right) \end{aligned} \tag{5.2}$$

where $\mathbf{c}^{(k)} = \nabla f(\mathbf{x}^{(k)})$ is the gradient of the objective function. In order to satisfy the inequality, the descent direction should satisfy the following condition:

$$\begin{aligned} &\mathbf{c}^{(k)T} \Delta\mathbf{x}^{(k)} < 0 \\ &\alpha_k \mathbf{c}^{(k)T} \mathbf{d}^{(k)} < 0 \end{aligned} \tag{5.3}$$

The search direction that satisfies Eq. (5.3) is called the descent direction. Knowing that the objective function increases in the direction of $\mathbf{c}^{(k)}$, the angle between $\mathbf{c}^{(k)}$ and $\mathbf{d}^{(k)}$ must be greater than 90°.

**Example 5.1 Descent direction**

An objective function is given as $f(x_1, x_2) = x_1^2 + 3x_1x_2 - 2x_2^2$. When $\mathbf{x} = \{2, 0\}$ is the location of the current design, plot the descent directions.

**Solution:**

At the current design, the gradient of the objective function becomes

$$\mathbf{c}^{(k)} = \begin{Bmatrix} 2x_1 + 3x_2 \\ 3x_1 - 4x_2 \end{Bmatrix} = \begin{Bmatrix} 4 \\ 6 \end{Bmatrix}$$

Therefore, the descent direction is those vectors that have an angle greater than 90°. Figure 5.2 shows possible descent directions. At $\mathbf{x} = \{2, 0\}$, the objective function has $f = 4$. The gradient $\mathbf{c}^{(k)}$ of the objective function is normal to the tangent plane. The object function increases along with the direction of the gradient. Then, all directions below the tangent plane are descent directions.

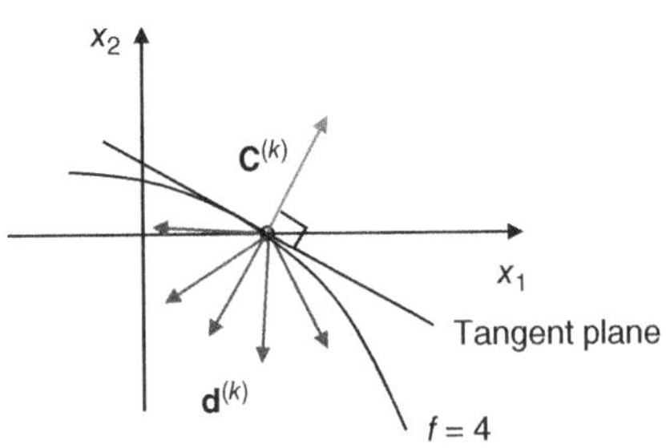


**Figure 5.2** Descent directions.

Since there are infinitely many descent directions, various optimization algorithms were developed based on how to determine the search direction $\mathbf{d}^{(k)}$. It seems that $\mathbf{d}^{(k)} = -\mathbf{c}^{(k)}$ is the best search direction as the objective function decreases fastest in this direction. However, we will be shown in the next section, it turns out that it is not an efficient direction. This is because the objective function is nonlinear, and the best descent direction may change at different designs. Different numerical optimization algorithms will be discussed in Sections 5.4 and 5.5.

### 5.3.2 Step-Size Termination Criterion

The determination of step size $\alpha_k$ is independent of the search direction. For the moment, let us assume that the search direction $\mathbf{d}^{(k)}$ is given. The goal of step-size determination is to find $\alpha_k$ that minimizes the objective function in the direction of $\mathbf{d}^{(k)}$. That is, the following 1D optimization problem needs to be solved for the step size:

$$\text{minimize} \quad \phi(\alpha_k) \equiv f\left(\mathbf{x}^{(k)} + \alpha_k \mathbf{d}^{(k)}\right) \tag{5.4}$$

As shown in Figure 5.3, the current design $\mathbf{x}^{(k)}$ is at the $y$-intercept $\alpha_k = 0$. The objective function has a negative slope at the current design because the search direction $\mathbf{d}^{(k)}$ is a descent direction. The goal is to find that step size $\alpha_k$ that minimizes $\phi(\alpha_k)$.

Similar to the optimality criteria, the minimum point must satisfy $\phi' = 0$ and $\phi'' > 0$. Therefore, the following condition can be obtained:

$$\phi' = \frac{\partial f}{\partial \mathbf{x}^{(k+1)}}^T \frac{\partial \mathbf{x}^{(k+1)}}{\partial \alpha} = \mathbf{c}^{(k+1)^T} \mathbf{d}^{(k)} = 0 \tag{5.5}$$

That is, the minimum point is when the search direction is perpendicular to the gradient of the objective function. The above condition is called the step-size termination criterion. The optimum step size $\alpha_k$ should satisfy this condition.

**Example 5.2 Linear search**

In Example 5.1, let the descent direction is chosen as $\mathbf{d}^{(k)} = \{-4, -6\}^T$ at the current design $\mathbf{x}^{(k)} = \{2, 0\}^T$. Find 1D line search to determine the step size $\alpha_k$ and new design $\mathbf{x}^{(k+1)}$.

**Solution:**

The step-size termination criterion in Eq. (5.5) can be written as

$$\mathbf{c}^{(k+1)^T} \mathbf{d}^{(k)} = -4(2x_1 + 3x_2) - 6(3x_1 - 4x_2) = 0$$

The new design point can be written as $\mathbf{x}^{(k+1)} = \mathbf{x}^{(k)} + \alpha_k \mathbf{d}^{(k)} = \{2 - 4\alpha_k, -6\alpha_k\}^T$. After substituting this point into the step-size termination criterion, we have

$$\mathbf{c}^{(k+1)^T} \mathbf{d}^{(k)} = -4(2(2 - 4\alpha_k) + 3(-6\alpha_k)) - 6(3(2 - 4\alpha_k) - 4(-6\alpha_k)) = 0$$

The above equation can be solved for $\alpha_k = 1.625$. Therefore, the new design point is

$$\mathbf{x}^{(k+1)} = \mathbf{x}^{(k)} + \alpha_k \mathbf{d}^{(k)} = \{2 - 4\alpha_k, -6\alpha_k\}^T = \{-4.5, -9.75\}^T$$

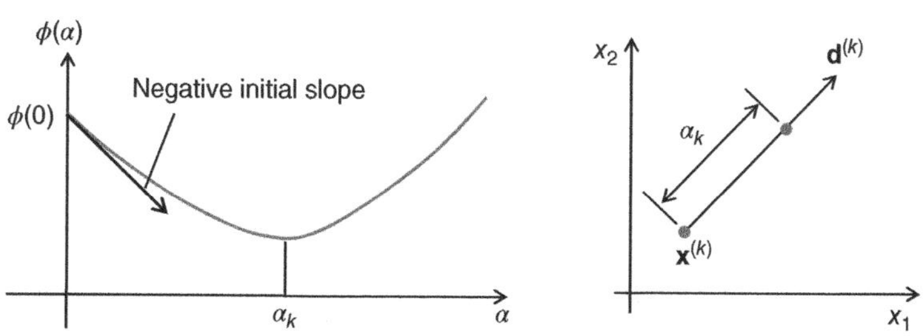


**Figure 5.3** Determination of step size in 1D line search.

Although $\mathbf{x}^{(k+1)}$ minimizes the objective function in the direction of $\mathbf{d}^{(k)}$, it is not the minimum point of the objective function. This can be verified by calculating the gradient of the objective function in this design, as

$$\mathbf{c}^{(k+1)} = \begin{Bmatrix} 2x_1 + 3x_2 \\ 3x_1 - 4x_2 \end{Bmatrix} = \begin{Bmatrix} -38.25 \\ 25.5 \end{Bmatrix}$$

Therefore, $\mathbf{x}^{(k+1)}$ does not satisfy the KKT condition, and the objective function can further be reduced by moving to a new descent direction.

Although it is best to choose the step size that satisfies the termination criterion, it is unnecessary to find the exact minimum point. This is because it is the minimum only in the direction of $\mathbf{d}^{(k)}$. Once we move to $\mathbf{x}^{(k)} + \alpha_k \mathbf{d}^{(k)}$, the objective function can be reduced in other directions as well. Therefore, instead of wasting computational resources in finding an accurate minimum in 1D line search, it might be more efficient to approximate it.

In this section, we assume that the search direction $\mathbf{d}^{(k)}$ is given and would like to determine the step size $\alpha_k$. There are many algorithms available to determine the step size. Some algorithms are based on reducing the intervals where the optimum step size exists, while other algorithms are approximating the objective function along the search direction as a simple polynomial function and finding the optimum step size from it. In this section, we briefly introduce an example of these two types of algorithms. For detailed discussions of step-size determination methods, readers are referred to the book by Arora [39].

### 5.3.3 Interval Reduction Method

The first group of algorithms starts from a range of $\alpha$, where the optimum $\alpha_k$ is within the range. It may not be straightforward to determine an appropriate range. A too-large range may slow the process of finding the optimum, while a too-small range may miss it. An important assumption in determining the range is that the objective function shows unimodal behavior within the interval. That is, the objective function initially decreases, and after the optimum step size $\alpha_k$, it increases again. Since different objective functions have different behaviors, it might be difficult to determine the range that works for general functions. As mentioned before, however, since finding the exact optimum $\alpha_k$ is not critical; the range is often determined heuristically.

Once the initial range is determined, the interval is gradually reduced by removing those subintervals that do not have optimum $\alpha_k$. This is done by computing the objective function at different $\alpha$'s and comparing their values. For example, the equal interval search method is to divide the range into $n$ intervals, to evaluate the function values starting from the first interval, and to stop when $\phi(\alpha_{j-1}) > \phi(\alpha_j)$ and $\phi(\alpha_{j+1}) > \phi(\alpha_j)$. Then, the range is reduced to $[\alpha_{j-1}, \alpha_{j+1}]$, and the process is repeated until the range becomes smaller than a threshold. This process is computationally expensive as it requires multiple simulations. The major disadvantage is that the samples in the previous interval cannot be used in the refined interval.

Among different interval reduction methods, the golden section search method is the most popular and efficient one. This method chooses the sample locations in such a way that the samples in the previous interval can be reused in the refined interval. It is a variable interval method, where the interval size is reduced in a constant ratio. With reference to Figure 5.4, let the interval length at $k$th iteration be $l^{(k)}$. The function values are evaluated at the two endpoints and two inside points. The two internal points $\alpha_3$ and $\alpha_4$ are located at the same distance from the two ends. The reduced interval is selected depending on function values at these sample locations. If $\phi(\alpha_3) < \phi(\alpha_4)$, then the

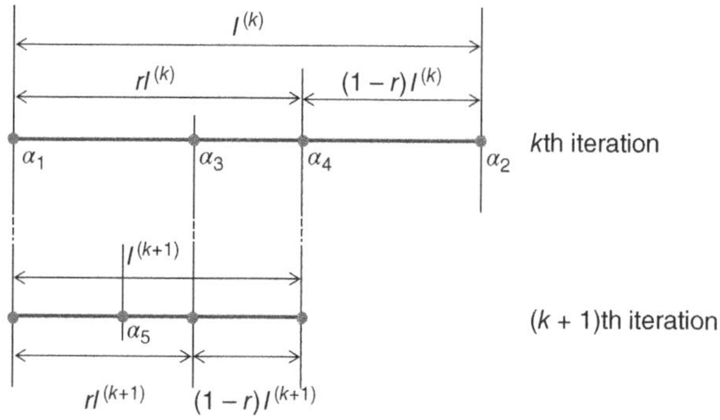


**Figure 5.4** Interval reduction in the golden section search method.

minimum lies between $\alpha_1$ and $\alpha_4$, and thus, the new interval becomes $l^{(k+1)} = \alpha_4 - \alpha_1$. If $\phi(\alpha_3) > \phi(\alpha_4)$, then the minimum lies between $\alpha_3$ and $\alpha_2$, and thus, the new interval becomes $l^{(k+1)} = \alpha_2 - \alpha_3$. Figure 5.4 shows the case of $\phi(\alpha_3) < \phi(\alpha_4)$. This process is repeated until the interval size is smaller than a predetermined threshold.

In the golden section search, the internal two points ($\alpha_3$ and $\alpha_4$) are chosen in such a way that these points can be reused in the next iteration with a reduced interval. As shown in Figure 5.4, $\alpha_3$ becomes the sample point in the next iteration. Therefore, only one additional point $\alpha_5$ is required in each iteration, and the interval is reduced to $l^{(k+1)} = rl^{(k)}$. In order to reuse $\alpha_3$, it is required that $rl^{(k+1)} = (1-r)l^{(k)}$. This requirement ends up being a quadratic equation $r^2 + r - 1 = 0$, whose positive root is $r = \left(-1 + \sqrt{5}\right)/2 = 0.618$. Thus, the two internal points are located at a distance of 0.618 l or 0.382 l from either end of the interval.

### 5.3.4 Quadratic Interpolation Method

The second group of algorithms approximates the objective function as a simple polynomial. For example, the quadratic interpolation method evaluates the objective function at three step sizes, $\alpha_1$, $\alpha_2$, and $\alpha_3$. Then, the function is approximated by a quadratic polynomial $\hat{\phi}(\alpha) = a\alpha^2 + b\alpha + c$. Since the polynomial has three coefficients, the three sample points are enough to determine the unknown coefficients. From the KKT condition in Chapter 4, the minimum point can be identified using the derivative information. That is, $\hat{\phi}'(\alpha) = 2a\alpha + b = 0$ and $\hat{\phi}''(\alpha) = 2a > 0$. From these two conditions, we can obtain the optimum step size $\alpha^* = -b/2a$ with $a > 0$ and $b < 0$. Figure 5.5 shows the quadratic approximation of the objective function and the approximate optimum step size.

**Figure 5.5** Quadratic approximation for step-size determination.

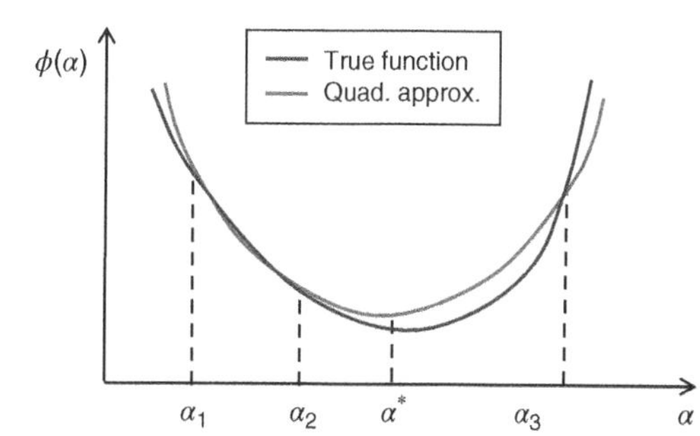

**Example 5.3 Quadratic interpolation method**
In Example 5.1, let the descent direction is chosen as $\mathbf{d}^{(k)} = \{-4, -6\}^T$ at the current design $\mathbf{x}^{(k)} = \{2, 0\}^T$. Use the quadratic interpolation method to determine the step size $\alpha_k$ and new design $\mathbf{x}^{(k+1)}$. Use the following three points to sample the objective function: $\alpha = 0$, 1, and 2.

**Solution:**
The new design point can be written as $\mathbf{x}^{(k+1)} = \mathbf{x}^{(k)} + \alpha_k \mathbf{d}^{(k)} = \{2 - 4\alpha_k, -6\alpha_k\}^T$. At the three sample points, the objective function has the following values: $\phi(0) = 4$, $\phi(1) = -32$, and $\phi(2) = -36$. These three samples are used to approximate a quadratic function $\hat{\phi}(\alpha) = a\alpha^2 + b\alpha + c = 16\alpha^2 - 52\alpha + 4$. Therefore, the optimum step size is $\alpha^* = -b/2a = 1.625$, which happens to be the same as the optimum step size in Example 5.2. This is because the original objective function is a quadratic polynomial. Therefore, the new design point is

$$\mathbf{x}^{(k+1)} = \mathbf{x}^{(k)} + \alpha_k \mathbf{d}^{(k)} = \{2 - 4\alpha_k, -6\alpha_k\}^T = \{-4.5, -9.75\}^T$$

When the objective function is a quadratic function of design variables, the step size of the line search can be calculated analytically. Consider the following objective function is a quadratic form:

$$f(\mathbf{x}) = \frac{1}{2}\mathbf{x}^T\mathbf{H}\mathbf{x} - \mathbf{x}^T\mathbf{b} \tag{5.6}$$

where $\mathbf{H}$ is the Hessian matrix and $\mathbf{b}$ is a constant vector. The gradient of the objective function can be written as $\mathbf{c} = \mathbf{H}\mathbf{x} - \mathbf{b}$. From the KKT condition, the minimum of the objective function is equivalent to solving the linear system of equations $\mathbf{H}\mathbf{x} = \mathbf{b}$. When the search direction $\mathbf{d}$ is given, the design can be updated to $\mathbf{x} + \alpha\mathbf{d}$. By substituting this new design into the objective function, we have

$$f(\mathbf{x} + \alpha\mathbf{d}) = \frac{1}{2}(\mathbf{x} + \alpha\mathbf{d})^T\mathbf{H}(\mathbf{x} + \alpha\mathbf{d}) - (\mathbf{x} + \alpha\mathbf{d})^T\mathbf{b} \tag{5.7}$$

By differentiating this with respect to $\alpha$, we can obtain the optimum step size as

$$\alpha = -\frac{\mathbf{d}^T\mathbf{c}}{\mathbf{d}^T\mathbf{H}\mathbf{d}} > 0 \tag{5.8}$$

Although this step size is only true for a quadratic function, it can be used for general nonlinear function as we can approximate them using a quadratic function when the design change is small.

## 5.4 Unconstrained Optimization Algorithms

When there are no constraints on the design problem, it is referred to as an unconstrained optimization problem. Even if most engineering problems have constraints, these problems can be transformed into unconstrained ones by using the penalty method or the Lagrange multiplier method. The unconstrained optimization problem sometimes contains the lower and upper limits of a design variable, since this type of constraint can be treated easily. The standard form of an unconstrained optimization problem can be written as

$$\text{minimize} \quad f(\mathbf{x}) \tag{5.9}$$

In the standard numerical optimization algorithms, the optimum design is found iteratively. Let the superscript $(k)$ be an iteration counter; the design at $k+1$th iteration is updated by

$x^{(k+1)} = x^{(k)} + \alpha_k \mathbf{d}^{(k)}$. In the previous section, the methods of determining the optimum step size $\alpha_k$ was discussed assuming that the search direction $\mathbf{d}^{(k)}$ is given. In this section, various methods of determining the search direction $\mathbf{d}^{(k)}$ will be discussed. As mentioned before, there are infinitely many search directions. Therefore, the performance of optimization algorithms depends on the selection of the search direction.

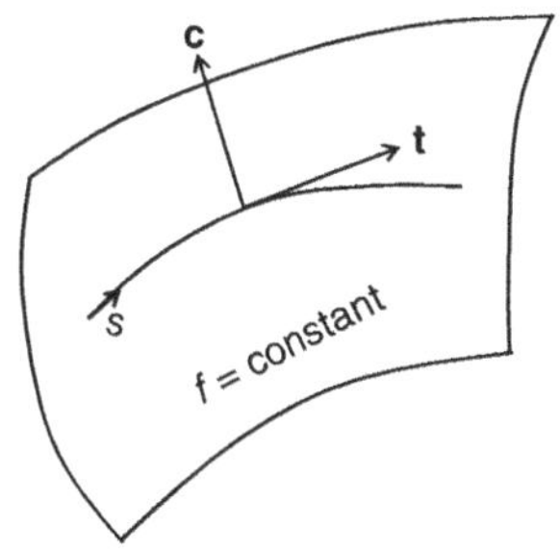


**Figure 5.6** Gradient of objective function.

### 5.4.1 Steepest Descent Method

Since the search direction needs to be a descent direction, it is intuitive to choose the direction that reduces the objective function the most. It turns out that the objective function increases fastest along the direction of its gradient. This is obvious from the definition of the gradient. It can also be shown that the gradient is normal to the hypersurface of $f(\mathbf{x})$ = constant. The surface shown in Figure 5.6 represents $f(\mathbf{x})$ = constant in 2D design space. Although we cannot plot the surface in a higher-dimensional space, it is called a hypersurface. On the surface, we can define a parametric curve $\mathbf{x}(s)$, where $s$ is a parametric coordinate. Then, a tangent vector $\mathbf{t}(s)$ of the curve can be defined as $\mathbf{t}(s) = d\mathbf{x}/ds$. That is, the derivative of the coordinate on the curve with respect to the parametric coordinate $s$. With the definition of the tangent vector, the derivative of the objective function with respect to $s$ can be defined as

$$\frac{df}{ds} = \frac{df}{d\mathbf{x}}^T \frac{d\mathbf{x}}{ds} = \mathbf{c}^T\mathbf{t} = 0 \tag{5.10}$$

This result is expected as the objective function remains constant along the curve. An important observation from this equation is that the gradient vector $\mathbf{c}$ is perpendicular to the tangent vector $\mathbf{t}$. Therefore, the normal vector to the hypersurface of $f(\mathbf{x})$ = constant is the direction that increases the objective function fastest.

With this observation, the steepest descent method chooses the negative of the gradient as a search direction:

$$\mathbf{d}^{(k)} = -\mathbf{c}^{(k)} \tag{5.11}$$

First, this search direction satisfies the descent condition: $\mathbf{c}^{(k)T}\mathbf{d}^{(k)} = -\|\mathbf{c}^{(k)}\|^2 < 0$. This seems the best direction for optimization, but it turns out that this is not a good strategy due to the nonlinearity of the objective function. Especially, the convergence of this strategy becomes slow near the optimum point. This is because the search directions in the consecutive two iterations are perpendicular. In order to show this, consider the step-size termination criterion:

$$\mathbf{c}^{(k+1)^T}\mathbf{d}^{(k)} = -\mathbf{c}^{(k+1)^T}\mathbf{c}^{(k)} = 0 \tag{5.12}$$

Therefore, $\mathbf{c}^{(k+1)}$ and $\mathbf{c}^{(k)}$ are perpendicular. This makes the search directions move in a zigzag pattern and slows the convergence. The reason for a slow convergence is that the search direction is determined based on the current information only, not using information from the previous iterations. In addition, only the first-order derivatives are used in determining the search direction.

**Example 5.4 Steepest descent method**

Consider an elliptical objective function $f(x_1, x_2) = x_1^2 + rx_2^2$. When the initial design is $\mathbf{x}^{(0)} = \{r, 1\}$, show the trajectory of design changes using the steepest descent method. The step size can be calculated using the step-size termination criterion in Eq. (5.5). Plot the contour of the objective function along with the trajectory of design change when $r = 10$.

**Solution:**

For the given objective function, the gradient vector can be calculated as

$$\mathbf{c}^{(k)} = \left\{2x_1^{(k)}, 2rx_2^{(k)}\right\}^T$$

Therefore, at the initial design, $\mathbf{x}^{(0)} = \{r, 1\}$, the gradient vector and the search direction become $\mathbf{c}^{(0)} = \{2r, 2r\}^T$ and $\mathbf{d}^{(0)} = -\mathbf{c}^{(0)}$. Then, at the next iteration, $k = 1$, the design becomes $\mathbf{x}^{(1)} = \{r - 2r\alpha, 1 - 2r\alpha\}$, and the gradient becomes $\mathbf{c}^{(1)} = \{2r - 4r\alpha, 2r - 4r^2\alpha\}^T$. Therefore, the step-size termination criterion can be used to determine the optimum step size:

$$\mathbf{c}^{(1)^T}\mathbf{d}^{(0)} = 2r(2r - 4r\alpha) + 2r\left(2r - 4r^2\alpha\right) = 0 \rightarrow \quad \alpha_0 = \frac{1}{r+1}$$

Therefore, the design at $k = 1$ can be determined as

$$\mathbf{x}^{(1)} = \{r - 2r\alpha, 1 - 2r\alpha\} = \frac{r-1}{r+1}\{r, -1\}$$

This process is repeated for $k = 2, 3, \ldots$ to obtain the general formulation of the updated design as

$$\mathbf{x}^{(k)} = \left(\frac{r-1}{r+1}\right)^k \left\{r, (-1)^k\right\}$$

The following Matlab code plots the contour of the objective function and the trajectory of design changes, as shown in Figure 5.7. For the quadratic objective function, it is possible to find the optimum design if the initial design is located along the axis of each design variable because then the initial descent direction is toward the optimum design $\mathbf{x}^* = \{0, 0\}$. However, if the initial design is not on the axis, the descent direction is misaligned with the optimum design. Therefore, although the updated designs move toward the optimum design, they show a zigzag pattern due to the orthogonal property of two consecutive descent directions.

```
r=10; x0=[r 1];
for k=1:10
x(k,:)=((r-1)/(r+1))^k*[r (-1)^k];
```

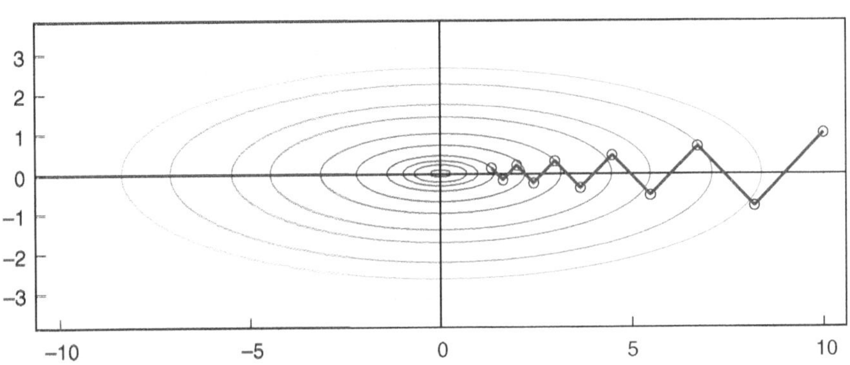


**Figure 5.7** Zigzag motion of the steepest descent method.

```
end
x = [x0; x];
figure(1);hold on; axis equal;
line(x(:,1),x(:,2),'marker','o')
%
x = linspace(-12,12);
y = linspace(-5,5);
[X,Y] = meshgrid(x,y);
Z = X.^2 + r*Y.^2;
contour(X,Y,Z,[0.1 0.5 1 2 5 10 20 30 50 70 100])
```

### 5.4.2 Conjugate Gradient Method

The slow convergence of the steepest descent method is because the method only uses the current gradient information. Due to the perpendicular nature of consecutive two gradients, the search directions move in a zigzag pattern, as shown by the solid blue lines in Figure 5.8. However, as shown by the dashed red arrow, the convergence behavior can be improved if the search direction is in the diagonal direction of the two descent directions.

The conjugate gradient method developed by Fletcher and Reeves [50] improves the rate of slow convergence in the steepest descent method by selecting a direction which takes into account not only the gradient direction but also the search direction in the previous iteration. By doing that, it is possible to develop algorithms that are guaranteed to converge to the minimum of an $n$-dimensional quadratic function in no more than $n$ iterations.

The difference in this method is the computation of search direction. The new descent direction is computed by

$$\mathbf{d}^{(k)} = -\mathbf{c}^{(k)} + \beta_k \mathbf{d}^{(k-1)} \tag{5.13}$$

and the first iteration is the same as the steepest descent method, $\mathbf{d}^{(0)} = -\mathbf{c}^{(0)}$. Because of the modification of the descent direction using the previous descent direction, the conjugate direction methods home at the minimum in the second move. The parameter $\beta_k$ controls how much the previous descent direction contributes to the current one. It is determined based on the magnitude of the gradients as

$$\beta_k = \left( \frac{\|\mathbf{c}^{(k)}\|}{\|\mathbf{c}^{(k-1)}\|} \right)^2 > 0 \tag{5.14}$$

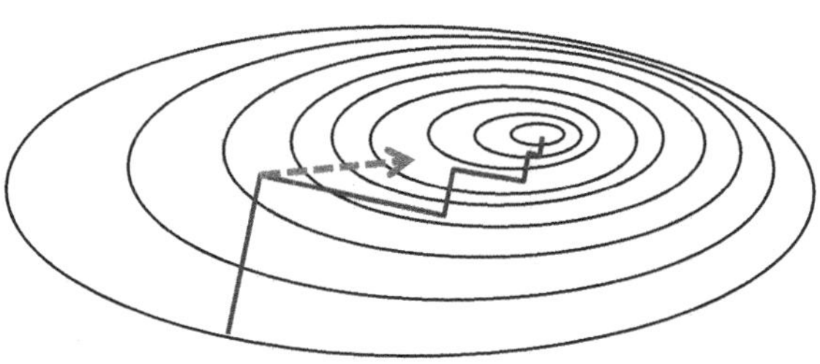

**Figure 5.8** Conjugate gradient direction.

This method tends to select the descent direction as a diagonal of the two orthogonal steepest descent directions, such that a zigzagging pattern can be eliminated. This method always has better convergence than the steepest descent method. Since the optimum design will satisfy the KKT condition, the parameter $\beta_k$ will be zero at the optimum design. It is noted that each move ends at the tangent to the function contour, because we minimize the function in the search direction. Most implementations approximate the function as a quadratic along the search direction, so that if the function is a quadratic, the search will end at the minimum in that direction. However, if the function is not quadratic, we are likely to stop at a distance from where the direction is tangent to the contour.

**Example 5.5 Conjugate gradient method**

Perform the first two iterations of the unconstrained optimization problem in Example 5.4 using the conjugate gradient method. Use the analytical step size shown in Eq. (5.8).

**Solution:**

The first iteration of the conjugate gradient method is identical to the steepest descent method. Therefore, $\mathbf{x}^{(0)} = \{10, 1\}$, $\mathbf{d}^{(0)} = -\mathbf{c}^{(0)} = \{-20, -20\}$. From Eq. (5.8), $\alpha_0 = -(\mathbf{d}^T\mathbf{c})/(\mathbf{d}^T\mathbf{H}\mathbf{d}) = 0.0909$. The updated design becomes $\mathbf{x}^{(1)} = \mathbf{x}^{(0)} + \alpha_0\mathbf{d}^{(0)} = \{8.1818, -0.8182\}$.

When $k = 1$: $\mathbf{c}^{(1)} = \{16.3636, -16.3636\}$, and

$$\beta_1 = \left(\frac{\|\mathbf{c}^{(1)}\|}{\|\mathbf{c}^{(0)}\|}\right)^2 = 0.6694$$

Therefore, the search direction in the conjugate gradient method becomes

$$\mathbf{d}^{(1)} = -\mathbf{c}^{(1)} + \beta_1\mathbf{d}^{(0)} = \{-29.7521, 2.9752\}$$

The step size $\alpha_1 = \mathbf{d}^T(\mathbf{b} - \mathbf{H}\mathbf{x})/(\mathbf{d}^T\mathbf{H}\mathbf{d}) = 0.2750$. The updated design becomes $\mathbf{x}^{(2)} = \mathbf{x}^{(1)} + \alpha_1\mathbf{d}^{(1)} = \{0, 0\}$, which is the optimum design. At this point, $\mathbf{c}^{(2)} = \{0, 0\}$. Since the objective function is a quadratic function, the conjugate gradient method converges in two iterations. Figure 5.9 shows the contour of the objective function and the trajectory of the design changes. The dark-colored lines are for the conjugate gradient method, while the light-colored lines are for the steepest descent method. By utilizing previous descent direction, the algorithm finds the optimum design in two iterations.

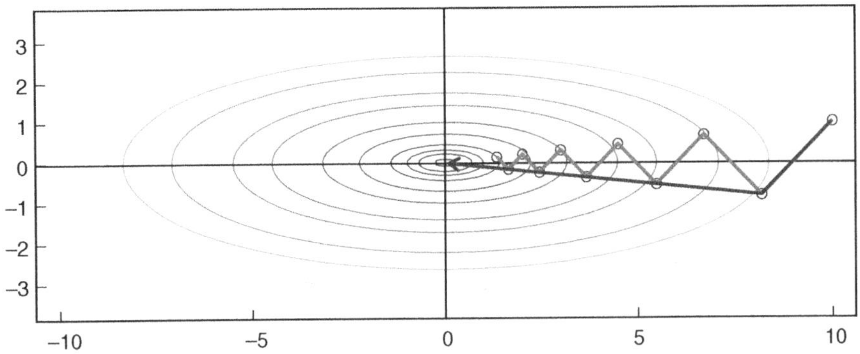


**Figure 5.9** Convergence of conjugate gradient method.

### 5.4.3 Newton Method

The previous two methods, the steepest descent and conjugate gradient methods, only use the first-order information (gradients) of the objective function to find the optimum design, which is called a first-order method. The convergence of these methods is generally slow and often requires many iterations to find the optimum design. Although the conjugate gradient method in Example 5.5 converged in two iterations, this happened because the objective function is a quadratic function. For a general nonlinear objective function, the conjugate gradient method may require many iterations as well.

In addition to the gradient information, Newton's method uses second-order information (Hessian) to approximate the objective function as a quadratic function of the design. The advantage of this method is that the design change $\Delta\mathbf{x}$ that can satisfy the KKT condition can be found by solving a linear system of equations. Therefore, if the objective function is a quadratic function, this method can find the optimum design in one iteration. For a general nonlinear objective function, this method shows a second-order convergence, which is much faster than the first-order convergence.

Let us consider the unconstrained optimization problem in Eq. (5.9). The design variable is updated at $k+1$th iteration by $\mathbf{x}^{(k+1)} = \mathbf{x}^{(k)} + \Delta\mathbf{x}^{(k)}$. Since $\mathbf{x}^{(k)}$ is already determined from $k$th iteration, the only variable is $\Delta\mathbf{x}^{(k)}$. Therefore, the objective function is approximated using the Taylor series expansion with respect to the previous design $\mathbf{x}^{(k)}$ as

$$\phi\left(\Delta\mathbf{x}^{(k)}\right) = f\left(\mathbf{x}^{(k)} + \Delta\mathbf{x}^{(k)}\right) \approx f\left(\mathbf{x}^{(k)}\right) + \mathbf{c}^{(k)^T}\Delta\mathbf{x}^{(k)} + \frac{1}{2}\Delta\mathbf{x}^{(k)^T}\mathbf{H}^{(k)}\Delta\mathbf{x}^{(k)} \tag{5.15}$$

where $\mathbf{H}^{(k)}$ is the Hessian matrix ($n \times n$), which is made of second-order derivatives. The above equation basically approximates the objective function at the new design as a quadratic function of $\Delta\mathbf{x}^{(k)}$. The goal is to find the design change $\Delta\mathbf{x}^{(k)}$ to minimize the objective function. One can use the KKT condition to find $\Delta\mathbf{x}^{(k)}$ that minimizes $\phi(\Delta\mathbf{x}^{(k)})$ as

$$\frac{\partial\phi}{\partial\Delta\mathbf{x}^{(k)}} = \mathbf{c}^{(k)} + \mathbf{H}^{(k)}\Delta\mathbf{x}^{(k)} = 0 \tag{5.16}$$

which yields the following design change:

$$\Delta\mathbf{x}^{(k)} = -\mathbf{H}^{(k)^{-1}}\mathbf{c}^{(k)} \tag{5.17}$$

In practice, since calculating the inverse of the Hessian matrix is expensive, the linear system of equations $\mathbf{H}^{(k)}\Delta\mathbf{x}^{(k)} = -\mathbf{c}^{(k)}$ is solved for $\Delta\mathbf{x}^{(k)}$. If the current estimated design $\mathbf{x}^{(k)}$ is sufficiently close to the optimum design, then Newton's method will show quadratic convergence.

Albeit its fast convergence, the major concern of Newton's method is how to calculate the Hessian matrix. The greater the number of design variables is, the greater the cost of computing $\mathbf{H}^{(k)}$ is. Although the Hessian information can be calculated using sensitivity analysis [49], it is not available in many application software. Therefore, Newton's method is often limited to cases when the functional form of the objective function is given in terms of design variables.

In addition, the Hessian matrix needs to be positive definite. Otherwise, the design change $\Delta\mathbf{x}^{(k)}$ from Eq. (5.17) can be unstable. The positive definite Hessian matrix means that the objective function should be convex (at least locally). From the descent direction, $\mathbf{c}^{(k)T}\Delta\mathbf{x}^{(k)} < 0$, the Hessian matrix must satisfy

$$\mathbf{c}^{(k)}\mathbf{H}^{(k)}\mathbf{c}^{(k)} > 0 \tag{5.18}$$

which is the definition of positive definiteness. Therefore, $\Delta\mathbf{x}^{(k)}$ is a descent direction when $\mathbf{H}^{(k)}$ is positive definite.

Lastly, Newton's method does not guarantee convergence. The method would find the optimum design if the starting design is close to the optimum design and the objective function is convex. Thus, several modifications are available. For example, the design update algorithm can be modified to include a step size by using a line search. In that case, the linear system of equations solves the search direction,

$$\Delta\mathbf{d}^{(k)} = -\mathbf{H}^{(k)^{-1}}\mathbf{c}^{(k)} \tag{5.19}$$

and the design can be updated as

$$\mathbf{x}^{(k+1)} = \mathbf{x}^{(k)} + \alpha_k \Delta\mathbf{d}^{(k)} \tag{5.20}$$

**Example 5.6 Newton's method**

Find the optimum design of the unconstrained optimization problem in Example 5.4 using Newton's method.

**Solution:**

Since the objective function is quadratic, the Hessian matrix is constant. With the initial design $\mathbf{x}^{(0)} = \{10, 1\}$, the gradient of the objective function is $\mathbf{c}^{(0)} = \{20, 20\}$. Therefore, the design change can be found by solving the following linear system of equations:

$$\mathbf{H}\Delta\mathbf{x}^{(0)} = \begin{bmatrix} 2 & 0 \\ 0 & 20 \end{bmatrix} \begin{Bmatrix} \Delta x_1^{(0)} \\ \Delta x_2^{(0)} \end{Bmatrix} = \begin{Bmatrix} -20 \\ -20 \end{Bmatrix} = -\mathbf{c}^{(0)}$$

The above equations can be solved for $\Delta\mathbf{x}^{(0)} = \{-10, -1\}$. Therefore, the updated design becomes $\mathbf{x}^{(1)} = \mathbf{x}^{(0)} + \Delta\mathbf{x}^{(0)} = \{0, 0\}$. With the new design $\mathbf{x}^{(1)}$, $\mathbf{c}^{(1)} = \{0, 0\}$, and therefore, it is the optimum design. Since the objective function is a quadratic function of design variables, Newton's method converges in one iteration.

### 5.4.4 Quasi-Newton Method

Although Newton's method has a quadratic convergence, the cost of computing the Hessian matrix and the lack of guaranteed convergence are drawbacks to this method. On the other hand, the steepest descent method only requires gradient information but has a problem of slow convergence. The main difference between these two methods is Hessian information, i.e., the second-order derivatives. In a similar way that the gradient information can be calculated using finite differences between two function values, the Hessian information can also be approximated using the difference of two gradients. Quasi-Newton methods are based on this concept and approximate the Hessian information using gradient information. These methods build an approximation to the Hessian matrix and update it after each gradient calculation. For a quadratic objective function, they are guaranteed to reach the optimum design in no more than $n$ iterations and to have an exact Hessian after $n$ iterations. Due to the approximate Hessian information, these methods show a slower convergence than that of Newton's method, but faster than that of the steepest descent method.

Quasi-Newton methods update the Hessian matrix starting from an identity matrix. An important property of Hessian update is that it needs to maintain the matrix positive definite when an exact line search is used. The calculation of search direction in Eq. (5.19) requires inverting the Hessian matrix or solving a linear system of matrix equations. Since the Hessian matrix is an approximate one, it would be more efficient if the inverse of the Hessian matrix were approximated directly. Therefore, the search direction in quasi-Newton methods can be determined by

$$\mathbf{d}^{(k)} = -\mathbf{H}^{(k)^{-1}}\mathbf{c}^{(k)}$$
$$\mathbf{d}^{(k)} = -\mathbf{A}^{(k)}\mathbf{c}^{(k)} \tag{5.21}$$

where $\mathbf{H}^{(k)}$ is an approximate Hessian matrix and $\mathbf{A}^{(k)}$ is an inverse of the approximate Hessian matrix. After the search direction is determined, the step size is determined using line search. Therefore, the main difference in quasi-Newton methods is how to update the Hessian or its inverse.

In order to show how the Hessian matrix is updated, the following vectors are defined first:

$$\mathbf{s}^{(k)} = \mathbf{x}^{(k+1)} - \mathbf{x}^{(k)} = \alpha_k \mathbf{d}^{(k)}$$
$$\mathbf{y}^{(k)} = \mathbf{c}^{(k+1)} - \mathbf{c}^{(k)} \tag{5.22}$$

where $\mathbf{s}^{(k)}$ is the change of design variables, and $\mathbf{y}^{(k)}$ is the change of gradients. Then, the quasi-Newton condition means that the curvature remains constant between two consecutive iterations. Consider the Taylor series expansion of the gradient vector, $\mathbf{c}^{(k+1)} = \mathbf{c}(\mathbf{x}^{(k)} + \mathbf{s}^{(k)}) = \mathbf{c}^{(k)} + \mathbf{H}^{(k)}\mathbf{s}^{(k)} +$ H. O. T. Therefore, the change of gradient can be approximated by $\mathbf{y}^{(k)} = \mathbf{c}^{(k+1)} - \mathbf{c}^{(k)} \approx \mathbf{H}^{(k)}\mathbf{s}^{(k)}$. By multiplying $\mathbf{s}^{(k)}$ on both sides, we can obtain the curvature at the current design as

$$\mathbf{s}^{(k)^T}\mathbf{y}^{(k)} = \mathbf{s}^{(k)^T}\mathbf{H}^{(k)}\mathbf{s}^{(k)} \tag{5.23}$$

Therefore, the quasi-Newton condition is to update the Hessian matrix while keeping the curvature constant. That is, $\mathbf{s}^{(k)T}\mathbf{H}^{(k+1)}\mathbf{s}^{(k)} = \mathbf{s}^{(k)T}\mathbf{H}^{(k)}\mathbf{s}^{(k)} = \mathbf{s}^{(k)T}\mathbf{y}^{(k)}$. Therefore, the following conditions can be written as

$$\mathbf{H}^{(k+1)}\mathbf{s}^{(k)} = \mathbf{y}^{(k)}$$
$$\mathbf{A}^{(k+1)}\mathbf{y}^{(k)} = \mathbf{s}^{(k)} \tag{5.24}$$

It is obvious that there are infinitely many ways of updating the Hessian matrix while satisfying the quasi-Newton condition in Eq. (5.24). Normally, the Hessian (or its inverse) matrix is updated using a vector at each iteration, starting from an identity matrix. For example, when a single vector $\mathbf{u}$ is used to update the Hessian matrix, it is called rank-1 update: $\mathbf{H}^{(k+1)} = \mathbf{H}^{(k)} + a_k\mathbf{u}\mathbf{u}^T$. Of course, the magic is to choose vector $\mathbf{u}$ such that the quasi-Newton condition is satisfied while the Hessian matrix is positive definite.

One of the most popular quasi-Newton methods is the Broyden–Fletcher–Goldfarb–Shanno (BFGS) algorithm [51], which uses a rank-2 update. It means that two vectors, $\mathbf{u}$ and $\mathbf{v}$, are used to update the Hessian matrix as

$$\mathbf{H}^{(k+1)} = \mathbf{H}^{(k)} + a_k\mathbf{u}\mathbf{u}^T + b_k\mathbf{v}\mathbf{v}^T \tag{5.25}$$

Among infinitely many possibilities, the first vector is chosen as $\mathbf{u} = \mathbf{y}^{(k)}$. Then, the quasi-Newton condition requires the following relation:

$$\mathbf{H}^{(k+1)}\mathbf{s}^{(k)} = \mathbf{H}^{(k)}\mathbf{s}^{(k)} + a_k\mathbf{y}^{(k)}\left(\mathbf{y}^{(k)^T}\mathbf{s}^{(k)}\right) + b_k\mathbf{v}\left(\mathbf{v}^T\mathbf{s}^{(k)}\right) = \mathbf{y}^{(k)} \tag{5.26}$$

In addition, the coefficients $a_k$ and $b_k$ are chosen to normalize the dot-product terms as $a_k = 1/\mathbf{y}^{(k)T}\mathbf{s}^{(k)}$ and $b_k = -1/\mathbf{v}^T\mathbf{s}^{(k)}$. Then, the quasi-Newton condition becomes

$$\mathbf{H}^{(k+1)}\mathbf{s}^{(k)} = \mathbf{H}^{(k)}\mathbf{s}^{(k)} + \mathbf{y}^{(k)} - \mathbf{v} = \mathbf{y}^{(k)} \tag{5.27}$$

This condition yields $\mathbf{v} = \mathbf{H}^{(k)}\mathbf{s}^{(k)}$. Therefore, the two vectors are determined. In summary, the Hessian matrix update of the BFGS method can be written as

$$\mathbf{H}^{(k+1)} = \mathbf{H}^{(k)} + \frac{\mathbf{y}^{(k)}\mathbf{y}^{(k)T}}{\mathbf{y}^{(k)T}\mathbf{s}^{(k)}} - \frac{\left(\mathbf{H}^{(k)}\mathbf{s}^{(k)}\right)\left(\mathbf{H}^{(k)}\mathbf{s}^{(k)}\right)^T}{\mathbf{s}^{(k)T}\mathbf{H}^{(k)}\mathbf{s}^{(k)}} \tag{5.28}$$

As mentioned before, this update can keep the Hessian matrix positive definite. In addition, if the objective function is quadratic, $\mathbf{H}^{(n)}$ will be the exact Hessian matrix.

As mentioned before, it would be unnecessary to update the Hessian matrix and invert it to calculate the search direction. Instead, the inverse of the Hessian matrix can be updated directly. The DFP (Davidon–Fletcher–Powell [52]) method approximates the inverse of the Hessian matrix using first-order sensitivity information. Similar to the Hessian update, the inverse of the Hessian matrix is updated by

$$\mathbf{A}^{(k+1)} = \mathbf{A}^{(k)} + a_k\mathbf{u}\mathbf{u}^T + b_k\mathbf{v}\mathbf{v}^T \tag{5.29}$$

By selecting $\mathbf{u} = \mathbf{s}^{(k)}$, the other vector $\mathbf{v} = \mathbf{A}^{(k)}\mathbf{y}^{(k)}$ can be determined from the quasi-Newton condition. Therefore, the inverse of the Hessian matrix is updated as

$$\mathbf{A}^{(k+1)} = \mathbf{A}^{(k)} + \frac{\mathbf{s}^{(k)}\mathbf{s}^{(k)T}}{\mathbf{y}^{(k)T}\mathbf{s}^{(k)}} - \frac{\left(\mathbf{A}^{(k)}\mathbf{y}^{(k)}\right)\left(\mathbf{A}^{(k)}\mathbf{y}^{(k)}\right)^T}{\mathbf{y}^{(k)T}\mathbf{A}^{(k)}\mathbf{y}^{(k)}} \tag{5.30}$$

The updated matrix $\mathbf{A}^{(k+1)}$ has similar properties to the updated Hessian matrix $\mathbf{H}^{(k+1)}$.

**Example 5.7 Quasi-Newton method**

Minimize $f(x_1, x_2) = 12x_1^2 + 4x_2^2 - 12x_1x_2 + 2x_1$ using the BFGS quasi-Newton method with the exact line search starting with the initial design $\mathbf{x}^{(0)} = \{-1, -2\}$.

**Solution:**

Although it is possible to calculate the exact Hessian matrix by differentiating the objective function twice, it is assumed that the Hessian matrix is unknown and needs to be updated. At the initial step, since $\mathbf{H}^{(0)} = \mathbf{I}$, the search direction is the same as that of the steepest descent method:

$$\nabla f = \begin{Bmatrix} 24x_1 - 12x_2 + 2 \\ 8x_2 - 12x_1 \end{Bmatrix}, \quad \mathbf{d}^{(0)} = -\mathbf{c}^{(0)} = \begin{Bmatrix} -2 \\ 4 \end{Bmatrix}$$

Using the search direction and step size, the design can be updated as

$$\mathbf{x}^{(1)} = \begin{Bmatrix} -1 \\ -2 \end{Bmatrix} + \alpha_0 \begin{Bmatrix} -2 \\ 4 \end{Bmatrix}$$

The objective function at the updated design becomes $f(\alpha_0) = 12(-1 - 2\alpha_0)^2 + 4(-2 + 4\alpha_0)^2 - 12(-1 - 2\alpha_0)(-2 + 4\alpha_0) + 2(-1 - 2\alpha_0)$. The value of $\alpha_0$ for which $f$ is the minimum is obtained from the condition $df/d\alpha_0 = 0$, or $\alpha_0 = 0.048077$. Therefore, the updated design and the gradient at that point are

$$\mathbf{x}^{(1)} = \begin{Bmatrix} -1.0961 \\ -1.8077 \end{Bmatrix}, \quad \mathbf{c}^{(1)} = \nabla f\left(\mathbf{x}^{(1)}\right) = \begin{Bmatrix} -2.6154 \\ -1.3077 \end{Bmatrix}$$

Then, the vectors $\mathbf{s}^{(0)}$ and $\mathbf{y}^{(0)}$ can be calculated as

$$\mathbf{s}^{(0)} = \mathbf{x}^{(1)} - \mathbf{x}^{(0)} = \begin{Bmatrix} -0.0961 \\ 0.1923 \end{Bmatrix}, \quad \mathbf{y}^{(0)} = \mathbf{c}^{(1)} - \mathbf{c}^{(0)} = \begin{Bmatrix} -4.6154 \\ 2.6923 \end{Bmatrix}$$

Therefore, the Hessian matrix in Eq. (5.28) is updated as

$$\mathbf{H}^{(1)} = \mathbf{H}^{(0)} + \frac{\mathbf{y}^{(0)}\mathbf{y}^{(0)T}}{\mathbf{y}^{(0)T}\mathbf{s}^{(0)}} - \frac{\left(\mathbf{H}^{(0)}\mathbf{s}^{(0)}\right)\left(\mathbf{H}^{(0)}\mathbf{s}^{(0)}\right)^T}{\mathbf{s}^{(0)T}\mathbf{H}^{(0)}\mathbf{s}^{(0)}} = \begin{bmatrix} 22.9538 & -12.5231 \\ -12.5231 & 7.7385 \end{bmatrix}$$

Then, the search direction is obtained from Eq. (5.21) as

$$\mathbf{d}^{(k)} = -\mathbf{H}^{(k)^{-1}}\mathbf{c}^{(k)} = \begin{Bmatrix} 1.7608 \\ 3.0186 \end{Bmatrix}$$

And the design at the next iteration becomes

$$\mathbf{x}^{(2)} = \begin{Bmatrix} -1.0961 \\ -1.8077 \end{Bmatrix} + \alpha_1 \begin{Bmatrix} 1.7608 \\ 3.0186 \end{Bmatrix}$$

The step size $\alpha_1$ is calculated from the condition $df/d\alpha_1 = 0$, to yield $\alpha_1 = 0.4332$, and

$$\mathbf{x}^{(2)} = \begin{Bmatrix} -0.3333 \\ -0.500 \end{Bmatrix}, \quad \mathbf{c}^{(2)} = \nabla f\left(\mathbf{x}^{(2)}\right) \approx \begin{Bmatrix} 0 \\ 0 \end{Bmatrix}$$

This implies convergence to the exact solution.

The quasi-Newton method is most popular in commercial optimization algorithms. The Matlab function `fminunc` uses a variant of Newton's method if the analytical gradient information is provided and BFGS if it has to calculate the gradient by finite differences. The variant of Newton's method used by `fminunc` is called a trust-region method. It constructs a quadratic approximation to the function and minimizes it in a box around the current point, with the size of the box adjusted depending on the success of the previous iteration.

### 5.4.5 Rate of Convergence

We used the order of convergence of the steepest descent method and Newton's method without formally defining it. In numerical analysis, the order of convergence and the rate of convergence of a convergent sequence are quantities that represent how quickly the sequence approaches its limit. In the context of optimization, the sequence means the history of design variables. Let $\{x^{(k)}\}$ be the history of design variables ($k = 1, 2, \ldots$) and let $x^*$ be the optimum design after convergence. Then the order of convergence means how fast the sequence $\{x^{(k)}\}$ converges to $x^*$. If the order of convergence is higher, then typically fewer iterations are necessary to converge to the optimum design.

A sequence $\{x^{(k)}\}$ is said to converge to $x^*$ with order $p$, where $p$ is the largest number such that

$$0 \le \lim_{k \to \infty} \frac{\left\|x^{(k+1)} - x^*\right\|}{\left\|x^{(k)} - x^*\right\|^p} < \infty \tag{5.31}$$

In addition, when a sequence has $p$ order of convergence, the rate of convergence is defined as

$$\beta = \lim_{k \to \infty} \frac{\left\|x^{(k+1)} - x^*\right\|}{\left\|x^{(k)} - x^*\right\|^p} \tag{5.32}$$

If $p = 1$, the sequence displays linear convergence, and $\beta$ must be less than 1 for the sequence to converge. In particular, if $\beta = 0$ when $p = 1$, then the sequence is called super-linear convergence. The steepest descent method shows linear convergence. A quadratic convergence is when $p = 2$. Newton's method shows a quadratic convergence when the initial design is close to the optimum design. The quasi-Newton methods show the order of convergence between 1 and 2.

**Example 5.8 Rate of convergence**
Calculate the order and rate of convergence of the sequence $x^{(k)} = a^k$, $a < 1$, and $x^* = 0$.

**Solution:**
First, assume that $p = 1$. Then, Eq. (5.31) becomes

$$0 \leq \lim_{k\to\infty} \frac{\|x^{(k+1)} - x^*\|}{\|x^{(k)} - x^*\|^1} = \frac{a^{k+1}}{a^k} = a < \infty$$

On the other hand, if $p = 2$ is assumed, then

$$0 \leq \lim_{k\to\infty} \frac{\|x^{(k+1)} - x^*\|}{\|x^{(k)} - x^*\|^2} = \frac{a^{k+1}}{a^{2k}} = \frac{1}{a^{k-1}} \to \infty$$

Therefore, the sequence shows linear convergence. With $p = 1$, $\beta = a$. Therefore, the rate of convergence is $a < 1$.

## 5.5 Constrained Optimization Using Unconstrained Algorithms

In the previous section, numerical algorithms for solving unconstrained optimization problems were discussed. Most engineering applications, however, have constraints that must be satisfied during the design optimization process. As shown in Figure 5.10, optimums design normally exists

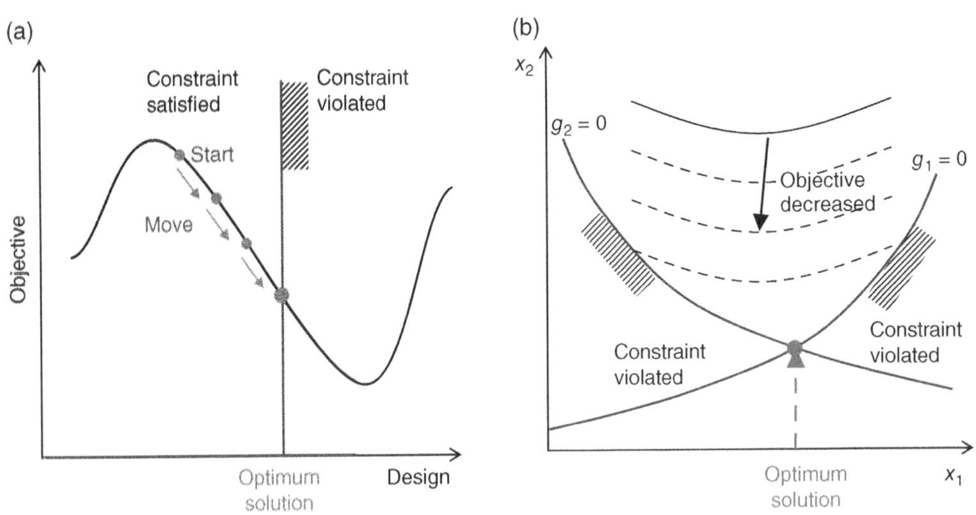


**Figure 5.10** The role of constraints on optimum design. (a) Single design variable and (b) two design variables.

on the boundary of constraints. These constraints play a critical role in determining the optimum design. In this section, numerical algorithms that can find the optimum design for constrained optimization problems will be discussed.

As we discussed in Section 4.4, not all constraints are active at the optimum design. All equality constraints should be active, while some inequality constraints may be inactive. Only those active constraints contribute to the optimum design. When all constraints are inactive, the constraint optimization problem can be considered as an unconstrained problem; it is assumed that at least one constraint should be active at the optimum design.

Numerical algorithms for constrained optimization can be grouped into two types. The first type converts the constrained optimization problem into an unconstrained optimization problem using either the Lagrange multiplier method or the penalty method and solves it using the numerical algorithms in the previous section. The second group solves the constrained optimization problem directly. We will discuss both types of algorithms in the following subsections.

### 5.5.1 Lagrange Multiplier Method

The Lagrange multiplier method, along with the penalty method in the next subsection, is a transformation method, where the constrained optimization problem is converted into an unconstrained optimization problem. Then, the problem can be solved using the unconstrained optimization algorithms in the previous section. Consider the constrained optimization problem in Eq. (4.3), which is rewritten here:

$$\begin{aligned} &\text{minimize } f(\mathbf{x}) \\ &\text{subject to } g_i(\mathbf{x}) \leq 0, \quad i = 1, \ldots, K \\ &\qquad h_j(\mathbf{x}) = 0, \quad j = 1, \ldots, M \\ &\qquad x_l^L \leq x_l \leq x_l^U, \quad l = 1, \ldots, n \end{aligned} \tag{5.33}$$

where $\mathbf{x} = \{x_1, x_2, \ldots, x_n\}^T$ is the vector of design variables, $f(\mathbf{x})$ is the objective function, $g_1(\mathbf{x}), \ldots, g_K(\mathbf{x})$ are inequality constraints, $h_1(\mathbf{x}), \ldots, h_M(\mathbf{x})$ are equality constraints, and $\mathbf{x}^L$ and $\mathbf{x}^U$ are lower and upper bounds of design variables, respectively. Since only active constraints contribute to the optimum design, let the number of active inequality constraints be $L < K$. Since the active inequality constraints are equivalent to equality constraints, the optimization problem can be considered as a constrained optimization with $M + L$ equality constraints.

The Lagrange multiplier method is based on the Lagrangian function defined in Eq. (4.36), which is rewritten here:

$$\text{minimize } L(\mathbf{x}, \boldsymbol{\mu}, \boldsymbol{\lambda}) = f(\mathbf{x}) + \sum_{j=1}^{M} \mu_j h_j(\mathbf{x}) + \sum_{i=1}^{L} \lambda_i g_i(\mathbf{x}) \tag{5.34}$$

It is noted that the slack variables are not used because the Lagrangian function is defined only with active inequality constraints. The minimization of the unconstrained optimization problem in Eq. (5.34) is equivalent to that of the constrained optimization problem in Eq. (5.33). The only difference is that in addition to design variables, the Lagrange multipliers are also considered unknown variables. Therefore, the objective function is the Lagrangian function, and the variables are design variables and Lagrange multipliers. It is noted that the Lagrange multipliers of inequality constraints have a sign restriction, which can easily be imposed using side constraints.

Although this seems a straightforward conversion, there are two challenges associated with the Lagrange multiplier method. The first challenge is that the number of active constraints varies during design iterations. Therefore, some Lagrange multipliers may disappear, while others are introduced in the middle of design iterations, which causes difficulty in initializing and updating them. It is possible that the Lagrange multipliers for all constraints are kept throughout all design iterations, and those multipliers associated with inactive constraints are set to a zero value. However, this still can cause a problem, especially when some constraints oscillate between active and inactive states. The second challenge is from the positive definite property of the Hessian matrix. Many Newton or quasi-Newton methods use the positive definite Hessian matrix or its approximation to enhance the convergence speed of design iterations. However, the second-order derivatives of the Lagrange multiplier are zero. Therefore, the diagonal terms of the Hessian matrix are zero, which makes the Hessian matrix positive semi-definite. Due to these two challenges, the Lagrange multiplier method is not popular in practical use.

### 5.5.2 Penalty Function Method

The penalty function method converts the constrained optimization problem to an unconstrained one by adding a penalty for violating constraints. A quadratic penalty is usually employed in order to make the penalty function smooth and preserve differentiability. The penalty for equality constraints is proportional to their square, while for inequality constraints, it is their square when they are positive and zero otherwise. Therefore, the loss function is defined as

$$\Phi(\mathbf{x},\omega)=f(\mathbf{x})+P(\mathbf{h},\mathbf{g},\omega)=f(\mathbf{x})+\omega\left\{\sum_{j=1}^{M}(h_j(\mathbf{x}))^2+\sum_{i=1}^{L}(g_i^+(\mathbf{x}))^2\right\} \tag{5.35}$$

where $\omega$ is a penalty parameter, and $g_i^+(\mathbf{x}) = \max(0, g_i(\mathbf{x}))$ is only nonzero when the inequality constraint is violated.

The disadvantage of a quadratic penalty is that it is very small for small violations, so the solution will tend to violate some constraints. This can be minimized by multiplying the penalty by a large parameter $\omega$, but that makes the problem numerically ill-conditioned. Therefore, the established procedure is to gradually increase $\omega$ as one gets closer to the solution. This is referred to as the sequential unconstrained minimization technique (SUMT). As the penalty parameter approaches infinity, the solution converges to the original optimization problem. The advantage of the penalty function method compared to the Lagrange multiplier method is that it does not introduce additional variables, but it has the disadvantage of numerical ill-conditioning of the Hessian matrix.

Because of this ill-conditioned Hessian matrix, the penalty function method is not popular anymore for gradient-based algorithms, but it is still popular for gradient-free global search algorithms. This is because most global search algorithms in Chapter 6 do not have a systematic way of including constraints. Therefore, the penalty function method is frequently used for them.

**Example 5.9 Penalty method**

Find the optimum design of the following constrained optimization problem by gradually increasing the penalty parameter from $\omega = 1$ to $\omega = 1{,}000$:

$$\text{minimize } f(\mathbf{x}) = x_1^2 + 10x_2^2$$
$$\text{subject to } h(\mathbf{x}) = 4 - x_1 - x_2 = 0$$

**Solution:**

In order to apply for the penalty function method, the following loss function is defined:

$$\Phi(\mathbf{x},\omega) = x_1^2 + 10x_2^2 + \omega(4 - x_1 - x_2)^2$$

Since this is a quadratic function, it is possible to find the optimum design analytically. From the KKT condition, the optimum design can be obtained as

$$x_1 = \frac{40\omega}{10 + 11\omega}, \quad x_2 = \frac{4\omega}{10 + 11\omega}$$

The following table shows the solution, the objective, and the loss function for a series of increasing $\omega$ values. The difference between $f$ and $\Phi$ is the penalty term. For low values of $\omega$, the constraint is violated a lot, the penalty is high, but the objective is small. As we increase the penalty parameter, the constraint violation decreases fast enough so that the total penalty actually decreases! For the highest penalty parameter, the sum of the variables is 3.966, so that the violation is smaller than 1%, but that may not be acceptable, in which case an even higher penalty parameter would be required.

| $\omega$ | $x_1$ | $x_2$ | $f$ | $\Phi$ |
|---|---|---|---|---|
| 1 | 1.905 | 0.1905 | 3.992 | 7.619 |
| 10 | 3.333 | 0.3333 | 12.220 | 13.333 |
| 100 | 3.604 | 0.3604 | 14.288 | 14.144 |
| 1,000 | 3.633 | 0.3633 | 14.518 | 14.532 |

The contour of the loss function when $\omega = 1$, as shown in Figure 5.11(a), is well-behaved. Even if we obtain an optimum design analytically, it is expected that numerical algorithms would converge easily as well. On the other hand, when $\omega = 1{,}000$, as shown in Figure 5.11(b), the contour shows a canyon at the constraint boundary. For a linear constraint, this may not be much of a problem, but if the constraint is nonlinear, an optimization algorithm that moves in straight lines would require a large number of iterations and may get stuck. Therefore, it is expected that numerical optimization with a large value of the penalty parameter may experience difficulty in convergence. For gradient-free algorithms, we may avoid this problem by using a penalty that is proportional to the violation instead of its square, so that the penalty parameter would not need to be very high.

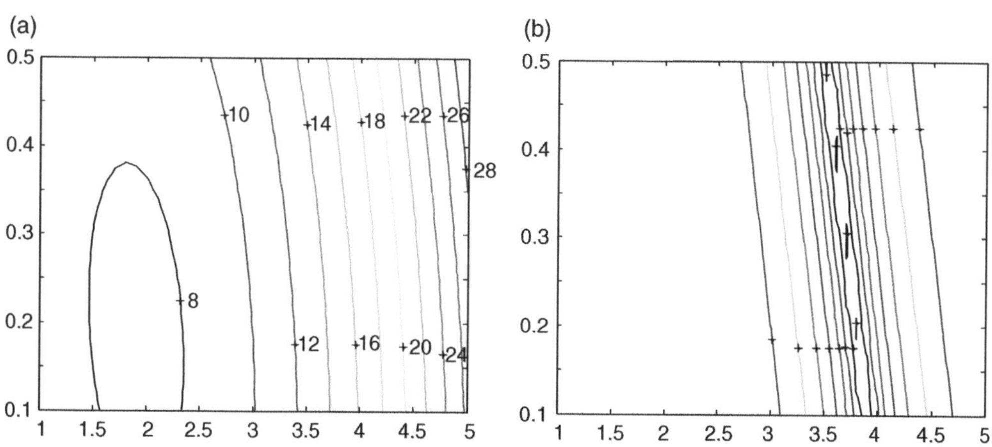


**Figure 5.11** Contours of the loss function with different penalty parameters. (a) $\omega$=1 and (b) $\omega$ = 1,000.

The contours in Figure 5.11 were obtained with the following Matlab script:

```
r=1;
x=linspace(1,5,41);    y=linspace(0.1,0.5,41);
[X,Y]=meshgrid(x,y);
Z=X.^2+10*Y.^2+r*(4-X-Y).^2;
cs=contour(X,Y,Z);    clabel(cs);
```

## 5.6 Constrained Optimization Using Direct Methods

In the previous section, constrained optimization problems were converted into unconstrained optimization problems using the Lagrange multiplier method and the penalty function method. However, in practice, it is popular to solve constrained optimization problems directly. The basic idea of direct methods is to calculate a search direction based on the gradients of both the objective function and the constraints. The direct methods for constrained optimization have an analogy of a person with a flashlight trying to go down a hill in a terrain with fenced areas. The flashlight can only provide limited information about slopes (gradients), and the person uses them to select a direction to move. Then, the person can move until the slope changes upward or the person meets with constraint boundaries.

Some algorithms move mostly away from constraint boundaries and stay inside the feasible domain. Since the optimum design is usually located on constraint boundaries, they approach the boundaries from the inside of the feasible domain. Some other algorithms follow the constraint boundaries. The gradient projection algorithm starts with a point on the boundary of the feasible domain and moves in the plane tangent to all the active constraints. The direction is the projection of the gradient on that plane. This move will end either when a new constraint is encountered or when the move takes it too far from the constraint boundaries due to their nonlinearity. Then there is a restoration move that brings it back to the constraint boundary.

Similar to the loss function in the previous section, the direct methods usually define a descent function or a merit function. The merit function must be reduced from one iteration to another and should be the same as the objective function $f(x)$ at the optimum design.

A practical difficulty associated with constraint boundaries is the oscillation between active/violated and inactive constraints. A constraint is considered inactive when $g(\mathbf{x}) < 0$, active when $g(\mathbf{x}) = 0$, and violated when $g(\mathbf{x}) > 0$. Oscillation frequently happens when the design is close to the constraint boundary. This oscillation can cause technical difficulties as the constraints may not be considered consistently at different iterations. In order to overcome this difficulty, inactive constraints that are close to the constraint boundaries are also included in the active constraint set. The $\epsilon$-active constraint set is defined as

$$I_\epsilon = \{i | g_i + \epsilon \geq 0\} \tag{5.36}$$

where $\epsilon > 0$ is a small positive constant to determine the value of the constraint that can be considered active. The $\epsilon$-active constraint set includes all active/violated constraints, as well as those constraints that are close to the constraint boundary.

### 5.6.1 Sequential Linear Programming (SLP) Method

The sequential linear programming (SLP) method approximates the nonlinear problem as a sequence of linear programming problems such that the simplex method may be used to find the solution to each iteration. By using function values and sensitivity information, the nonlinear problem in Eq. (5.33) is linearized in a similar way to Taylor's expansion method in the first order. Let the linear approximation of the objective function be $f(\mathbf{x}^{(k+1)}) \approx f(\mathbf{x}^{(k)}) + \mathbf{c}^T\Delta\mathbf{x}$, that of equality constraint $h_i(\mathbf{x}^{(k+1)}) \approx h_i(\mathbf{x}^{(k)}) + \nabla h_i^T \Delta\mathbf{x}$, and that of active inequality constraint $g_j(\mathbf{x}^{(k+1)}) \approx g_j(\mathbf{x}^{(k)}) + \nabla g_j^T \Delta\mathbf{x}$. In order to simplify notations, the gradient vectors are defined as $\nabla h_i = \mathbf{n}_i$ and $\nabla g_j = \mathbf{a}_j$. Also, the gradient matrix is defined as $\mathbf{N} = [\mathbf{n}_1, ..., \mathbf{n}_M]$ and $\mathbf{A} = [\mathbf{a}_1, ..., \mathbf{a}_L]$. Then, the constrained optimization problem in Eq. (5.33) at iteration $k$ can be written as

$$
\begin{aligned}
&\text{minimize } \mathbf{c}^T\Delta\mathbf{x} \\
&\text{subject to } \mathbf{n}_i^T\Delta\mathbf{x} + h_i = 0, \quad i = 1, ..., M \\
&\mathbf{a}_j^T\Delta\mathbf{x} + g_i \le 0, \quad j = 1, ..., K \\
&\Delta x_l^L \le \Delta x_l \le \Delta x_l^U, \quad l = 1, ..., n
\end{aligned}
\tag{5.37}
$$

That is, the original nonlinear optimization problem with respect to $\mathbf{x}$ is converted into a linear optimization problem with respect to $\Delta\mathbf{x}$. There are many efficient algorithms to solve linear optimization problems, such as the simplex method. Once the design increment $\Delta\mathbf{x}$ is solved, the design is updated by $\mathbf{x}^{(k+1)} = \mathbf{x}^{(k)} + \Delta\mathbf{x}$. Therefore, the SLP method sequentially linearizes the nonlinear optimization problem, where its name comes from. It is noted that gradient information is critical to construct the linear optimization problem.

### 5.6.2 Quadratic Programming (QP) Subproblem

In solving the linear optimization problem in Eq. (5.37) for $\Delta\mathbf{x}$, the move limits (side constraints or step size constraints) are critically important for convergence. Consider the move limits of two-dimensional design variables, as shown in Figure 5.12(a). The center circle is the current design $\mathbf{x}^{(k)}$, and the square is the move limits. As can be shown in the figure, the length of the design change $\Delta\mathbf{x}$ is large toward the corner of the move limits. Therefore, design tends to change along the corners. In order to prevent such a bias, the move limits can be given in the circular domain, as shown in Figure 5.12(b). In this case, all design changes are of the same length, and there is no preferred direction.

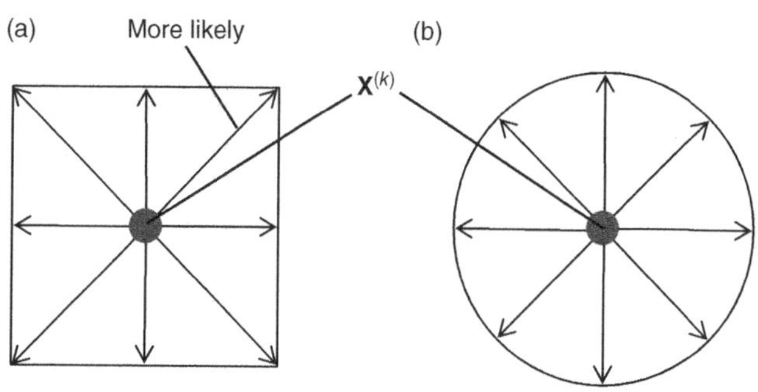


**Figure 5.12** Box-shaped move limits versus equal-size move limits. (a) Rectangular move limits and (b) equal-size move limits.

In order to apply the equal-size design change, the linearized optimization problem in Eq. (5.37) is modified as

$$\begin{aligned}&\text{minimize } \mathbf{c}^T\mathbf{d}\\&\text{subject to } \mathbf{N}^T\mathbf{d} = -\mathbf{h}\\&\qquad\qquad \mathbf{A}^T\mathbf{d} \le -\mathbf{g}\\&\qquad\qquad \frac{1}{2}\mathbf{d}^T\mathbf{d} \le \xi^2\end{aligned} \tag{5.38}$$

Instead of the design change $\Delta\mathbf{x}$, the search direction $\mathbf{d}$ is used as an additional line search can be added to find the optimum design change. The last term in Eq. (5.38) imposes the step-size constraint, where the length of the design change should be less than a pre-determined threshold. The only difficulty in Eq. (5.38) is that the constraints are nonlinear, and many linear programming algorithms, such as the simplex method, cannot be used.

Instead of solving Eq. (5.38), it is possible to show that the following quadratic programming (QP) problem is equivalent to the linearized constrained optimization with equal-size constraints in Eq. (5.38):

$$\begin{aligned}&\text{minimize } \mathbf{c}^T\mathbf{d} + \frac{1}{2}\mathbf{d}^T\mathbf{d}\\&\text{subject to } \mathbf{N}^T\mathbf{d} = -\mathbf{h}\\&\qquad\qquad \mathbf{A}^T\mathbf{d} \le -\mathbf{g}\end{aligned} \tag{5.39}$$

Since the objective function of the QP is convex and the feasible set of linear constraints is a convex set, Eq. (5.39) is a convex optimization problem, and a local optimum becomes the global optimum design. There are many numerical algorithms to solve for the QP in Eq. (5.39). This QP can be effectively solved, for example, by using the Kuhn–Tucker condition and the simplex method. Starting from the identity matrix, the Hessian matrix $\mathbf{H}$ is updated at each iteration by using the aforementioned methods in unconstrained optimization algorithms. The advantage of solving Eq. (5.39) in this way is that for a positive definite $\mathbf{H}$, the problem is convex and the solution is unique. Moreover, this method does not require the move limit as in SLP.

### 5.6.3 Constrained Steepest Descent Method

In the steepest descent algorithm detailed in Section 5.4, the descent direction $\mathbf{d}$ is obtained from the gradient of the objective function; $\mathbf{d} = -\mathbf{c}$. When constraints exist, this descent direction has to be modified in order to include their effect. In the constrained steepest descent method, this is accomplished by defining the following descent function with violated constraints:

$$\Phi(\mathbf{x}, \omega, \mathbf{h}, \mathbf{g}) = f(\mathbf{x}) + \omega_k V(\mathbf{h}, \mathbf{g}) \tag{5.40}$$

where $V(\mathbf{h}, \mathbf{g})$ is the maximum constraint violation and $\omega_k$ is a penalty parameter. The penalty parameter is user-defined but has to be larger than the absolute sum of all Lagrange multipliers. The maximum constraint violation is defined as

$$V(\mathbf{h}, \mathbf{g}) = \max\left\{0, |h_i|, g_j, j \in I_\epsilon\right\} \tag{5.41}$$

**Table 5.2** Procedure of the constrained steepest descent algorithm.

| Step | Procedure | Comment |
|---|---|---|
| 1 | Set $\mathbf{x}^{(0)}$, $k = 0$, $\omega_0 = 1$ | Initial design must be given |
| 2 | Compute $f(\mathbf{x}^{(k)})$, $h_i(\mathbf{x}^{(k)})$, $g_j(\mathbf{x}^{(k)})$, $\mathbf{c}$, $\mathbf{N}$, $\mathbf{A}$, $V$ | |
| 3 | Using QP subproblem, solve for $\mathbf{d}^{(k)}$ | |
| 4 | Check for convergence $\|\mathbf{d}^{(k)}\| < \epsilon_2$ and $V < \epsilon_1$ | Stop if converged |
| 5 | Update $\omega_k$ | |
| 6 | Line search: $\mathbf{x}^{(k+1)} = \mathbf{x}^{(k)} + \alpha_k \mathbf{d}^{(k)}$ | |
| 7 | Increase $k = k + 1$, go to Step 2 | Design iteration |

If constraints are violated, then these constraints are added to the objective function using a penalty method. The gradient of the descent function combines the effects of the objective function and the violated constraint functions. Table 5.2 summarizes the procedure of the constrained steepest descent algorithm.

### 5.6.4 Feasible Direction Method

The feasible direction method is designed to allow design movement within the feasible region in each iteration. Based on the previous design, the updated design reduces the objective function and remains in the feasible region. Since all designs are feasible, a design at any iteration can be used, even if it is not an optimum design. Since this method uses the linearization of functions as in SLP, it is difficult to maintain nonlinear equality constraints. Thus, this approach is used exclusively for inequality constraints. Search direction **d** can be found by solving the following linear subproblem:

$$\begin{aligned}&\text{minimize } \bar{f} = f_0 + \mathbf{c}^T\mathbf{d}\\&\text{subject to } \bar{g}_j = g_{j0} + \mathbf{a}_j^T\mathbf{d} \le 0\end{aligned} \tag{5.42}$$

Knowing that $g_{j0} \le 0$, the direction that satisfies $\mathbf{a}_j^T\mathbf{d} \le 0$ is a feasible direction, i.e., the direction maintains the constraints satisfied. In addition, the direction that satisfies $\mathbf{c}^T\mathbf{d} \le 0$ is a useful direction as it reduces the objective function. Therefore, the goal is to find the direction that satisfies both the feasible and useful directions. This can be achieved by defining a new variable $\beta = \max\left\{\mathbf{c}^T\mathbf{d}, \mathbf{a}_j^T\mathbf{d}\right\}$ and solve the following optimization problem:

$$\begin{aligned}&\text{minimize } \beta(\mathbf{d})\\&\text{subject to } \mathbf{a}_j^T\mathbf{d} \le \beta\\&\qquad\qquad \mathbf{c}^T\mathbf{d} \le \beta\\&\qquad\qquad -1 \le \|\mathbf{d}\| \le 1\end{aligned} \tag{5.43}$$

After finding a direction **d** that can reduce cost function and maintain feasibility, a line search is used to determine the step size.

### 5.6.5 Constrained Quasi-Newton Method

In the case of unconstrained optimization in Section 5.5, it was shown that the Newton method has a quadratic convergence, while the steepest descent method only shows linear convergence. It was also shown that since the Hessian information is expensive to calculate, quasi-Newton methods were used to approximate the Hessian or its inverse matrix. A similar approach can also be taken in constrained optimization problems.

The constrained quasi-Newton methods, or sequential QP methods, are based on solving the nonlinear KKT conditions using Newton's method. Since Newton's method requires the Hessian matrix, it is approximated using the quasi-Newton method. In order to explain the constrained quasi-Newton method, consider the following constrained optimization problem with equality constraints alone:

$$\begin{aligned}&\text{minimize } f(\mathbf{x})\\&\text{subject to } h_i(\mathbf{x}) = 0\end{aligned} \tag{5.44}$$

whose Lagrangian function becomes $L(\mathbf{x}, \boldsymbol{\lambda}) = f(\mathbf{x}) + \sum \lambda_i h_i(\mathbf{x})$. The KKT optimality condition can be written in the form of a nonlinear equation. First, let us define the combined variables $\mathbf{y} = \{\mathbf{x}, \boldsymbol{\lambda}\}^T$. Then the KKT condition can be written as

$$\begin{cases}\nabla_{\mathbf{x}} L = \nabla_{\mathbf{x}} f + \mathbf{N}\boldsymbol{\lambda} = 0\\ \nabla_{\boldsymbol{\lambda}} L = \mathbf{h} = 0\end{cases} \quad \rightarrow \quad \mathbf{F}(\mathbf{y}) = \begin{Bmatrix}\nabla_{\mathbf{x}} f + \mathbf{N}\boldsymbol{\lambda}\\ \mathbf{h}\end{Bmatrix} = 0 \tag{5.45}$$

The constrained quasi-Newton method solves the above nonlinear equations of the KKT condition using Newton's method. From Taylor series expiation, $\mathbf{F}(\mathbf{y}^{(k+1)}) = \mathbf{F}(\mathbf{y}^{(k)}) + (\nabla_{\mathbf{y}}\mathbf{F})^T\Delta\mathbf{y}^{(k)} = 0$. By replacing $\Delta\mathbf{y} = \{\mathbf{d}, \Delta\boldsymbol{\lambda}\}^T$, this linearized incremental equation can be written as

$$\begin{bmatrix}\nabla_{\mathbf{xx}} L & \mathbf{N}\\ \mathbf{N}^T & 0\end{bmatrix}\begin{Bmatrix}\mathbf{d}^{(k)}\\ \boldsymbol{\lambda}^{(k+1)}\end{Bmatrix} = -\begin{Bmatrix}\nabla_{\mathbf{x}} f\\ \mathbf{h}\end{Bmatrix} \tag{5.46}$$

In the above equation, $\Delta\boldsymbol{\lambda}^{(k)} = \boldsymbol{\lambda}^{(k+1)} - \boldsymbol{\lambda}^{(k)}$ is used to replace the unknown variables $\Delta\boldsymbol{\lambda}^{(k)}$ with $\boldsymbol{\lambda}^{(k+1)}$. This is the $k$th iteration to find the search direction and the Lagrange multiplier for the KKT condition.

In practice, instead of solving Eq. (5.46), an equivalent quadratic problem is solved. It is left as an exercise problem to show that the following quadratic problem is equivalent to solving Eq. (5.46), as

$$\begin{aligned}&\text{minimize } \mathbf{c}^T\mathbf{d} + \frac{1}{2}\mathbf{d}^T[\nabla_{\mathbf{xx}} L]\mathbf{d}\\&\text{subject to } \mathbf{N}^T\mathbf{d} = -\mathbf{h}\end{aligned} \tag{5.47}$$

As discussed before, the Hessian matrix is expensive to calculate. Therefore, the constrained quasi-Newton method uses the BFGS method to approximate it.

When there are both equality and inequality constraints, the equivalent quadratic problem is defined using equality constraints and active inequality constraints as

$$\begin{aligned}&\text{minimize } \mathbf{c}^T\mathbf{d} + \frac{1}{2}\mathbf{d}^T[\nabla_{\mathbf{xx}} L]\mathbf{d}\\&\text{subject to } \mathbf{N}^T\mathbf{d} = -\mathbf{h}\\&\qquad\qquad \mathbf{A}^T\mathbf{d} \le -\mathbf{g}\end{aligned} \tag{5.48}$$

Once the search direction $\mathbf{d}^{(k)}$ is found by solving the QP problem, the line search algorithm can be used to find the step size $\alpha_k$ and update the design $\mathbf{x}^{(k+1)} = \mathbf{x}^{(k)} + \alpha_k\mathbf{d}^{(k)}$.

## 5.7 Matlab Optimization Toolbox

Matlab provides various optimization functions, as shown in Table 5.3. Among various optimization algorithms, unconstrained optimization problems can be solved using `fminunc` and `fminsearch`. For constrained optimization problems with nonlinear equality and inequality constraints, `fmincon` is the most common function. In this section, brief explanations of these functions with their algorithms are introduced along with several examples.

**Example 5.10 Constrained optimization using Matlab**
Solve the following constrained optimization problem using the Matlab `fmincon` function:

$$\text{minimize } f(\mathbf{x}) = e^{x_1}\left(4x_1^2 + 2x_2^2 + 4x_1x_2 + 2x_2 + 1\right)$$
$$\text{subject to } g_1(\mathbf{x}) = x_1x_2 - x_1 - x_2 + 1.5 \leq 0$$
$$g_2(\mathbf{x}) = -x_1x_2 - 10 \leq 0$$

**Solution:**
In order to call `fmincon`, it would be necessary to make m-files for the objective function and constraints. First, create the following m-file named `objfun.m`:

```
function f = objfun(x)
f = exp(x(1))*(4*x(2)^2 + 2*x(2)^2 + 4*x(1)*b(2) + 2*x(2) + 1);
```

Also, create the following m-file named `confun.m`

```
function [c, ceq] = confun(x)
% Nonlinear inequality constraint
c = [1.5 + x(1)*x(2) - x(1) - x(2);
     -x(1)*x(2) - 10];
% Nonlinear equality constraints
ceq = [];
```

Then, optimization can be invoked by using the following Matlab script:

```
x0 = [-1,1]; % Make a starting guess at the solution
options = optimset('LargeScale','off');
```

**Table 5.3** Matlab optimization functions.

| Type | Function |
|---|---|
| Minimization of a scalar function | `fminbnd` |
| Unconstrained minimization | `fminunc, fminsearch` |
| Linear programming | `linprog` |
| Quadratic programming | `quadprog` |
| Constrained minimization | `fmincon` |
| Goal attainment | `fgoalattain` |
| Minmax | `fminmax` |
| Semi-infinite minimization | `fseminf` |
| Binary integer programming | `bintprog` |

```
[x, fval, exitflag, output, lambda, grad, hessian] = ...
fmincon(@objfun,x0,[],[],[],[],[],[],@confun, options)
```

First, design variables are stored in the array $x$. The function `objfun.m` calculates the objective function value and returns it in the variable `f`. On the other hand, the function `confun.m` calculates an array of inequality constraints `c` and an array of equality constraints `ceq`. In this particular optimization problem, there are two inequality constraints and no equality constraints. Therefore, `c` is $2 \times 1$ array, while `ceq` is null. With `objfun.m` and `confun.m` being available, the Matlab script assigns the initial design to x0, defines optimization options in options, and calls the function `fmincon`. The function `optimset` assigns various options that control the optimization process. Table 5.4 summarizes the available options for Matlab optimization functions, which include algorithms, convergence criteria, result display, etc. For detailed explanations of the `optimset` function, the users are referred to the Matlab manual.

The function `fmincon` in this example returns the following outputs:

```
x =      -3.4049     1.1135
fval =     -0.1494
exitflag =      1
output =
                iteration: 13
                funcCount: 45
            constrviolation: 0
                 stepsize: 1.2743e-05
                algorithm: 'interior-point'
            firstorderopt: 4.0000e-07
             cgiterations: 0
                  message: 'Local minimum found that satisfies the
constraints. Optimization completed because the objective function is
non-decreasing in feasible directions, to within the value of the
optimality tolerance, and constraints are satisfied to within the
value of the constraint tolerance. <stopping criteria details>
Optimization completed: The relative first-order optimality measure,
4.000016e-07, is less than options.OptimalityTolerance = 1.000000e-06,
and the relative maximum constraint violation, 0.000000e+00, is less
than options.ConstraintTolerance = 1.000000e-06.'

        lambda =
                    eqlin: [0x1 double]
                 eqnonlin: [0x1 double]
                  ineqlin: [0x1 double]
                    lower: [2x1 double]
                    upper: [2x1 double]
               ineqnonlin: [2x1 double]

        grad =
            -0.0015
             0.0579

        hessian =
             0.1464     0.2038
             0.2038     0.3980
```

**Table 5.4** Available optimset options.

| Option | Data | Meaning |
|---|---|---|
| Display | 'off' 'iter' 'final' 'notify' | Level of display |
| GradObj | 'on' 'off' | Objective gradient |
| Jacobian | 'on' 'off' | Constraint gradient |
| LargeScale | 'on' 'off' | Algorithm |
| MaxFunEvals | Integer | Max. number of function evaluations |
| MaxIter | Integer | Max. number of optimization iterations |
| TolFun | Real | Convergence tolerance for objective |
| TolX | Real | Convergence tolerance for design variables |

The function `fmincon` returns various outputs. First, the variable `x` contains optimum design variables, and the variable `fval` contains the value of optimum objective function. The values of constraints at the optimum design can be obtained by calling `confun.m` as `[c, ceq] = confun(x)`. The meaning of variable `exitflag` = 1 is explained in Table 5.5, where the optimum is achieved by satisfying the optimality criterion. When the `exitflag` is 0, −1, or −2, the optimization iteration stops without convergence. The variable output contains various information regarding the optimization process, such as the total number of iterations and the number of function evaluations. The variable lambda contains the Lagrange multiplier for all constraints. The variables grad and Hessian contain the gradient and Hessian information of the objective function at the optimum design.

**Table 5.5** Interpretation of exitflag in Matlab optimization.

| exitflag | Meaning |
|---|---|
| 1 | First-order optimality measure was less than `options.TolFun` and maximum constraint violation was less than `options.TolCon`. |
| 2 | Change in x was less than `options.TolX`. |
| 3 | Change in the objective function value was less than `options.TolFun`. |
| 4 | The magnitude of the search direction was less than `2*options.TolX` and constraint violation was less than `options.TolCon`. |
| 5 | Magnitude of directional derivative in search direction was less than `2*options.TolFun` and maximum constraint violation was less than `options.TolCon`. |
| 0 | Number of iterations exceeded `options.MaxIter` or number of function evaluations exceeded `options.FunEvals`. |
| −1 | The algorithm was terminated by the output function. |
| −2 | No feasible point was found. |

**Example 5.11 Optimization of a ring problem**
We will examine the behavior of the function `fmincon` for a simple example of minimizing a quadratic function in a ring. Solve the following constrained optimization problem when (a) $a = 10$ and (b) $a = 1.1$. Compare the differences between the two cases in terms of

$$\text{minimize } f(\mathbf{x}) = x_1^2 + ax_2^2$$
$$\text{subject to } 10^2 \le x_1^2 + x_2^2 \le 20^2$$

**Solution**

a) Since the objective function is an elliptical shape, the outer ring constraint will not be active, but it may slow down convergence by limiting the straight-line moves. As long as $a > 1$, the optimum on the inner circle will go to $x_2 = 0$. For the first case, we choose $a = 10$, which would make it easy for fmincon to find the optimum, even if we start far away at $\mathbf{x}_0 = (1, 10)$. The following two Matlab functions, `quad2.m` and `ring.m`, are defined for the objective and constraint functions, respectively,

```
function f=quad2(x)
    global a
    f=x(1)^2+a*x(2)^2;
end
%
function [c,ceq]=ring(x)
   c(1)=10^2-x(1)^2-x(2)^2;
   c(2)=x(1)^2+x(2)^2-20^2;
   ceq=[];
end
```

It is noted that the inequality constraint is split into two inequality constraints in order to use the standard form of constraints. The following Matlab script is used to call the `fmincon` function:

```
global a
x0=[1,10];a=10;
[x,fval,exitflag,output,lambda]=fmincon(@quad2,x0,[],[],[],[],[],
[],@ring)
```

The output from `fmincon` tells us that it satisfied convergence criteria based on the lack of progress on the objective function and constraint satisfaction. In both cases, this is based on tolerances that we can change with the `optimset` function, which uses default values.

```
x =      10.0000     0.0000
fval =   100.0000
exitflag =       1
output =
                        iterations: 9
                         funcCount: 36
                   constrviolation: 0
                          stepsize: 9.9662e-06
                         algorithm: 'interior-point'
                     firstorderopt: 2.0004e-06
                      cgiterations: 2
```

With the calling sequence we had, it gives us the objective function value of 100 and the optimum $\mathbf{x} = (10, 0)$. The optimization finished successfully (`exitflag=1`), and it took 9 iterations and 36 function evaluations. This is remarkable since we did not provide `fmincon` with a routine to calculate derivatives of objective function and constraints and it had to calculate them by finite differences. So, with 9 iterations, 18 of the function evaluations were used to calculate derivatives. This means that the average number of function evaluations per one dimensional search was only 2.

b) Setting $a = 1.1$, we make the advantage of reducing $x_2$ and increasing $x_1$ much smaller. This causes fmincon to slow down a lot. It took 36 iterations with 216 function evaluations. The optimization converges to the same optimum design as with case (a). Note that each iteration now takes about four function evaluations!

```
x =    10.0000     0.0000
fval =   100.0000
exitflag =      1
output =
              iterations: 36
               funcCount: 216
         constrviolation: 0
                stepsize: 9.9694e-06
               algorithm: 'interior-point'
           firstorderopt: 2.0002e-06
            cgiterations: 19
```

## 5.8 Practical Suggestions for Numerical Optimization

Although various numerical optimization algorithms are presented in this chapter, there are several things that the users need to pay attention to in order to make the optimization solution process robust. These are applicable to all algorithms in this chapter.

1) It is always good to normalize the design space to $0 \le x_k \le 1$. Mathematically, it is possible to solve design variables with different lower and upper bounds. However, when the ranges of design variables are significantly different, a numerical difficulty can occur. For example, let us consider the case where both Young's modulus and Poisson's ratio are design variables. In the MKS unit system, Young's modulus of metal is in the order of $10^{11}$ Pa, while the Poisson's ratio is $0 < \nu < 0.5$. Such a huge difference makes it difficult for numerical algorithms to find the optimum design for both variables. Therefore, it would be better to normalize all design variables with the same range:

$$\hat{x} = \frac{x - x_{\min}}{x_{\max} - x_{\min}} \tag{5.49}$$

where $x_{\min}$ and $x_{\max}$ are, respectively, the lower and upper bounds of the design variable. Many commercial optimization algorithms internally normalize design variables. When a design variable does not have lower and/or upper bounds, it is better to select very low or high values for the bounds.

2) Similar to the normalization of design variables, it is also better to normalize constraints. This is because different constraints have different scales. For example, in solid mechanics, displacement constraints are in the order of $10^{-5}$ m, while stress constraint is in the order of $10^8$ Pa. Such a huge difference in scale can cause difficulty in satisfying the constraints. A violation of $10^3$ is ignorable for stress constraint, but it is a huge violation of the displacement constraint. Therefore, it would be necessary to scale all constraints such that their violation and margin can be interpreted in the same way. For example, when a constraint is given in the form of $g(\mathbf{x}) \leq 0$, it would be desirable to scale them such that $g = -0.1$ means a 10% margin, and $g = +0.1$ corresponds to a 10% violation. When a constraint is given in the form of $c(x) \leq c_{\text{limit}}$, the normalized constraint can be written as

$$g(\mathbf{x}) = \frac{c(\mathbf{x})}{c_{\text{limit}}} - 1 \leq 0 \tag{5.50}$$

Similar normalization can be possible for greater-than-equal-to type constraints. When a constraint bound is zero, an appropriate scale should be chosen by the users.
3) When the users have many algorithms to choose, it is better to choose an optimization algorithm that computes Lagrange multipliers. In Chapter 4, the optimality condition of a constrained optimization problem is formulated using Lagrange multipliers. These Lagrange multipliers are not only important to find an optimum design but also useful to understand the role of constraints on the optimum design. If a Lagrange multiplier is zero at the optimum design, the corresponding constraint is inactive. If a Lagrange multiplier at the optimum design is small, the influence of the corresponding constraint is not significant. On the other hand, if a Lagrange multiplier is high, the constraint is important. That is, if the constraint bound is changed slightly, the optimum design can change significantly. Therefore, it is good practice to check the Lagrange multipliers at the optimum design to understand what constraints play an important role in determining the optimum design.
4) Most optimization algorithms in this chapter require gradient (i.e., derivative) information of the objective and constraint functions, which is why they are categorized as gradient-based optimization. There are systematic ways of calculating gradient information [49], which is called design sensitivity analysis. They require an additional computer program that depends on the objective and constraint functions. In many cases, design sensitivity analysis can calculate the gradient information with much cheaper computational costs. However, since it requires significant modification of the response analysis program, not all computer programs have the capability of calculating the gradient information. Instead, the gradient information is often calculated using a finite difference method, where the design variable is slightly perturbed, and the response is calculated again. Then, using the difference between the two response values, the gradient is approximated, as

$$\nabla g(\mathbf{x}) \approx \frac{g(\mathbf{x} + \Delta\mathbf{x}) - g(\mathbf{x})}{\Delta\mathbf{x}} \tag{5.51}$$

Indeed, this is the definition of the derivative as $\Delta\mathbf{x} \to 0$. The above method of calculating the gradient is referred to as the forward finite difference method because the design is perturbed in the positive direction of the design variable. Gradient calculation using a finite difference method is computationally expensive because each design variable is perturbed one at a time, and the response is calculated again. For example, if there are $n$ design variables, it requires

$n + 1$ response analyses to calculate the gradient information. In fact, gradient calculation is a major part of the computational cost in optimization. However, the forward finite difference in Eq. (5.51) can cause biased gradient information as the design is perturbed in the positive direction. If the computational cost is affordable, it is recommended to use the central finite difference method, where the sensitivity is calculated by

$$\nabla g(\mathbf{x}) \approx \frac{g(\mathbf{x} + \Delta\mathbf{x}) - g(\mathbf{x} - \Delta\mathbf{x})}{2\Delta\mathbf{x}} \tag{5.52}$$

The central finite difference method is expensive as it requires perturbing the design in both positive and negative directions. When there are $n$ design variables, the central finite difference method requires $2n + 1$ response analyses to calculate the gradient information. However, the calculated gradient is unbiased.

5) Finite element analysis programs are often used to calculate the objective and constraint functions. When the size of the model is large, it is difficult to solve a large matrix equation using a direct method. Instead, iterative methods are often used to solve a large system of equations. The iterative methods often yield less accurate solutions than the direct method. Therefore, if iterative analysis methods are used for the analysis, it would be better to use an extra stringent convergence criterion during the optimization if you can afford to. Otherwise, you can get poor finite-difference derivatives and bogus local minima.
6) Optimization algorithms assume that the analysis can produce results anywhere in the design space. In practice, however, it is possible that the analysis may not be successful in some portion of the design space. Since the analysis cannot produce results in this region, the optimization algorithm may think that the objective function value is low in this region, which is wrong. Therefore, it would be necessary to carefully check the final designs with more accurate analysis codes or more refined models.
7) It would not be smart to solve an optimization problem without having any prior estimation or expectation of the optimum design. It is dangerous to use the optimization program as if it were a black box. Before using optimization, it is crucial to gain some insight into the problem by studying carefully the results of an analysis of a representative or intuitive design.
8) It would be a bad choice if nonlinear optimization algorithms were used to solve linear problems. It is true that nonlinear algorithms can solve linear problems. However, nonlinear optimizers are not as robust as linear optimizers and have more difficulty in calculating Lagrange multipliers. Therefore, if both the objective function and constraints are linear, it is always better to use linear optimization algorithms.
9) When optimization algorithms stop convergence iterations, it does not always mean that the optimum design has been found. There are many other reasons that the optimization algorithms stop without converging to the optimum design. Optimization algorithms can stop iterations for many reasons. For example, optimization algorithms stop when the maximum number of function evaluations is reached. Also, they stop when the maximum number of iterations is reached. The best way to confirm local optima is to check the Kuhn–Tucker conditions. In practice, this might not be an easy task. In the case of Matlab optimization algorithms, the reason for stopping the algorithms is stored in the EXITFLAG variable. When EXITFLAG is negative, it means that the optimization algorithms stopped without finding local optima.

In presenting optimization results, the guiding principle is one that applies to almost any document that is long enough, so some readers may not want to read everything. The document should provide summaries that allow the reader to get the important information in the document without reading everything and indicate where in the rest of the document additional details on each topic may be found. When presenting the solution to an optimization problem, the following information should be included:

1) Summary of design variables, objective function, and constraints
2) Method of solution
3) Solution, including minimizer, the value of the objective function, the indication of which constraints are critical, and if they are not exactly zero, what is the margin or degree of violation
4) List of the other materials included with the solution, which allows the reader to get additional details.

## Exercise

**5.1** Explain the differences and commonalities of steepest descent, conjugate gradients, Newton's method, and quasi-Newton methods for unconstrained minimization.

**5.2** Minimize $f(x_1, x_2) = 12x_1^2 + 4x_2^2 - 12x_1x_2 + 2x_1$ using the conjugate gradient method, starting with the initial design $\mathbf{x}^{(0)} = \{-1, -2\}$.

**5.3** Calculate the order and rate of convergence of the sequence $x^{(k)} = 1/k$ and $x^* = 0$.

**5.4** With an extremely robust algorithm, we can find a very accurate solution with a penalty function approach by using a very high $\omega$. However, at some high value, the algorithm will begin to falter, either taking a very large number of iterations or not reaching the solution. Test `fminunc` and `fminsearch` on Eq. (5.9) starting from $\mathbf{x}_0 = [2, 2]$. Start with $\omega = 1{,}000$ and increase.

**5.5** Show that the KKT condition of the quadratic programming problem in Eq. (5.47) is equivalent to the KKT condition in Eq. (5.46).

**5.6** Use `fminunc` to minimize the Rosenbrock Banana function and compare the trajectories of `fminsearch` and `fminunc` starting from (−1.2,1), with and without the routine for calculating the gradient. Plot the three trajectories.

**5.7** Minimize a quadratic objective function with a ring constraint using the Matlab `fmincon` function. Using the bounds $r_i = 10$ and $r_o = 20$. Is feasible domain convex?

$$\underset{x_1, x_2}{\text{minimize}} \; f(x_1, x_2) = x_1^2 + 10x_2^2$$

$$\text{subject to } \; r_i^2 \le x_1^2 + x_2^2 \le r_o^2 s$$

**5.8** For the ring problem in Example 5.11 with $a = 10$, can you find a starting point within a circle of radius 30 around the origin that will prevent fmincon of finding the optimum?

**5.9** Solve the problem of minimizing the surface area of the cylinder subject to a minimum volume constraint as an inequality constraint. Do also with Matlab by defining non-dimensional radius and height using the cubic root of the volume.

**5.10** A 10-bar truss structure shown in the figure is under two loads, $P_1$ and $P_2$. The design goal is to minimize the weight, $W$, by varying the cross-sectional areas, $A_i$, of the truss members. The stress of the member should be less than the allowable stress with the safety factor. For manufacturing reasons, the cross-sectional areas should be greater than the minimum value. Input data are summarized in the table. Find optimum design using the `fmincon` Matlab function. Use initial cross-sectional area $A_i = 0.5$ in$^2$. Plot objective history (objective value versus iteration). Discuss EXITFLAG. Discuss why some members have minimum area. In order to calculate the member forces, the following terms need to be defined first:

$$a_{11} = \left(\frac{1}{A_1} + \frac{1}{A_3} + \frac{1}{A_5} + \frac{2\sqrt{2}}{A_7} + \frac{2\sqrt{2}}{A_8}\right), \quad a_{12} = a_{21} = \frac{1}{A_5}$$

$$a_{22} = \left(\frac{1}{A_2} + \frac{1}{A_3} + \frac{1}{A_5} + \frac{1}{A_6} + \frac{2\sqrt{2}}{A_9} + \frac{2\sqrt{2}}{A_{10}}\right)$$

$$b_1 = \sqrt{2}\left(\frac{P_2}{A_1} - \frac{P_1 + 2P_2}{A_3} - \frac{P_2}{A_5} - \frac{2\sqrt{2}(P_1 + P_2)}{A_7}\right), \quad b_2 = \left(-\frac{\sqrt{2}P_2}{A_4} - \frac{\sqrt{2}P_2}{A_5} - \frac{4P_2}{A_9}\right)$$

Then, member forces are defined as

$$N_8 = \frac{a_{22}b_1 - a_{12}b_2}{a_{11}a_{22} - a_{12}a_{21}}, \quad N_{10} = \frac{b_2a_{11} - a_{21}b_1}{a_{11}a_{22} - a_{12}a_{21}}, \quad N_1 = P_2 - \frac{1}{\sqrt{2}}N_8$$

$$N_2 = -\frac{1}{\sqrt{2}}N_{10}, \quad N_3 = -P_1 - 2P_2 - \frac{1}{\sqrt{2}}N_8, \quad N_4 = -P_2 - \frac{1}{\sqrt{2}}N_{10}$$

$$N_5 = -P_2 - \frac{1}{\sqrt{2}}N_8 - \frac{1}{\sqrt{2}}N_{10}, \quad N_6 = -\frac{1}{\sqrt{2}}N_{10}$$

$$N_7 = \sqrt{2}(P_1 + P_2) + N_8, \quad N_9 = \sqrt{2}P_2 + N_{10}$$

| **Parameters** | **Values** |
|---|---|
| Dimension, $b$ | 360 in. |
| Safety factor, $S_F$ | 1.5 |
| Load, $P_1$ | 66.67 kips |
| Load, $P_2$ | 66.67 kips |
| Density, $\rho$ | 0.1 lb/in$^3$ |
| Modulus of elasticity, $E$ | $10^4$ ksi |
| Allowable stress, $\sigma_{\text{allowable}}$ | 25 ksi* |
| Initial area $A_i$ | 1.0 in$^2$ |
| Minimum cross-sectional area | 0.1 in$^2$ |

*For Element 9, allowable stress is 75 ksi.

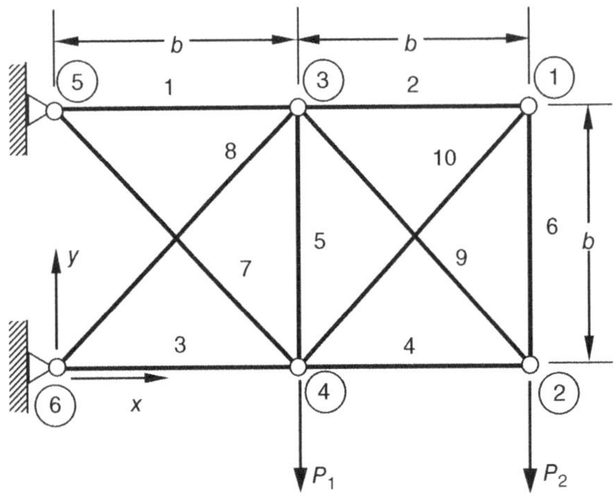
b
b
5
1
3
2
1
8
10
5
6
b
y
7
9
3
4
6
x
4
2
$P_1$
$P_2$

# 6

# Global Search Optimization Algorithms

## 6.1 Introduction

The gradient-based numerical optimization algorithms that we discussed in Chapter 5 are local search algorithms where a design is continuously improved/updated based on derivatives. For differentiable functions, it has been shown that even if the derivatives are calculated by finite difference methods, gradient-based algorithms are better than ones that do not use derivatives. However, gradient-based algorithms are limited to continuously differentiable functions. If the objective function or constraint functions are not smooth or have some discontinuities, we have to rely on gradient-free optimization algorithms.

Another challenge of gradient-based algorithms is that they find local optimum designs. Since design moves from an initial design to an improved design, the local search algorithms tend to converge on the local optimum that is close to the initial design. Therefore, different initial designs may end up with different local optima. Indeed, it is often suggested to repeat the gradient-based optimization algorithms with different initial designs in order to increase the chance of finding a global optimum design. However, the lack of capability in exploring the entire design space is considered a major drawback of gradient-based optimization algorithms.

In order to overcome the requirements of smoothness and local optima, gradient-free optimization algorithms have gained a lot of attention recently. These algorithms are relatively easy to implement as they do not require gradient or Hessian information and the line search for step-size determination. In addition, since they explore the entire design space, they have a better chance of finding the global optimum. However, most gradient-free algorithms are computationally expensive as they require thousands of function evaluations. In addition, these algorithms often come with many parameters to be tuned, and the performance strongly depends on these parameters. Therefore, an algorithm that works great for one optimization problem may not work well for other problems. Typically, these algorithms are developed based on some physical phenomena. Some examples are genetic algorithms, simulated annealing algorithms, and particle swarm optimization.

In the literature, these algorithms are often referred to as global optimization algorithms, but they are referred to as global search algorithms in this text. This is because there is no guarantee of finding the global optimum design. They are indeed structured random searches, which increases the chance of finding global or near-global optimum designs. These methods tend to be numerically robust as they generate a population of designs and choose the best one. An important aspect of this group of algorithms is that they can provide a number of good designs instead of a single optimum design.

Some global search algorithms are deterministic, while most others are probabilistic. The Nelder–Mead sequential simplex algorithm in Section 6.2 and the DIRECT method in Section 6.3 are two

*Surrogate Modeling and Optimization: Theories, Applications, and Limitations*, First Edition. Nam-Ho Kim.

Companion website: www.wiley.com/go/kim/surrogate_modeling

examples of deterministic algorithms. These algorithms are deterministic in the sense that the same optimum design can be found when the algorithm is used with the same initial condition. However, they still search globally to find the best design. Most other algorithms are stochastic in their nature, where different results are expected if the algorithm is repeated. These algorithms include the genetic algorithm in Section 6.4, the particle swarm algorithm in Section 6.5, and the simulated annealing algorithm in Section 6.6.

Depending on how they generate population and do not get trapped in local optima, probabilistic optimization algorithms can be categorized into three groups. The first group is based on pure random search, including the hill-climbing and taboo search algorithms. These algorithms essentially jump around the design space and accept designs with some probability. The second group is based on evolutionary algorithms, such as the genetic algorithm, which try to imitate biological evolution. The last one tries to imitate swarm intelligence, particle swarm optimization, which will cover essentially imitates birds and bees, while ant colony optimization imitates ants.

The major bottleneck of global search algorithms is computational cost. The cost of local optimization algorithms usually increases somewhere between linearly and quadratically proportional to the number of variables. However, the global search algorithms are often considered as NP-hard [53], which means that as the dimension of a problem increases, the cost of solving it increases faster than any polynomial order, which means that the computational cost increases exponentially.

Another interesting observation in global search algorithms is the no-free-lunch theorem [54], where for certain types of mathematical problems, the computational cost of finding a solution, averaged over all problems in the class, is the same for any solution method. In global optimization, it means that no single algorithm will perform well on all problems. The performance of global search algorithms changes significantly by changing their parameters. Therefore, it is possible to make an algorithm perform well for a specific problem, but it may not work efficiently for other problems. This is a great opportunity for engineers who can use problem-specific knowledge to tailor algorithms to their problems, something that they can rarely do for local optimization algorithms. However, it is difficult to claim that one algorithm is better than the other.

The complexity of a function can be categorized in the perspective of difficulty associated with finding the global optimum. Thomas Weise [55] illustrates different characteristics of optimization problems of increasing challenge. The function in Figure 6.1(a) shows a very benign function, where local minima are very shallow, and most local optimization algorithms may be able to find the global minimum. Figure 6.1(b) shows many significant local optima, where it is likely that gradient-based algorithms may find the closest local optimum from the initial design. However, a good global algorithm may find the global optimum in such a problem. Gradient-based algorithms can still find the global optimum by repeating the algorithm with different initial designs. Figure 6.1(c) shows a challenging problem in the sense that most optimization algorithms may get trapped in the wide flat region, as we do not know the functional form of the entire design space. In Figure 6.1(d), the global optimum is hidden in a narrow region. Even a good global search optimizer may take forever to find the optimum.

## 6.2 Nelder–Mead Sequential Simplex Algorithm

The first gradient-free algorithm that we will discuss in this section is the Nelder–Mead sequential simplex algorithm, which does not require the objective function to be differentiable or even continuous. The algorithm was initially invented by Spendley et al. [56] and then refined by British mathematicians Nelder and Mead [57]. It is considered the simplest algorithm in a random search

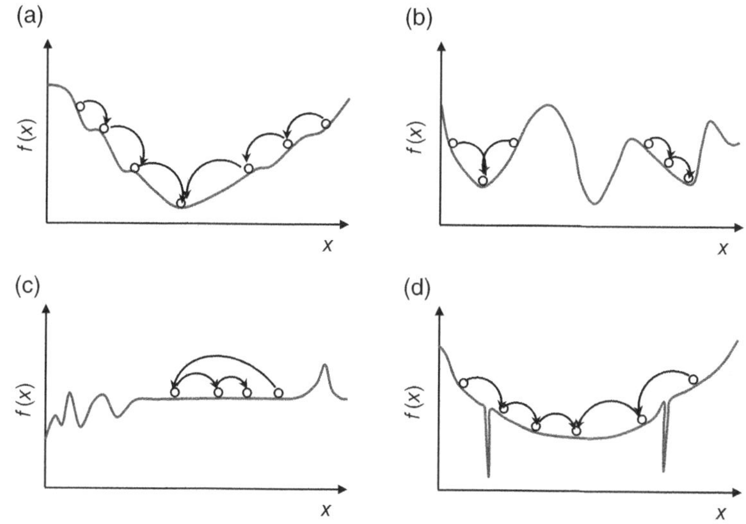


**Figure 6.1** Complexity of functions and global optimization. (a) Benign function, (b) many significant local optima, (c) function with a wide flat region, and (d) global optima in a narrow region.

by adding simple logic, which is referred to as a structured random search. It is easy to implement, robust, and computationally efficient. This algorithm is currently implemented in the Matlab function fminsearch. Matlab refers to a paper on its convergence properties in one or two dimensions [58].

The algorithm has random search characteristics as it is based on generating population in the design space. However, it is not purely a random search because it moves the design in a direction that can reduce the objective function. This sounds like a gradient-based algorithm, but it moves the design without requiring gradient information.

The Nelder–Mead sequential simplex algorithm utilizes the geometry of a simplex, which is the simplest body in $N$ dimensional design space with $N + 1$ vertices (triangle in 2D and tetrahedron in 3D). The initial vertices are randomly generated in the design space, but the performance of the algorithm is relatively insensitive to these initial choices. The objective function is evaluated at each vertex. The basic idea of the Nelder–Mead algorithm is to move away from the worst point in the direction of the other points. Then, the algorithm is composed of the following four simple logics to update the design:

- **Reflection:** The point with the highest objective function value is reflected on the opposite side (preserve volume, nondegeneracy). Specifically, the worst point moves in the direction of the center of gravity of the other points. Most of the time, this point may be an improvement on the worst point, but not on the best point. In that case, the new vertex is the point that is the mirror image of the worst point on the other side.
- **Expansion:** If the reflected point is the best design (better than the objective functions at all vertices), this means that the reflection direction is promising and the reflected point should go further along that direction, i.e., expand the simplex further (possibly the optimum is further in that direction).
- **Contraction:** If the reflected point is worse than all the existing points, it means the reflected point possibly goes too far, and it would be necessary to contract the simplex (possible valley floor).

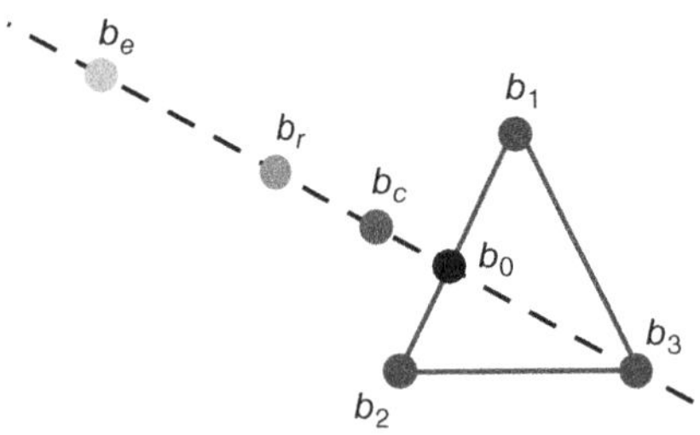


**Figure 6.2** Nelder–Mead sequential simplex algorithm in two-dimensional design space.

- **Reduction:** If the contracted point is still worse, reduce the size of the simplex (possibly the best point is within the simplex).

Figure 6.2 shows an illustration of the Nelder–Mead algorithm for a two-dimensional design space. The initial simplex is the triangle composed of $b_1 - b_2 - b_3$. At these three vertices, the objective function has the highest value at $b_3$, which is reflected against $b_1 - b_2$. The reflected point is $b_r$. If the objective value at $b_r$ is between that of $b_1$ and $b_2$, it becomes a new vertex, replacing $b_3$. If the objective value at $b_r$ is less than that of $b_1$ and $b_2$, it is further extended to $b_e$. If the objective value at $b_r$ is greater than that of $b_1$ and $b_2$, it is contracted to $b_c$. If the objective function value at the contracted point $b_c$ is greater than that of $b_1$ and $b_2$, reduce the size of the simplex by half.

With reference to Figure 6.2, the Nelder–Mead sequential simplex algorithm in $N$-dimensional design space starts with $N+1$ randomly selected vertices of a simplex. The objective functions at all vertices are evaluated and ordered by function values: $f(\mathbf{b}_1) \le f(\mathbf{b}_2) \ldots \le f(\mathbf{b}_{N+1})$. That is, $\mathbf{b}_1$ is the best design and $\mathbf{b}_{N+1}$ is the worst design for the given simplex. Then, the following algorithm is repeated until the size of the simplex is less than a threshold.

1) Calculate $\mathbf{b}_0$ at the center of gravity of all the points except for $\mathbf{b}_{N+1}$

$$\mathbf{b}_0 = \frac{1}{N}\sum_{i=1}^{N} \mathbf{b}_i$$

2) (Reflection) Reflect the worst point $\mathbf{b}_{N+1}$ about $\mathbf{b}_0$ ($\alpha = 1$)

$$\mathbf{b}_r = \mathbf{b}_0 + \alpha(\mathbf{b}_0 - \mathbf{b}_{N+1})$$

If $\mathbf{b}_r$ is better than the second worst, but not the best, use it to replace with the worst and go back to Step 1 ($\mathbf{b}_{N+1} = \mathbf{b}_r$)

3) (Expansion) If $\mathbf{b}_r$ is the best, i.e., $f(\mathbf{b}_r) < f(\mathbf{b}_1)$, move further in the direction of $\mathbf{b}_r$ ($\gamma = 2$)

$$\mathbf{b}_e = \mathbf{b}_0 + \gamma(\mathbf{b}_0 - \mathbf{b}_{N+1})$$

If $f(\mathbf{b}_e) < f(\mathbf{b}_r)$, then replace $\mathbf{b}_{N+1}$ by $\mathbf{b}_e$. Otherwise, replace $\mathbf{b}_{N+1}$ by $\mathbf{b}_r$. Go back to Step 1.

4) (Contraction) If $\mathbf{b}_r$ is the worst, i.e., $f(\mathbf{b}_r) \ge f(\mathbf{b}_N)$, move less in the direction of $\mathbf{b}_r$ ($\rho = 0.5$)

$$\mathbf{b}_c = \mathbf{b}_0 + \rho(\mathbf{b}_0 - \mathbf{b}_{N+1})$$

If $f(\mathbf{b}_c) < f(\mathbf{b}_{N+1})$, then replace $\mathbf{b}_{N+1}$ by $\mathbf{b}_c$.

5) (Reduction) If $f(\mathbf{b}_c) \ge f(\mathbf{b}_{N+1})$, contract all points about the best one ($\sigma = 0.5$)

$$\mathbf{b}_i = \mathbf{b}_1 + \sigma(\mathbf{b}_i - \mathbf{b}_1), i = 2, \ldots, N+1$$

Go back to Step 1.

**Example 6.1 Minimization of the banana function**

Consider the minimization of the Rosenbrock function, commonly called a banana function:

$$\text{minimize } f(\mathbf{x}) = 100\left(x_2 - x_1^2\right)^2 + (1 - x_1)^2$$

It is known that the function has its minimum at $\mathbf{x}^* = (1, 1)$ with $f(\mathbf{x}^*) = 0$. With the initial design at $(-1.2, 1.0)$, perform 20 optimization iterations using the Matlab fminsearch function (Nelder–Mead sequential simplex method).

**Solution:**

First, the following Matlab function banana is stored in the banana.m file:

```
function [y]=banana(x)
global z1 z2 yg count
y=100*(x(2)-x(1)^2)^2+(1-x(1))^2;
z1(count)=x(1); z2(count)=x(2);
yg(count)=y;
count=count+1;
```

The banana function not only codes the Banana function, but it also stores the coordinates and the function values every time it is called. This allows us to plot the history of design evolution. The following Matlab script calls `fminsearch` with a limit on the number of function evaluations set at 20

```
global z1 z2 yg count
count =1;
options=optimset('MaxFunEvals',20)
[b,fval] = fminsearch(@banana,[-1.2, 1],options)
mat=[z1;z2;yg]
```

The array mat stores the history of design variables and objective function:

```
mat =
  Columns 1 through 8
   -1.200    -1.260    -1.200    -1.140    -1.080    -1.080    -1.020    -0.960
    1.000     1.000     1.050     1.050     1.075     1.125     1.1875    1.150
   24.20     39.64     20.05     10.81      5.16      4.498     6.244     9.058

  Columns 9 through 16
   -1.020    -1.020    -1.065    -1.125    -1.046    -1.031    -1.007    -1.013
    1.125     1.175     1.100     1.100     1.119     1.094     1.078     1.113
    4.796     5.892     4.381     7.259     4.245     4.218     4.441     4.813
```

The first two rows are design variables, $x_1$ and $x_2$, while the last row is the objective function. In order to construct the initial simplex (triangle in 2D problem), `fminsearch` perturbs the initial design by 5% in each design direction. Figure 6.3 shows the first seven design points generated by `fminsearch` function in Matlab. The first three points are the initial simplex. Here the first reflection, $\mathbf{x}^{(4)}$, is very successful, in that the value of the function, at 10.81, is better than any of the previous points, so we double the distance and get an even better result of 5.16 at $\mathbf{x}^{(5)}$. At the step, the simplex is composed of $\mathbf{x}^{(5)}$, $\mathbf{x}^{(3)}$, and $\mathbf{x}^{(2)}$.

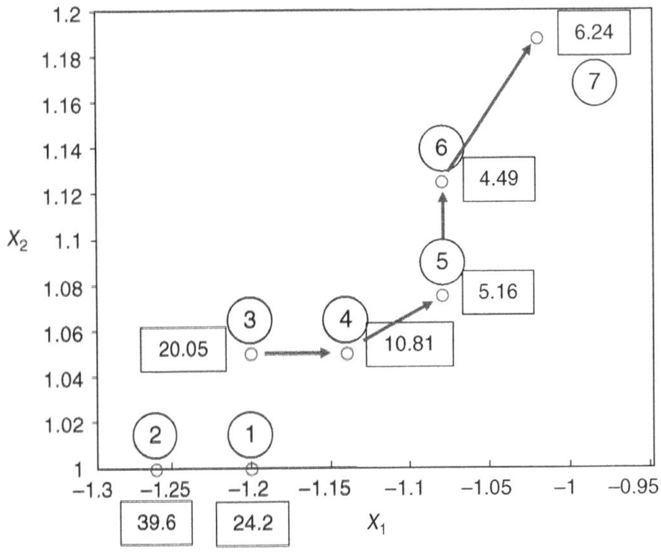


**Figure 6.3** First seven design points generated by `fminsearch` in Matlab.

The second iteration again starts with a reflection of the worst point $f(\mathbf{x}^{(2)}) = 24.2$ about the middle of the line connecting the other two points, $\mathbf{x}^{(5)}$ and $\mathbf{x}^{(3)}$. This gives us an improvement to 4.49 at $\mathbf{x}^{(6)}$, so we try expansion again to $\mathbf{x}^{(7)}$. However, this time we do not get improvement, so the simplex for the next iteration will include $\mathbf{x}^{(6)}$, $\mathbf{x}^{(5)}$, and $\mathbf{x}^{(3)}$.

## 6.3 DIRECT Method

Although the Nelder–Mead sequential simplex method is a gradient-free algorithm, it has a limitation to be called a global search algorithm as it starts with an initial simplex and updates the simplex based on function values. If a function has many local optima, it tends to find a local optimum depending on the size and location of the initial simplex. A global search algorithm should have the capability to explore the entire design space for a possible, better optimum design.

In this section, we will discuss a new algorithm, called DIviding RECTangles (DIRECT) algorithm, originally proposed by Jones et al. [59]. The DIRECT algorithm is an example of systematic exploration of the design space by dividing it into regions and then into subregions. The regions would be sections of a line in 1D, rectangles in 2D, and cuboids in 3D. In higher dimensions, they are referred to as hyperboxes. Compared to other global search algorithms, the DIRECT algorithm is purely deterministic, which means that the algorithm will produce the same optimum design even if it runs multiple times. In order to understand the DIRECT optimization algorithm, it would be necessary to discuss Lipschitzian optimization first, followed by how the DIRECT algorithm extends it.

### 6.3.1 Lipschitzian Optimization

Lipschitzian optimization was originally proposed by Shubert [60], which is designed for seeking the global minimum of a function. It is named after Rudolf Otto Sigismund Lipschitz, who was a

German mathematician and professor at the University of Bonn from 1864. Lipschitzian optimization is a predecessor of the DIRECT optimization algorithm that divides the space into boxes, samples at the center of the boxes, and divides boxes further based on the function's Lipschitz constant. The Lipschitz constant is an upper bound on the rate of change of the function, and it can be used even for functions that are not even differentiable.

A function $f: D \in R^n \rightarrow R$ is called Lipschitz continuous if there exists a positive constant $K \in R^+$ such that

$$|f(x) - f(x')| \leq K|x - x'|, \quad \forall x, x' \in D \tag{6.1}$$

where the lowest $K$ is called the Lipschitz constant. This basically excludes those functions whose absolute value of the slope approaches infinity. The upper bound of the slope is the Lipschitz constant. As long as the objective function is a bounded function of design variables, it is Lipschitz continuous. For $x \in [a, b]$, replacing $a$ and $b$ with $x'$ in the previous equation obtains

$$\begin{aligned} f(x) &\geq f(a) - K(x - a) \\ f(x) &\geq f(b) + K(x - b) \end{aligned} \tag{6.2}$$

In the first equation, the right-hand side is the straight line from $f(a)$ with a slope of $-K$. In the second equation, the right-hand side is the straight line from $f(b)$ with a slope of $+K$. Therefore, the function value is above the two straight lines. As shown in Figure 6.4(a), the intersection of these two lines is the lowest possible value if the function is decreased by a rate of $K$ from both ends. Note that given $K$ the function cannot be possibly lower than the intersection value. The intersection point can be calculated as

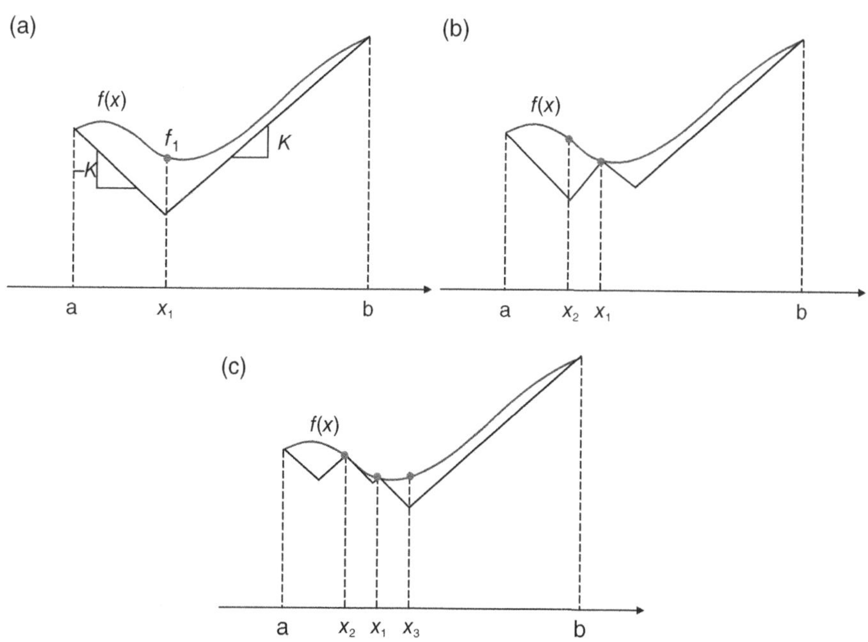


**Figure 6.4** Lipschitzian optimization of 1D function. (a) First iteration, (b) second iteration with intervals $(a, x_1)$, and (c) third iteration with intervals $(x_1, b)$.

$$x_1(a,b,f,K) = \frac{a+b}{2} + \frac{f(a)-f(b)}{2K}$$
$$f_1(a,b,f,K) = \frac{f(a)+f(b)}{2} - \frac{K(b-a)}{2} \tag{6.3}$$

Lipschitz optimization repeats this process by reducing the intervals. As shown in Figure 6.4(b), the process is repeated with two intervals that were created by adding $x_1$; that is, $(a, x_1)$ and $(x_1, b)$. Figure 6.4(b) shows that the process predicts a lower bound on the function in $(a, x_1)$ that is better than the one in $(x_1, b)$. Therefore, we select the intersection there and get $x_2$. We can repeat the process at the additional two intervals that were created by adding $x_2$, that is $(a, x_2)$ and $(x_2, x_1)$. Among all intersection points, we choose the lowest one as $x_3$, as shown in Figure 6.4(c).

The Lipschitzian optimization algorithm is simple and straightforward and yet provides a robust global search capability. The algorithm is deterministic, which means there is no need for multiple runs. In addition, the algorithm requires a few tuning parameters. The only parameter it needs is the Lipschitz constant $K$, which depends on the function. On the other hand, the determination of the Lipschitz constant may not be an easy task, as the global behavior of the function is unknown. In practice, the Lipschitz constant affects the convergence speed and the global searching capability. A large $K$ can help global search, but it slows down the convergence speed. On the other hand, a small $K$ may end up searching for local optima. Lastly, the algorithm may have a computational complexity for high dimensions. Since the algorithm requires evaluating function values at all corners of the design space, it requires a large number of function evaluations $\sim O(2^n)$ for a high-dimensional design space.

### 6.3.2 DIRECT in 1D

The direct algorithm is inspired by Lipschitzian optimization. The algorithm divides the design space into hyperboxes (lines in 1D, rectangles in 2D, cuboids in 3D, etc.) and samples the function at the center of each hyperbox. Based on Jones et al. [59], another challenge in Lipschitzian optimization is the determination of the Lipschitzian constant $K$. In the DIRECT algorithm, instead of finding the intersection of two lines as in Figure 6.4(a), the function value is evaluated at the center of the region. In addition, in order to recycle the center of the region in subsequent subregions, the region of design is divided into three subregions, and the function values are evaluated at the center of each subregion. As shown in Figure 6.5(a), the center point P in the initial region is recycled after

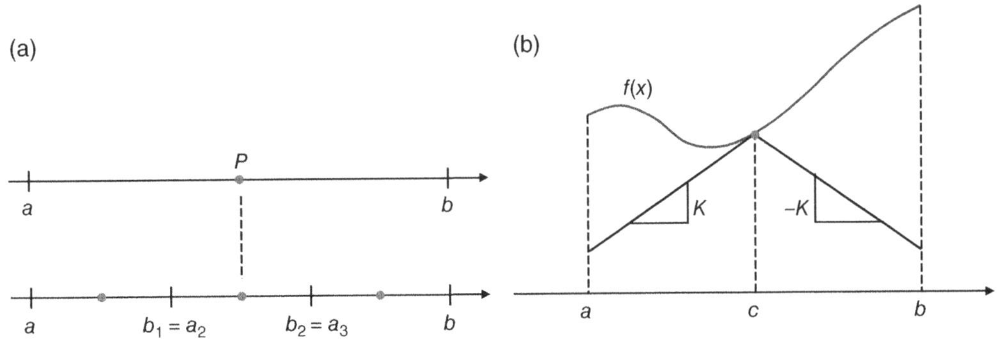


**Figure 6.5** The DIRECT algorithm in 1D design space. (a) Evaluating function at the center of intervals and (b) recycling center point after dividing into three subregions.

the region is divided into three subregions. From the perspective of Lipschitzian optimization, the DIRECT algorithm uses the following Lipschitz bound:

$$f(x) \geq f(c) + K(x - a) \quad \text{for } x \leq c$$
$$f(x) \geq f(c) - K(x - c) \text{ for } x \geq c \tag{6.4}$$

Since the function value at the center of the region is used instead of the two ends, it is unnecessary to find the intersection of the two lines, as shown in Figure 6.5(b). Also, it is unnecessary to define the Lipschitz constant $K$ that is the upper bound of the slope of the function.

### 6.3.3 DIRECT Algorithm

The DIRECT algorithm starts from one region of the entire design space and sequentially subdivides the region. At a given iteration, the design space is divided into subregions, and the objective function is evaluated at the center of each subregion. Figure 6.6 illustrates an example of the subdivision process of the DIRECT algorithm in a two-dimensional design space. The horizontal axis is $x_1$ and the vertical axis is $x_2$. First, the design space is normalized such that the box is square. The grey boxes represent the regions that need to be subdivided, and the circles represent the location of function evaluation. Initially, a single box is used for the entire design space, and the function is evaluated at the center of the design space, as shown in Figure 6.6(a). In the implementation of Jones et al. [59], the rule of the subdivision is such that the longer dimension is subdivided into three subregions. When a box is a square, the region is subdivided into three regions in the direction of $x_2$ first, and then, only the center region is subdivided into three regions in the direction of $x_1$, as shown in Figure 6.6(b). Such a scheme only requires four additional function evaluations. After comparing all five function values, it is determined that the bottom box is further subdivided. The rule of selecting the regions to be subdivided will be discussed later. In this case, the subdivision is only performed in the direction of $x_1$ because the bottom box is rectangle and has a longer length in the direction of $x_1$. In the next iteration, as shown in Figure 6.6(c), both the top rectangle and the bottom-center square are selected. In the optimization term, subdividing a large box corresponds to exploring the design space, while subdividing a small box corresponds to refining the design space. The top rectangle is subdivided into three squares in the direction of $x_1$, while the bottom-center square is subdivided into both $x_1$ and $x_2$ directions, as shown in Figure 6.6(d). In the subsequent evaluations, the top-center and bottom-right boxes will be further subdivided in the next iteration.

A critical question of the DIRECT algorithm is which boxes should be subdivided. As mentioned before, this subdivision is used to explore the design space and refine the optimum design. In the DIRECT algorithm, this determination is done using a graph of box size versus objective function,

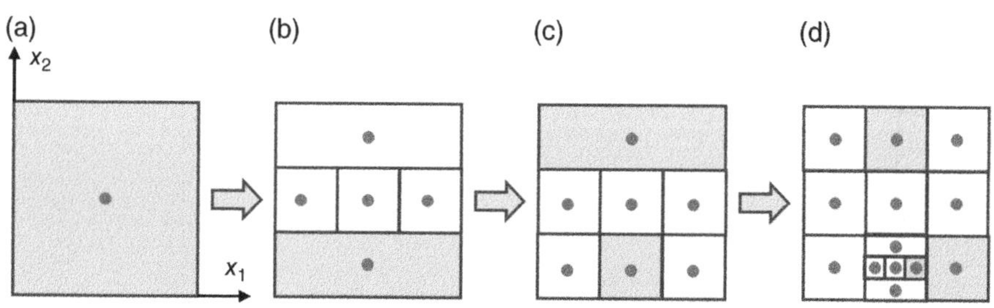


**Figure 6.6** DIRECT algorithm in 2D design space. (a) Initial single box, (b) first subdivision, (c) second subdivision, and (d) third subdivision.

as shown in Figure 6.7. In the graph, the box size represents the largest dimension of a box, while the ordinate is the value of the objective function at the center of the box. With reference to Figure 6.7(a), if the smallest box with the lowest objective function is subdivided, it can look for a better design within the current best box, which corresponds to refining the current optimum design. On the other hand, the lowest objective functions of larger boxes may have higher values, but they have a chance to yield a better design if they are subdivided, which corresponds to exploring the design space.

In the DIRECT algorithm, subdivisions are made at those points that have the best possibility of improving the optimum design. These points can be selected by constructing a convex hull of box sizes as shown in Figure 6.7(b). Those points on the convex hull are subdivided, and function values are evaluated at the center of the new boxes. If a sufficient number of iterations are performed, every region in the design space will eventually be divided into a small boxes. Therefore, the algorithm is guaranteed to find the global optimum even if the objective function has a difficult shape to find the global optimum design. In practice, however, it will require a huge number of function evaluations.

Figure 6.7 only shows a two-dimensional illustration. However, the number of boxes (i.e., the number of function evaluations) increases exponentially with dimension. For example, consider the case of 10 design variables. Having 10 regions in each design variable is not considered a high resolution. Even with such a coarse resolution, the number of function evaluations can reach $10^{10}$, which is practically impossible for most functions. Therefore, there is a serious question as to how

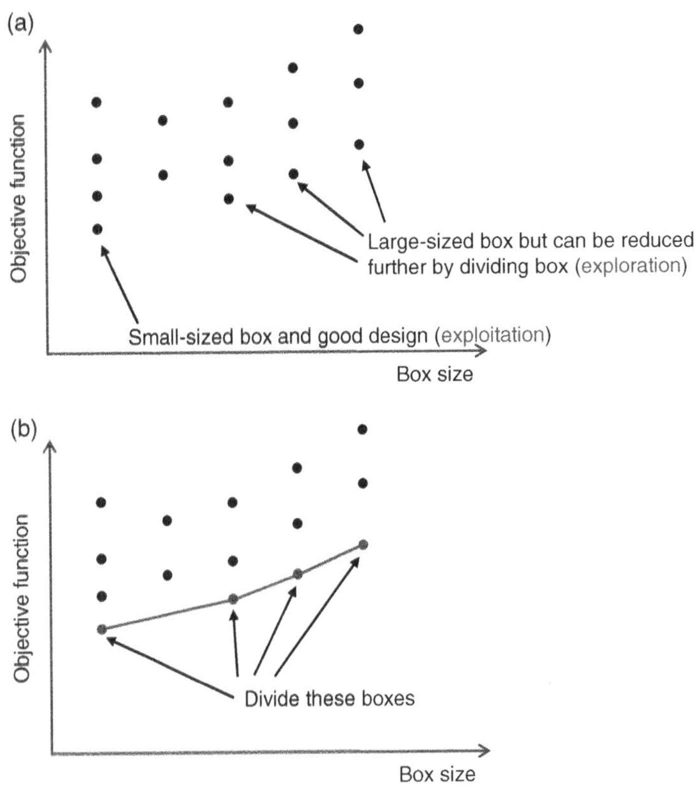


**Figure 6.7** Determination of subdivision boxes in the DIRECT algorithm. (a) Exploration versus exploitation and (b) convex hull of box sizes.

practical DIRECT is for a high-dimensional problem, which is discussed by Cox et al. [61]. Therefore, this algorithm is appropriate for an optimization problem with a small number of design variables.

The Matlab code of the DIRECT algorithm from Dr. Daniel Finkel is available on the website (https://ctk.math.ncsu.edu/Finkel_Direct/). The code is based on the algorithm published by Jones et al. [59]. The website also has user's guide as well as sample optimization problems.

**Example 6.2 Optimization of Goldstein–Price function**

Optimize the following Goldstein–Price function in the design space of $-2 \le x_1,\ x_2 \le 2$ using the DIRECT algorithm:

$$\text{minimize } f(\mathbf{x}) = \left(1 + (x_1 + x_2 + 1)^2 * \left(19 - 14x_1 + 3x_1^2 - 14x_2 + 6x_1x_2 + 3x_2^2\right)\right) \\ *\left(30 + (2x_1 - 3x_2)^2\left(18 - 32x_1 + 12x_1^2 + 48x_2 - 36x_1x_2 + 27x_2^2\right)\right)$$

**Solution:**

The Goldsten–Price function has its global minimum at $(x_1, x_2) = (0, -1)$ with a minimum value $f_{\min} = 3.0$. In order to solve the optimization problem, it would be necessary to define the objective function as an m-file. The following gp.m file is defined:

```
function value = gp(x)
x1 = x(1); x2 = x(2);
value =(1+(x1+x2+1).^2.*(19-14.*x1+3.*x1.^2-14.*x2+6.*x1.*x2+3.*x2.^2))...
.*(30+(2.*x1-3.*x2).^2.*(18-32.*x1+12.*x1.^2+48.*x2-36.*x1.*x2+27.*x2.^2));
```

The above gp.m file will calculate the objective function value at a given design $(x_1, x_2)$. Then, the following Matlab code will define the optimization problem and call the Direct.m Matlab function to find the optimum design:

```
clear all;
bounds = [-2 2;-2 2];                % lower- and upper-bounds of design
options.testflag  = 1;                        % Global minimum is known
options.globalmin = 3;                    % The value of known global min
options.showits   = 1;                        % Show iteration history
options.tol       = 0.01;                        % Stopping criterion
Problem.f = 'gp';                             % Objective function name
[fmin,xmin,hist] =Direct(Problem,bounds,options);% Calling DIRECT algorithm
plot(hist(:,2),hist(:,3))                   % Plot iteration statistics
xlabel('Fcn Evals'); ylabel('f_{min}');
```

The optimum design is stored at xmin, and the optimum objective value is at fmin. It turns out $x_{\min} = (0, -1.0005)$ and $f_{\min} = 3.0001$, which is within the tolerance. Figure 6.8(a) shows the contour of the objective function, which shows $(x_1, x_2) = (0, -1)$ as the global optimum design. Figure 6.8(b) shows the history of the objective function versus the number of function evaluations. Although optimization took only 14 iterations, it requires a total of 191 function evaluations. As boxes are refined, more function evaluations are required at a single iteration.

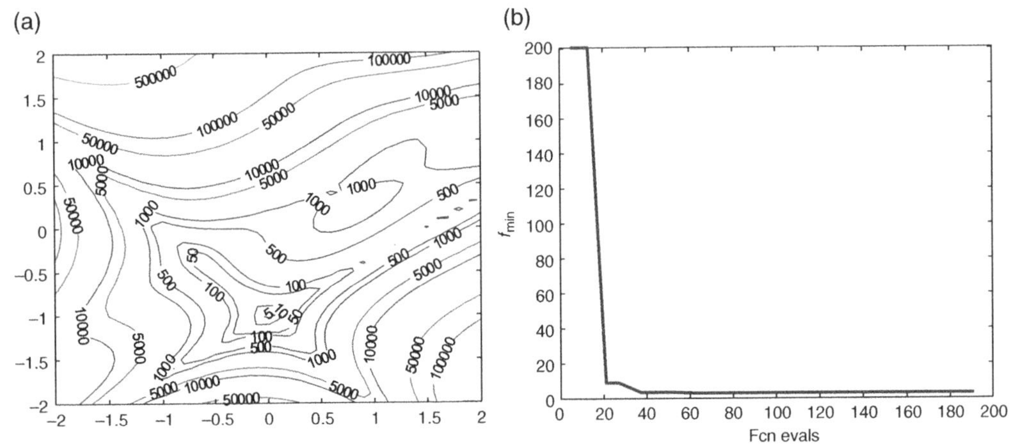

**Figure 6.8** Contour of objective function and optimization history of Example 6.2. (a) Contour of the objective function and (b) history of the objective function.

## 6.4 Genetic Algorithms

The sequential simplex method and the DIRECT method in the previous sections have the global searching capability, but they are deterministic in the sense that they will produce the same optimum design when the algorithm is repeated. That is, there is no randomness in the searching process. This is not a desirable feature of global searching algorithms, as randomness can increase the chance of finding the global optimum. Therefore, many global searching algorithms have randomness imbedded in their iteration process. Among many global searching algorithms, genetic algorithms are the most popular ones. Genetic algorithms were developed in 1975 by John Holland from the University of Michigan [62]. However, similar algorithms were developed in Europe around the same time with the name of evolutionary strategies. The main difference is that genetic algorithms are usually associated with discrete variables, while evolutionary strategies with continuous ones.

Genetic algorithms are inspired by Darwin's principle of evolution, where a population of individuals can adapt to its environment because individuals that possess traits that make them less vulnerable than others are more likely to have descendants and therefore to pass on their desirable traits to the next generation. One can think of this process of adaptation as an optimization process that probabilistically creates fitter individuals through selection and recombination of good characters. Genetic algorithms are simplified computer models of evolution, where the environment is emulated by the objective function to maximize/minimize, and the design variables play the role of the individuals. Moving from parents to children corresponds to one iteration in optimization, and the goodness is measured in terms of the objective function. The key ideas that distinguish genetic algorithms from traditional search algorithms are the use of a population of designs, the mating of pairs of designs to create child designs in a process called crossover, and the use of mutations (i.e., randomness) to enhance exploration. Together they create an optimization process that has a strong random component, so every time the algorithms are run, they produce different optimum designs.

### 6.4.1 Representation of Design

An important difference between genetic algorithms and conventional engineering optimization is the representation of design. In engineering design, a design variable is represented using a number, either continuous or discrete. In genetic algorithms, a design variable is stored in the chain of DNA (deoxyribonucleic acid) chromosomes, often using binary digits. Each design encodes a particular candidate structure in the form of one or several chromosomes, which are strings of finite length. A design variable is represented in the form of a string of binary numbers. Each chromosome can have a value of 0 or 1. The mapping between binary numbers and the state of the design variable is called coding. As an example, let a discrete design variable have four different states: red, green, blue, and yellow. In order to represent this design variable, two chromosomes should be enough as they can represent four different states: 00, 01, 10, and 11. Therefore, by coding 00 as red, 01 as green, 10 as blue, and 11 as yellow, the design variable can have all possible states. Although binary coding is most common, real number coding is also possible but requires special treatment.

The same binary digits can be used to store integer variables in a similar way that a computer stores numbers in the form of a binary numeral system. Figure 6.9 shows an eight-digit binary number. In the binary system, each bit represents an increasing power of 2, with the rightmost bit representing $2^0$, the next representing $2^1$, then $2^2$, and so on. The value of a binary number is the sum of the powers of 2 represented by each "1" bit. For example, the binary number 100101 is converted to decimal form as follows:

$$100101_2 = \left(1 \times 2^5\right) + \left(0 \times 2^4\right) + \left(0 \times 2^3\right) + \left(1 \times 2^2\right) + \left(0 \times 2^1\right) + \left(1 \times 2^0\right) = 37$$

As expected, the eight-digit binary numbers can represent up to 256 states of a design variable. A great number of bits are required to represent a design variable with a large range.

**Example 6.3 Binary coding**

Four integer variables are used for the following optimization problem definition

$$\begin{aligned}&\text{minimize } f(\mathbf{x}), \quad \mathbf{x} = \{x_1, x_2, x_3, x_4\}\\&\{0 \le x_1, x_4 \le 15\}, \quad \{0 \le x_2 \le 7\}, \quad \{0 \le x_3 \le 3\}\end{aligned}$$

When $x_1 = 6$, $x_2 = 5$, $x_3 = 3$, and $x_4 = 11$, represent the integer design variables using binary coding.

**Solution:**

Since $x_1$ and $x_4$ are between 0 and 15, all their states can be represented using four bits. Similarly, $x_2$ needs three bits and $x_3$ needs two bits. Therefore, all four design variables are represented using 13 bits, as shown in Figure 6.10.

| | | | | | | | | |
|---|---|---|---|---|---|---|---|---|
| Value of cell | $2^7$ | $2^6$ | $2^5$ | $2^4$ | $2^3$ | $2^2$ | $2^1$ | $2^0$ |
| | 128 | 64 | 32 | 16 | 8 | 4 | 2 | 1 |
| Binary number | 0 | 0 | 1 | 0 | 0 | 1 | 0 | 1 |

**Figure 6.9** Representation of 37 in the form of binary numbers.

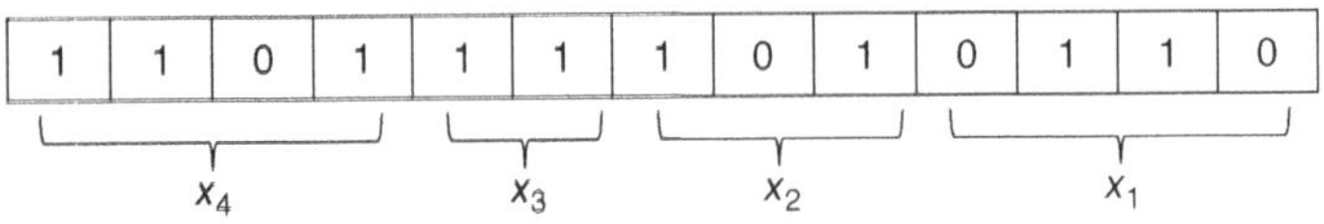


**Figure 6.10** Representation of four integer variables.

Many engineering optimization problems include real-valued design variables. It is a challenge to store a continuous variable using binary coding. Instead of storing continuous variables, binary coding stores the variable with a resolution or interval. Let a continuous design variable have a lower and upper bound, $x^L \le x \le x^U$. We want to store this variable with a resolution of $x^{\mathrm{inc}}$. Then the number $m$ of the required digits can be found from

$$2^m \ge \frac{x^U - x^L}{x^{\mathrm{inc}}} + 1 \tag{6.5}$$

For example, if $\{0.01 \le x \le 1.81\}$ with $x^{\mathrm{inc}} = 0.001$, the required digits should be

$$2^m \ge \frac{1.81 - 0.01}{0.001} + 1 = 1801 \tag{6.6}$$

Therefore, the smallest integer $m = 11$ and the actual intervals should be $x^{\mathrm{inc}} = (1.81 - 0.01)/2^{11} = 0.00088$. It is noted that the interval is adjusted such that the entire 11 digits are used to represent the range. As another example, let a design variable be the height of a stiffener, which can change from 1 to 2 in. with 0.1-in. intervals. Even if the design variable can have 11 different values, binary coding requires four digits with 16 values. Therefore, the entire range is divided into 15 intervals, each 1/15 in. wide. Therefore, 0000 denotes 1″, 0001 denotes 16/15″, ..., 1111 denotes 2″. Genetic algorithms, like most global search algorithms, are not effective in getting high-precision designs. It is better to go for a coarse grid of real values.

A common application of discrete variables is the stacking sequence optimization of composite materials, where the sequence of four different ply angles, 0°, ±45° or 90°, are determined to achieve the best performance of the composite panel. Since the ply thickness is fixed due to the manufacturing process, the design problem is selecting the number of plies and selecting the ply angles for each ply from the four choices. Coding the stacking sequence of a composite laminate that has a finite number of possible angles is typically done by assigning a number to each angle, or even just using the angle itself. For the convenience of explanation, natural coding will be used instead of common binary coding. In natural coding, the four possible ply angles are assigned as $(0° \to 1, 45° \to 2, -45° \to 3, 90° \to 4)$. If binary coding is used, then two bits can be used instead. Therefore, the following stacking sequence can be represented using natural coding as

$$(45/-45/90/0)_s \to (2/3/4/1)$$

The subscribed "*s*" stands for symmetric stacking, where the total stacking sequence is (45/ −45/ 90/0/0/90/ −45/45). When the laminate is symmetric, only one half is coded.

There is often also a balance condition that requires that the laminate has an equal number of positive and negative angles. This means that the number of 45° plies need to be equal to the number of −45° plies. This is a difficult constraint to enforce, and an easy way out is to require that plies come in groups of two: $(0°_2 \to 1, \pm 45° \to 2,\ 90°_2 \to 3)$. This, however, may mean wasted plies for 0° or 90° directions.

### 6.4.2 Genetic Operators

A key factor of genetic algorithms is working with a population of designs that can mate and create child designs. Among the population, two parents are selected based on their goodness (fitness) of their objective function values. Then, child designs are created using genetic operations, which take some genes from one parent and some from a second parent. Figure 6.11 shows four basic genetic operators.

The first genetic operator is selection, where good parents are chosen to produce children in the next generation. The basic idea is to pass designs with higher fitness to the next generation. This selection operation increases the population of high-fitness designs, while removing low-fitness designs. Therefore, the mean fitness of the population is improved.

The crossover operator exchanges some genes from the two parents to produce children. This corresponds to the case when a child takes a part of their gene from one parent and the remaining part from the other parent. First, some chromosomes of the two parents are sliced and exchanged. This operation diversifies the population, while the mean fitness is preserved. If the two parent designs are good (i.e., having good fitness value), the chance is high that their child design can produce better fitness. Selection plus crossover form together an exploitation mechanism that looks for a combination of good designs. Randomness can be included in the selection process of the two parents and in determining the crossover location. The genetic representation of designs with a fixed chromosome length makes the crossover convenient, as their parts can be easily aligned.

Mutation changes a chromosome randomly and adds an element of exploration. These two operators are used in practically every genetic algorithm. Permutation reverses the order of a portion of chromosomes. In the case of stacking sequence optimization of composite material, it preserves the in-plane properties and changes only the bending properties. Addition/deletion adds/removes one chromosome from a design. This operation is useful when the design space needs to be expanded to satisfy design constraints.

**Example 6.4 Crossover operation**

In the stacking sequence design of a composite material, two parents are given in $[\pm 45_2/0_4/90_2]s$ and $[\pm 45_4/0_2]s$ with five-digit genes. When the crossover occurred at the second digit, calculate their child designs.

**Solution:**

For coding, let us assign $0_2 \rightarrow 1, \pm 45 \rightarrow 2, 90_2 \rightarrow 3$. Also, we count the sequence from left to right. Then, the two parents can be represented as

Parent 1: [2/2/1/1/3]

Parent 2: [2/2/2/2/1]

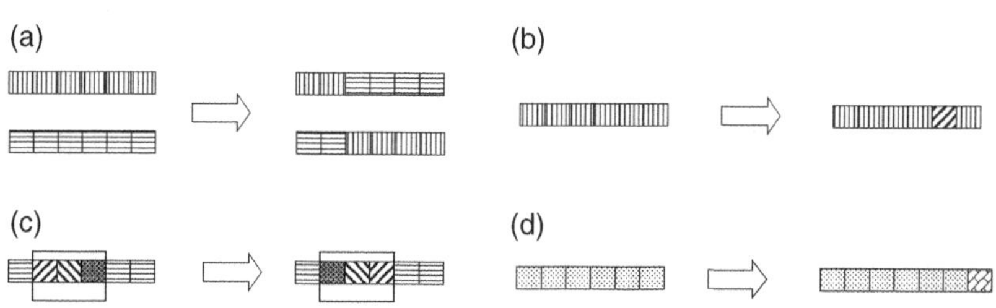


**Figure 6.11** Genetic operations. (a) Crossover, (b) mutation, (c) permutation, and (d) addition/deletion.

After the crossover at the second digit, the two children become

Child 1: [2/2/1/2/1]

Child 2: [2/2/1/1/3]

Therefore, the stacking sequences of the two children are $[\pm45_2/0_2/\pm45/0_2]s$ and $[\pm45_2/0_4/90_2]s$.

**Example 6.5 Mutation operation**

The first child in Example 6.4 has a mutation at the fourth digit, where a random value of 3 is assigned, followed by a permutation between 2-3-4 digits. Calculate the resulting stacking sequence.

**Solution:**

From Example 6.4, Child 1 has a design of [2/2/1/2/1]. After a mutation at the fourth digit, it becomes [2/3/1/2/1]. Then, a permutation between 2-3-4 digits yields [2/2/1/3/1]. This design corresponds to a stacking sequence of $[\pm45_2/0_2/90_2/0_2]s$.

### 6.4.3 Procedure of Genetic Algorithms

Figure 6.12 illustrates the basic procedure of genetic algorithms using the genetic operators. The algorithms start with a population of designs, and the fitness of each design is calculated based on its objective function and possibly constraint satisfaction. Then, parent designs are selected based on their fitness but with randomness thrown in. From the illustrative figure, the best (vertical line pattern) design was selected three times, as was the second best (inclined line pattern). The third-best design (horizontal line pattern) was selected twice, and the worst design (dotted pattern) not at all. Once parents are selected, children are created through genetic operations. In the figure, only crossover is used to create children. After all children are created, they become parents in the next generation (iteration). This process is continued until the stopping criteria are satisfied.

The procedure of genetic algorithms can be summarized as follows:

- Create a random initial population
- Generate the next population based on the following steps

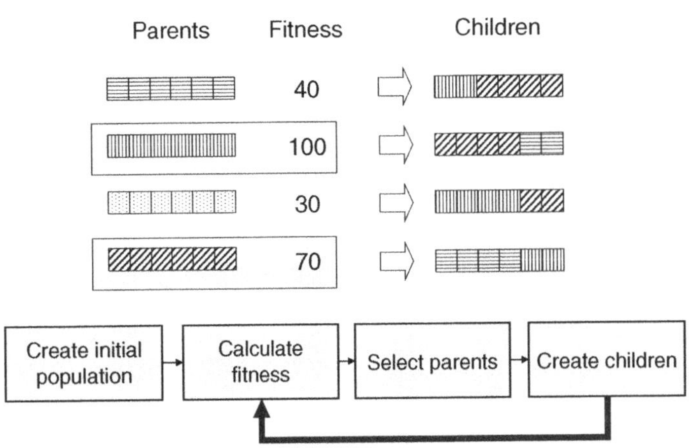


**Figure 6.12** Illustration of the procedure of genetic algorithms.

- Scores each design of the current population based on its *fitness*
- Selects members, called *parents*, based on their fitness
- Choose lower/higher fitness members as *elite*. These elite members are passed to the next population
- Produces *children* from the parents using *genetic operations*
- Replaces the current population with children to form the next generation

• Repeat the process until the *stopping criteria* are satisfied

The procedure of genetic algorithms starts from creating the initial population of designs randomly. In the case of Figure 6.12, the initial population consists of four designs with six digits each. In order to generate random designs, random number generators in common computer programs can be used. For example, Matlab uses the rand function to generate random numbers from a uniform distribution$\sim U[0, 1]$. However, most random number generators in computer programs are pseudo-random based on the initial seed. If the same initial seed is used, the same sequence of random numbers is generated. The initial seed may reflect some internal computer state, time of day, or some other data available to the generator. It is selected so that it is practically random, and different results are expected at different trials. However, seeds are useful when we want to generate the same sequence of random numbers. In most cases, it is necessary to interpret/transform random numbers to design values.

Once a population of designs is randomly generated, the next step is to calculate the fitness of each design. In unconstrained optimization, the fitness corresponds to the objective function to minimize. In the case of a constrained optimization problem, the following three-term augmented objective function is commonly used:

$$f^* = f + \mathrm{p}v - \mathrm{b}m + \mathrm{sign}(v)\Delta \tag{6.7}$$

where $f$ is the objective function, $v$ is the maximum violation of constraints, $m$ is the minimum margin of satisfying constraints, and p, b, and $\Delta$ are constants. By having a large value of p, the objective function is penalized proportional to the violated constraints. In addition, there is a small penalty $\Delta$ if there are any violated constraints, to give a preference to designs that satisfy the constraints over those who have very small violations. Finally, there is a term that rewards the margin with respect to the constraint with a bonus. This is important when there are multiple designs that have the same value of the objective function, but some satisfy the constraint with a larger margin.

Some objective functions vary in a wide range, while others vary in a small range. In order to make a robust algorithm, it would be necessary to select the fitness so that the difference between the best design and the poorest design is large, even if their objective function values are close. This prevents stagnation in evolution as we are near the optimum. Accentuating the difference may be done by normalizing it as

$$\text{fitness} = \frac{|f^* - f^*_{\max}|}{|f^*_{\min} - f^*_{\max}|} \tag{6.8}$$

Normalized fitness is insensitive to the range of objective function.

The two most common ways of selecting parents are the roulette wheel and the tournament. In this section, only the roulette wheel approach will be explained. Tournament selection is based on selecting randomly two individuals from the population and picking the best one as one parent, then repeating to select the second parent. Some genetic algorithms pursue elitist strategies, where the top design in a generation is always selected to be moved to the next generation. This guarantees monotonic progress, but it has not been proven to lead to faster convergence. In fact, putting too much preference to the top design means focusing on refining the local optimum but limiting exploration of a better design far away from the current good one. Finally, often genetic algorithms have no twin rule, in that both parents cannot be identical.

In a roulette wheel approach, each design occupies a slice of the roulette wheel that is proportional to its fitness. Spinning the roulette wheel is simulated by generating a random number from a uniform distribution$\sim U[0, 1]$. Figure 6.13(a) shows an illustration of roulette wheel selection with six fitness values, {0.62, 0.60, 0.65, 0.61, 0.57, 0.64}. Since the six fitnesses are close, the slices allocated to the six designs are close in size. Therefore, if they are randomly selected (arrows in the figure), they all have a similar probability of being selected. If, on the other hand, the fitness is made proportional to the reverse rank, then the top design will get a slice of 6/21 of the wheel, the second best 5/21, and the poorest design 1/21. Therefore, the following reverse rank {3/21, 5/21, 1/21, 4/21, 6/21, 2/21} is used to make the roulette wheel, as shown in Figure 6.13(b). The figure also shows the effect of generating six random numbers and selecting individuals based on these numbers. With the original fitness, each design is selected once. With the reverse rank-based fitness, the first- and second-best designs are selected twice, while the third and fourth are selected once, and the fifth and sixth are not selected at all.

Once parents are selected, genetic operators are applied to create child designs. The process of moving from parents to children is referred to as a generation, which is equivalent to iteration in optimization. While conventional optimization updates a design to a better design, genetic algorithms modify a population of designs via selection and genetic operations, by which the population evolves toward an optimal design. Once child designs are created, they become a new population, and the process is repeated until the algorithm satisfies the stopping criteria.

Stopping criteria determines when to stop optimization iteration. Although, the users can determine when and how to stop optimization iteration, below are common stopping criteria that are widely used.

- Stop when the maximum *number* of generations reaches
- Stop after running for the maximum computational *time* reaches
- Stop when the *best* fitness becomes less than or equal to the limit
- Stop when the *relative change* in fitness is less than the tolerance
- Stop if there is *no improvement* during an interval of time
- Stop if the average *relative change* in fitness is less than the tolerance.

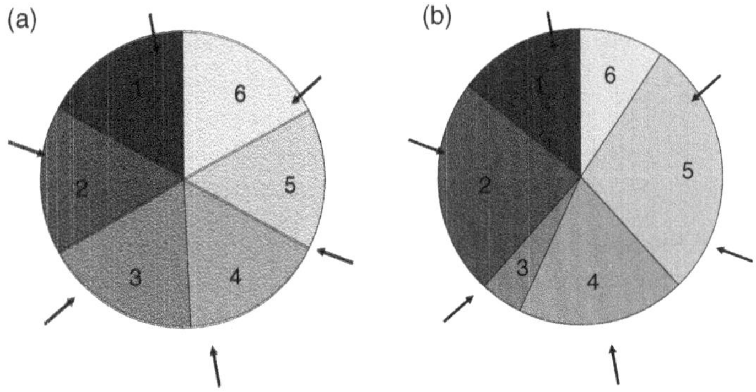


**Figure 6.13** Example of the roulette wheel selection method. (a) Proportional to fitness value and (b) proportional to the reverse rank.

### 6.4.4 Genetic Algorithm in Matlab

Genetic algorithms can be applied to solve optimization problems that are not well suited for standard optimization algorithms, including problems in which the objective function is discontinuous, nondifferentiable, stochastic, or highly nonlinear. Matlab provides the `ga` function for minimizing an objective function with linear and nonlinear constraints using genetic algorithms. The `ga` function can be called as

```
[x,fval,exitflag,output,population,scores] = ...
ga(fun,nvars,A,b,[],[],lb,ub,nonlcon,IntCon,options)
```

The `ga` function takes various input parameters. The `fun` is the function handle of the fitness function, `nvars` is the number of design variables, A and b are for linear inequality constraints in the form of $\mathbf{Ax} \leq \mathbf{b}$, lb and ub are lower and upper bounds of design variables, `nonlcon` is the function handle of the nonlinear inequality constraints, `IntCon` is for integer variables, and options are optimization options.

Once the `ga` algorithm is finished, it returns with various output results. The variable `x` includes the optimum design, `fval` is the value of the objective function at the optimum design, population includes the final population of designs, and scores include the fitness function values of the population. When the `exitflag` is positive, it means the algorithm stopped normally, while when it is negative, it means the algorithm stopped due to erroneous reasons.

**Example 6.6 Optimization of a banana function using genetic algorithms**

The Rosenbrock function is frequently used for testing optimization algorithms. The functional has two design variables, as

$$f = 100\left(x_1^2 - x_2\right)^2 + (1 - x_1)^2 \tag{6.9}$$

It has a minimum value of zero at design $\mathbf{x} = \{1, 1\}$. Find the optimum design of the Rosenbrock function using the Matlab `ga` function. Use the range of design variables $-3 \leq x_1, x_2 \leq 3$.

**Solution:**

Figure 6.14(a) shows the plot of the Rosenbrock function with the optimum design at $\mathbf{x} = \{1, 1\}$. Although the function varies over a wide range [0,15,000], it does not vary near the optimum design. Therefore, this function is difficult to identify the optimum design. Since the function is quite steep, it

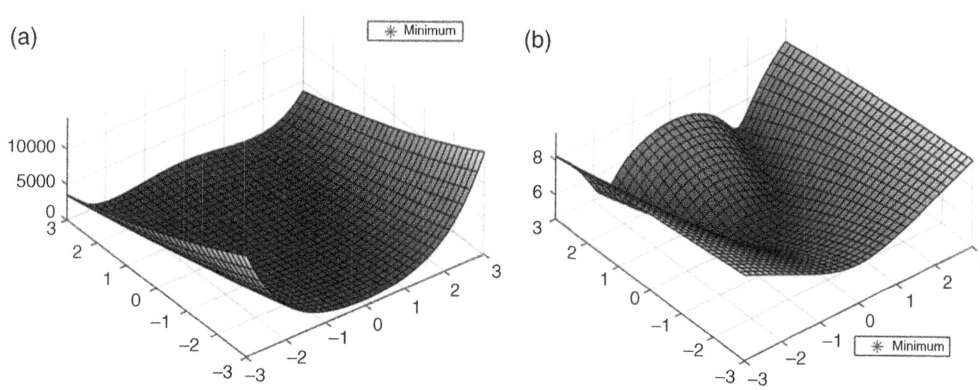


**Figure 6.14** Plot of Rosenbrock function in logarithm. (a) Rosenbrock function and (b) modified Rosenbrock function.

would be easier to visualize it if the logarithm of the function is plotted instead. Figure 6.14(b) shows the plot of the modified Rosenbrock function $\hat{f} = \log\left(100 + 100(x_1^2 - x_2)^2 + (1 - x_1)^2\right)$. A value of 100 is added to the function in order to prevent the function from approaching negative infinity. The figure also shows the optimum design at $\mathbf{x} = \{1, 1\}$.

Since this is unconstrained optimization, the objective function is used as the fitness function. To minimize the fitness function using ga, the following fitness function Rosenbrock.m is created:

```
function y = Rosenbrock(x)
y = 100 * (x(1)^2 - x(2))^2 + (1 - x(1))^2;
```

For given input design variables $\mathbf{x} = \{x_1, x_2\}$, the Rosenbrock function returns the value of the objective function. The following Matlab script defines the lower and upper bounds of the design and calls the ga function.

```
rng default       % For reproducibility
ndv = 2;
lb = [-3,-3];
ub = [3,3];
[x,fval] = ga(@Rosenbrock,ndv,[],[],[],[],lb,ub)
```

The rng function sets the seed of the random number generator default so that the same result can be obtained. After launching the ga function, the optimization stops with the following output:

```
Optimization terminated: maximum number of generations exceeded.
x = 1.5083    2.2781
fval = 0.2594
```

The x returned by the solver is the best point in the final population computed by ga. The fval is the value of the function Rosenbrock evaluated at point x. ga did not find an especially good solution. The maximum generation in Matlab is $100 * ndv = 200$. Therefore, the algorithm stops after 200 generations. It turns out that the optimum design is far away from the global optimum. In fact, 200 generations are often not enough to find a good optimum design. In order to allow more generations, the following optimoptions function is used to change the maximum allowed generations:

```
options = optimoptions('ga','MaxGenerations',10000);
[x,fval] = ga(@Rosenbrock,ndv,[],[],[],[],lb,ub,[],[],options)
```

With the increased number of generations, the ga algorithm stops with the following output:

```
Optimization terminated: average change in the fitness value less than
options.FunctionTolerance.
x = 1.0354    1.0721
fval = 0.0013
```

At this time, the algorithm stops because the average fitness value does not change much. Now, the optimum design is much closer to the global optimum.

### 6.4.5 When to Use Genetic Algorithm?

Genetic algorithms are computationally expensive compared to deterministic algorithms. Consequently, their domain of application is problems that would be difficult to solve by gradient-based algorithms or problems that would be out of range for these algorithms. Some of the main advantages of genetic algorithms are the following:

- Genetic algorithms do not require continuity or differentiability of the objective function. They do not need gradients.
- They are robust: they are very insensitive to noise.
- They are modular and therefore portable; because the evolutionary mechanism is separate from the problem representation, they can be transferred from problem to problem.
- They are particularly efficient in discrete and combinatorial problems, which are typically difficult to solve by conventional algorithms; they can be used for problems involving both discrete and continuous variables.
- Because they explore the design space with populations, genetic algorithms are "naturally" amenable to parallelization.

## 6.5 Particle Swarm Optimization

Many global search algorithms mimic the natural behaviors of animals or insects [63]. Among them, a particle swarm algorithm is introduced by Kennedy and Eberhart [64] in 1995, inspired by social behavior and movement dynamics of insects, birds, and fish. Particle swarm optimization imitates the basic strategy of a swarm of bees seeking food. The search direction relies on remembered good locations from the past. For each member of the swarm, the search direction is a compromise between the best point visited by the swarm and the best point visited by the individual.

Similar to genetic algorithms, this algorithm is also based on a population of designs. However, instead of creating a new population, the particle swarm algorithm moves the existing designs to better ones. Since this movement to a new design is continuous, this algorithm has the advantage of handling continuous design variables. Similar to other global search algorithms, this algorithm also has a stochastic property in search. The algorithm is an efficient and robust global search capability. There is some tuning of its parameters that is needed for peak efficiency, but commonly used values often give reasonably good results. The performance of the algorithm is comparable to that of genetic algorithms, and it is highly suitable for parallel computation. The algorithm has successfully been applied to a wide variety of problems, including neural networks, structural optimization, and shape and topology optimization.

Figure 6.15 illustrates an example of a three-bar truss with two cross-sectional area design variables. Initially, 20 particles are randomly distributed in the design space. Each particle will have a position, which is the value of the design variables. As the search progresses, the swarm coalesces at the presumed optimum, as shown in the bottom right figure.

In the particle swarm optimization algorithm, a particle means a design, and the collection of particles is the population. Let us assume that $n$ be the dimension of design space and $m$ be the number of particles (i.e., designs). These particles are represented by $\mathbf{x}^i = \{x_1, x_2, \ldots, x_n\}^i$, $i = 1, \ldots, m$. The particle swarm algorithm considers the design as a location of a particle. A design at a point moves to a different location using a design change. Let subscript $k$ represent the current design iteration. Then, the location of a $i$th particle at the next iteration can be updated as

$$\mathbf{x}^i_{k+1} = \mathbf{x}^i_k + \Delta\mathbf{x}^i_{k+1} \tag{6.10}$$

where $\Delta\mathbf{x}^i_{k+1}$ is the design change of particle $i$ at iteration $k+1$. The computational cost of the particle swarm algorithm is proportional to the number of particles and the number of iterations: $m \times k$.

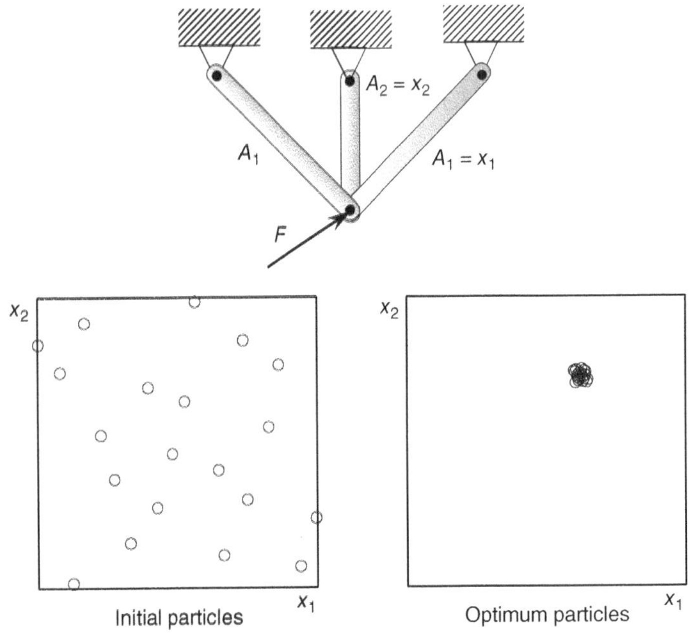


**Figure 6.15** Initial and optimum particles of particle swarm optimization.

The main idea of the particle swarm algorithm is to choose the design change such that the new design can improve the objective function and move toward the global optimum design. In the gradient-based optimization algorithms in Chapter 5, the design change is calculated based on the gradient information. In the particle swarm algorithm, the design change is calculated based on individual and group experiences of the best design. Let $\mathbf{p}^i$ be the best design of particle $i$ up to the current iteration, and $\mathbf{p}^g$ be the best design of the group up to the current iteration. The former represents an individual experience, while the latter represents the group experience. Therefore, the design change can be directed toward these well-experienced designs. Therefore, the design change can be calculated as

$$\Delta\mathbf{x}_{k+1}^i = w\Delta\mathbf{x}_k^i + c_1 r_1\left(\mathbf{p}^i - \mathbf{x}_k^i\right) + c_2 r_2\left(\mathbf{p}^g - \mathbf{x}_k^i\right) \tag{6.11}$$

where $w$, $c_1$, and $c_2$ are algorithmic parameters, which determine the importance of each term. The first term $w\Delta\mathbf{x}_k^i$ represents the inertial effect, where the particle tends to move in the direction of the previous iteration. A typical value of $w$ (the inertia fraction) is 0.5. To prevent divergence of the algorithm, the inertia fraction must be smaller than 1. The second term $c_1 r_1\left(\mathbf{p}^i - \mathbf{x}_k^i\right)$ is based on moving toward $\mathbf{p}^i$, the best past position of the $i$th particle. The size of the move is random, controlled by a random number $r_1$, uniformly distributed in [0, 1]. The multiplier $c_1$ is used to control the mean of the move distance. A typical value is $c_1 = 2$, in which case the mean of the move is at $\mathbf{p}^i$, and there is a 50% chance of overshooting or undershooting it. The third term $c_2 r_2\left(\mathbf{p}^g - \mathbf{x}_k^i\right)$ is exactly the same structure as the second term, but the target is $\mathbf{p}^g$, which is the group (swarm) best. Again, a typical value for $c_2$ is also 2. With all three terms present, we get motion in an intermediate direction between the three directions of the three terms.

The procedure of the particle swarm optimization algorithm can be summarized as follows:

1) Initialize population
   a) Set algorithmic parameters $w$, $c_1$, $c_2$
   b) Set $i = 1$, $k = 1$
   c) Randomly generate particles $\mathbf{x}_k^i$
2) Optimization iteration
   a) Evaluate fitness value $f_k^i$ using design particle $\mathbf{x}_k^i$
   b) If $f_k^i < f_{\text{best}}^i$, then $f_{\text{best}}^i = f_k^i$, $\mathbf{p}^i = \mathbf{x}_k^i$
   c) If $f_k^i < f_{\text{best}}^g$, then $f_{\text{best}}^g = f_k^i$, $\mathbf{p}^g = \mathbf{x}_k^i$
   d) If the stopping condition is satisfied, go to 3.
   e) Update particle design change $\Delta\mathbf{x}_{k+1}^i$
   f) Update particle position vector $\mathbf{x}_{k+1}^i$
   g) $i = i + 1$. If $i > m$ then increment $k = k + 1$, set $i = 1$.
   h) Go to 2(a).
3) Report results and terminate the algorithm
   Similar to most population-based optimization algorithms, it would be hard to determine the stopping criteria of the particle swarm algorithm. It is partly because there are no optimality criteria for these kinds of algorithms. In addition, there is no mathematical theory available to distinguish the global optimum from local optima. With these difficulties, the convergence of the particle swarm optimization algorithm can be defined in two different ways. (i) All particles converged to a point in the design space, which may or may not be the optimum design, and (ii) the individual optimum $\mathbf{p}^i$ and the swarm's best-known position $\mathbf{p}^g$ does not change for many iterations, regardless of how the swarm behaves. In contrast to the gradient-based algorithms, the first corresponds to the case when the design change is less than a threshold, while the second corresponds to the case when the change of the objective function is less than a threshold.

   Matlab supports the particle swarm optimization based on the original algorithm of Kennedy and Eberhart [64], with modifications suggested in Mezura-Montes and Coello Coello [65] and in Pedersen [66]. The particle swarm algorithm begins by creating the initial particles and assigning them initial velocities (design change). It evaluates the objective function at each particle location and determines the best (lowest) function value and the best location. It chooses new velocities based on the current velocity, the particles' individual best locations, and the best locations of their neighbors. It then iteratively updates the particle locations (the new location is the old one plus the velocity, modified to keep particles within bounds), velocities, and neighbors.

   The Matlab function for the particle swarm algorithm is `particleswarm`. The following calling convention can be used:

```
[x,fval,exitflag,output] = particleswarm(fun,nvars,lb,ub,options)
```

The `particleswarm` function takes various input parameters. The `fun` is the function handle of the fitness function, `nvars` is the number of design variables, `lb` and `ub` are lower and upper bounds of design variables, and options are optimization options. As mentioned before, the algorithm depends on the number of particles, which can be changed using the optimization option:

```
options = optimoptions('particleswarm','SwarmSize',50);
```

Once the particle swarm algorithm is finished, it returns with various output results. The variable $x$ includes the optimum design, and `fval` is the value of the objective function at the optimum design. When the `exitflag` is positive, it means the algorithm stopped normally, while when it is negative, it means the algorithm stopped due to erroneous reasons. The output structure includes the summary of the solution process, including the number of iterations, the number of function evaluations, and the stopping conditions.

**Example 6.7 Optimization of a banana function using particle swarm algorithm**
Optimize the Rosenbrock function in Example 6.6 using the particle swarm algorithm in Matlab. Try with different swarm sizes and discuss the optimum design, iterations, the number of function evaluations, and stopping criteria.

**Solution:**
The Rosenbrock problem in Example 6.6 has two design variables in the range of $-3 \leq x_1, x_2 \leq 3$. The objective function is available in Rosenbrock.m file, and the following Matlab script can be used to solve the optimization problem using the particle swarm algorithm with 10 particles:

```
rng default  % For reproducibility
nvars = 2;
lb = [-3,-3];
ub = [3,3];
options = optimoptions('particleswarm','SwarmSize',10);
[x,fval,exitflag,output] = particleswarm(@Rosenbrock,nvars,lb,ub,options)
```

The optimization algorithm stopped because the relative change in the objective value is less than the tolerance of $1 \times 10^{-1}$. The following outputs are returned from the algorithm:

```
x =      0.9683     0.9375
fval =      0.0010
exitflag =       1
output =
       iterations: 67
        funccount: 680
```

The optimum design is close to the global optimum design $\mathbf{x}^* = [1, 2]$ with a minimum objective function value of 0. The algorithm is converged after 67 iterations. Since there are 10 particles, each iteration takes 10 function evaluations. Including 10 initial function evaluations, the total number of function evaluations becomes 680.

The same optimization algorithm can be used with 50 particles by changing the option 'SwarmSize' to 50. At this time, the algorithm stopped after consuming the maximum function evaluations of 2,000.

```
x =      1.0308     1.0622
fval =    9.6164e-04
exitflag =       1
output =
       iterations: 39
       funccount: 2000
```

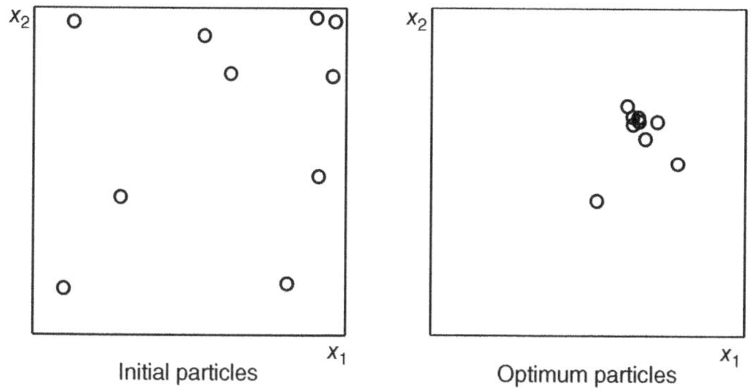


**Figure 6.16** Initial and optimum particle distributions of Rosenbrock optimization.

Although the final design and final objective function are close to the previous ones, it only performed 39 iterations. This is because each iteration requires 50 function evaluations. Figure 6.16 shows the particle locations at the initial and optimum designs of the Rosenbrock function.

## 6.6 Simulated Annealing Optimization

The particle swarm optimization in the previous section is based on the social behavior of a bird flock or fish school. The other group of global search algorithms is based on a statistical process commonly found in nature. Among them, the simulated annealing algorithm in this section is motivated by studies in statistical mechanics, which deal with the equilibrium of a large number of atoms in solids and liquids by gradually decreasing temperature. During the solidification of metals or the formation of crystals, for example, a number of solid states with different internal atomic or crystalline structures that correspond to different energy levels can be achieved depending on the rate of cooling. If the system is cooled too rapidly, it is likely that the resulting solid state would have a small margin of stability because the atoms will assume relative positions in the lattice structure to reach an energy state, which is only locally minimal. In order to reach a more stable, globally minimum energy state, the process of annealing is used in which the metal is reheated to a high temperature and cooled slowly, allowing the atoms enough time to find positions that minimize a steady-state potential energy. It is observed in the natural annealing process that during the time spent at a given temperature, it is possible to have the system jump to a higher energy state temporarily before the steady state is reached. This characteristic of the annealing process makes it possible to achieve near-global minimum energy states.

A computational algorithm that simulates the annealing process was proposed by Metropolis et al. [67] and is referred to as the Metropolis algorithm. Simulated annealing efficiently predicts the behavior of a collection of atoms in equilibrium at a given temperature. At a given temperature $T$, the algorithm perturbs the position of an atom randomly and computes the resulting change in the energy $\Delta E$ of the system. If the energy $E$ at the perturbed state is lower than that of the initial state, $\Delta E < 0$, then the new position of the atom is accepted. On the other hand, if the perturbed position causes an increase in the energy $\Delta E \geq 0$, the new position is accepted with a probability inversely proportional to the energy increase. Accepting a higher energy state with probability is the key mechanism to explore the design space for a possible better design far away from the current design.

First, the positive energy change is defined as $\Delta E_+ = \max(0, \Delta E)$. The probability of acceptance, $P(\Delta E)$, of a higher energy state is computed as

$$P(\Delta E) = e^{\left(-\frac{\Delta E_+}{T}\right)} \tag{6.12}$$

Note that when $\Delta E < 0$, $P(\Delta E) = e^0 = 1$. Therefore, the probability is in the range of $0 < P(\Delta E) \leq 1$. If the energy is decreased, it will be accepted with a probability of one. On the other hand, if the energy is increased, it will be accepted with a probability of $P(\Delta E) < 1$. The probability is inversely proportional to the positive energy change. If the temperature of the system is high, then the probability of acceptance of a higher energy state increases. If, on the other hand, the temperature is close to zero, then the probability of acceptance becomes very small. The decision to accept or reject is made by randomly selecting a number $u \sim U(0, 1)$ and comparing it with $P(\Delta E)$. If $u < P(\Delta E)$, then the perturbed state is accepted, if $u \geq P(\Delta E)$, the state is rejected.

The temperature $T$ becomes a control parameter and is used to first "melt" the system at a high temperature. At each temperature, a pool of atomic structures is generated by randomly perturbing positions until a steady-state energy level is reached (commonly referred to as thermal equilibrium). The iterative simulation is then run at this temperature until the system is considered to be in equilibrium, and only then is the temperature reduced by a small fraction. This cooling and equilibrium process continues until no further improvement becomes possible, and the system is considered "frozen" or crystallized, with the design parameters at this condition optimal. The temperature reduction sequence and number of iterations allowed for the system to reach equilibrium are considered analogues to an annealing schedule.

The analogy between the simulated annealing and the optimization of functions with many variables was established recently by Kirkpatrick et al. [68] and Cerny [69]. By replacing the energy state with an objective function $f$, and using variables $x$ for the configurations of the particles, we can apply the Metropolis algorithm to optimization problems. The method requires only function values. The move in the design space from one point, $x^i$ to another $x^j$ causes a change in the objective function, $\Delta f^{ij}$. The temperature $T$ now becomes a control parameter that regulates the convergence of the process. Important elements that affect the performance of the algorithm are the selection of the initial value of the "temperature," $T_0$, and how to update it. In addition, the number of iterations (or combinations of design variables) needed to achieve "thermal equilibrium" must be decided before the $T$ can be reduced. These parameters are collectively referred to as the "cooling schedule."

The procedure of the simulated annealing algorithm can be summarized as follows:

- Set initial design $\mathbf{x} = \mathbf{x}_0$, objective function $f(\mathbf{x})$, the maximum number of iterations $M$, and initial temperature $T_0$
- Loop for cooling schedule
  - Calculate temperature $T$ from the cooling schedule
  - Loop for thermal equilibrium
    - Generate a new candidate design $\mathbf{x}_{new}$ randomly near the current design $\mathbf{x}$
    - Calculate the objective function $f(\mathbf{x}_{new})$, and $\Delta f = f(\mathbf{x}_{new}) - f(\mathbf{x})$
    - Calculate the probability of acceptance $P(\Delta f) = e(-\Delta f_+/T)$ with $\Delta f_+ = \max(0, \Delta f)$ and generate a random sample $u \sim U(0, 1)$
    - If $P(\Delta f) \geq u$, set $\mathbf{x} = \mathbf{x}_{new}$
  - End of the thermal equilibrium loop

- End of the cooling schedule loop
- Finish the algorithm with the optimum design **x** and the optimum objective function $f(\mathbf{x})$.

From the algorithm, it is obvious that the algorithm has to determine the maximum number of iterations $M$, initial temperature $T_0$, and the cooling schedule.

The definition of the cooling schedule begins with the selection of the initial temperature $T_0$. If a low value of $T_0$ is used, the algorithm would have a low probability of reaching a global minimum. The initial value of $T_0$ must be high enough to permit virtually all moves in the design space to be acceptable, so that almost a random search is performed. Typically, $T_0$ is selected such that the acceptance ratio $X$ (defined as the ratio of the number of accepted moves to the total number of proposed moves) is approximately $X_0 = 0.95$ [70]. Johnson et al. [71] determined $T_0$ by calculating the average increase in the objective function, $\overline{\Delta f}^{(+)}$, over a predetermined number of moves and solved

$$X_0 = e^{\left(-\frac{\overline{\Delta f}^{(+)}}{T_0}\right)} \tag{6.13}$$

leading to

$$T_0 = \frac{\overline{\Delta f}^{(+)}}{\ln\left(X_0^{-1}\right)} \tag{6.14}$$

However, since $\overline{\Delta f}^{(+)}$ is difficult to estimate, the above formula will be an approximation.

Once the initial temperature is set, a number of moves in the variable space are performed by perturbing the design. The number of moves at a given temperature must be large enough to allow the solution to escape from a local minimum. One possibility is to move until the value of the objective function does not change for a specified number, $M$, of successive iterations. Another possibility suggested by Aarts [72] for discrete-valued design variables is to make sure that every possible combination of design variables in the neighborhood of a steady-state design is visited at least once with a probability of $P$. That is, if there are $S$ neighboring designs, then

$$M = S\ln\left(\frac{1}{1-P}\right) \tag{6.15}$$

where $P = 0.99$ for $S > 100$, and $P = 0.995$ for $S < 100$. For discrete-valued variables, there are often many options for defining the neighborhood of the design. One possibility is to define it as all the designs that can be obtained by changing one design variable to its next higher or lower value. A broader immediate neighborhood can be defined by changing more than one design variable to its next higher or lower value. For an $n$-variable problem, the immediate neighborhood has

$$S = 3^n - 1 \tag{6.16}$$

Once convergence is achieved at a given temperature, generally referred to as thermal equilibrium, the temperature is reduced, and the process is repeated.

Many different schemes have been proposed for updating the temperature (cooling schedule). A frequently used rule is a constant cooling update

$$T_{k+1} = \alpha T_k, \quad k = 0, 1, \ldots, K \tag{6.17}$$

where $0.5 \leq \alpha \leq 0.95$. Nahar [73] fixes the number of decrement steps $K$ and suggests the determination of the values of the $T_k$ experimentally. It is also possible to divide the interval $[0, T_0]$ into a fixed $K$ number of steps and use

$$T_k = \frac{K-k}{K} T_0, \quad k = 0, 1, \ldots, K \tag{6.18}$$

The number of intervals typically ranges from 5 to 20.

The use of simulated annealing for structural optimization has been quite recent. Elperin [74] applied the method to the design of a 10-bar truss problem, where member cross-sectional dimensions were to be selected from a set of discrete values. Kincaid and Padula [75] used it for minimizing the distortion and internal forces in a truss structure. A six-story 156-member frame structure with discrete valued variables was considered by Balling and May [76]. Optimal placement of active and passive members in a truss structure was investigated by Chen et al. [77] to maximize the finite-time energy dissipation to achieve increased damping properties.

An important difference from iterative improvement procedures such as gradient-based methods is that the Metropolis method need not get stuck, since transitions out of a local optimum are always possible at a nonzero temperature. Like genetic algorithms, this population-based method is relatively insensitive to noise and does not require gradient information. A major drawback to this method, however, is the problem dependency of algorithm parameters. That is, it is important to use an annealing schedule that is tailored to the type of optimization problem being solved; otherwise, the optimizer will perform poorly. Various statistical methods have been proposed, all with limited success, to obtain optimum cooling rates to avoid entrapment in local optima.

Matlab provides the `simulannealbnd` function for solving a minimization problem using the simulated annealing algorithm. The calling convention is similar to other optimization algorithms:

```
[x,fval,exitflag,output] = simulannealbnd(fun,x0,lb,ub,options)
```

The Matlab implementation of `simulannealbnd` is for solving unconstrained optimization problems with bounds. The annealing schedule can be provided as an option, including the initial temperature in `InitialTemperature` and the cooling schedule in `TemperatureFcn`.

**Example 6.8 Simulated annealing algorithm in Matlab**

Find the minimum of a function $f(\mathbf{x}) = (4 - 2.1x_1^2 + x_1^4/3)x_1^2 + x_1x_2 + (-4 + 4x_2^2)x_2^2$ with the lower and upper bounds of design $-64 \leq x_1,\ x_2 \leq 64$ using the `simulannealbnd` function of Matlab.

**Solution:**

To implement the objective function calculation, the Matlab file obj.m has the following code:

```
function f = objfun(x)
x1 = x(1);
x2 = x(2);
f = (4-2.1.*x1.^2+x1.^4./3).*x1.^2+x1.*x2+(-4+4.*x2.^2).*x2.^2;
```

The objective function computes the scalar value of the objective function and returns it in its single output argument `f`. To minimize the objective function using `simulannealbnd`, pass a function handle to the objective function and a starting point `x0` as the second argument. For reproducibility, set the random number stream.

```
x0 = [0.5 0.5];    % Starting point
lb = [-64 -64];
ub = [64 64];
rng default % For reproducibility
[x,fval,exitFlag,output] = simulannealbnd(@objfun,x0,lb,ub)
```

The outputs from Matlab are as follows:

```
x = -0.0896     0.7127
fval = -1.0316
exitFlag = 1
output =
      iterations: 2465
       funccount: 2484
         message: 'Optimization terminated: change in best function
                  value less than options.FunctionTolerance.'
        rngstate: [1x1 struct]
     problemtype: 'boundconstraints'
     temperature: [2x1 double]
       totaltime: 0.5566
```

It shows that the optimization converged as the change in the objective function is less than the tolerance. The optimum design was $\mathbf{x}^* = (-0.0896, 0.7127)$, where the optimum objective function was $-1.0316$. The algorithm took 2,465 iterations with 2,484 function evaluations. When the algorithm stopped, the cooling temperatures for both variables were $(0.2058 \times 10^{-1}, 0.2207 \times 10^{-3})$, which was stored in `output.temperature`.

## Exercise

**6.1** For a 2D problem, the current simplex has $f(0, 0) = 1, f(1, 0) = 2, f(0, 1) = 3$. (i) Where will you evaluate $f$ next? (ii) If the next two points gave us function values of 4 and 5, respectively, where will you evaluate the functions next? (iii) If instead of scenario (ii), the next point (reflection point) gave us a function value of 0, where will you evaluate the function next?

**6.2** The first four points selected by the Nelder–Mead algorithm are $(x_1, x_2, f) = 1:(0, 0, 3),\ 2:(0, 1, 7),\ 3:(1, 1, 6),\ 4:(1, 0, 2)$. Calculate where the algorithm will select the next point, also naming the operation that leads to this point (i.e., reflection, expansion, contraction, or shrinkage).

**6.3** Track and plot the first two iterations of fminsearch on Himmelblau's function starting from (1,1). The form of Himmelblau's function is defined as

$$f(x_1, x_2) = \left(x_1^2 + x_2 - 11\right)^2 + \left(x_1 + x_2^2 - 7\right)^2$$

The function has one local maximum at $\mathbf{x} = (-0.270845, -0.923939)$ with $f_{max} = 181.617$ and four identical local minima at $\mathbf{x}$ = (3.0, 2.0), (−2.805118, 3.131312), (−3.779310, −3.283186), (3.584428, −1.848126).

**6.4** Consider the following pairs of objective functions to be minimized: (3, 9), (4, 1), (10, −3), (−7, 11), (−8, 14), (7, −17), (6, 7), (5, 8). (i) Divide them into ranks. (ii) Calculate the crowding distance of the points in the second rank.

**6.5** Global optimization balances exploration and exploitation. How is that reflected in genetic algorithms?

**6.6** What are all possible child designs of $[0_2/\pm 45/90]s$ and $[\pm 45_2/0]s$ that are balanced and symmetric with uniform crossover?

**6.7** Use the Matlab `fminsearch` function to minimize the function $f = x_1^2 + (x_1 - x_2)^2$ starting at the point (10, 1). Report the first 23 function evaluations and identify each with the Nelder–Mead operation that yields each point (i.e., reflection, contraction, expansion, etc.).

# Part III

# Advanced Topics in Surrogate Modeling

# 7

# Kriging Surrogate–Gaussian Process Model

## 7.1 Introduction

In Part I of this book, we discussed polynomial response surfaces, where a model form is composed of known basis functions with unknown coefficients. These coefficients were determined using linear regression, which means that the surrogate approximation converges to the mean of the trend of samples. This surrogate works well when samples are obtained from experiments with random measurement noise. In such a case, regression to the mean is a desirable feature. When the model form is correct and random noise is normally distributed, this surrogate can filter out random noise and converge to the mean trend of samples.

Nowadays, computer simulations are often used to generate samples, and surrogate models are used to approximate the response of those simulations. This is because high-fidelity computer simulations are as expensive as experiments. Although computer simulations also have random noises from truncation errors, their magnitudes are in general ignorable compared to that of bias errors, which are caused by numerical methods. Bias errors have different characteristics from random noises. They are not random and have the same or similar values everywhere. Therefore, it makes sense to fit the computer simulation results as if there is no random noise. The bias errors can be considered separately after fitting the computer simulations. In such a case, it would be better to interpolate samples from computer simulation, not regress them. In addition, the model form from computer simulation is usually complicated because they are often the solution of partial differential equations. Therefore, it would be difficult to find the correct form of the model in advance. These characteristics are actually opposite to those of polynomial response surfaces (PRSs). Therefore, we need a new surrogate model with different assumptions from those of PRS.

The Kriging surrogate assumes that samples are correct, but the model form is unknown or uncertain. The Kriging surrogate was invented in the 1950s by Daniel Krige (Professor at the University of the Witwatersrand, Republic of South Africa), a South African geologist, for the purpose of predicting the distribution of minerals on the basis of samples [78]. Later, the surrogate model was formalized by a French engineer, Georges Matheron, in 1960s [79]. It is known as a Kriging surrogate to engineers, but it is better known as a Gaussian process (GP) model for statisticians. This is because the samples are assumed to be multivariate Gaussian random variables. That is, the vector of function values $f(x)$ is just one sample from a multivariate Gaussian distribution of dimension equal to the number of observed coordinates $n_y$. It became very popular for fitting surrogates to expensive computer simulations only about 40~50 years later. This is at least partly because it is an expensive surrogate in terms of computation, and it had to wait until computers

*Surrogate Modeling and Optimization: Theories, Applications, and Limitations*, First Edition. Nam-Ho Kim.
 Published 2026 by John Wiley & Sons, Inc.
Companion website: www.wiley.com/go/kim/surrogate_modeling

became fast enough. However, still it is assumed that the cost of building a Kriging surrogate is ignorable compared to that of calculating samples.

Linear regression was invented at a time when fitting a surrogate was computationally onerous, even though it only involves solving a system of linear equations. As computer resources have become cheaper, more expensive surrogates have been proposed and adopted. For moving least squares surrogate [80], we need to form and solve the system of equations at every prediction point. This surrogate approximates local effects better because nearby points have more influence than points that are far away. Radial basis surrogates [81] create the same effect of nearby points having a larger effect by using kernel functions that decay away from sample points. Matlab's radial basis neural networks perform optimization to decide which of the sample points should serve as shape functions and which should not. However, the spread of the shape functions (how fast they decay from the sample points) is provided by the user. The Kriging surrogate goes one step further in that it uses optimization to calculate the spread, and it allows the spread to be different in different directions. This requires solving an optimization problem to find these spreads, and in addition, each surrogate evaluation is not cheap, especially when there is a large number of sample points. Although the Kriging surrogate model is the most expensive surrogate, it has the highest chance of being the most accurate for a given problem. However, based on the comprehensive study by Goel et al. [9], the chance of being the best surrogate is still less than 50%. Therefore, it is important to study the Kriging surrogate thoroughly such that the users understand the advantages and disadvantages of the surrogate.

From the perspective of surrogate modeling, it is important to understand the conceptual difference between polynomial response surfaces and Kriging surrogates. Section 7.2 explains the fundamental difference between the two types of surrogates. In particular, it is important to understand the concept of correlation in the Kriging surrogate. Another important concept is the difference between interpolating samples and regression. Section 7.3 develops a Kriging surrogate model using samples. The Kriging surrogate model requires identifying several hyperparameters, which determine the prediction quality of the surrogate model. In addition, uncertainty plays an important role in the Kriging surrogate. Section 7.4 summarizes issues related to the Kriging surrogate. The readers should focus on how an uncertainty model in Kriging is different from that of polynomial response surfaces. Section 7.5 extends the one-dimensional surrogate to higher dimensions. The Kriging surrogate in Section 7.3 assumes that samples are accurate, and Kriging surrogate interpolates samples. When samples include additional random noise, it is necessary to use Gaussian process regression or Kriging with nuggets, which is presented in Section 7.6.

## 7.2 Kriging Philosophy

As mentioned earlier, the basic assumptions of Kriging are opposite to those of polynomial response surfaces. In polynomial response surfaces, it is assumed that the form of the true function is known in terms of a linear combination of polynomials, while the samples have random noises that are normally distributed with zero mean and variance $\sigma^2$. In addition, the distribution of noise is independent and identically distributed, i.e., uncorrelated. These samples are used to find the unknown coefficients that minimize the root-mean-squared errors at the sample points.

Kriging starts from different assumptions. First, it assumes that the form of the true function is unknown or uncertain, except for the form of correlation between the values of the function at nearby points. The value of the function at a point is correlated to the values at neighboring points based on their separation in different directions. The correlation is strong with nearby points and weak with faraway points. This is a valid assumption if the function is continuous. The correlation strength does not change based on the location, only the separation distance between points, and decays as they are further apart. Because of this assumption, Kriging is a smooth predictor. If two points are close, then their predictions are also close. Because the correlation decays proportional to the distance, Kriging is a local surrogate with shape functions that are similar to those used for radial basis surrogates, except that the decay can be different in different directions. The concept of correlation is the major strength of Kriging surrogates, but at the same time, it can be a major weakness. Due to correlation, it is possible to interpolate samples smoothly without requiring some knowledge about the true model. But the main difficulty is how to find the right (or at least an appropriate) correlation. In the coming sections, therefore, it is advised that the readers pay attention to correlation.

Second, Kriging assumes that samples are accurate, i.e., noise-free. Therefore, the surrogate passes through the samples, and the errors are zero at the sample locations. Because of this, Kriging is called an interpolating surrogate. At unsampled locations, however, the surrogate prediction is uncertain as the model itself is uncertain. This is an important property of the Kriging surrogate, contrasting to polynomial response surfaces. Because of this interpolating property, the root-mean-squared errors cannot be used to find model parameters. There is a version of Kriging when samples have noise, which is often called Gaussian process regression or Kriging with nuggets. This will be discussed in Section 7.6.

The Kriging surrogate is one of the regression-based methods, which is linear regression like the least squares method in Chapter 2. The difference between GP and ordinary linear regression is whether the correlation in errors between a regression function and samples is considered or not. It is assumed that errors are independent and identically distributed in ordinary linear regression, while they are assumed to be correlated in Kriging.

### 7.2.1 Correlation Between Two Random Variables

Correlation is an important concept in Kriging, which is directly related to the covariance of two random variables. The covariance of two random variables, $X$ and $Y$, can be defined as

$$\text{cov}(X, Y) = \mathbb{E}[(X-\mu_X)(Y-\mu_Y)] = \mathbb{E}[XY] - \mu_X\mu_Y \tag{7.1}$$

where $\mathbb{E}[\cdot]$ is the expected value of a random variable, and $\mu_X$ and $\mu_Y$ are the mean values of $X$ and $Y$, respectively. The covariance of a random variable itself is the square of the standard deviation, which is also called variance. Note that the variance of a random variable is the second moment around its mean, or the square of the standard deviation: $\text{Var}[X] = \text{cov}(X, X) = \sigma_X^2$. The covariance of a pair of random variables is the mixed moment of the two variables. It is noted that the covariance between a random variable and a deterministic variable is zero. This can easily be seen by setting $X = \mu_X$ in Eq. (7.1).

Using covariance, the correlation of two random variables can be defined as

$$R(X, Y) = \frac{\text{cov}(X, Y)}{\sigma_X\sigma_Y} \tag{7.2}$$

where $\sigma_X$ and $\sigma_Y$ are the standard deviations of $X$ and $Y$, respectively. Note that $-1 \le R(X, Y) \le 1$. The correlation measures their propensity to move together. A correlation of 1 means that they increase and decrease in perfect synchrony. The most common case is when $X = \alpha Y$ with a positive constant $\alpha$. Similarly, a correlation of −1 also means perfect synchrony, but in the opposite direction, $X = \alpha Y$ with a negative constant $\alpha$. Finally, a correlation of 0 means that they are not related. The common case is when the random variables are independent, but not all uncorrelated variables are independent. Zero correlation relates to the first moment, but there may be dependence through higher-order effects.

The correlation between two variables is considered to be strong if the absolute value of the correlation coefficient is greater than 0.75. However, the definition of a "strong" correlation can vary from one field to the next. A moderate correlation is considered when $0.5 \leq |R(X, Y)| \leq 0.75$, and the correlation is weak when $0.25 \leq |R(X, Y)| \leq 0.5$.

**Example 7.1 Correlation Between Functions**

The fitting process in Kriging needs to estimate how fast the correlation between function values decays with distance. This depends on the wavelength of the function; that is, the correlation quickly decays for a function with a short wavelength. In order to check how fast the correlation decays for function sin($x$) that has a wavelength of $2\pi$, (i) generate 10 random numbers, $x_i$, $i = 1, \ldots, 10$, between 0 and 10; (ii) move $x_i$ by a small amount, $x_i^{\text{near}} = x_i + 0.1$, and a large amount, $x_i^{\text{far}} = x_i + 1.0$; and (iii) calculate correlations between $y = \sin(x_i)$ and $y = \sin\left(x_i^{\text{near}}\right)$ and between $y = \sin(x_i)$ and $y = \sin\left(x_i^{\text{far}}\right)$.

**Solution:**

Different random numbers may be generated, but the following Matlab code can be used to generate random numbers and calculate correlations:

```
x=10*rand(1,10);
xnear=x+0.1; xfar=x+1;
y=sin(x);
ynear=sin(xnear)
yfar=sin(xfar)
rnear=corrcoef(y, ynear)      %rnear=0.9894;
rfar=corrcoef(y, yfar)        %rfar=0.4229
```

The Matlab function `corrcoef` calculates the correlation coefficient between two sets of samples. The following table shows 10 samples of random numbers along with $y = \sin(x_i)$ and $y = \sin\left(x_i^{\text{near}}\right)$ and $y = \sin\left(x_i^{\text{far}}\right)$.

| x | 8.147 | 9.058 | 1.267 | 9.134 | 6.324 | 0.975 | 2.785 | 5.469 | 9.575 | 9.649 |
|---|---|---|---|---|---|---|---|---|---|---|
| ynear | 0.923 | 0.264 | 0.980 | 0.190 | 0.140 | 0.880 | 0.254 | −0.655 | −0.248 | −0.319 |
| y | 0.957 | 0.359 | 0.955 | 0.287 | 0.040 | 0.828 | 0.349 | −0.727 | −0.150 | −0.222 |
| yfar | 0.274 | −0.592 | 0.765 | −0.651 | 0.863 | 0.919 | −0.600 | 0.185 | −0.913 | −0.941 |

It is clear that the pattern between `y` and `ynear` is similar, but not between `y` and `yfar`, as shown in Figure 7.1. That is, `y` and `ynear` are strongly correlated, but not for `y` and `yfar`. Indeed, the correlation coefficient between `y` and `ynear` is 0.9894, while it is 0.4229 between `y` and `yfar`. This reflects the change in function values, as illustrated by the pair of points marked in grey cells. In Kriging, finding the rate of correlation decay is part of the fitting process. This example shows that with a wavy function, the correlation decays to about 0.4 over one-sixth of the wavelength $(1 : 2\pi)$. More precisely, the correlation becomes $e^{-1} = 0.3679$ when $x_{\text{far}} = x + 1.1906$, which corresponds to $1.1906/2\pi = 0.1895$ times the wavelength.

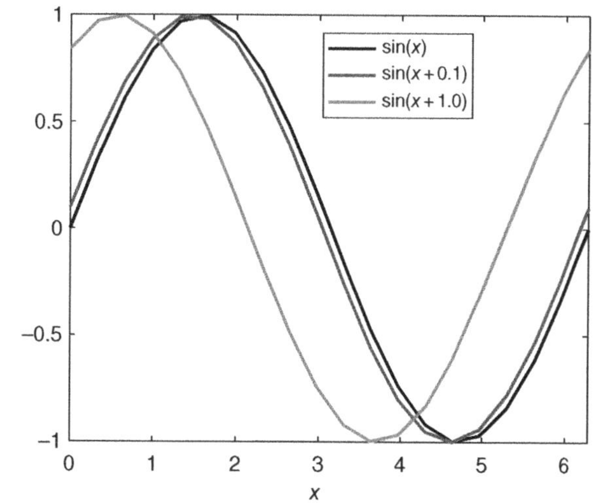


**Figure 7.1** High and low correlation of function sin(*x*).

### 7.2.2 Kriging Surrogate Approximation

In Kriging, a simulation output, $\hat{y}(\mathbf{x})$, is composed of a global function output, $\boldsymbol{\xi}(\mathbf{x})^T\boldsymbol{\beta}$, that often takes the form of constant or polynomials and its local departure, $s(\mathbf{x})$ as

$$\hat{y}(\mathbf{x}) = \boldsymbol{\xi}(\mathbf{x})^T\boldsymbol{\beta} + s(\mathbf{x}) \tag{7.3}$$

where $\mathbf{x}$ is a vector of input variables, $\boldsymbol{\xi}(\mathbf{x})$ is a $n_p \times 1$ vector of bases associated with $n_p \times 1$ vector of global function parameters/coefficients, $\boldsymbol{\beta}$, and $s(\mathbf{x})$ is a local departure, i.e., an error between the global function and Kriging prediction. In practice, a simple polynomial function, or a constant, is used for the global function. For example, when the global function is assumed to be a linear polynomial, $\beta_0 + \beta_1 x$ in one-dimensional Kriging, $\boldsymbol{\xi}(x) = \{1, x\}^T$ and $\boldsymbol{\beta} = \{\beta_0, \beta_1\}^T$. Since the global function determines the overall trend of the Kriging prediction, it is also referred to as a trend function.

Figure 7.2 illustrates Kriging approximation of a one-dimensional function. The global function is a linear polynomial, and the difference between Kriging prediction and the global function is defined as a local departure. As mentioned earlier, since the samples are accurate, the Kriging approximation passes through the samples and there is no uncertainty at sample locations. However, the prediction between samples is uncertain, which is represented using a Gaussian distribution. Therefore, the local departure is not a deterministic function, but a statistical process. This is why statisticians call the Kriging surrogate a Gaussian process model. The Kriging prediction, shown as the grey curve in the figure, is nothing but the mean of the Gaussian process.

One may think that Eq. (7.3) is similar to Eq. (2.3) of Chapter 2, but they are significantly different in terms of the fundamental assumptions. The polynomial-based least squares method in Eq. (2.3) assumes that the model form $\hat{y}(\mathbf{x}) = \boldsymbol{\xi}(\mathbf{x})^T\boldsymbol{\beta}$ is correct, but the samples have errors $\epsilon_k$ that are normally distributed as $N(0, \sigma^2)$. These errors are assumed to be statistically independent; that is, the error in one point is not related to the error in other points. The regression process finds the model parameters to minimize these errors in the sense of mean-squared-error. On the other hand, Kriging assumes that the samples are correct, but the model form is uncertain. Since the samples are correct, Kriging passes through all the samples. That is, the model predictions and samples match: $y_k = \hat{y}(\mathbf{x}_k) = \boldsymbol{\xi}(\mathbf{x}_k)^T\boldsymbol{\beta} + s(\mathbf{x}_k)$. At prediction points, the uncertainty or error in Kriging can be

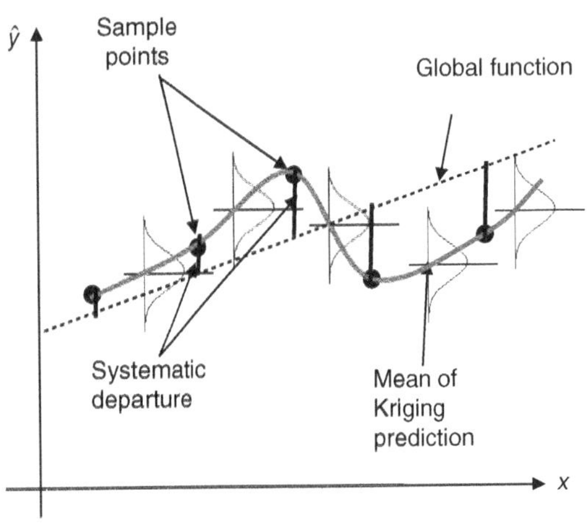


**Figure 7.2** Kriging approximation with samples.

described using the local departure, $s(\mathbf{x})$, in Eq. (7.3), which is a realization of the Gaussian stochastic process with zero mean, variance $\sigma^2$, and nonzero covariance as

$$s(\mathbf{x}) \sim N(0, \sigma^2) \tag{7.4}$$

The standard deviation $\sigma$ of the Gaussian process can be estimated using the differences between samples and the global function. The standard deviation is small when the samples are dense. The samples are considered to be dense when there are more than enough points to describe the waviness of the function. In order to be dense, the distance from one point to the nearest one should be an order of magnitude less than the wavelength of the function.

The key concept of Kriging is that the functional form of $\hat{y}(\mathbf{x})$ is unknown, except for the form of correlation between the value of the functions at nearby points. This correlation controls the smoothness of Kriging predictions. In particular, the correlation depends only on the separation distance between points and decays as they are further apart. The correlation matrix between $n_y$ samples can be defined as

$$\mathbf{R} = [R(\mathbf{x}_k, \mathbf{x}_l)], \quad k, l = 1, \ldots, n_y \tag{7.5}$$

where $R(\mathbf{x}_k, \mathbf{x}_l)$ is the correlation between two sample points, as defined in Eq. (7.2). Note that the diagonal components of $\mathbf{R}$ are one as is from its definition in Eq. (7.2). The correlation is strong between two points nearby and weak between two points far away. The correlation between a prediction point, $\mathbf{x}$, and sample points can be defined as

$$\mathbf{r}(\mathbf{x}) = [R(\mathbf{x}_k, \mathbf{x})], \quad k = 1, \ldots, n_y \tag{7.6}$$

which is a $n_y \times 1$ vector that represents the correlation between the sample points $\mathbf{x}_k$ and prediction point $\mathbf{x}$, which makes different magnitudes in departure at different prediction points.

Because of the correlation in the local departure, Kriging has an outstanding feature that the simulated output, denoted by the dashed curve in Figure 7.3, passes through the set of samples $\left[\mathbf{x}_{1:n_y}, \mathbf{y}_{1:n_y}\right]$. When the prediction point (a new input), $x^{\text{new}}$, is located at a sample point A, $x_i$ ($i = 1, 2, \ldots, n_y$), the magnitude of departure is the same as the difference between samples, $y_i$ ($i = 1, 2, \ldots, n_y$), and the global function output (the global function is constant in the figure); hence, the prediction output $\hat{y}(x_i)$ becomes samples, $y_i$. When the prediction point deviates from the samples (point B), the magnitude of departure is changed based on the correlation between point B and the sample points. As a result, if the prediction point is in between sample points, the simulated

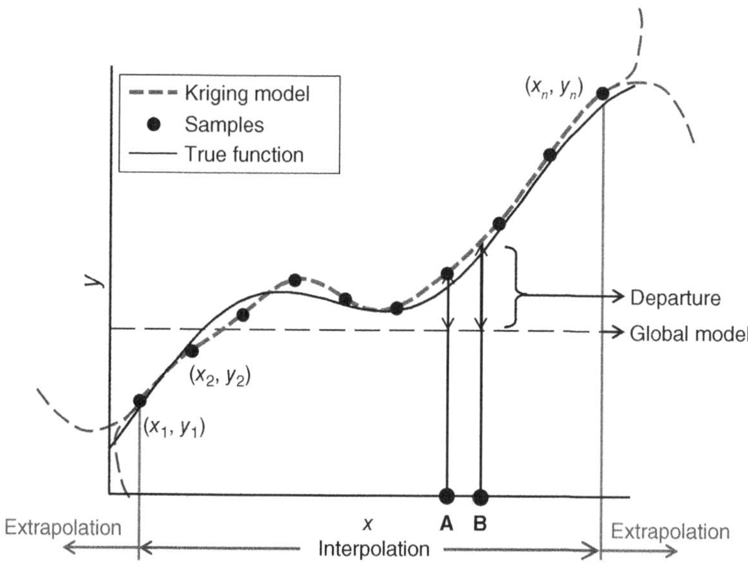


**Figure 7.3** Illustration of Kriging surrogate model.

output smoothly interpolates the samples. If, however, the prediction point moves away from sample points, which is extrapolation, the influence of this departure is reduced as correlation decreases and the Kriging prediction becomes close to the global function. Therefore, the case of extrapolation is not so different from an ordinary linear regression, which follows the global model. For interpolation, the correlation between sample points is an important factor to characterize Kriging's attribute, which is determined by selecting an appropriate type of correlation function and estimating its hyperparameters. The hyperparameters in Kriging control the effect of correlation between two points, which affects the smoothness of the surrogate model. The hyperparameters are usually determined using an optimization algorithm based on the samples. A typical correlation function and estimating hyperparameters will be discussed later in this section.

Kriging has several different names based on the global function. The most general version of Kriging is called universal Kriging, which uses a general polynomial function as a global function. However, in most cases, linear or quadratic polynomials are used for the global function. When a constant is used for the global function, it is referred to as an ordinary Kriging; that is, the parameter $\beta$ is a scalar. The constant is estimated using samples, such as the mean of the samples or maximum likelihood estimate. A simple Kriging is when the constant global function is assumed to be known.

### 7.2.3 Correlation Model

Kriging assumes that predictions are correlated inversely proportional to the distance. This correlation is captured by the local departure term. The local departure has a zero mean, $\mathbb{E}[s(\mathbf{x})] = 0$, and the covariance of two samples as

$$\text{cov}\left[s\left(\mathbf{x}^{(i)}\right), s\left(\mathbf{x}^{(j)}\right)\right] = \sigma^2\phi\left(\theta, \mathbf{x}^{(i)}, \mathbf{x}^{(j)}\right) \tag{7.7}$$

where $\sigma^2 = \text{cov}[s(\mathbf{x}), s(\mathbf{x})]$ is the variance of the local departure and $\phi$ is the correlation function. The correlation function (also called kernel function) is a crucial ingredient for a Kriging surrogate, since it contains the assumptions about the approximation function. Generally speaking, it describes the "similarity" between observations and new points, i.e., how similar such points are, depending on the distance between the input points. Input points that are close to each other are expected to have similar outputs or responses. The correlation function has a hyperparameter $\theta$ that determines the shape of correlation. When $\theta$ is large, the correlation effect reaches to a region far away from the point, and thus, the prediction tends to be smooth. On the other hand, when $\theta$ is small, the correlation effect decays fast and reaches only a nearby region. Therefore, predictions show fluctuation. Also, the correlation function depends on two sample locations, $\mathbf{x}^{(i)}$ and $\mathbf{x}^{(j)}$. However, as mentioned before, the correlation depends on the distance between two samples, not their locations. Therefore, these two sample locations appear as a relative distance.

It would be good to be reminded that the input design space is a multidimension. When the correlation at all input variables is the same, it is referred to as isotropic correlation. In such a case, only a single hyperparameter $\theta$ is enough to define the correlation model as

$$\phi\left(\theta, \mathbf{x}^{(i)}, \mathbf{x}^{(j)}\right) = \phi\left(\theta, \|\mathbf{x}^{(i)} - \mathbf{x}^{(j)}\|\right) \tag{7.8}$$

It is noted that the correlation depends only on the distance between the two variables $\|\mathbf{x}^{(i)} - \mathbf{x}^{(j)}\|$.

Since different input variables may represent different physical variables, the true function may have different behavior in different input variables. Therefore, the correlation may not be the same in different input variables. In such a case, it makes more sense to use different hyperparameters, $\theta_k$, for different input variables, which is referred to as anisotropic correlation, defined as

$$\phi\left(\boldsymbol{\theta}, \mathbf{x}^{(i)}, \mathbf{x}^{(j)}\right) = \prod_{k=1}^{n} \phi_k\left(\theta_k, \left|x_k^{(i)} - x_k^{(j)}\right|\right) \tag{7.9}$$

where $\boldsymbol{\theta} = \{\theta_1, ..., \theta_n\}^T$ is the vector of hyperparameters. Figure 7.4 illustrates an example of isotropic and anisotropic correlation functions in two-dimensional design space (see Exercise Problem 7.4).

There are different types of correlation models available, including Gaussian, radial basis (or squared exponential), rational quadratic, neural network, Matern, periodic, constant, linear functions, and also a combination of these (Rasmussen and Williams [82]). Even though the quality of the Kriging simulation depends on the correlation function, the Gaussian correlation model is common in surrogate modeling. The Gaussian correlation between point $\mathbf{x}$ and point $\mathbf{s}$ can be defined as

$$\phi(\boldsymbol{\theta}, \mathbf{x}, \mathbf{s}) = \prod_{k=1}^{n} \exp\left[-\left(\frac{x_k - s_k}{\theta_k}\right)^2\right] \tag{7.10}$$

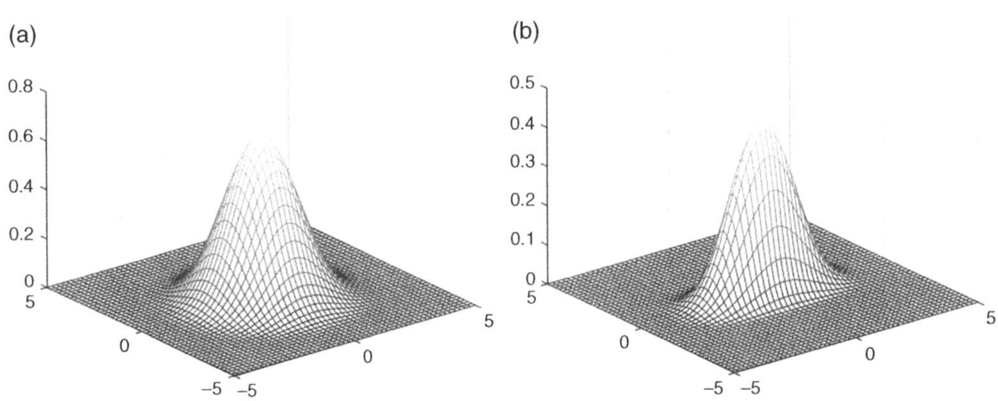


**Figure 7.4** Isotropic versus anisotropic correlation functions in two dimensions. (a) Isotropic correlation and (b) anisotropic correlation.

where $\theta_k$ is a hyperparameter that is a scale parameter in this function, which is used to control the smoothness of the function. When all hyperparameters are identical, the anisotropic correlation becomes an isotropic correlation. That is, when $\boldsymbol{\theta} = \{\theta, \ldots, \theta\}^T$, the correlation becomes

$$\phi(\boldsymbol{\theta}, \mathbf{x}, \mathbf{s}) = \prod_{k=1}^{n} \exp\left[-\left(\frac{x_k - s_k}{\theta}\right)^2\right] = \exp\left[-\frac{1}{\theta^2}\sum_{k=1}^{n}(x_k - s_k)^2\right] = \exp\left[-\frac{\|\mathbf{x}-\mathbf{s}\|^2}{\theta^2}\right]$$

A small $\theta$ means a fast decay of the correlation with distance. The correlation should decay to about 0.4 at one-sixth of the wavelength, $l$ (see Example 7.1). Since $e^{-1} = 0.3679$ is close to 0.4, the hyperparameter must satisfy $(l/6\theta)^2 \approx 1$, which yields $\theta \approx l/6$. That is, if a function has a wavelength $l$, then the hyperparameter should be close to $\theta \approx l/6$. For example, for $y = \sin(x)$, $\theta \approx 1$, while for $y = \sin(5x)$, $\theta \approx 0.2$. More precisely, $(0.1895l/\theta)^2 = 1$. Therefore, $\theta = 0.1895l$.

Let us consider the case where there are $n_y$ sample points, $(\mathbf{x}^{(i)}, y_i)$, with $n$-dimensional input variables $x_k^{(i)}, k = 1, \ldots, n$, and $y_i = y(\mathbf{x}^{(i)})$. Given the hyperparameters $\theta_k$, the covariance matrix of the samples can be defined as

$$\operatorname{cov}\left(y_i, y_j\right) = \sigma^2 \exp\left[-\sum_{k=1}^{n}\left(\frac{x_k^{(i)} - x_k^{(j)}}{\theta_k}\right)^2\right] = \sigma^2 R_{ij}, \quad i, j = 1, \ldots, n_y \tag{7.11}$$

where the correlation matrix $\mathbf{R} = [R_{ij}]$ is also defined from the covariance matrix. In defining the correlation matrix, it is assumed that the variance $\sigma^2$ of the local departure is constant over the design space, which is called stationary covariance. It means that the wavelength of the function is the same for the entire design space. The stationary covariance depends only on the relative position of its two inputs.

In the process of fitting the samples, we will also need to estimate the hyperparameters $\theta_k$ and variance $\sigma^2$. The variance will be small when the data is dense. Dense means that we have more than enough points to describe the waviness of the function. That means that the distance from one point to the nearest one should be an order of magnitude less than the wavelength of the function.

**Example 7.2 Correlation Matrix**

Five samples are equally spaced in the one-dimensional design space $x \in [0, 4]$. Calculate the Gaussian correlation matrix when (i) $\theta = 2.0$ and (ii) $\theta = 0.1$. Calculate the determinant of the correlation matrices.

**Solution:**

When hyperparameter $\theta$ is small, the correlation decays fast, and thus the affecting distance of one sample is short. On the other hand, when $\theta$ is large, samples affect other samples at a distance. The following Matlab script calculates the correlation matrix $\mathbf{R}_1$ when $\theta = 2.0$ and $\mathbf{R}_2$ when $\theta = 0.1$.

```
x=linspace(0,4,5);
[col,raw] = meshgrid(x,x);
dist=abs(col-raw);
theta1=2.0;
R1=exp(-(dist/theta1).^2)
det(R1)
theta2=0.1;
R2=exp(-(dist/theta2).^2)
det(R2)
```

$$\mathbf{R}_1 = \begin{bmatrix} 1.000 & 0.779 & 0.368 & 0.105 & 0.018 \\ 0.779 & 1.000 & 0.779 & 0.368 & 0.105 \\ 0.368 & 0.779 & 1.000 & 0.779 & 0.368 \\ 0.105 & 0.368 & 0.779 & 1.000 & 0.779 \\ 0.018 & 0.105 & 0.368 & 0.779 & 1.000 \end{bmatrix}, \quad \mathbf{R}_2 = \begin{bmatrix} 1 & 0 & 0 & 0 & 0 \\ 0 & 1 & 0 & 0 & 0 \\ 0 & 0 & 1 & 0 & 0 \\ 0 & 0 & 0 & 1 & 0 \\ 0 & 0 & 0 & 0 & 1 \end{bmatrix}$$

It turned out that $\mathbf{R}_1$ is a fully populated matrix, while $\mathbf{R}_2$ is an identity matrix. In addition, the determinants of correlation matrices are $\det(\mathbf{R}_1) = 0.0032$ and $\det(\mathbf{R}_2) = 1.0$. Therefore, when the hyperparameter $\theta$ is small, the correlation converges to an identity matrix, which means that there is no correlation between samples. On the other hand, when $\theta$ is large, the determinant of the correlation matrix approaches zero, and all samples have significant correlation. In this case, it might be difficult to calculate the inverse of the correlation matrix as it becomes a singular matrix. It would be a good exercise to find the relationship between the hyperparameter and the determinant of the correlation matrix (see Exercise Problem 7.6).

## 7.3 Kriging Surrogate Model

In this section, a Kriging surrogate model is developed based on its model form given in Eq. (7.3) with $n_y$ sample points. The development of a Kriging surrogate is equivalent to determining the global function and local departure. They are determined from the assumptions given in the previous section. First, the global function parameters, $\boldsymbol{\beta}$, in Eq. (7.3) and variance of samples with respect to the global function, $\sigma^2$, in Eq. (7.4) can be estimated by using maximum likelihood estimation (MLE). Once the global function is determined, the local departure is calculated by minimizing the unbiased mean squared errors. In addition, it is necessary to determine the correlation between sample points and prediction points. Lastly, we will discuss the uncertainty associated with the Kriging prediction.

### 7.3.1 Global Function and Distribution of Errors

As explained in Eqs. (7.3) and (7.4), the errors between the global function and sample points are assumed to be normally distributed. At $n_y$ sample points, the errors can be defined as

$$\begin{Bmatrix} e_1 \\ e_2 \\ \vdots \\ e_{n_y} \end{Bmatrix} = \begin{Bmatrix} y_1 \\ y_2 \\ \vdots \\ y_{n_y} \end{Bmatrix} - \begin{bmatrix} \xi_1(\mathbf{x}_1) & \xi_2(\mathbf{x}_1) & \cdots & \xi_{n_p}(\mathbf{x}_1) \\ \xi_1(\mathbf{x}_2) & \xi_2(\mathbf{x}_2) & \cdots & \xi_{n_p}(\mathbf{x}_2) \\ \vdots & \vdots & \ddots & \vdots \\ \xi_1(\mathbf{x}_{n_y}) & \xi_2(\mathbf{x}_{n_y}) & \cdots & \xi_{n_p}(\mathbf{x}_{n_y}) \end{bmatrix} \begin{Bmatrix} \beta_1 \\ \beta_2 \\ \vdots \\ \beta_{n_p} \end{Bmatrix}$$

$$\mathbf{e} = \mathbf{y} - \mathbf{X}\boldsymbol{\beta} \tag{7.12}$$

where $\mathbf{X}$ is the design matrix, the values of the basis vector at all sample points. It is noted that the above errors depend on the global function parameters, $\boldsymbol{\beta}$.

The assumption in Kriging is that the error $\mathbf{e}$ follows a Gaussian distribution with a zero mean, variance $\sigma^2$, and correlation between sample points. In linear regression, the parameters are determined to minimize the mean-squared-errors at sample points. In Kriging, these parameters are determined using a likelihood function. The likelihood is a measure of how the distribution of the global prediction, $\mathbf{X\beta}$, is close to that of the samples, $\mathbf{y}$. This is basically the same with the statistical view of linear regression in Section 2.8. The only difference is that in linear regression, the correlation was not considered, but Kriging uses the correlation. From the assumption that the errors (local departures) follow a Gaussian distribution, the likelihood is in the form of the probability density function (PDF) of a normal distribution. That is, the likelihood is the value of probability density of obtaining samples $\mathbf{y}$ for given parameters, $\boldsymbol{\beta}$ and $\sigma^2$. Therefore, the joint PDF of correlated $n_y$ sample points can be used to define the likelihood as

$$f(\mathbf{y} \mid \boldsymbol{\beta}, \sigma^2) = \frac{1}{\sqrt{(2\pi\sigma^2)^{n_y}|\mathbf{R}|}} \exp\left( - \frac{(\mathbf{y}-\mathbf{X}\boldsymbol{\beta})^T\mathbf{R}^{-1}(\mathbf{y}-\mathbf{X}\boldsymbol{\beta})}{2\sigma^2} \right) \tag{7.13}$$

which is the multiplication of $n_y$ Gaussian distributions. This is the likelihood function of $\mathbf{y}$ with given $\boldsymbol{\beta}$ and $\sigma^2$ with a correlation matrix, $\mathbf{R}$. In the above equation, $|\mathbf{R}|$ is the determinant of the correlation matrix. The goal is to find the parameters, $\boldsymbol{\beta}$ and $\sigma^2$, which can maximize the likelihood. When the global function $\mathbf{X\beta}$ is close to sample $\mathbf{y}$, the likelihood has its maximum, but the correlation between samples plays a role. Therefore, the likelihood in Eq. (7.13) is a function of $\boldsymbol{\beta}$, $\sigma^2$ and $\mathbf{R}$. For the moment, we will assume that the correlation matrix $\mathbf{R}$ is given and will be determined later.

The main difference between the likelihood function in Eq. (7.13) and that of linear regression in Section 2.8 is the inverse of correlation matrix. However, this difference can yield a significantly different estimate. In linear regression, the likelihood function is maximized when the MSE, $(\mathbf{y}-\mathbf{X}\boldsymbol{\beta})^T(\mathbf{y}-\mathbf{X}\boldsymbol{\beta})$ is minimum. In the case of Kriging, the maximum location of the likelihood function depends on the relationship between the difference $(\mathbf{y}-\mathbf{X}\boldsymbol{\beta})$ and the inverse of correlation matrix $\mathbf{R}^{-1}$. That is, the maximum location of the likelihood function may not be achieved when the difference becomes minimum. In fact, the maximum likelihood can be obtained even when the difference $(\mathbf{y}-\mathbf{X}\boldsymbol{\beta})$ is large by modifying the inverse of the correlation matrix (see Exercise Problem 7.7). Therefore, in Kriging surrogates, the maximum likelihood estimate is to find the hyperparameters that make the differences in samples and the correlation matrix consistent. It is interesting to note that the differences $(\mathbf{y}-\mathbf{X}\boldsymbol{\beta})$ are in the output space, while the correlation is defined in the input space.

The parameters, $\boldsymbol{\beta}$ and $\sigma^2$, can be found by maximizing the likelihood function. For the purpose of handling an algebraic function instead of an exponential function, a logarithm of the likelihood is often used. Since logarithm is a monotonically increasing function, it may change the value of likelihood, but it will not change the parameter values that yield the maximum value. The logarithmic likelihood or log-likelihood is given as

$$\ln\left[f(\mathbf{y}\mid\boldsymbol{\beta},\sigma^2)\right] = -\frac{n_y}{2}\ln(2\pi) - \frac{n_y}{2}\ln\left(\sigma^2\right) - \frac{1}{2}\ln|\mathbf{R}| - \frac{(\mathbf{y}-\mathbf{X}\boldsymbol{\beta})^T\mathbf{R}^{-1}(\mathbf{y}-\mathbf{X}\boldsymbol{\beta})}{2\sigma^2} \tag{7.14}$$

The first term is not important for the purpose of finding hyperparameter $\theta$. The second term penalizes the likelihood for a large variance of local departure. The third term corresponds to a penalty term that increases proportionally to a model's complexity. In Kriging, a model is considered complex when correlation decays fast. When all samples are uncorrelated, the third term has its minimum, i.e., $\ln|\mathbf{R}| = 0$. As more and more samples are correlated, determinant $|\mathbf{R}|$ approaches zero, and it increases the log-likelihood function. Therefore, maximizing the log-likelihood function prefers to a more correlated model. The last term in Eq. (7.14) corresponds to a penalty term for a model's failure to fit observed samples. As the differences between samples and the global function increase, the log-likelihood is decreased. Maximizing the log-likelihood function is a trade-off between fitting performance and simplicity of the Kriging model. That is, in order to have a simple model, variance $\sigma^2$ and determinant $|\mathbf{R}|$ need to be decreased, which will increase the numerator of the last term due to $\mathbf{R}^{-1}$ term.

The maximum location can be found by differentiating Eq. (7.14) with respect to $\boldsymbol{\beta}$ and $\sigma^2$ as

$$\frac{\partial \ln f}{\partial \boldsymbol{\beta}} = \frac{\mathbf{X}^T\mathbf{R}^{-1}(\mathbf{y}-\mathbf{X}\boldsymbol{\beta})}{\sigma^2} = 0$$

$$\frac{\partial \ln f}{\partial \sigma^2} = -\frac{n_y}{2}\frac{1}{\sigma^2} + \frac{(\mathbf{y}-\mathbf{X}\boldsymbol{\beta})^T\mathbf{R}^{-1}(\mathbf{y}-\mathbf{X}\boldsymbol{\beta})}{2\sigma^4} = 0 \tag{7.15}$$

Therefore, the estimated parameters that maximize the likelihood function can be obtained by solving the above equation for $\boldsymbol{\beta}$ and $\sigma^2$ as

$$\hat{\boldsymbol{\beta}} = \left(\mathbf{X}^T\mathbf{R}^{-1}\mathbf{X}\right)^{-1}\left\{\mathbf{X}^T\mathbf{R}^{-1}\mathbf{y}\right\}$$

$$\hat{\sigma}^2 = \frac{\left(\mathbf{y}-\mathbf{X}\hat{\boldsymbol{\beta}}\right)^T\mathbf{R}^{-1}\left(\mathbf{y}-\mathbf{X}\hat{\boldsymbol{\beta}}\right)}{n_y - n_p} \tag{7.16}$$

Note that the estimated parameter, $\hat{\boldsymbol{\beta}}$, is in the same form as Eq. (2.14) except for the inverse of the correlation matrix, and the degree of freedom $n_y - n_p$ is used instead of the number of samples $n_y$ in $\hat{\sigma}^2$ for the unbiased estimation, which is basically the same with Eq. (2.20), except for the inverse of correlation matrix. Different from polynomial response surfaces, the difference between $\boldsymbol{\beta}$ and $\hat{\boldsymbol{\beta}}$ is not important in Kriging surrogates. In polynomial response surfaces, $\hat{y}(\mathbf{x}) = \boldsymbol{\xi}(\mathbf{x})^T\hat{\boldsymbol{\beta}}$ is the final form of the surrogate approximation. In Kriging, however, it only defines the global function. Therefore, the difference is compensated by the local departure term $s(\mathbf{x})$. Therefore, the true function in Eq. (7.3) can also be written as $y(\mathbf{x}) = \boldsymbol{\xi}(\mathbf{x})^T\hat{\boldsymbol{\beta}} + s(\mathbf{x})$. It is also noted that the estimated parameters depend on the correlation matrix $\mathbf{R}$, which will be discussed later. If there is no correlation between samples, matrix $\mathbf{R}$ will be an identity matrix, and Eq. (7.16) becomes identical to linear regression of polynomial response surfaces.

By substituting the expression of variance $\hat{\sigma}^2$ into the last term in Eq. (7.14), the last term becomes a constant $(n_y - n_p)/2$. Since the first term is also constant, it would only be necessary to consider the second and third terms in Eq. (7.14) to maximize the log-likelihood function, which will be discussed in the following section.

In deriving Eq. (7.16), it is assumed that the wavelength of the function is constant in all regions of input space. In such a case, the covariance matrix that is found using the initial samples is constant throughout the input space, which is called stationary covariance. For nonstationary covariance, readers are referred to Xiong et al. [83].

When the main purpose of the surrogate is for estimating function in interpolating region, the choice of global function might not be important as the local departure will model such a way that the prediction passes through sample points and prediction between sample points are determined based on correlation between sample points. However, the choice of global function becomes important when the purpose of surrogate is to predict a function at the extrapolation region. As discussed in Chapter 2, it is inevitable to have a large extrapolation region in a high-dimensional input space. In the extrapolation region, the correlation between sample points is quickly diminished as the extrapolation distance increases and the prediction goes back to the global function. Therefore, it would be important to choose the global function that is close to the true functional behavior. Without having the knowledge of functional behavior in the extrapolation region, Kriging prediction can have a significant limitation. Along with determining the correlation structure, this will be the critical issue for validity of Kriging predictions.

**Example 7.3 Effect of Correlation on the Global Function**

The correlation between samples can affect the fitting process. Consider fitting a linear polynomial response surface using three samples in Example 2.1 of Chapter 2. The three samples were $(x_i, y_i) = (0, 0), (1, 1), (2, 0)$, and the model form was $\hat{y}(x, \boldsymbol{\beta}) = \beta_1 + \beta_2 x$. The linear regression yields a constant function $\hat{y}(x, \boldsymbol{\beta}) = 1/3$. The linear regression corresponds to the case when the correlation matrix is an identity matrix. When the correlation coefficient between two samples $x_2$ and $x_3$ is 0.5, calculate the two unknown coefficients.

**Solution:**

First, the correlation matrix and its inverse need to be calculated as

$$\mathbf{R} = \begin{bmatrix} 1 & 0 & 0 \\ 0 & 1 & 0.5 \\ 0 & 0.5 & 1 \end{bmatrix}, \quad \mathbf{R}^{-1} = \begin{bmatrix} 1 & 0 & 0 \\ 0 & 1.333 & -0.667 \\ 0 & -0.667 & 1.333 \end{bmatrix}$$

Therefore, the matrix and vector that are required in the regression equation in Eq. (7.16) can be obtained as

$$\mathbf{X}^T\mathbf{R}^{-1}\mathbf{X} = \begin{bmatrix} 2.333 & 2 \\ 2 & 4 \end{bmatrix}, \quad \mathbf{X}^T\mathbf{R}^{-1}\mathbf{y} = \begin{Bmatrix} 0.667 \\ 0 \end{Bmatrix}$$

Therefore, the regression equation $\hat{\boldsymbol{\beta}} = \left(\mathbf{X}^T\mathbf{R}^{-1}\mathbf{X}\right)^{-1}\left\{\mathbf{X}^T\mathbf{R}^{-1}\mathbf{y}\right\}$ can be solved for regression coefficients $\hat{\boldsymbol{\beta}} = \{0.5, -0.25\}^T$. Therefore, the global function becomes $\hat{y}(x) = 0.5 - 0.25x$. Figure 7.5 compares the regression results with and without the correlation. Since samples $x_2$ and $x_3$ are correlated, the regression chose a slope that follows the trend of these two samples. As the correlation coefficient of samples $x_2$ and $x_3$ increases, the magnitude of the slope increases. For example, when the correlation coefficient becomes 0.9, the slope changes from −0.25 to −0.75. However, linear regression with correlation increases the root-mean-squared error from 0.8165 to 0.9014. In the same way, the variance in Eq. (7.16) is increased from 0.6667 to 1.0.

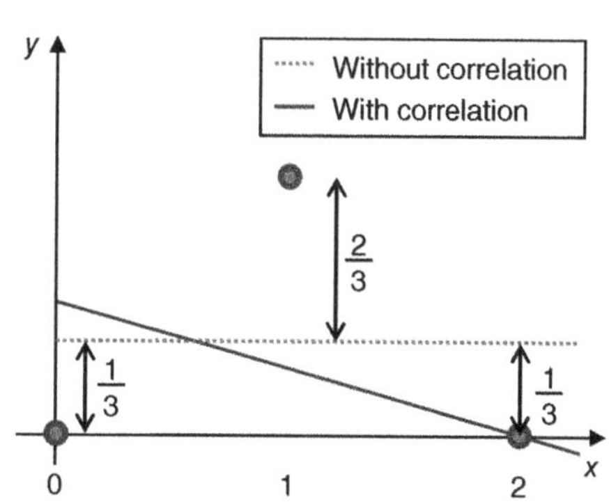


**Figure 7.5** Linear approximation of three samples with and without correlation.

**Example 7.4 Comparison with PRS**
Consider the linear PRS fit using three samples in Example 2.1, where the three samples were given as $(x, y) = (0, 0), (1, 1), (2, 0)$. Recall that different error measures yielded different fit results: $\hat{y} = 1/3$ with $e_{RMS}$, $\hat{y} = 0.5$ with $e_{max}$, and $\hat{y} = 0$ with $e_{av}$. Calculate the linear global function $\hat{y} = \beta_1 + \beta_2 x$ using the maximum likelihood estimate, given in Eq. (7.16). Use the hyperparameter $\theta = 1$.

**Solution:**
First, the correlation matrix and its inverse need to be calculated. When $\theta = 1$, they are

$$\mathbf{R} = \begin{bmatrix} 1 & 0.3679 & 0.0183 \\ 0.3679 & 1 & 0.3679 \\ 0.0183 & 0.3679 & 1 \end{bmatrix}, \quad \mathbf{R}^{-1} = \begin{bmatrix} 1.1781 & -0.4921 & 0.1594 \\ -0.4921 & 1.3620 & -0.4921 \\ 0.1594 & -0.4921 & 1.1781 \end{bmatrix}$$

Therefore, the matrix and vector that are required in the regression equation in Eq. (7.16) can be obtained as

$$\mathbf{X}^T\mathbf{R}^{-1}\mathbf{X} = \begin{bmatrix} 2.0689 & 2.0689 \\ 2.0689 & 4.1062 \end{bmatrix}, \quad \mathbf{X}^T\mathbf{R}^{-1}\mathbf{y} = \begin{Bmatrix} 0.3779 \\ 0.3779 \end{Bmatrix}$$

Therefore, the regression equation $\hat{\boldsymbol{\beta}} = (\mathbf{X}^T\mathbf{R}^{-1}\mathbf{X})^{-1}\{\mathbf{X}^T\mathbf{R}^{-1}\mathbf{y}\}$ can be solved for regression coefficients $\hat{\boldsymbol{\beta}} = \{0.1827, 0.0\}^T$. Therefore, the global function becomes $\hat{y}(x) = 0.1827$. Figure 7.6 (a) shows the linear global function along with linear PRS fits with the three different error measures. It is clear that the stationary condition of the likelihood function yields a different estimation from PRS fits. This is due to the contribution of the correlation matrix.

As explained before, when the hyperparameter is small, the correlation matrix approaches an identity matrix, which then becomes equivalent to linear regression. Figure 7.6(b) shows the constant term in the global function $\hat{y}(x) = \beta_1$ as a function of hyperparameter $\theta$. When $\theta \rightarrow 0$, the constant term $\beta_1 \rightarrow 1/3$, which is the same as the coefficient from linear regression. However, as $\theta$ increases, the constant term decreases. It is also noted that the constant term is insensitive to the hyperparameter when $\theta \leq 0.5$, but it decreases rapidly as the hyperparameter increases. It is interesting to note that when $\theta > 1.2$, the constant term becomes negative, which is difficult to

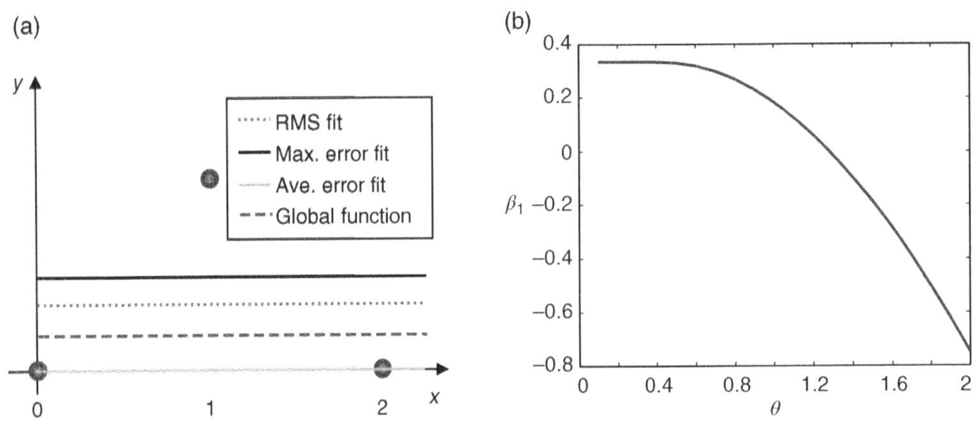


**Figure 7.6** Comparison of the global function with PRS fits. (a) Global function and PRSs with different error measures and (b) global function constant with different hyperparameters.

justify as all samples are non-negative. Therefore, the global function may fail to represent the global trend of samples when a wrong hyperparameter is estimated.

The main reason that the global function fails to represent the trend of samples is related to the fundamental assumption in Kriging. In Kriging, it is assumed that the samples are correlated, and the correlation structure is given in the form of the correlation matrix. If the three samples are uncorrelated, it is expected that all samples are random, and the same constant $\beta_1 = 1/3$ with linear regression can be found in the case of the RMS error. However, when the correlation is present, $\beta_1 = 1/3$ determines that the errors of the three samples $\mathbf{e} = \{-1/3, 2/3, -1/3\}^T$ seems uncorrelated because two of them are negative and the other is positive. When correlation is strong among samples, $\beta_1$ becomes negative so that all errors become positive.

### 7.3.2 Local Departure

In the previous subsection, the global function parameters, $\hat{\boldsymbol{\beta}}$, and variance, $\hat{\sigma}^2$, were estimated using maximum likelihood estimation, whose results are given in Eq. (7.16). Therefore, at this point, the form of the global function in Eq. (7.3) is determined. Now then, the local departure term $s(\mathbf{x})$ in Eq. (7.3) is considered, which eventually makes a Kriging simulation function. From the property that Kriging simulation passes sample points, Kriging simulation can be expressed by a linear combination of samples and weight functions. Then, the weight functions can be found by minimizing the mean-squared-errors (MSE) between outputs from a Kriging simulation function and the true function.

First, the error between the Kriging simulation and the true function is defined as

$$\epsilon(\mathbf{x}) = \hat{y}(\mathbf{x}) - y(\mathbf{x}) = \mathbf{w}(\mathbf{x})^T\mathbf{y} - y(\mathbf{x}) \tag{7.17}$$

where $\hat{y}(\mathbf{x})$ and $y(\mathbf{x})$ are, respectively, the Kriging simulation function and the true function. The Kriging simulation function can be expressed as the dot product of a $n_y \times 1$ vector of weight functions, $\mathbf{w}(\mathbf{x})$, and a $n_y \times 1$ vector of samples, $\mathbf{y}$ (i.e., $\mathbf{w}(\mathbf{x})^T\mathbf{y}$). The goal is to determine the weight function that minimizes the error in Eq. (7.17). Even if the Kriging simulation function is shown in the form of $\hat{y}(\mathbf{x}) = \mathbf{w}(\mathbf{x})^T\mathbf{y}$, this form will be expressed in terms of the global function and local departure as shown in Eq. (7.3).

From the assumption that there is no error in samples, the samples can be expressed in the form of the Kriging simulation function in Eq. (7.3). That is, $\mathbf{y} = \mathbf{X}\hat{\boldsymbol{\beta}} + \mathbf{s}$, where $\mathbf{s} = \{s(\mathbf{x}_1), s(\mathbf{x}_2), \ldots, s(\mathbf{x}_{n_y})\}^T$ is the vector of local departures at sample points. Note that $s(\mathbf{x})$ is a function, while $\mathbf{s}$ is a vector of $s(\mathbf{x})$ at all sample points. Similarly, the true function can also be expressed as $y(\mathbf{x}) = \boldsymbol{\xi}(\mathbf{x})^T\hat{\boldsymbol{\beta}} + s(\mathbf{x})$. Therefore, Eq. (7.17) can be expressed as

$$\begin{aligned} \epsilon(\mathbf{x}) &= \mathbf{w}(\mathbf{x})^T\left\{\mathbf{X}\hat{\boldsymbol{\beta}} + \mathbf{s}\right\} - \left(\boldsymbol{\xi}(\mathbf{x})^T\hat{\boldsymbol{\beta}} + s(\mathbf{x})\right) \\ &= \left(\mathbf{w}(\mathbf{x})^T\mathbf{X} - \boldsymbol{\xi}(\mathbf{x})^T\right)\hat{\boldsymbol{\beta}} + \mathbf{w}(\mathbf{x})^T\mathbf{s} - s(\mathbf{x}) \end{aligned} \tag{7.18}$$

In Eq. (7.18), $\left(\mathbf{w}(\mathbf{x})^T\mathbf{X} - \boldsymbol{\xi}(\mathbf{x})^T\right)\hat{\boldsymbol{\beta}}$ and $\mathbf{w}(\mathbf{x})^T\mathbf{s} - s(\mathbf{x})$ are the global and the departure error terms, respectively. To keep the global function unbiased, a constraint for the weight function is demanded so that the global error term is zero. Since the global function parameters $\hat{\boldsymbol{\beta}}$ cannot be all zero, the following equation is obtained as the constraint:

$$\mathbf{X}^T\mathbf{w}(\mathbf{x}) - \boldsymbol{\xi}(\mathbf{x}) = 0 \tag{7.19}$$

Now the MSE of Eq. (7.18) becomes

$$\text{MSE} = \mathbb{E}\left[\epsilon(\mathbf{x})^2\right] = \mathbb{E}\left[\left(\mathbf{w}^T\mathbf{s} - s\right)^2\right] = \mathbb{E}\left[\mathbf{w}^T\mathbf{s}\mathbf{s}^T\mathbf{w} - 2\mathbf{w}^T\mathbf{s}s + s^2\right] \tag{7.20}$$

It is noted that the weight function $\mathbf{w}(\mathbf{x})$ is deterministic, it can be moved out of the expectation operator. By using the definition of the second central moment (the variance) and covariance[1], the MSE can be obtained as

$$\text{MSE} = \sigma^2\left(\mathbf{w}^T\mathbf{R}\mathbf{w} - 2\mathbf{w}^T\mathbf{r} + 1\right) \tag{7.21}$$

which represents uncertainty in simulated Kriging outputs, and thus, it becomes the variance of the Kriging simulation.

The weight function can be found by minimizing the MSE in Eq. (7.21), while satisfying the unbiased constraint in Eq. (7.19). The method of Lagrange multipliers, which is an optimization method with equality constraints, is employed to find $\mathbf{w}$ by minimizing the MSE in Eq. (7.21) with the constraint in Eq. (7.19). The Lagrange function is defined by introducing a Lagrange multiplier, $\boldsymbol{\lambda}$ that is a $n_p \times 1$ vector in this case:

$$L(\mathbf{w}, \boldsymbol{\lambda}) = \sigma^2\left(\mathbf{w}^T\mathbf{R}\mathbf{w} - 2\mathbf{w}^T\mathbf{r} + 1\right) - \boldsymbol{\lambda}^T\left(\mathbf{X}^T\mathbf{w} - \boldsymbol{\xi}\right) \tag{7.22}$$

Minimizing the Lagrange function is equivalent to minimizing the MSE in Eq. (7.21) while satisfying the unbiasedness constraint in Eq. (7.19). The minimum can be found by differentiating Eq. (7.22) with respect to $\mathbf{w}$ and $\boldsymbol{\lambda}$ as

$$\frac{\partial L(\mathbf{w}, \boldsymbol{\lambda})}{\partial \mathbf{w}} = 2\sigma^2(\mathbf{R}\mathbf{w} - \mathbf{r}) - \mathbf{X}\boldsymbol{\lambda} = 0 \tag{7.23}$$

$$\frac{\partial L(\mathbf{w}, \boldsymbol{\lambda})}{\partial \boldsymbol{\lambda}} = \mathbf{X}^T\mathbf{w} - \boldsymbol{\xi} = 0 \tag{7.24}$$

By solving Eq. (7.23) for $\mathbf{w}$, the following equation is obtained:

$$\mathbf{w} = \mathbf{R}^{-1}\mathbf{r} + \mathbf{R}^{-1}\mathbf{X}\frac{\boldsymbol{\lambda}}{2\sigma^2} \tag{7.25}$$

To obtain $\boldsymbol{\lambda}/2\sigma^2$, Eq. (7.25) is substituted into Eq. (7.24) and solved for $\boldsymbol{\lambda}/2\sigma^2$ as

$$\frac{\boldsymbol{\lambda}}{2\sigma^2} = \left(\mathbf{X}^T\mathbf{R}^{-1}\mathbf{X}\right)^{-1}\left\{\boldsymbol{\xi} - \mathbf{X}^T\mathbf{R}^{-1}\mathbf{r}\right\} \tag{7.26}$$

Therefore, the weight vector, $\mathbf{w}$, can be obtained with Eqs. (7.25) and (7.26), which is used for the Kriging simulation in Eq. (7.17) as

$$\begin{aligned}
\hat{y}(\mathbf{x}) &= \mathbf{w}(\mathbf{x})^T\mathbf{y} \\
&= \left(\mathbf{R}^{-1}\mathbf{r} + \mathbf{R}^{-1}\mathbf{X}\frac{\boldsymbol{\lambda}}{2\sigma^2}\right)^T\mathbf{y} \\
&= \mathbf{r}^T\mathbf{R}^{-1}\mathbf{y} + \frac{\boldsymbol{\lambda}^T}{2\sigma^2}\mathbf{X}^T\mathbf{R}^{-1}\mathbf{y} \\
&= \mathbf{r}^T\mathbf{R}^{-1}\mathbf{y} + \left(\boldsymbol{\xi}^T - \mathbf{r}^T\mathbf{R}^{-1}\mathbf{X}\right)\underline{\left(\mathbf{X}^T\mathbf{R}^{-1}\mathbf{X}\right)^{-1}\left\{\mathbf{X}^T\mathbf{R}^{-1}\mathbf{y}\right\}}
\end{aligned} \tag{7.27}$$

1 The covariance between different inputs are as follows: cov[$s(\mathbf{x})$, $s(\mathbf{x})$] = $\sigma^2$, cov[$s(\mathbf{x}_k)$, $s(\mathbf{x})$] = $\sigma^2\mathbf{r}$, cov[$s(\mathbf{x}_k)$, $s(\mathbf{x}_l)$] = $\sigma^2\mathbf{R}$, $k$, $l = 1, 2, \ldots, n_y$

Note that the symmetric property of the correlation matrix is used. In Eq. (7.27), the underlined term is the estimated parameter, $\hat{\boldsymbol{\beta}}$ in Eq. (7.16). Finally, the equation is obtained for Kriging simulation as

$$\hat{y}(\mathbf{x}) = \xi(\mathbf{x})^T\hat{\boldsymbol{\beta}} + \mathbf{r}(\mathbf{x})^T\mathbf{R}^{-1}\left(\mathbf{y} - \mathbf{X}\hat{\boldsymbol{\beta}}\right) \tag{7.28}$$

where $\mathbf{y}$ is a $n_y \times 1$ sample vector, $\mathbf{X}$ is a $n_y \times n_p$ design matrix, $\mathbf{R}$ is a $n_y \times n_y$ correlation matrix between the sample points, and $\hat{\boldsymbol{\beta}}$ is a $n_p \times 1$ vector of global function parameters/coefficients, which are determined based on the samples, while $\mathbf{r}(\mathbf{x})$ is a $n_y \times 1$ correlation vector between simulation point and the measurement points. In Eq. (7.28), the first term, $\xi(\mathbf{x})^T\hat{\boldsymbol{\beta}}$, represents the global function term and the second term is the local departure. To calculate the departure at a new input, the differences between the samples and global function outputs, $\left(\mathbf{y} - \mathbf{X}\hat{\boldsymbol{\beta}}\right)$, are weighted and summed based on the correlation term, $\mathbf{r}(\mathbf{x})^T\mathbf{R}^{-1}$. In order to rewrite the prediction in Eq. (7.28) into a similar form with linear regression, the sample-dependent term in the local departure is defined as $\boldsymbol{\alpha} = \mathbf{R}^{-1}\left(\mathbf{y} - \mathbf{X}\hat{\boldsymbol{\beta}}\right)$. Then the prediction can be rewritten as

$$\hat{y}(\mathbf{x}) = \xi(\mathbf{x})^T\hat{\boldsymbol{\beta}} + \mathbf{r}(\mathbf{x})^T\boldsymbol{\alpha} \tag{7.29}$$

The computational cost of the Kriging surrogate is much higher than that of other surrogate models, albeit it is still considered ignorable compared to the cost of calculating each sample. The computational cost of calculating parameters, $\hat{\boldsymbol{\beta}}$ and $\hat{\sigma}^2$, is almost the same as that of polynomial response surfaces. However, the computational cost of the local departure is expensive as it requires (i) calculating the correlation matrix and its inverse for all samples, (ii) calculating the correlation vector $\mathbf{r}(\mathbf{x})$ at every prediction point, and (iii) multiplying with the inverse of the correlation matrix. This computational cost increases rapidly along with the number of samples. For a large $n_y$, it becomes impractical to calculate the inverse of the correlation matrix, $\mathbf{R}^{-1}$. Since the last term $\boldsymbol{\alpha} = \mathbf{R}^{-1}\left(\mathbf{y} - \mathbf{X}\hat{\boldsymbol{\beta}}\right)$ only depends on the samples, a linear system of equations, $\mathbf{R}\boldsymbol{\alpha} = \left(\mathbf{y} - \mathbf{X}\hat{\boldsymbol{\beta}}\right)$, is solved for vector $\boldsymbol{\alpha}$ instead of calculating the inverse of the correlation matrix. As the number of samples increases, or equivalently, as the correlation between samples becomes strong, the correlation matrix converges to a singular matrix as shown in Example 7.2. Therefore, numerical difficulty can be experienced with a large number of samples.

In Eq. (7.29), only $\xi(\mathbf{x})$ and $\mathbf{r}(\mathbf{x})$ are a function of prediction point $\mathbf{x}$; all other terms are fixed for a given set of samples. Therefore, Kriging is similar to linear regression, which is a linear combination of unknown coefficients multiplied by known basis functions. However, there are two important differences. First, the basis function, $\mathbf{r}(\mathbf{x})$, is unknown, in that the correlation matrix, $\mathbf{R}$, needs to be estimated from the samples (in the next subsection). Second, the coefficients are not found by minimizing the root-mean-squared-errors (RMSE) at the sample points; in fact, RMSE is zero because the prediction interpolates the samples exactly.

**Example 7.5 Weight Function**

In Eq. (7.17), the Kriging prediction is expressed as a linear combination of weight functions and samples: $\hat{y}(\mathbf{x}) = \mathbf{w}(\mathbf{x})^T\mathbf{y}$. With the three samples in Example 7.3, plot the weight function $\mathbf{w}(x)$ as a function of $x$ when $\theta = 1$ and discuss its characteristics.

**Solution:**

From Eq. (7.27), the weight function can be written as

$$\mathbf{w}(\mathbf{x})^T = \mathbf{r}^T\mathbf{R}^{-1} + \left(\xi^T - \mathbf{r}^T\mathbf{R}^{-1}\mathbf{X}\right)\left(\mathbf{X}^T\mathbf{R}^{-1}\mathbf{X}\right)^{-1}\left\{\mathbf{X}^T\mathbf{R}^{-1}\right\} \tag{7.30}$$

where $\xi = \{1, x\}^T$ and

$$\mathbf{r}(x) = \left\{e^{-(0-x)^2}, e^{-(1-x)^2}, e^{-(2-x)^2}\right\}^T$$

The following Matlab code is used to plot Figure 7.7:

```
y=[0 1 0]';
X=[1 0;1 1;1 2];
R=[   1      exp(-1) exp(-4)
    exp(-1)    1     exp(-1)
    exp(-4) exp(-1)     1];
Rinv=inv(R);
A=X'*Rinv*X;
Ainv=inv(A);
for i=1:51
 x(i)=0.04*(i-1);
 r = [exp(-((0-x(i))/th)^2);exp(-((1-x(i))/th)^2);exp(-((2-x(i))/th)^2)];
 xi= [1; x(i)];
 w(i,:)=r'*Rinv + (xi'-r'*Rinv*X)*Ainv*X'*Rinv;
end
plot(x,w(:,1),x,w(:,2),x,w(:,3))
legend('w_1','w_2','w_3')
```

It turns out that the weight functions satisfy the Kronecker delta property of $w_i(x_j) = \delta_{ij}$, where $\delta_{ij} = 1$ when $i = j$ and zero otherwise. The weight functions can be considered shape functions that interpolate samples. Since $y_1 = y_3 = 0$ for this example, the Kriging prediction is the same as $w_2(x)$.

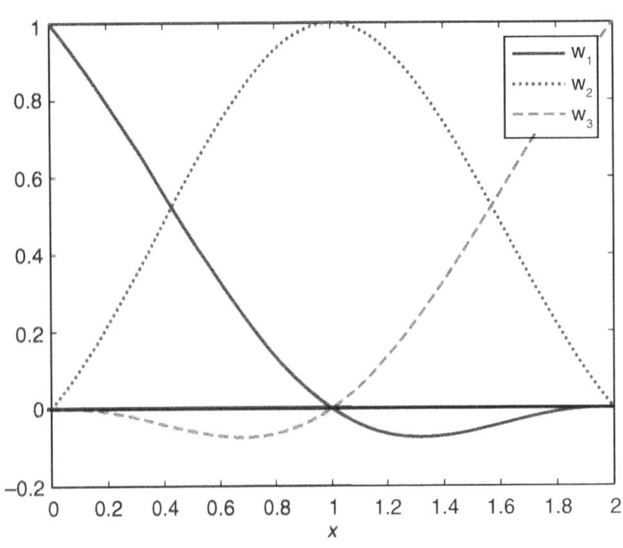


**Figure 7.7** Weight functions of three samples.

The plot of weight function in Figure 7.7 provides a lot of similarity with other types of interpolation methods, such as radial basis functions and reproducing kernel particle methods [84].

**Example 7.6 Ordinary Kriging**

With the three samples in Example 7.3, derive the ordinary Kriging model where the global function is a constant. When the Gaussian correlation function in Eq. (7.11) is used with $\theta^2 = 1, 10, 0.02$, determine the Kriging prediction $\hat{y}(x)$ when the correlation matrix is an identity matrix. Plot the Kriging prediction and compare it with the global function.

**Solution:**

Using the three samples, the correlation matrix at samples and the correlation vector at the prediction point can be defined as

$$\mathbf{r}(x) = \begin{Bmatrix} \exp\left(-x^2/\theta^2\right) \\ \exp\left(-(1-x)^2/\theta^2\right) \\ \exp\left(-(2-x)^2/\theta^2\right) \end{Bmatrix}, \quad \mathbf{R} = \begin{bmatrix} 1 & \exp\left(-\theta^{-2}\right) & \exp\left(-4\theta^{-2}\right) \\ \exp\left(-\theta^{-2}\right) & 1 & \exp\left(-\theta^{-2}\right) \\ \exp\left(-4\theta^{-2}\right) & \exp\left(-\theta^{-2}\right) & 1 \end{bmatrix}$$

In the case of an ordinary Kriging, the global function is constant. Using $\mathbf{X} = [1\ 1\ 1]^T$ and $\mathbf{y} = [0\ 1\ 0]^T$, the constant global function can be determined as

$$\hat{\beta} = \left(\mathbf{X}^T\mathbf{R}^{-1}\mathbf{X}\right)^{-1}\left\{\mathbf{X}^T\mathbf{R}^{-1}\mathbf{y}\right\} = \begin{cases} 0.1827, & \text{when } \theta^2 = 1 \\ -2.7340, & \text{when } \theta^2 = 10 \\ 0.3333, & \text{when } \theta^2 = 0.02 \end{cases}$$

As explained before, when hyperparameter $\theta$ is too large, the global trend function becomes negative. On the other hand, when the hyperparameter is too small, the global trend function becomes the mean of samples, which is the same as linear regression.

$$\hat{y}(x) = \hat{\beta} + \mathbf{r}(x)^T\mathbf{R}^{-1}\left(\mathbf{y} - \mathbf{1}\hat{\beta}\right)$$

Figure 7.8 shows the plot of Kriging prediction along with the global function with the three different hyperparameter values, $\theta^2 = 1, 10, 0.02$. It is confirmed that the Kriging prediction passes

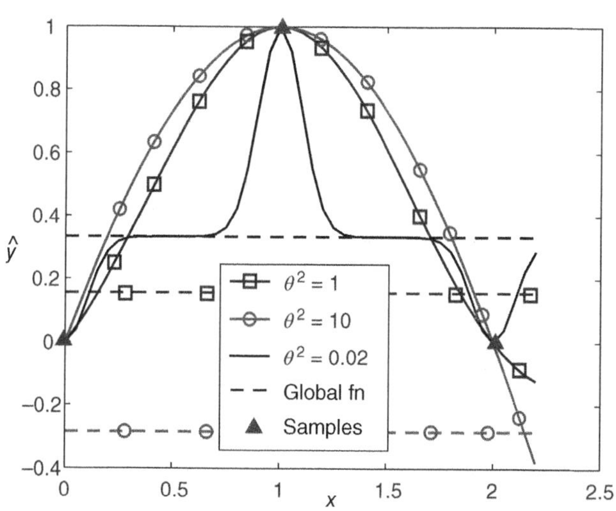


**Figure 7.8** Ordinary Kriging prediction with three samples.

through all sample points irrespective of the hyperparameter. With a large hyperparameter ($\theta^2 = 10$), the Kriging prediction is smooth (curve with circles) because the correlation between samples is strong for all samples. This is achieved by moving the global function to the negative value of $\hat{\beta} = -2.734$, such that the differences $(\mathbf{y} - \mathbf{1}\hat{\beta})$ become all positive. When the hyperparameter is small ($\theta^2 = 0.02$), the correlation is almost zero and the correlation matrix is close to the identity matrix. In such a case, the effect of one sample quickly decays and the majority of Kriging prediction follows the global function. Based on the discussion in the correlation model, the wavelength of the Kriging model would be about $l \approx 6\theta = 0.85$, which is consistent to the figure. Note that the curves are actually the mean prediction of the Gaussian process.

### 7.3.3 Hyperparameters and Likelihood Function

It is noted that both the parameters, $\hat{\boldsymbol{\beta}}$ and $\hat{\sigma}^2$, and the Kriging prediction in Eq. (7.28) depend on the correlation matrix $\mathbf{R}$, which in turn depends on the hyperparameter $\theta$. So far, we derived the Kriging surrogate model assuming that the hyperparameters are known. In general, the hyperparameters are determined via an optimization algorithm by maximizing a log-likelihood (or minimizing a negative log-likelihood) function corresponding to the error between global function outputs and samples, which is given in Eq. (7.14). The following equation is obtained by omitting the first and fourth[2] terms in Eq. (7.14) since they are constant with respect to the correlation matrix, i.e., the hyperparameters (Toal et al. [85]). Therefore, the hyperparameter, $\theta$, can be found by maximizing the following equation:

$$\theta = \arg\max\left[-\frac{n_y}{2}\ln\left(\hat{\sigma}^2\right) - \frac{1}{2}\ln|\mathbf{R}|\right] \tag{7.31}$$

The optimum value of $\theta$ can be obtained by maximizing Eq. (7.31) or minimizing the equivalent function of Eq. (7.31) as

$$\theta = \arg\min\left[\ln\left(\hat{\sigma}^{2(n_y - n_p)}|\mathbf{R}|\right)\right] \tag{7.32}$$

In this equation, the degree of freedom, $n_y - n_p$ is used instead of the number of samples $n_y$ for the purpose of unbiased estimation in variance.

The reduced negative log-likelihood function in Eq. (7.32) works well for most cases, but there are two cases in which it is impossible to find the hyperparameter that minimizes the negative log-likelihood function. The first case is when the number of samples is not enough to represent the functional behavior. In such a case, the negative log-likelihood function continues to increase with minimum value being $\theta = 0$, which basically removes correlations between samples. If this happens, Kriging predictions will follow the global function except for the sample points. This issue will be discussed in the next section.

The second case is when the global function is the same as the true function. In that case, the variance $\hat{\sigma}^2 = \left(\mathbf{y} - \mathbf{X}\hat{\boldsymbol{\beta}}\right)^T \mathbf{R}^{-1}\left(\mathbf{y} - \mathbf{X}\hat{\boldsymbol{\beta}}\right)/\left(n_y - n_p\right) = 0$, and thus, the term $\hat{\sigma}^{2(n_y - n_p)}|\mathbf{R}| = 0$. Therefore, the log-likelihood term goes to negative infinity. In such a case, it is not possible to find a hyperparameter that maximizes the likelihood function. However, any hyperparameter $\theta$ that does

2 Even if the fourth term includes the inverse of correlation matrix, it will be canceled out when $\sigma^2$ is substituted with $\hat{\sigma}^2$ in Eq. (7.16).

not make the correlation matrix singular would be valid. Also, since $\hat{\sigma}^2 = 0$, prediction uncertainty cannot be calculated. Therefore, it is not a good idea to use a complex form of global function in Kriging. That is why ordinary Kriging is popular in practice, where a constant global function is used. However, the choice of global function becomes important in the extrapolation region, which will be discussed in the next section.

**Example 7.7 Likelihood Function**

For the ordinary Kriging model in Example 7.6, plot the negative log-likelihood as a function of hyperparameter $\theta$ and estimate the hyperparameter that minimizes the negative log-likelihood.

**Solution:**

In order to calculate the negative log-likelihood function in Eq. (7.32), it would be necessary to calculate the variance $\hat{\sigma}^2$ in Eq. (7.16) and the determinant of the correlation matrix at different values of $\theta$. Also, in the case of the ordinary Kriging, $n_y = 3$ and $n_p = 1$. The following Matlab code can be used to plot the negative log-likelihood function:

```
y=[0 1 0]';
X=[1 1 1]';
for k=1:100
 th(k)=0.02*k;
 e1=exp(-(1/th(k))^2); e2=exp(-(2/th(k))^2);
 R=[1 e1 e2; e1 1 e1; e2 e1 1];
 Rinv=inv(R);
 detR=det(R);
 beta=(X'*Rinv*y)/(X'*Rinv*X);
 ym=y-X*beta;
 Like(k)=log((ym'*Rinv*ym/2)^4*detR);
end
plot(th,Like)
```

Figure 7.9(a) shows the plot of the negative log-likelihood function. It turns out that the function stays constant when $\theta \le 0.4$ and monotonically increases when $\theta \ge 0.4$. As shown in the figure, it is

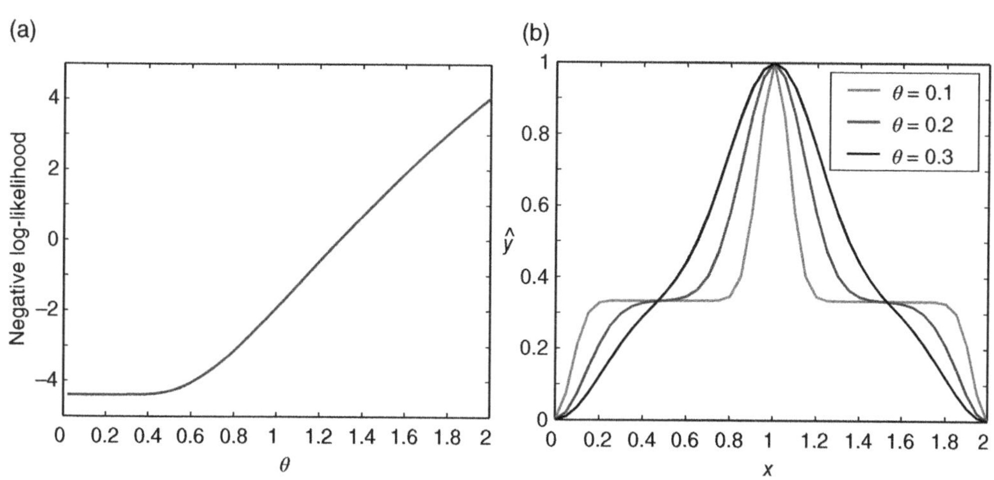


**Figure 7.9** Negative log-likelihood function and Kriging predictions with three samples. (a) Negative log-likelihood function and (b) Kriging predictions.

difficult to identify the minimum of the likelihood function, and thus, the hyperparameter. When the negative log-likelihood function monotonically increases after the initial flat region. It is an indicator that the number of samples is too small to represent the variation of the function properly. If the number of samples is reasonable, the negative log-likelihood function should decrease after the initial flat region. In order to check the effect of hyperparameter in the flat region, Figure 7.9(b) plots the Kriging predictions with three different values of hyperparameter. It looks like the three predictions are significantly different, while there is no difference in the likelihood function. This is one of the challenges in Kriging that the optimum hyperparameter is often difficult to identify. In practice, minimizing the negative log-likelihood function may not find the optimum hyperparameter when the number of samples is not enough. More than five samples per wavelength of the function are required such that the optimum hyperparameter can approximate the function reasonably.

As shown in Example 7.7, identifying the optimum hyperparameter by maximizing the likelihood function is the major challenge in Kriging surrogate modeling. First, it would be beneficial to understand why the likelihood function has a flat region, which makes it difficult to find an optimum. In order to understand this, let us consider the log-likelihood in Eq. (7.31). When $\theta$ decreases, the correlation decays fast and the correlation matrix becomes an identity matrix. Therefore, $|\mathbf{R}| = 1$. From Eq. (7.16), the variance $\hat{\sigma}^2 = \left(\mathbf{y} - \mathbf{X}\hat{\boldsymbol{\beta}}\right)^T \mathbf{R}^{-1}\left(\mathbf{y} - \mathbf{X}\hat{\boldsymbol{\beta}}\right) / \left(n_y - n_p\right)$ also becomes independent of the hyperparameter. Therefore, if a value of hyperparameter makes the correlation matrix identity, the log-likelihood function remains constant, which happens when $\theta \leq 0.4$ in Example 7.7. Therefore, it is important to limit the hyperparameter not to make the correlation matrix close to the identity matrix. In practice, it is recommended that $|\mathbf{R}| < 0.95$.

The next question is then how to determine the right hyperparameter when the likelihood shows a flat region. Unfortunately, there is no good way to determine the right hyperparameter in Example 7.7 because the number of samples is too small to determine the right hyperparameter that can best represent the distribution of samples. In general, there always is a flat region when $\theta$ becomes small. But with a proper number of samples, the flat region should not be the best log-likelihood because all samples are uncorrelated in this region. There should be a lower value of log-likelihood.

**Example 7.8 Likelihood Function with More Samples**

In Example 7.7, it was shown that the three samples were not enough to determine the hyperparameter. Use the following five samples to plot the hyperparameter and find the one that minimizes the negative log-likelihood using the graph: $(x_i, y_i) = (0, 0), (0.5, 0.7071), (1, 1), (1.5, 0.7071), (2, 0)$. Plot the Kriging predictions with three different hyperparmeters including the best one and compare them.

**Solution:**

The same procedure with Example 7.7 is used. In the case of the ordinary Kriging, $n_y = 5$ and $n_p = 1$. The following Matlab code can be used to plot the negative log-likelihood function:

```
x=[0 0.5 1 1.5 2];
y=[0 0.7071 1 0.7071 0]';
X=[1 1 1 1 1]';
ny=size(x,2);
```

```
R=zeros(ny,ny);
for k=1:100
 th(k)=0.02*k;
 for i=1:ny
  for j=1:ny
   R(i,j)=exp(-((x(i)-x(j))/th(k))^2);
  end
 end
 Rinv=inv(R);
 detR=det(R);
 beta(k)=(X'*Rinv*y)/(X'*Rinv*X);
 ym=y-X*beta(k);
 Like(k)=log((ym'*Rinv*ym/4)^8*detR);
end
figure; plot(th,Like)
```

Figure 7.10(a) shows the negative log-likelihood as a function of the hyperparameter. Different from the case with three samples, the case of five samples shows a clear global minimum of the likelihood function when $\theta = 0.54$ and a local minimum at $\theta = 1.68$. In addition, the likelihood function shows a flat region when $\theta < 0.2$. The flat region of negative log-likelihood at small values of $\theta$ is where the correlation matrix becomes the identity matrix and $|\mathbf{R}| = 1$. This means that correlations between samples disappear too quickly, which is not a desired condition. In Figure 7.10 (b), it is clear that the prediction with $\theta = 0.1$ is not a reasonable one as the correlation disappears too fast. Even if $\theta = 0.54$ is the global minimum, its prediction is not significantly different from that of $\theta = 1.68$. However, the two predictions become significantly different in the extrapolation region. In addition, the uncertainty distribution will be significantly different, which will be shown in the later section.

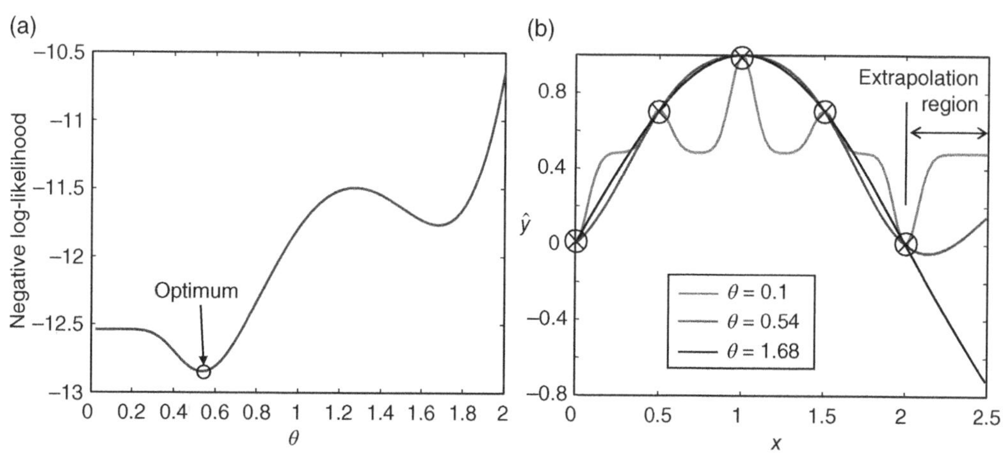


**Figure 7.10** Negative log-likelihood function and Kriging predictions with five samples. (a) Negative log-likelihood function and (b) Kriging predictions.

## 7.4 Issues in Determining Hyperparameters

Fitting a Kriging surrogate model boils down to determining hyperparameter $\theta$ because all other parameters depend on it. It is commonly accepted that the optimum hyperparameter can be determined to maximize the likelihood function in Eq. (7.13), which states that the differences between samples and global functions at sample locations follow a multivariate Gaussian distribution with correlation **R**. Since this is the fundamental assumption of Kriging surrogate modeling, the performance depends on the hyperparameter. In practice, it is not straightforward to determining the optimal hyperparameter. This section discusses issues related to determining the optimal hyperparameter and provides possible remedies to achieve an acceptable hyperparameter that can be used for the purpose of prediction.

### 7.4.1 Lower and Upper Bounds of Hyperparameter

Both the negative log-likelihood functions in Figures 7.9 and 7.10 show a flat region when the hyperparameter approaches zero. As explained before, this is due to the fact that the correlation matrix becomes an identity matrix, where all samples are uncorrelated. Since this is not the case that we want to consider (otherwise, it will be the same as linear regression), it would be necessary to limit the lower bound of the hyperparameter not including this region. In general, it would be necessary to limit the lower bound based on the minimum distance between samples. That is,

$$\theta \geq \theta_{\min} = \frac{\text{min.sample distance}}{\sqrt{2}} \tag{7.33}$$

Note that when $\theta = \theta_{\min}$, correlation between two samples becomes $\rho = \exp(-2) = 0.135$ and the determinant of the correlation matrix with two samples becomes 0.982.

Although Eq. (7.33) directly limits the lower bound of the hyperparameter, its choice is arbitrary. In addition, when many samples are at a large distance, while a single sample is at a small distance, this criterion may not be accurate to prevent an uncorrelated sample. It would be better to control the correlation matrix directly instead of the hyperparameter because the effect of the hyperparameter occurs through the correlation matrix. As discussed before, when the hyperparameter is too small, correlations between samples decay quickly, and the correlation matrix becomes an identity matrix. When this happens, the predictions mostly follow the global function and show sharp variations near the sample points, which are not desirable predictions. Therefore, it would be desirable to limit the hyperparameter such that the correlation matrix cannot be an identity matrix. As discussed in Example 7.7, it would be necessary to limit the hyperparameter such that

$$\theta \geq \theta_{\min} = \arg\min|\mathbf{R}(\theta)| \leq 0.95 \tag{7.34}$$

This condition would be more difficult to impose than Eq. (7.33) because the determinant of the correlation matrix is nonlinear function of hyperparameter.

The optimization process of finding the optimum hyperparameter that can minimize the negative log-likelihood may anticipate difficulty when the hyperparameter becomes too large. When a variable does not have an upper bound, an optimization algorithm searches for the optimum variable over a wide range of the variable's domain. As shown in Figure 7.11, when the hyperparameter increases, the likelihood function becomes unstable. When the value of hyperparameter increases, the correlation between samples increases and the reciprocal condition number of the correlation matrix decreases to zero. Therefore, calculating the inverse of the correlation matrix cannot be reliable. The likelihood function in the figure is calculated using the same method in Example 7.8 with

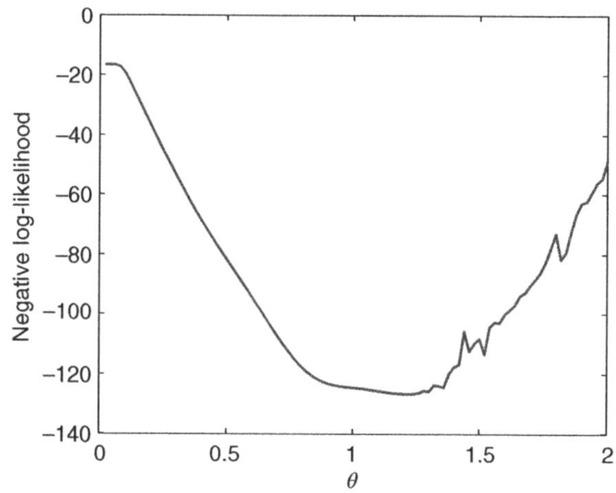


**Figure 7.11** Instability of the negative log-likelihood with a large $\theta$.

13 equally spaced samples (see Exercise Problem 7.12) from true function $y = \sin(\pi x)$. Initially the negative log-likelihood function has a flat region where correlation between samples disappears, followed by minimum at $\theta = 1.2$. This optimum value of hyperparameter is consistent with the previous discussion $\theta = 0.1895 \times 2\pi = 1.19$. After that, the negative log-likelihood starts increasing and becomes unstable when $\theta \geq 1.3$. This happens because the correlation matrix becomes singular, and the calculation of its inverse involves a large numerical error. In fact, the log-likelihood function becomes a complex variable. Therefore, it is necessary to limit the upper bound of the hyperparameter such that the inverse of the correlation matrix can be calculated without numerical difficulty. In Matlab implementation, if the reciprocal condition number of the correlation matrix `rcond(R)` is less than the floating-point relative accuracy `eps`, then the inverse matrix cannot be calculated with a desired accuracy. Therefore, the upper bound of the hyperparameter can be defined as

$$\theta \leq \theta_{\max} := \arg\min \mathrm{rcond}(\mathbf{R}(\theta)) \geq \mathrm{eps} \tag{7.35}$$

The relative accuracy `eps` in Matlab with 64-bit personal computers is about $2.224 \times 10^{-16}$.

### 7.4.2 Finding Optimum Value of Hyperparameter

Once the lower and upper bounds of the hyperparameter are determined, the optimal hyperparameter can be found within this range by minimizing the negative log-likelihood function. Three possible scenarios can be considered. First, the negative log-likelihood function starts decreasing after the initial flat region and has a minimum within the range of the hyperparameter. In such a case, the negative log-likelihood function in Eq. (7.32) can be minimized using conventional unconstrained optimization algorithms in Chapter 5. As shown in Figure 7.10(a), it is possible that the negative log-likelihood function has multiple local optima. Since the gradient-based optimization algorithms in Chapter 5 can only find a local optimum, it is possible to run optimization multiple times with different initial values. However, as discussed previously, a negative log-likelihood function normally has an initial flat region, followed by a decreasing trend as the hyperparameter increases. Therefore, it would be a good strategy to start with the lower bound in Eq. (7.33) and gradually increase the hyperparameter until it reaches the minimum value.

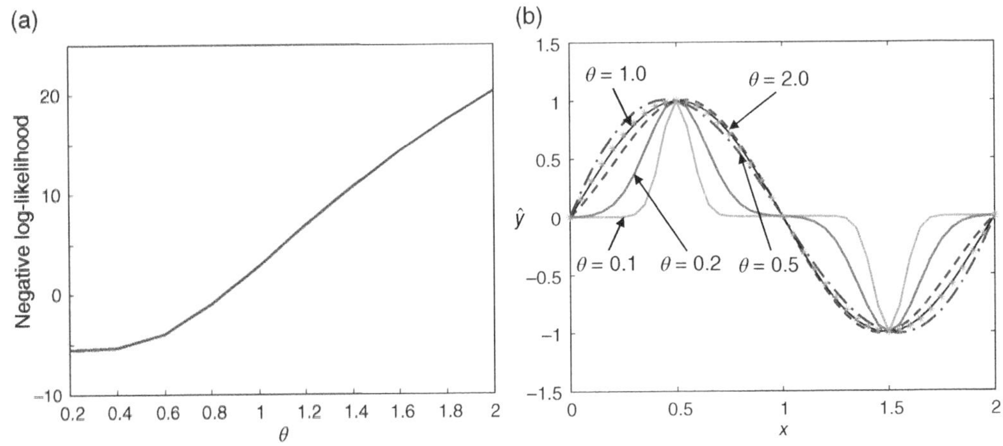


**Figure 7.12** Kriging surrogate with a too-small number of samples. (a) Negative log-likelihood function and (b) Kriging predictions with different hyperparameters.

Second, the negative log-likelihood function continues to increase after the initial flat region. This happens when the number of samples is too small to describe the variation of the function. In the case of $y = \sin(\pi x)$, it would require six samples within the period of the function to describe the fluctuation of the function. When five equally spaced samples are used, the negative log-likelihood function continues to increase as shown in Figure 7.12(a). The five samples are calculated from true function $y = \sin(\pi x)$ (see Exercise Problem 7.13). Figure 7.12(b) shows Kriging predictions with different values of hyperparameter. The star markers correspond to the true function values. Although the negative log-likelihood function continues to increase, the best fit can be obtained with $\theta = 1.0$. However, the negative log-likelihood function does not provide any clue for this optimal value. In this case, the hyperparameter that makes the negative log-likelihood minimum is the lower bound in Eq. (7.33). However, as shown in Figure 7.12(b), the Kriging surrogate with the lower-bound value of hyperparameter does not provide reasonable predictions as it removed correlation between samples. The theoretical study in correlation shows that the optimal hyperparameter should be $\theta = 0.1895 \times 2\pi = 1.19$. Therefore, when the number of samples is too small, it is impossible to determine the optimal hyperparameter solely based on the likelihood function. There should be additional information to determine the optimal hyperparameter. Good news is that the difference in predictions for different values of hyperparameter are not significantly different once the negative log-likelihood function passes the initial flat region. However, this relief cannot be true in the extrapolation region, which will be shown later.

The last scenario is when the negative log-likelihood function continues to decrease within the range of the hyperparameter. In such a case, it is suggested to use the hyperparameter that makes the negative log-likelihood minimum within the range. For example, Figure 7.13(a) shows the negative log-likelihood function of $y = \sin(\pi x)$ versus the hyperparameter with a different number of samples (see Exercise Problem 7.14). All the negative log-likelihood functions show the initial flat region when the hyperparameter is too small. When the number of samples is too small, such as $n_y = 6$, the negative log-likelihood function barely shows decrease after the initial flat region. This indicates that the number of samples is too small to represent the variation of the function. After $n_y = 8$, the negative log-likelihood function shows a clear minimum after the initial flat region. However, instability starts after $n_y = 12$, and it is difficult to identify the minimum negative log-likelihood function after $n_y = 13$. This happens because correlation between samples is too strong,

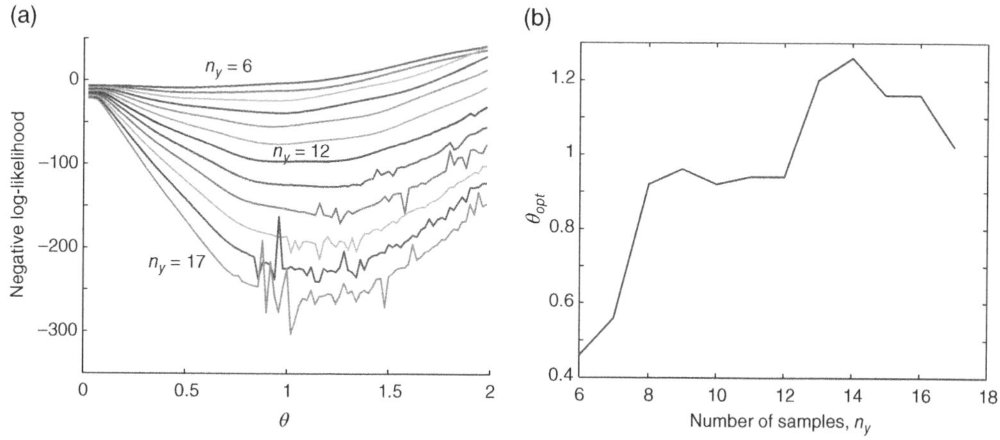


**Figure 7.13** Effect of the number of samples. (a) Negative log-likelihood function and (b) optimum hyperparameter.

and the correlation matrix becomes singular. Figure 7.13(b) shows that a meaningful optimum hyperparameter can be found between $8 \le n_y \le 12$, where the optimum hyperparameter was $\theta_{opt} = 0.92$. When the negative log-likelihood function does not have a clear minimum within the range, it is suggested to use the minimum value within the range given in Eqs. (7.34) and (7.35).

Unfortunately, it is difficult to estimate what value of the hyperparameter can produce the ill-conditioned correlation matrix in advance. This is related to the number of samples, distribution of samples, and dimension of the input variables. For practical applications, it would be a good strategy to start with the lower bound of the hyperparameter by gradually increasing it to find the optimum value. When the reciprocal condition number of correlation matrix `rcond(R)` is less than the floating-point relative accuracy `eps` before obtaining the minimum, it is suggested to use the largest hyperparameter so far as the optimum value.

### 7.4.3 Issues Related to the Number of Samples

Although the optimum value of the hyperparameter can be found using optimization algorithms, it would be beneficial to investigate the behavior of the likelihood function for different numbers of samples. Similar to the $y = \sin(\pi x)$ function in Figure 7.11, the design space is fixed $x \in [0, 1]$. Then, the number of samples gradually increases from 6 to 14 in equal intervals. Figure 7.14(a) shows the plots of the negative log-likelihood as a function of the hyperparameter $\theta$ for different numbers of samples (see Exercise Problem 7.16). It is interesting to note that the value of the hyperparameter that minimizes the likelihood function gradually decreases as the number of samples increases, as shown in Figure 7.14(b). Since the design space $x \in [0, 1]$ is fixed, the distance between samples decreases as the number of samples increases. Therefore, it is expected that the ratio between the optimum hyperparameter and the distance between samples may have a relationship. Based on the graph in Figure 7.14(b), when $7 \le n \le 12$, this relationship can be written as $\theta \approx 3.2d$, where $d$ is the distance between adjacent samples.

An important observation in this section is that Kriging surrogate anticipates difficulty when the number of samples is either too small or too large. When the number of samples is too small, it is impossible to determine the optimal hyperparameter that minimizes the negative log-likelihood function. In this case, the minimum of negative log-likelihood function is when samples do not

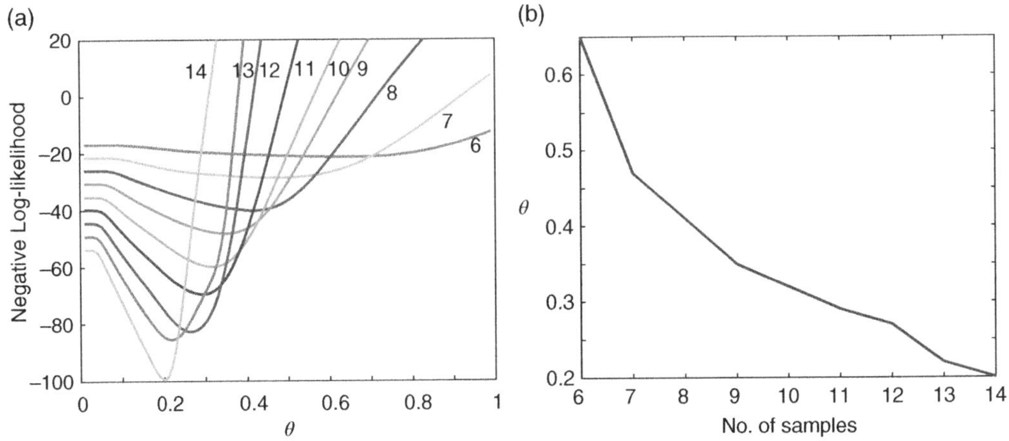


**Figure 7.14** Relationship between the hyperparameter and number of samples. (a) Negative log-likelihood versus hyperparameter and (b) optimal hyperparameter versus the number of samples.

have any correlation. On the other hand, when the number of samples is too large, the correlation matrix becomes singular before the negative log-likelihood function reaches its minimum. This is expected because it is difficult to interpolate too many samples. It would be technically difficult to determine an appropriate number of samples that are enough to follow the trend of the function while not causing singularity of the correlation matrix.

When the number of samples is too large, it is possible to use a reduced number of samples for fitting the Kriging surrogate. In fact, Matlab implementation of Gaussian process regression `fitrgp` has an option to decide how the correlation matrix is calculated. When `FitMethod="-none"`, no optimization is performed to find the hyperparameter. Instead, the initial value of hyperparameter is used for prediction. When `FitMethod="exact"`, all samples are used to calculate the correlation matrix. This is the same as what we discussed in this section. When `FitMethod="sd"`, the correlation matrix is generated using a subset of data. Since a subset of data is used, the Kriging surrogate cannot interpolate all data. When a Kriging surrogate does not interpolate all data, it is called Gaussian process regression (GPR), which will be discussed in the later section. When too many samples are available, it would be better to regress samples instead of interpolating all samples.

### 7.4.4 Computational Cost of Kriging Surrogate

Another important issue in determining the hyperparameter is related to the cost of fitting the Kriging surrogate. The expressions for determining $\hat{\beta}, \hat{\sigma}$, and $\hat{\theta}$ requires the inverse of the correlation matrix. Since the dimension of the correlation matrix is the same as the number of samples, the dimension of the correlation matrix increases proportional to the number of samples. This can cause a significant computational burden, especially for high-dimensional problems with a large number of samples. In particular, the optimization problem in determining the hyperparameter $\hat{\theta}$ is an iterative process where the inverse of the correlatin matrix is required at every iteration. After fitting the surrogate, the prediction stage in Eq. (7.28) also requires the inverse of the correlation matrix. However, since the correlation matrix is independent of the prediction point, it can be calculated once. On the other hand, the polynomial response surface requires solving a matrix equation once, which is in the dimension of the number of coefficients, not the number of samples. Therefore, the Kriging surrogate is considered one of the most expensive surrogate models. In Section 7.5, we will talk about numerical methods to relieve the computational difficulties associated with calculating the inverse of the correlation matrix.

As discussed before, maximizing the negative log-likelihood function is a challenging optimization problem because the correlation matrix in the likelihood function becomes singular or close to singular. Instead of using the maximum likelihood approach, cross-validation can also be used. The cross-validation method leaves out one of the points, fits the rest with the parameters, and calculates the error at the left-out point. This process is repeated for every point to produce a total error measure. Then, the set of parameters that yields the lowest cross-validation error is chosen. The optimization problem to solve for maximum likelihood or minimum cross-validation error in Kriging regression is often ill-conditioned, which can lead to either a poor fit or poor estimate of the prediction variance. Poor estimate of the prediction variance can be checked by comparing it to the cross-validation error. A poor fit can often be found in Kriging regression when the curvature changes significantly near sample points.

For example, Figure 7.15(a) shows a simple quadratic polynomial $y(x) = x^2 + 5x - 10$, where the curvature changes relatively quickly. Nine sample points are generated from the polynomial (only a

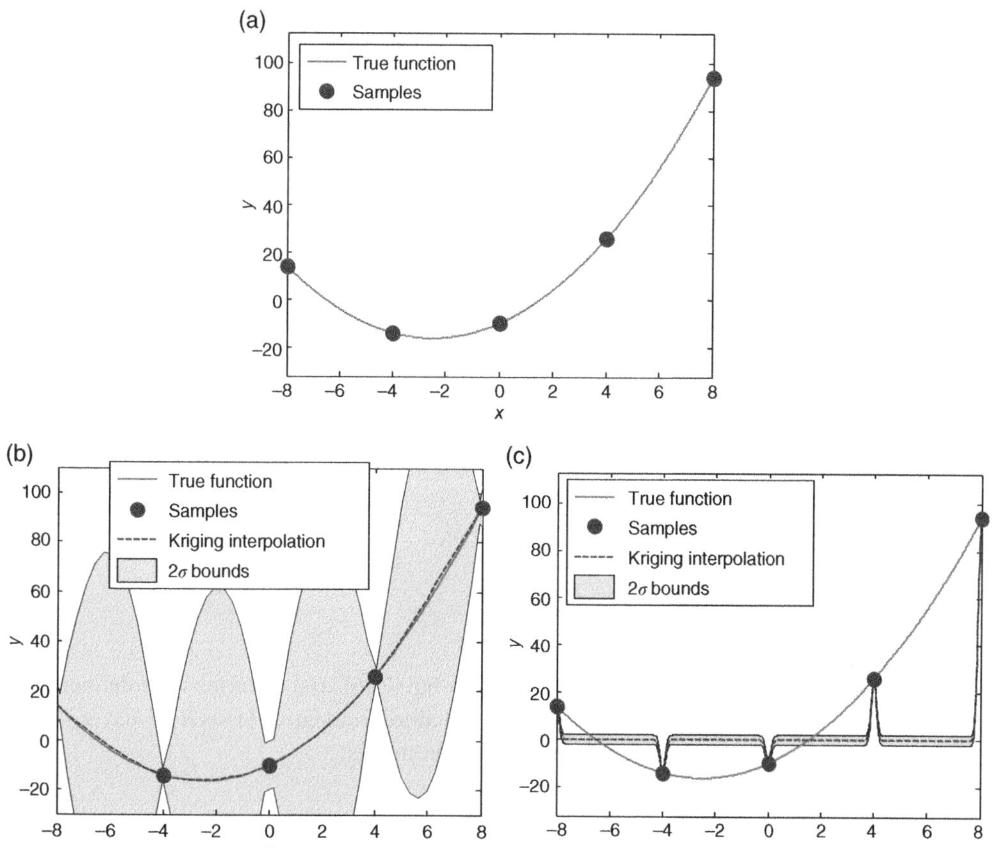


**Figure 7.15** Effect of hyperparameters in the Kriging surrogate. (a) True function and sample locations, (b) Kriging prediction with too-large hyperparameters, and (c) Kriging prediction with too-small hyperparameters.

portion of samples are shown in the figure), and the Kriging surrogate is used to fit the samples with a constant global function $\xi(x) = 0$. Figure 7.15(b) shows Kriging regression with an unreasonably large hyperparameter, $\theta$, while Figure 7.15(c) shows the case with a too-small hyperparameter. The grey area is the $\pm\sigma$ confidence intervals of the Kriging prediction. We will discuss the confidence intervals of Kriging approximation in the next section. Note that there is no uncertainty in the Kriging prediction at sample locations. When $\theta$ is too large, the fit is good and smooth, but the estimated prediction variance is too large. On the other hand, when $\theta$ is too small, the fit is poor, and the prediction intervals fail to cover the true function. Therefore, it is always recommended to visualize by plotting the Kriging prediction with its uncertainty.

**Example 7.9 Kriging Prediction and Hyperparameter**

Using the Kriging simulation, perform the following steps with the following five samples: $(x_i, y_i) = (0, 1), (5, 0.99), (10, 0.99), (15, 0.94), (20, 0.95)$.

a) Calculate the global function parameter for a constant form and the variance of samples with respect to the global function. Use the Gaussian correlation function given in Eq. (7.11) and assume $\theta = 5.2$.
b) Obtain an optimum value of the hyperparameter, $\theta$, graphically.
c) Calculate the Kriging simulation results at $x = 10, 14$ after obtaining the global and departure terms separately.

**Solution:**

This problem can be solved by using the given equations in this section. Before solving the problems, the samples and design matrix need to be defined as

$$\mathbf{x} = [0 \quad 5 \quad 10 \quad 15 \quad 20]^T$$

$$\mathbf{y} = [1 \quad 0.99 \quad 0.99 \quad 0.94 \quad 0.95]^T$$

$$\mathbf{X} = [1 \quad 1 \quad 1 \quad 1 \quad 1]^T$$

Note that the global function is defined as a constant in the problem, which means the global function outputs will be the same for any input and there will be one parameter. That is, $\boldsymbol{\xi}(x) = \{1\}$ and $\boldsymbol{\beta} = \{\beta\}$. In Matlab, these terms are defined as

```
x=[0 5 10 15 20]';                  % input variable
y=[1 0.99 0.99 0.94 0.95]';         % samples
X=ones(5,1);                        % design matrix
ny=length(y); np=size(X,2);
```

a) The equations for solving this problem are given in Eq. (7.16), and all terms were defined previously except for the correlation matrix, $\mathbf{R}$, that is obtained from Eq. (7.11) as $\mathbf{R} = [R(x_k, x_l)]$. The following Matlab commands calculate the correlation matrix:

```
th=5.2;
for k=1:ny; for l=1:ny;
 R(k,l)=exp(-(norm(x(k,:)-x(l,:))/th)^2);
end; end;
```

and the results become

$$\mathbf{R} = \begin{bmatrix} 1 & 0.3967 & 0.0248 & 0.0002 & 0 \\ 0.3967 & 1 & 0.3967 & 0.0248 & 0.0002 \\ 0.0248 & 0.3967 & 1 & 0.3967 & 0.0248 \\ 0.0002 & 0.0248 & 0.3967 & 1 & 0.3967 \\ 0 & 0.0002 & 0.0248 & 0.3967 & 1 \end{bmatrix}$$

The diagonal elements should be one because they represent the correlation between themselves, e.g., $\mathbf{R}(1, 1)$ and $\mathbf{R}(2, 2)$ are correlations between $x_1$ and $x_1$ and between $x_2$ and $x_2$, respectively. On the other hand, the off-diagonal elements are determined based on the distance between two different inputs, e.g., $\mathbf{R}(1, 2)$ and $\mathbf{R}(2, 1)$ are calculated from $x_1$ and $x_2$, which makes the correlation matrix symmetric.

The global function parameter and variance of samples with respect to the global function can be calculated by using Eq. (7.16) as

$$\hat{\beta} = \left(\mathbf{X}^T\mathbf{R}^{-1}\mathbf{X}\right)^{-1}\left\{\mathbf{X}^T\mathbf{R}^{-1}\mathbf{y}\right\} = 3.0989^{-1} \times 3.0226 = 0.9754$$

$$\hat{\sigma}^2 = \frac{\left(\mathbf{y}-\mathbf{X}\hat{\boldsymbol{\beta}}\right)^T\mathbf{R}^{-1}\left(\mathbf{y}-\mathbf{X}\hat{\boldsymbol{\beta}}\right)}{n_y - n_p} = 7.28 \times 10^{-4}, \quad \hat{\sigma} = 0.0270$$

```
Rinv=inv(R);
betaH=(X'*Rinv*X)\(X'* Rinv*y);
sigmaH=sqrt(1/(ny-np)*((y-X*betaH)'*Rinv*(y-X*betaH)));
```

b) During the process in part (a), the value of $\theta = 5.2$ is given. In fact, this is the optimum result, which can be obtained by minimizing Eq. (7.32). Note that the two variables, $\hat{\sigma}$ and $\mathbf{R}$ are dependent on $\theta$ as shown in Eqs. (7.11) and (7.16) and how to calculate these variables are shown in (a). Since this is a one-parameter optimization problem, it can be solved approximately by plotting the objective function (Eq. (7.32)) at different values of $\theta$. By calculating Eq. (7.32) for $\theta > 0$, Figure 7.16 can be obtained. In the figure, the dot marker represents the optimum point,

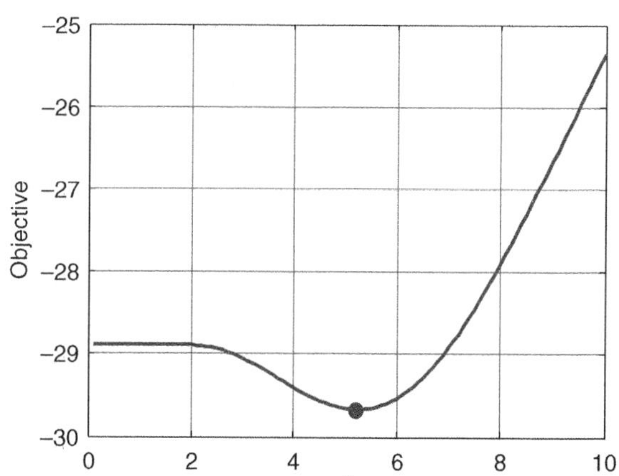


**Figure 7.16** Objective function (log-likelihood) with respect to $\theta$.

from which the optimum value of the hyperparameter is approximately obtained as $\theta_{\text{opt}} = 5.2$. The following Matlab code can be used to calculate Figure 7.16.

```
th=zeros(20,1); Obj=zeros(20,1);
for i=1:20
 th(i)=0.5*i;
 for k=1:ny; for l=1:ny;
  R(k,l)=exp(-(norm(x(k,:)-x(l,:))/th(i))^2);
 end; end;
 Rinv=inv(R);
 betaH=(X'*Rinv*X)\(X'* Rinv*y);
 sigmaH=sqrt(1/(ny-np)*((y-X*betaH)'*Rinv*(y-X*betaH)));
 Obj(i)=log(sigmaH^(2*(ny-np))*det(R));
end
plot(th,Obj,'linewidth',2); grid on;
```

c) Since the global function is a constant, the design vector $\boldsymbol{\xi} = \{1\}$ and there is a single function parameter. Therefore, the global term $\boldsymbol{\xi}^T\hat{\boldsymbol{\beta}}$ in Eq. (7.28) becomes a constant: $\xi\hat{\beta} = 1 \times 0.9754$. For the departure term, $\mathbf{r}^T\mathbf{R}^{-1}\left(\mathbf{y} - \mathbf{X}\hat{\boldsymbol{\beta}}\right)$ in Eq. (7.28), the correlation vector between $x = 10$, 14 and the sample points should be calculated as:

$$\mathbf{r} = \{R(x_k, x)\}, \quad k = 1, \ldots, n_y$$

and thus at $x = 10$;

$$\mathbf{r} = \{0.0248 \quad 0.3967 \quad 1 \quad 0.3967 \quad 0.0248\}^T$$

$$\mathbf{r}^T\mathbf{R}^{-1}\left(\mathbf{y} - \mathbf{X}\hat{\boldsymbol{\beta}}\right) = 0.0146$$

Note that since $x = 10$ is the same as one of the measurement points (the third point), the correlation vector, **r**, is identical to the third column of correlation matrix, **R**.

In the same manner, at $x = 14$;

$$\mathbf{r} = \{0.0007 \quad 0.05 \quad 0.5534 \quad 0.9637 \quad 0.2641\}^T$$

$$\mathbf{r}^T\mathbf{R}^{-1}\left(\mathbf{y} - \mathbf{X}\hat{\boldsymbol{\beta}}\right) = -0.0272$$

The correlation vectors can be calculated with the following Matlab code:

```
xNew=10; %or xNew=14
for k=1:ny; r(k,1)=exp(-(norm(x(k,:)-xNew)/th)^2); end;
gpDepar=r'*Rinv*(y-X*betaH);
```

Therefore, the Kriging simulation results in Eq. (7.28) become as follows:

$$\hat{y}(10) = 0.9754 + 0.0146 = 0.99$$

$$\hat{y}(14) = 0.9754 - 0.0272 = 0.9482$$

The results are shown as blue star markers in Figure 7.17(a), which are departures from the global function. By repeating the aforementioned process for $x \in [-5, 25]$, the red solid curve in Figure 7.17(a) is obtained. Note that the Kriging simulation results pass through the samples and make a smooth curve.

Figure 7.17(b) shows the Kriging simulation results when an inappropriate hyperparameter is used. Too-small value, $\theta = 0.5$ in this case, reduces the correlation between inputs. When the

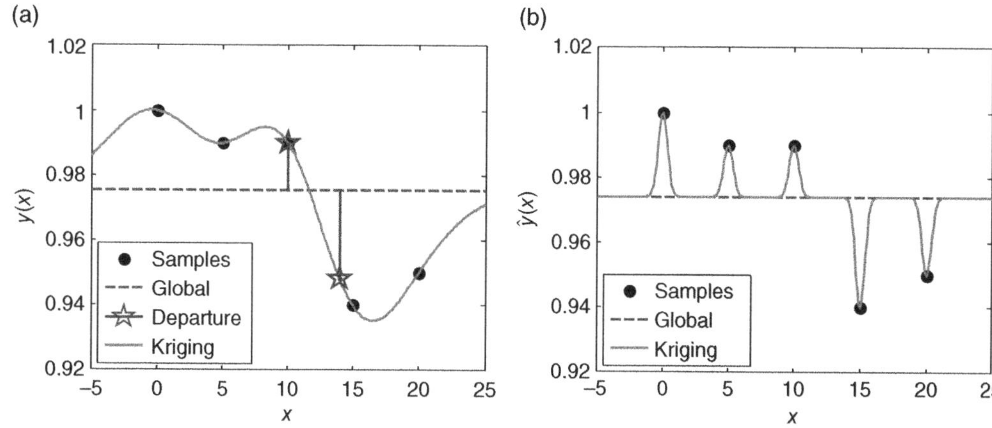


**Figure 7.17** Kriging predictions with five samples. (a) With $\theta = 5.2$ and (b) with $\theta = 0.5$.

correlation is reduced, Kriging regression approaches a linear regression and thus follows the global function. Both the ends and in-between sample points are overlapped with the global function. But still, the Kriging regression pass through the sample points since the distance between the same points is zero, which makes the correlation value one from Eq. (7.11). It is given as an exercise problem to consider the opposite case, i.e., when the hyperparameter is very large.

### 7.4.5 Hyperparameter and Extrapolation Accuracy

Note that since the exponential form of correlation decays away from the sample points when the prediction point is far from all sample points, Kriging predictions will converge to the global function $\boldsymbol{\xi}(\mathbf{x})^T\hat{\boldsymbol{\beta}}$. Therefore, Kriging is not good for extrapolation. As shown in Figure 7.12, the prediction accuracy of Kriging surrogate is less sensitive to the value of hyperparameter. However, the hyperparameter can play an important role in prediction accuracy in the extrapolation region. The following example shows the effect of the hyperparameter in the extrapolation region.

**Example 7.10 Kriging Prediction and Hyperparameter**

Sometimes the negative log-likelihood function may not have a minimum within the range of the hyperparameter. Generate equally spaced 20 samples from $y = \sin x + 0.2x$ in the range of $x \in [0, 2\pi]$ and fit the samples using the ordinary Kriging surrogate. Plot the negative log-likelihood function versus hyperparameter $\theta$. Also plot Kriging surrogate predictions at various values of $\theta$ in the range of $x \in [-\pi, 3\pi]$ and compare the prediction accuracy in the interpolation and extrapolation regions.

**Solution:**

The following Matlab script is used to plot the negative log-likelihood function versus hyperparameter $\theta$ and Kriging predictions with different hyperparameters. First, 20 equally spaced samples are generated in the range of $x \in [0, 2\pi]$, and true function $y = \sin x + 0.2x$ is used to generate output samples. For the hyperparameter, 10 equally spaced hyperparameters are generated between $\theta \in [0.1, 2.0]$. For predictions, equally spaced prediction points are defined in the range of $s \in [-\pi, 3\pi]$. The outer loop gradually increases the hyperparameter, which stops when the reciprocal condition number of correlation matrix becomes less than the floating-point relative accuracy `eps`.

```
clear; ny=20; nth=10;
x=linspace(0,2*pi,ny)'; thmin=0.1;
y=sin(x)+0.2*x;
th=linspace(thmin,2.0,nth)';
s=[-pi:0.05:3*pi]';
X=ones(ny,1);
for k=1:nth                              % Loop for increasing hyperparameter
 for i=1:ny, for j=1:ny                                  % Correlation matrix
  R(i,j)=exp(-((x(i)-x(j))/th(k)).^2);
 end, end
 if rcond(R) < eps, break; end          % Check for reciprocal cond. No.
 Rinv=inv(R);
 beta=inv(X'*Rinv*X)*(X'*Rinv*y);             % Global coefficient beta
 ym=y-beta;
 Like(k)=log((ym'*Rinv*ym/(ny-1))^(ny-1)*det(R));   % Neg. log-likelihood
 for i=1:ny
  r(:,i)=exp(-((s-x(i))/th(k)).^2);      % Correlation at prediction point
 end
 yhat(:,k)=beta+r*Rinv*ym;                           % Kriging predictions
 figure(1); plot(s,yhat); hold on;
 fprintf("k=%2d th=%5.3f beta=%5.3f Like=%8.3e\n",k,th(k),beta,Like(k))
end
plot(x,y,'*',s,sin(s)+0.2*s,'k-'); hold off;
figure(2); plot(th(1:size(Like,2)),Like','b*-');
```

The hyperparameter is gradually increased from $\theta = 0.1$ to 2.0. However, the increment stops at $\theta = 1.58$, where the reciprocal condition number of the correlation matrix is smaller than eps. Figure 7.18(a) plots the negative log-likelihood function versus hyperparameter $\theta$. The negative log-likelihood function continues to decrease until the correlation matrix becomes singular. Figure 7.18(b) plots Kriging predictions with different values of hyperparameter. The dashed curve

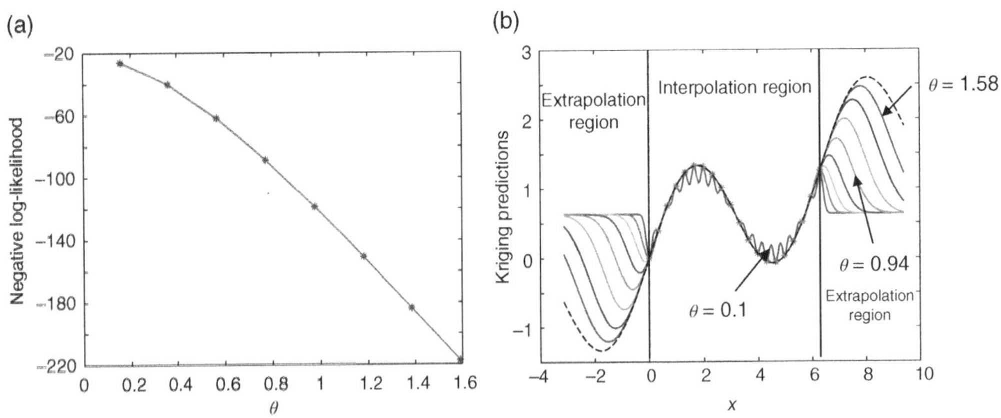


**Figure 7.18** Kriging predictions of $y = \sin x + 0.2x$. (a) Negative log-likelihood function and (b) Kriging predictions at interpolation and extrapolation regions.

is the true function. It turns out that when $\theta = 0.1$, the predictions oscillate between samples since the correlation decays quickly. However, when $\theta > 0.36$, all predictions are almost the same in the interpolation region. Therefore, even if the negative log-likelihood function is not minimum, Kriging predictions are still accurate. However, predictions are significantly different in the extrapolation region. When the value of hyperparameter is small, the prediction at the extrapolation region quickly converged to the global function. When the value of hyperparameter is large, the effect of correlation continues to extend the extrapolation region. The larger the hyperparameter is, the closer the predictions to the true function (dashed curve). Therefore, it is better to use the upper bound of the hyperparameter when the minimum of log-likelihood function cannot be found within the lower and upper bounds of the hyperparameter.

### 7.4.6 Uncertainty in Kriging Predictions

An important attribute for Kriging compared to other surrogate models is that it allows to estimate uncertainty in prediction points. This attribute is similar to the polynomial response surface (PRS) but in a different context. In PRS, the fitting process has uncertainty due to the lack of samples, which causes uncertainty in estimated coefficients, and thus, in prediction. This is from the basic assumption of PRS, where the model form is accurate, but the coefficients need to be estimated using a finite number of samples. Therefore, the uncertainty is distributed over the entire design space as shown in Figure 2.14 of Chapter 2. It is noted that some samples are out of the confidence intervals, which indicates that the noises of some samples are out of the intervals. As the number of samples increases, the estimated coefficients converge to the true coefficient, i.e., the prediction uncertainty is reduced.

In Kriging, on the other hand, the uncertainty comes from the uncertain model form, not from samples. Since a sample is a realization of the true model, there is no uncertainty at the sample location. However, uncertainty is present between samples, which is represented as a Gaussian distribution. In the viewpoint of uncertainty, the Kriging prediction in Eq. (7.28) is nothing but a mean prediction of a Gaussian process. The variance of the Kriging simulation is introduced in the form of MSE in Eq. (7.21). In order to understand the prediction variance, consider a Kriging prediction in Figure 7.19(a), where $x$ is an input variable, and $\hat{y}_{\text{KRG}}(x)$ is Kriging prediction in a red solid curve. In order to represent the prediction uncertainty, a Gaussian distribution is plotted with a dashed curve at $x = 0.8$. The 95% confidence intervals of this distribution are the upper and lower bounds of the

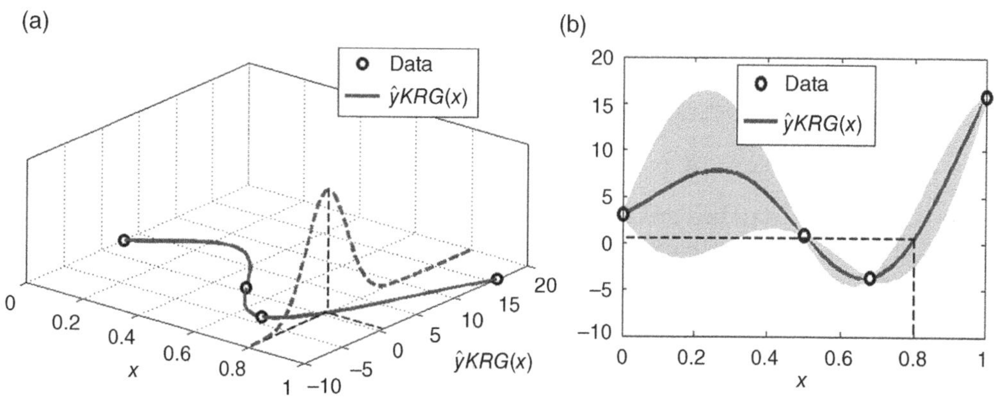


**Figure 7.19** Kriging surrogate and prediction variance. (a) Kriging predictions and (b) prediction confidence intervals.

shaded area in Figure 7.19(b) at $x = 0.8$. This prediction uncertainty is zero at sample locations, which is why the shaded area disappears at the sample locations. As the prediction point moves away from the sample locations, the prediction uncertainty increases. In the figure, the prediction uncertainty is large around $x = 0.2$. Since the uncertainty disappears at a sample location, the uncertainty can significantly be reduced if an additional sample is added at $x = 0.2$.

The uncertainty in the Kriging simulation follows a Gaussian distribution with mean at the deterministic model in Eq. (7.28) and the variance given in Eq. (7.21) as

$$Y(\mathbf{x}) \sim N\left(\boldsymbol{\xi}^T\hat{\boldsymbol{\beta}} + \mathbf{r}^T\mathbf{R}^{-1}\left(\mathbf{y} - \mathbf{X}\hat{\boldsymbol{\beta}}\right), \hat{\sigma}^2\left(\mathbf{w}^T\mathbf{R}\mathbf{w} - 2\mathbf{w}^T\mathbf{r} + 1\right)\right) \tag{7.36}$$

Note that the true variance of the Kriging simulation is unknown but estimated from the limited number of samples. When the variance is estimated from a finite number of samples, the distribution of mean follows Student's t-distribution. Therefore, Eq. (7.36) is modified with the Student's t-distribution with a degree of freedom, $n_y - n_p$ as

$$Y(\mathbf{x}) \sim \boldsymbol{\xi}^T\hat{\boldsymbol{\beta}} + \mathbf{r}^T\mathbf{R}^{-1}\left(\mathbf{y} - \mathbf{X}\hat{\boldsymbol{\beta}}\right) + t_{n_y - n_p}\hat{\sigma}\sqrt{\mathbf{w}^T\mathbf{R}\mathbf{w} - 2\mathbf{w}^T\mathbf{r} + 1} \tag{7.37}$$

where $t_{n_y - n_p}$ is the Student's t-distribution with $n_y - n_p$ degrees-of-freedom. $Y(\mathbf{x})$ represents the distribution of simulation outputs that is caused by the uncertainty in function parameters. Due to the assumption in Kriging, the above uncertainty does not include any noise in samples. Equation (7.37) can be used to calculate the confidence intervals of Kriging predictions. The confidence intervals, however, can be considered as prediction intervals at the same time because the simulation results pass through measurement points. Therefore, the prediction uncertainty for Kriging is performed based on Eq. (7.37), whose intervals are called prediction intervals.

Using the weight function given in Eq. (7.30), the prediction variance can be rewritten as

$$\hat{\sigma}_y^2(\mathbf{x}) = \hat{\sigma}^2\left[1 - \mathbf{r}(\mathbf{x})^T\mathbf{R}^{-1}\mathbf{r}(\mathbf{x}) + \left(\boldsymbol{\xi}(\mathbf{x}) - \mathbf{X}^T\mathbf{R}^{-1}\mathbf{r}(\mathbf{x})\right)^T\left(\mathbf{X}^T\mathbf{R}^{-1}\mathbf{X}\right)^{-1}\left(\boldsymbol{\xi}(\mathbf{x}) - \mathbf{X}^T\mathbf{R}^{-1}\mathbf{r}(\mathbf{x})\right)\right] \tag{7.38}$$

When the global function is constant (ordinary Kriging), the design matrix $\mathbf{X}$ becomes a column vector of ones, $\mathbf{X} \equiv \mathbf{1} = [1, 1, \ldots, 1]^T$. Also, the basis vector $\boldsymbol{\xi}(\mathbf{x}) = 1$. Therefore, the prediction variance can be simplified as

$$\hat{\sigma}_y^2(\mathbf{x}) = \hat{\sigma}^2\left[1 - \mathbf{r}(\mathbf{x})^T\mathbf{R}^{-1}\mathbf{r}(\mathbf{x}) + \frac{\left(1 - \mathbf{1}^T\mathbf{R}^{-1}\mathbf{r}(\mathbf{x})\right)^2}{\mathbf{1}^T\mathbf{R}^{-1}\mathbf{1}}\right] \tag{7.39}$$

It is interesting to note that the prediction variance is zero at sample points in Kriging. In order to show this, the prediction point is chosen at a sample point; $\mathbf{x} = \mathbf{x}_k$. From the expression of the Lagrange multiplier in Eq. (7.26), it can be shown that

$$\boldsymbol{\xi}(\mathbf{x}_k) - \mathbf{X}^T\mathbf{R}^{-1}\mathbf{r}(\mathbf{x}_k) = \mathbf{0} \tag{7.40}$$

This is because $\mathbf{R}^{-1}\mathbf{r}(\mathbf{x}_k)$ becomes a unit vector whose $k$th component is one and all other components are zero, and the $k$th column of design matrix $\mathbf{X}$ is nothing but $\boldsymbol{\xi}(\mathbf{x}_k)$. Therefore, the Lagrange multiplier vanishes at sample locations; $\boldsymbol{\lambda}(\mathbf{x}_k) = 0$. With zero Lagrange multiplier at the $k$th sample point, the interpolating weight $\mathbf{w}$ in Eq. (7.25) becomes a unit vector whose $k$th component is one and all other components are zero.[3] Finally, the MSE at the $k$th sample point becomes

3 This is the interpolation property of the weight function, shown in Example 7.5.

$$\hat{\sigma}_y^2(\mathbf{x}_k) = \hat{\sigma}^2\left(\mathbf{w}^T\mathbf{R}\mathbf{w} - 2\mathbf{w}^T\mathbf{r} + 1\right) = 0 \tag{7.41}$$

In the above equation, $\mathbf{w}^T\mathbf{R}\mathbf{w} = 1$ because the diagonal component of the correlation matrix is one, and $\mathbf{w}^T\mathbf{r} = 1$ because the $k$th component of correlation vector is also one. Therefore, at sample points, the prediction variance becomes zero and there is no uncertainty at sample points.

**Example 7.11 Uncertainty in Kriging Prediction**

Calculate 90% confidence intervals at $x = 10$, 14 for the problem given in Example 7.9.

**Solution:**

To calculate the prediction interval, Eq. (7.37) can be used, and the mean of the Kriging regression is obtained in Example 7.9. For the uncertainty part, the weight function $\mathbf{w}$ is given in Eq. (7.25), and then the standard deviation of the prediction uncertainty, $\hat{\sigma}_y = \hat{\sigma}\sqrt{\mathbf{w}^T\mathbf{R}\mathbf{w} - 2\mathbf{w}^T\mathbf{r} + 1}$ can be calculated as

```
xi=1;
w=Rinv*r+Rinv*X*((X'*Rinv*X)\(xi'-X'*Rinv*r));
ySigmaH=sigmaH*sqrt(w'*R*w-2*w'*r+1);
```

Since there is no uncertainty at measurement points, $\hat{\sigma}_y$ should be zero at $x = 10$, and the 5th and 95th percentiles are the same as the mean value, 0.99 (given in Example 7.9). On the other hand, $\hat{\sigma}_y$ is calculated as 0.0041 at $x = 14$, and the confidence intervals can be obtained in two ways based on the inverse CDF of Student's t-distribution or the random samples from the t-distribution. The following is Matlab script to calculate the confidence intervals.

```
% using the inverse calculation
gpMean=0.9482; % from Example 7-9
PI=[ gpMean + tinv(0.05,ny-np)*ySigmaH, ...
     gpMean + tinv(0.95,ny-np)*ySigmaH]
% using the random samples
ns=5e3;
tDist=trnd(ny-np,1,ns);
yHat=gpMean+tDist*ySigmaH;
PI=prctile(yHat,[5 95])
```

The results are listed in the following table, and the dashed curves in Figure 7.20 are obtained by repeating the process for $x \in [-5, 25]$. Note that the confidence intervals are zero at sample points, and they increase in the extrapolation region.

| $x_{new}$ | 5 percentiles | 95 percentiles | 90% P.I. |
|---|---|---|---|
| $x = 10$ | 0.99 | 0.99 | 0 |
| $x = 14$ | 0.9394 | 0.9570 | 0.0176 |

The uncertainty in Kriging predictions can be represented by prediction intervals. The meaning of 95% prediction intervals is that the probability of the true function being located within these intervals is 95%. This requirement of prediction intervals can mislead the purpose of uncertainty quantification because any poor predictions can include the true function by increasing the level

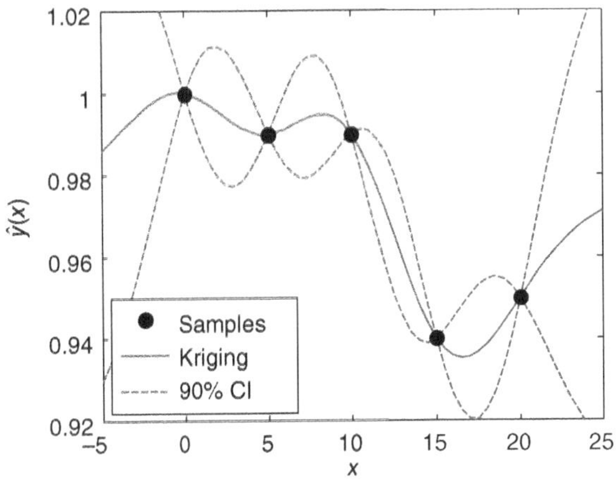


**Figure 7.20** Uncertainty in Kriging regression with five samples.

of uncertainty. Therefore, it is important to make the intervals as small as possible while they can include the true function with 95% probability.

In Example 7.10, it has been shown that Kriging predictions are insensitive to the values of hyperparameter in the interpolation region. However, uncertainty in predictions is sensitive to the hyperparameter. An extreme example is shown in Figure 7.15(b) and (c). In Figure 7.15(b), the value of hyperparameter is too high such that the 95% prediction intervals cover a wide range. In such a case, even if the prediction intervals cover the true function, it may not be useful because of a too high level of uncertainty. On the other hand, Figure 7.15(c) shows an opposite case where the level of uncertainty is too low. When the level of uncertainty is too low, prediction intervals cannot cover the true function.

**Example 7.12 Prediction Interval Versus Hyperparameter**

Consider true function $y = \sin \pi x + 0.2x$ in Example 7.10. Ten equally spaced samples are generated in the range of $x \in [0, 2\pi]$ using the true function. Plot Kriging predictions and 90% prediction intervals in the range of $s \in [-\pi, 3\pi]$ with the following values of hyperparameter: $\theta = 0.1, 0.5, 4.36$, and 5.5. Note that $\theta = 4.36$ is the optimal hyperparameter that minimizes the negative log-likelihood function. Discuss how prediction accuracy and uncertainty depend on the hyperparameter.

**Solution:**

The following Matlab script is used to calculate Kriging predictions and prediction intervals with different values of hyperparameter. The first part of the script is to set up variables and generate samples. Variable `x` and `y` store samples, while variable `s` stores prediction points. For ordinary Kriging, the design matrix `X` becomes a column vector of ones.

```
ny=10; nth=4;
x=linspace(0,2*pi,ny)'; s=[-pi:0.05:3*pi]';
y=sin(x)+0.2*x;
th=[0.1; 0.5; 4.36; 5.5];
X=ones(ny,1);
```

Once the initial setup is finished, the next part of the script is to loop for different values of hyperparameter. For a given hyperparameter, the correlation matrix and its inverse are calculated. The loop stops when the reciprocal condition number is less than `eps`.

```
for k=1:nth
 for i=1:ny, for j=1:ny
  R(i,j)=exp(-((x(i)-x(j))/th(k)).^2);
end, end
if rcond(R) < eps, break; end
Rinv=inv(R);
```

The coefficient of global function $\beta$ and standard deviation $\sigma$ of samples with respect to the global function in Eq. (7.16) are calculated. In addition, the correlations between prediction points and samples are stored in variable `r`. Note that the dimension of variable `r` is the number of prediction points × the number of samples. Then, Kriging predictions are calculated using Eq. (7.28). Also, the prediction variance can be calculated using Eq. (7.21).

```
beta=inv(X'*Rinv*X)*(X'*Rinv*y);
sigmaH=ym'*Rinv*ym/(ny-1);
ym=y-beta;
for i=1:ny
 r(:,i)=exp(-((s-x(i))/th(k)).^2);
end
yhat(:,k)=beta+r*Rinv*ym;
A=X'*Rinv*X; Ainv=inv(A);
 for i=1:size(s,1)
  w=r(i,:)*Rinv + (1-r(i,:)*Rinv*X)*Ainv*X'*Rinv;
  ySigmaH(i,k)=sigmaH*sqrt(w*R*w'-2*w*r(i,:)'+1);
end
```

We plot Kriging predictions and prediction intervals in a single figure for a given value of hyperparameter. In the following Matlab script, `yhat(:,k)` is Kriging predictions with the $k$th hyperparameter. In addition to Kriging predictions, samples and true function are also plotted in the same figure. In order to plot prediction intervals, Matlab function `patch` is used, which plots polygons. The first series of data are for the lower intervals, while the second series are for the upper intervals. In order to make a continuous polygon, the order of upper intervals is reversed using `flipud` function. The lower intervals are `tinv(0.05,ny-1)*ySigmaH(:,k)`, while the upper intervals are `tinv(0.95,ny-1)*ySigmaH(:,k))`. These are 5th and 95th percentiles of Student's t-distribution.

```
 figure; hold on;
 plot(s,yhat(:,k),x,y,'*',s,sin(s)+0.2*s,'r-','LineWidth',1);
 patch([s;flipud(s)],[yhat(:,k)+tinv(0.05,ny-1)*ySigmaH(:,k); ...
        flipud(yhat(:,k)+tinv(0.95,ny-1)*ySigmaH(:,k))], ...
        'k','FaceAlpha',0.1);
 fprintf("k=%2d th=%5.3f beta=%5.3f Like=%8.3e, detR=%8.3e\n",...
          k,th(k),beta,Like(k),det(R))
end
```

Figure 7.21 shows Kriging predictions (blue curve), 10 samples (asterisk markers), true function (red curve), and 90% prediction intervals (grey area) with different values of hyperparameter. When $\theta = 0.1$, Kriging predictions are unsmooth because correlations between samples disappear. Most

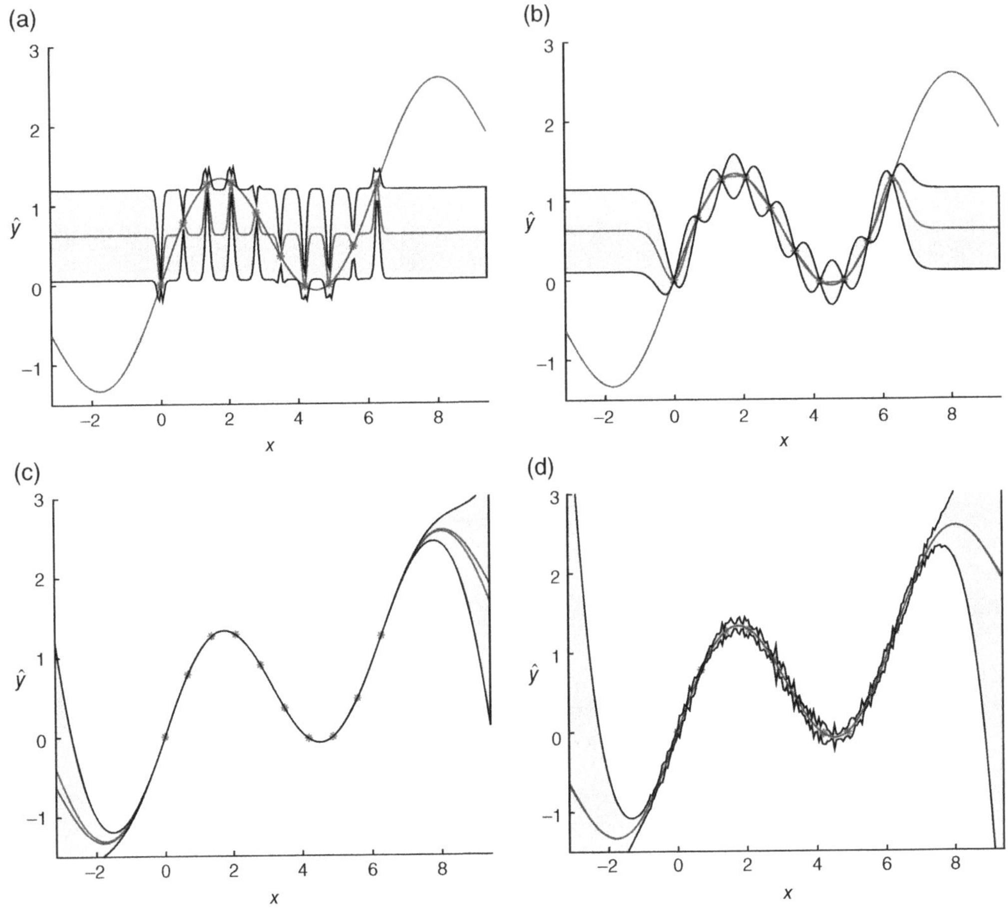


**Figure 7.21** Prediction intervals with different values of hyperparameter. (a) $\theta = 0.1$, (b) $\theta = 0.5$, (c) $\theta = 4.36$, and (d) $\theta = 5.5$.

predictions follow the global function, which is constant. When $\theta = 0.5$, Kriging predictions in the interpolation region are accurate but show large uncertainty between samples. In addition, the predictions quickly converge to the global function in the extrapolation region, and 90% prediction intervals failed to cover the true function. Therefore, the main issue for a nonoptimal hyperparameter is uncertainty at prediction points and inaccuracy in the extrapolation region. When the optimum hyperparameter is used as shown in Figure 7.21(c), not only predictions but also their uncertainties are significantly reduced. In addition, predictions are relatively accurate at the extrapolation region, albeit prediction intervals increase as prediction points move away from the last sample location. When the value of hyperparameter is too high as shown in Figure 7.21 (d), Kriging predictions are accurate, but choppy variation of prediction intervals appears in prediction points. This is partly because the reciprocal condition number of the correlation matrix becomes too small. In addition, prediction intervals in the extrapolation region significantly increase compared to those with the optimal hyperparameter. It is interesting to note that the mean predictions in the extrapolation are more accurate than those with the optimal hyperparameter. This is because the effect of samples reaches longer with a higher value of hyperparameter.

It can be concluded that the effect of hyperparameter is more on prediction uncertainty than mean prediction itself.

### 7.4.7 Effect of Global Function

So far, we have only focused on ordinary Kriging where the global function is a constant function. Since the local departure is the difference between the samples and the global function, the global function can affect the fitting process of Kriging and prediction accuracy. It is natural to ask what a good global function is for the given set of samples. That is, for a given set of samples, is there any way to determine the optimal global function?

As shown in Example 7.6, in the case of ordinary Kriging, the value of constant global function depends on samples and correlation between them. Different from the polynomial response surface, the constant global function is not the mean of samples. When the correlation distance is small, the constant global function approaches the mean of samples. But as the correlation becomes strong, the constant global function deviates from the mean of samples. If Eq. (7.16) and Eq. (2.14) are compared, the difference comes from $\mathbf{R}^{-1}$. If $\mathbf{R}^{-1}$ becomes an identity matrix, these two equations are identical. That is, when correlation between samples disappears, the global function of Kriging becomes identical to the mean prediction of polynomial response. When correlation between samples varies, the constant global function changes in order to match the distribution of samples with correlation. Therefore, the role of global function is not to follow the trend of samples but to make the distribution of samples consistent to the given correlation. In that sense, it is not a good idea to call the global function to be a trend function. In maximum likelihood estimate, the global function is determined so that the differences, $\mathbf{y} - \mathbf{X}\boldsymbol{\beta}$, become as close as a Gaussian distribution with correlation matrix $\mathbf{R}$, which is determined by the hyperparameter and distance between samples.

In the case of simple Kriging, the constant global function is often fixed to zero, which may not be consistent to the maximum likelihood estimate in Eq. (7.16), and yet, simple Kriging has the same level of accuracy with ordinary Kriging (see Exercise Problem 7.18). Therefore, we can expect that the coefficient $\boldsymbol{\beta}$ that satisfies the maximum likelihood may not be critical at least in Kriging prediction in the interpolation region. Therefore, it is possible to determine the coefficients not from Eq. (7.16), but using regular regression in Chapter 2.

In ordinary Kriging, the global function is a constant function. It would be interesting to see the effect of nonconstant global function on Kriging predictions and prediction uncertainty. In order to study the effect of nonconstant global function, eight equally spaced samples are generated from true function $y = \sin x + 0.2x$. The eight samples are fitted using (i) ordinary Kriging and (ii) universal Kriging with a linear global function. Figure 7.22(a) shows Kriging predictions (blue curve) with a constant global function (dashed line), i.e., ordinary Kriging. It turns out that the constant global function becomes $\beta = 0.6283$ and the optimal hyperparameter becomes $\theta_{\text{opt}} = 3.778$. Kriging predictions are accurate compared to the true function (red curve) in the interpolation region. Also, prediction uncertainty is ignorable. However, Kriging predictions converge to the global function in the extrapolation region because correlation decays quickly. In addition, the level of prediction uncertainty significantly increases in the extrapolation region. Figure 7.22(b) shows Kriging predictions with a linear global function. The global function turns out to be $\boldsymbol{\xi}^T\boldsymbol{\beta} = -0.3496 + 0.3113x$, and the optimal hyperparameter becomes $\theta_{\text{opt}} = 3.333$. (see Exercise Problem 7.19). The values of hyperparameter for the constant and linear global functions are close. Also, the prediction accuracy and uncertainty are similar in the interpolation region. However, the major difference occurs in the

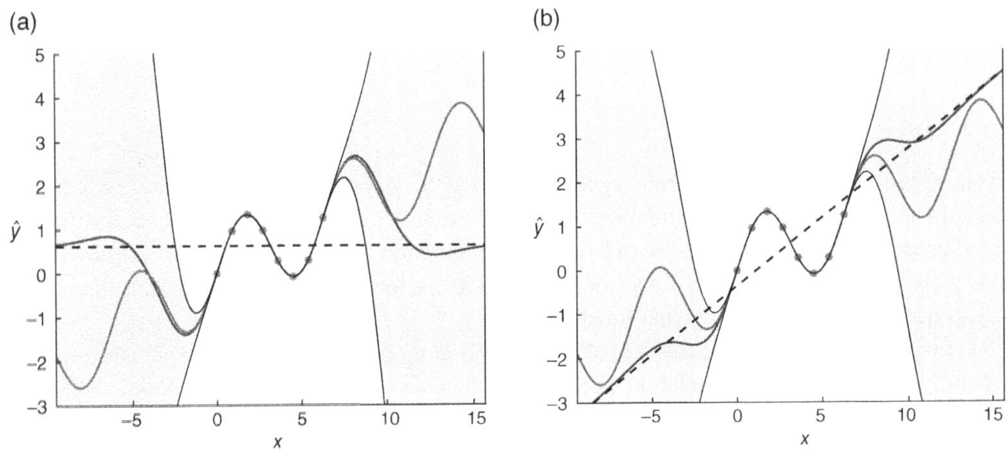


**Figure 7.22** Kriging predictions with different global functions. (a) Ordinary Kriging and (b) linear global function.

extrapolation region, where Kriging predictions converge to the global function. Therefore, the effect of global function is significant in the extrapolation region.

## 7.5 Numerical Implementation of Kriging Surrogate

It is possible to implement the Kriging surrogate as shown in the examples in the previous section, but it would be computationally inefficient. For example, the computational cost of calculating the inverse of correlation matrix is proportional to $n_y^3$, with $n_y$ being the number of samples. On top of that, the Kriging formulation requires a double inverse of matrices, such as in the term $(\mathbf{X}^T\mathbf{R}^{-1}\mathbf{X})^{-1}$. Therefore, the Matlab scripts that were shown in the examples in the previous section work well when the number of samples is relatively small, in the range of $n_y < 20$. However, the number of samples may increase drastically as the dimensions of input variables increase. In addition, as the number of dimensions and the number of samples increase, the correlation matrix approaches a singular matrix. Therefore, a robust and efficient numerical method is required for large-scale Kriging surrogate modeling. The numerical method used in this section is originally from Design and Analysis of Computer Experiments (DACE) Matlab toolbox [86]. It is noted that Kriging is the most expensive surrogate model, albeit it is inexpensive compared to expensive computer simulations. Therefore, we will use various numerical methods to make the surrogate calculation robust and efficient.

So far, we explained the Kriging surrogate using one-dimensional examples. Most engineering designs have many design variables, and engineers want to approximate the functional relationship between quantity-of-interest (QoI) and design variables using surrogate models. Therefore, the real benefit of a surrogate model is its approximation capability of high-dimensional functions. Most characteristics of a one-dimensional Kriging surrogate can be applicable to a multidimensional Kriging surrogate. The main differences of multidimensional approximation from the one-dimensional one is the format of sample pairs, the choice of the global function, and the form of the correlation function.

Before we develop a multidimensional Kriging surrogate, it would be necessary to discuss samples first. Let the dimension of input variables be $n$. Then the vector of input variables can be defined as $\mathbf{x} = \{x_1, x_2, ..., x_n\}^T$. Also, sample pairs are defined as $(\mathbf{x}_i, y_i)$, $i = 1, ..., n_y$. The process of obtaining the pairs of samples can be described as follows.

a) The locations of input samples are determined: $\mathbf{x}_i$, $i = 1, ..., n_y$. Various methods of design of experiments in Chapter 3 can be used for this purpose. When computer simulations are used to generate samples in high dimensions, the Lartin hypercube sampling method in Section 3.4 is often used because of its space-filling characteristics.
b) It is recommended that the ranges of input variables are normalized using their mean and standard deviation. Let $\boldsymbol{\mu}_x = \{\mu_{x1}, \mu_{x2}, ..., \mu_{xn}\}^T$ be the mean of input samples, and $\boldsymbol{\sigma}_x = \{\sigma_{x1}, \sigma_{x2}, ..., \sigma_{xn}\}^T$ be the standard deviation of input samples. Then, input samples are normalized by

$$x'_{i,j} = \frac{x_{i,j} - \mu_{xi}}{\sigma_{xi}} \tag{7.42}$$

where $x_{i,\,j}$ is the $j$th sample of the $i$th input variable.

For given input samples $\mathbf{x}_i$, the samples of output QoI $y_i$ are calculated. This process can use analytical functions, numerical simulations, or experiments. The output QoI samples are also normalized using their mean $\mu_y$ and standard deviation $\sigma_y$ as

$$y'_i = \frac{y_i - \mu_y}{\sigma_y} \tag{7.43}$$

Normalization can stabilize the optimization process of finding hyperparameters because distances can be measured in terms of standard deviation.

Once all samples are defined, the next step is to determine the correlation function. As discussed in Section 7.2, it is possible to use an isotropic correlation as shown in Figure 7.4(a). However, it is possible that the function may have different behaviors in different input variables. In the case of two-dimensional function $y(x_1, x_2) = x_2 \sin x_1$, for example, the function varies linearly along $x_2$ direction, while it is a periodic function in the $x_1$ direction. Therefore, it makes more sense to use an anisotropic correlation function in Figure 7.4(b). Therefore, there is the same number of hyperparameters with the number of input variables, defined as $\boldsymbol{\theta} = \{\theta_1, \theta_2, ..., \theta_n\}^T$. Finally the correlation function is defined as in Eq. (7.10). For example, in the case of two-dimensional input variables, the correlation between point $\mathbf{x}$ and point $\mathbf{s}$ can be defined as

$$\phi(\boldsymbol{\theta}, \mathbf{x}, \mathbf{s}) = \exp\left[-\left(\frac{x_1 - s_1}{\theta_1}\right)^2\right] \exp\left[-\left(\frac{x_2 - s_2}{\theta_2}\right)^2\right] = \exp\sum_{i=1}^{2}\left[-\left(\frac{x_i - s_i}{\theta_i}\right)^2\right] \tag{7.44}$$

It is noted that the distance between two points is measured in each variable's direction. Compared to one-dimensional correlation, the two-dimensional correlation decays fast as it accumulates decay in each variable. For example, when the hyperparameter is the same as the distance between two points, one-dimensional correlation is $e^{-1} = 0.368$, while in the case of two dimensions, it becomes $e^{-2} = 0.135$. Therefore, it is expected that the values of hyperparameters in multi-dimensions are higher than that of one dimension.

The hyperparameters in correlation need to be determined using maximum likelihood estimate. First, the polynomial form of trend function is defined. Although it is possible to use higher-order polynomials, it is normally satisfactory to use constant, linear, or quadratic polynomials because higher-order polynomials come with many unknown parameters. An important aspect is that it is not desirable to introduce too many parameters to be determined. Instead of maximizing the likelihood function in Eq. (7.13), a reduced form of negative log-likelihood in Eq. (7.32) is minimized.

Various optimization algorithms can be used for this purpose. Since the likelihood function can show complex behaviors with numerous local minima, gradient-free globally searching algorithms can be more attractive. As we discussed in Chapter 6, Nelder-mead simplex algorithm or genetic algorithms are good candidates. Once the optimum hyperparameters are found, the next step is to calculate surrogate predictions and prediction variances at prediction points.

In this section, a 100-line Matlab code for fitting a Kriging surrogate model is introduced. The code includes three Matlab functions: `Kriging`, `fitKRG`, and `trend`. All these three functions can be combined in a single file `Kriging.m`. The three files can also be saved in separate files: `Kriging.m`, `fitKRG.m`, and `trend.m`.

```
function [yp, ysig, nb]=Kriging(xs,ys,xp,trPRS,noise)
% INPUTS  xs: ny x ndim sample points
%         ys: ny x 1 samples
%         xp: np x ndim prediction points
%         trPRS: polynomial order of trend function (0, 1, 2)
%         noise: 1 for noisy samples, 0 for no noise
% OUTPUTS yp: np x 1 mean predictions
%         ysig: np x 1 prediction standard deviations
%         nb: # of coefficients in trend function
%
 [ny,ndim]=size(xs);nvar=ndim+1;np=size(xp,1);% Size of samples&predictions
 meanx=mean(xs);stdx=std(xs);xs=(xs-meanx)./stdx;% Scaling sample locations
 meany=mean(ys); stdy=std(ys);ys=(ys-meany)/stdy;          % Scaling samples
 xp=(xp-meanx)./stdx;                          % Scaling prediction points
 %
 nij=ny*(ny-1)/2; l=0;              % # of upper-triangular component of R
 diff=zeros(nij,ndim);                         % Distance between samples
 ij=zeros(nij,2);                            % Index of correlation matrix
 for i=1:ny-1
   l=l(end)+(1:ny-i);
   ij(l,:)=[repmat(i,ny-i,1),(i+1:ny)'];
   diff(l,:)=repmat(xs(i,:),ny-i,1)-xs(i+1:ny,:);
 end
 %
 % Estimating hyper-parameters using max. likelihood estimate
 par=struct('xs',xs,'ys',ys,'d',diff,'ij',ij,'trPRS',min(trPRS,2));
 hyp0 = [ny^(-1/4)*ones(ndim,1); min(0.1,noise)];
 lb=[0.01*ones(ndim,1);0];ub=[6*ones(ndim,1);noise];% Lower- & upper-bounds
 options=optimset('Display','off','TolFun',1e-6,'DiffMinChange',0.001);
 hypOpt=fmincon(@(hyp) fitKRG(hyp,par),hyp0,[],[],[],[],lb,ub,[],options);
 %hypOpt = ga(@(hyp) fitKRG(hyp,par),nvar,[],[],[],[],lb,ub,[],options)';
 %hypOpt = simulannealbnd(@(hyp) fitKRG(hyp,par),hyp0,lb,ub,options);
 hypmle=hypOpt(1:size(xs,2)); t=max(0,hypOpt(end));
 fprintf('Hyperparam:%8.3g\n',hypmle.*stdx');
 if noise ~= 0, fprintf('Noise frac:%8.3g\n',tau); end
 %
 [~,fit]=fitKRG(hypOpt,par);                      % Fitting Kriging surrogate
 %
 xi = trend(xp,trPRS); nb=size(xi,2);     % Trend basis vector at np points
 xrep=repmat(xp',ny,1); xrep=transpose(reshape(xrep,ndim,np*ny));
 d=xrep-repmat(xs,np,1);                 % Distance b/w predictions & samples
```

```
  r = (1-t)*exp(sum(-(d./hypmle').^2,2)); % Correlation b/w preds & samples
  r = reshape(r,ny,np);
  yp=xi*fit.beta + (fit.alpha*r)';                      % Kriging predictions
  yp=yp*stdy+meany;                                   % Unscaling predictions
  %
  rt=fit.C\r;
  q=fit.G\(xi'-fit.Ft'*rt);
  sigma2=fit.sigma2.*stdy.^2;                              % Process variance
  ysig=sqrt(sigma2.*(1-sum(rt.^2,1)+sum(q.^2,1))');        % Prediction STD
  fprintf('Total std      :%8.3g\n',sqrt(sigma2)*stdy);
  if noise~=0, fprintf('Noise std :%8.3g\n',sqrt(tau*sigma2)*stdy); end
 end
 %
 function [negLogLK,fit]=fitKRG(xvar,par)
 % Negative log-likelihood estimate for given hyper-parameter
 %
 % INPUTS  hyp          : Hyper-parameter
 %         par          : Model parameters structure
 % OUTPUTS negLogLK     : Negative log-likelihood function
 %         fit          : Necessary parameters for function evaluation
 %
  hyp=xvar(1:size(par.xs,2)); t=xvar(end);
  H = trend(par.xs,par.trPRS);                              % Design matrix
  ny=size(par.ys,1); nb=size(H,2);     % ny: # of samples, nb: # of coeff.
  r = (1-t).*exp(sum(-(par.d./hyp(:)').^2, 2));  % Correlation b/w samples
  idx=find(r>0);
  ii=(1:ny)';
  mu=(10+ny)*eps;                    % Small diagonal to prevent singularity
  R=sparse([par.ij(idx,1);ii],[par.ij(idx,2);ii],[r(idx);ones(ny,1)+mu]);
  [C,~]=chol(R);                       % Cholesky decomposition R = C'*C
  C=C';
  Ft=C\H;
  Yt=C\par.ys;
  [Q,G]=qr(Ft,0);
  b=G\(Q'*Yt);                           % Coefficients of trend function
  rho=(Yt-Ft*b);                                      % rho = Cinv*(y-Xb)
  alpha=rho'/C;                                   % alpha = (y-Xb)'*Rinv
  detR = sum(log(full(diag(C)).^2));  % Determinant of correlation matrix
  sigma2=sum(rho.^2)/(ny-nb);                           % Process variance
  negLogLK=detR + (ny-nb)*log(sigma2);            % Negative log-likelihood
  fit = struct('sigma2',sigma2,'beta',b,'alpha',alpha,'C',C,'Ft',Ft,'G',G);
end
 %
 function H=trend(x,order)
 % Basis vector of trend function
 %
  [m,n]=size(x);
  if order==0,     H=ones(m,1);           % Constant trend function basis
  elseif order==1, H=[ones(m,1), x];        % Linear trend fucntion basis
  elseif order==2                         % Quadratic trend function basis
  nn=(n+1)*(n+2)/2;
```

```
93        H=[ones(m,1) x zeros(m,nn-n-1)];
94        j=n+1; q=n;
95        for k=1:n
96         H(:,j+(1:q))=repmat(x(:,k),1,q).*x(:,k:n);
97         j=j+q; q=q-1;
98       end
99     end
100 end
```

Function `Kriging`: The Matlab function takes sample pairs `(xs,ys)`, prediction points `xp`, the order to trend function `trPRS`, and the indicator of noise in data `noise` as inputs. The function produces predictions `yp` corresponding points `xp`, prediction standard deviation `ysig`, and number of coefficients `nb` of trend function as output. The matrix `xs` of sample locations is `ny×ndim` matrix, with `ny` being the number of samples and `ndim` being the dimensions of the input variables. The vector `ys` of samples is `ny×1` vector with sample values. The matrix `xp` of prediction points is `np×ndim` matrix, with `np` being the number of prediction points. In order to build a Kriging surrogate, the user also needs to specify the order of trend function. The current implementation allows up to quadratic polynomials. Therefore, `trPRS` should be 0, 1, or 2. When samples include noise, `noise = 1`, otherwise use `noise = 0`. The Kriging function returns with the vector `yp` of predictions at np points and the vector `ysig` of prediction standard deviations at `np` points. The function also returns the number of coefficients `nb` in the trend function. In the case of trPRS = 0, nb = 1, when trPRS = 1, nb = 1 + ndim, and when trPRS = 2, nb = (ndim + 1)(ndim + 2)/2.

For the purpose of a stable and robust process, it would be good practice to normalize all samples. Lines 11~14 normalize sample locations and predictions using the mean and standard deviation of the samples. It is important to note that it is necessary to denormalize the predictions and prediction variances before they are returned (Lines 45 and 49). In Kriging, correlation between samples is determined using distances. Even if correlation depends on hyperparameters, the distances remain the same. Therefore, it would be efficient to store distances between samples, which are calculated in Lines 16~23. Since distances between two samples are symmetric, only the upper triangular portion is stored in the sparse matrix format. The array `ij` stores the index, while the array `diff` stores distances. The distances are measured in different dimensions.

Hyperparameters are determined in Lines 26~35. First, all necessary variables are stored in the parameter structure `par` so that they can be transferred to the `fitKRG` function (Line 26). In current implementation, hyperparameters are $\boldsymbol{\theta} = \{\theta_1, \theta_2, \ldots, \theta_{\text{ndim}}, \tau\}^T$, where $\tau$ is the fraction of noise variance. Fitting a Kriging surrogate with noise in samples will be discussed in Section 7.6. For the moment, it is enough to set `noise = 0`, such that $\tau = 0$ throughout optimization. The initial values of hyperparameters `hyp0` are set to be $n_y^{-1/4}$ in Line 27. For noise fraction, the initial value is 0.1 when samples include noise. Otherwise, the initial value is zero. The lower and upper bounds of hyperparameters are established in Line 28. An optimum hyperparameters were sought in the range of [0.01, 6]. It means hyperparameters can increase up to six times the standard deviation of input variables. On the other hand, the noise fraction is sought in the range of [0,1] when samples include noise, or [0,0] when samples do not include noise. The `fmincon` function is used to minimize the negative log-likelihood calculated in the `fitKRG` function by changing hyperparameter `hyp` (Line 30). The `fmincon` function uses a gradient-based constrained optimization algorithm. The optimum values of hyperparameters are stored in `hypmle` for hyperparameter and `tau` for noise fraction (Line 33). When printing the optimum hyperparameters, they need to be denormalized to the original variables (Line 34).

Matlab provides different optimization algorithms. Although the `fmincon` function is robust, different algorithms can be used. For example, a genetic algorithm can be used by uncommenting Line 31, or a simulated annealing algorithm can be used by uncommenting Line 32.

Once the hyperparameters are optimized, the `fitKRG` function is used again to calculate the Kriging structure `fit` for predictions and prediction variance (Line 37). It is noted that during optimization, the negative log-likelihood from `fitKRG` is used, while for the purpose of Kriging predictions, the structure `fit` is required from `fitKRG`. Prediction points are stored in the `xp` matrix, whose dimension is $n_p \times \text{ndim}$. Then, the `trend` function is used to calculate the basis functions at prediction points (Line 39). Then, the next task is to calculate correlation between $n_p$ prediction points and $n_y$ sample points. In order to do this, the `xp` matrix is copied `ny` times, and the copied matrix is rearranged to $\text{ndim} \times n_p n_y$ dimensions (Line 40). Then the distances between $n_p$ prediction points and $n_y$ sample points are stored in the matrix `d` (Line 41), which is used to calculate correlation `r` (Line 42). The correlation is reshaped to a $n_y \times n_p$ matrix in Line 43. The surrogate predictions are calculated using (Line 44)

$$\hat{y}(\mathbf{x}_i) = \boldsymbol{\xi}(\mathbf{x}_i)^T\hat{\boldsymbol{\beta}} + \mathbf{r}(\mathbf{x}_i)^T\boldsymbol{\alpha}, \quad i = 1, \dots, n_p \tag{7.45}$$

Since these predictions are normalized, they need to be denormalized in Line 45.

The prediction variance can be calculated using Eq. (7.38). In order to calculate the $\mathbf{r}(\mathbf{x})^T\mathbf{R}^{-1}\mathbf{r}(\mathbf{x})$ term, the modified correlation vector is defined $\mathbf{r}_t(\mathbf{x}) = \mathbf{C}^{-1}\mathbf{r}(\mathbf{x})$ in Line 47. Then, $\mathbf{r}(\mathbf{x})^T\mathbf{R}^{-1}\mathbf{r}(\mathbf{x}) = \mathbf{r}_t(\mathbf{x})^T\mathbf{r}_t(\mathbf{x})$. The last term in Eq. (7.38) is computationally expensive due to inverse of matrices. In order to make the computation efficient, we use the variables defined in the `fitKRG` function: $\mathbf{C}$, $\mathbf{G}$, and $\mathbf{F}_t$ (please refer to function `fitKRG` for these variables). Then, the term related to the inverse can be converted into $\mathbf{X}^T\mathbf{R}^{-1}\mathbf{X} = \mathbf{F}_t^T\mathbf{F}_t = \mathbf{G}^T\mathbf{Q}^T\mathbf{Q}\mathbf{G} = \mathbf{G}^T\mathbf{G}$, where $\mathbf{Q}$ is an orthonormal matrix: $\mathbf{Q}^T\mathbf{Q} = \mathbf{I}$. Then a temporary variable is defined as $\mathbf{q}(\mathbf{x}) = \mathbf{G}^{-1}(\boldsymbol{\xi}(\mathbf{x}) - \mathbf{X}^T\mathbf{R}^{-1}\mathbf{r}(\mathbf{x}))$ (Line 48). Therefore, the last term in Eq. (7.38) becomes $\mathbf{q}(\mathbf{x})^T\mathbf{q}(\mathbf{x})$. The process variance $\hat{\sigma}^2$ is denormalized in Line 49, and the prediction standard deviation is the square root of prediction variance in Line 50. The process standard deviation $\hat{\sigma}$ is printed in Line 51, and the standard deviation of estimate noise $\hat{\sigma}_n$ is printed in Line 52. When there is no noise in samples, $\hat{\sigma}_n = 0$.

Function `fitKRG`: Function `fitKRG` calculates negative log-likelihood for the given values of hyperparameters. The main purpose of this function is to calculate the likelihood function for given values of hyperparameters, such that the hyperparameters that can maximize the likelihood can be estimated, i.e., maximum likelihood estimate. This function also calculates other variables that are necessary to calculate Kriging prediction and prediction variance. Function `fitKRG` takes the vector `xvar` of optimization variables and the structure `par` of the model database. Then, function `fitKRG` returns with the negative log-likelihood `negLogLK` and Kriging structure `fit` of surrogate parameters. The negative log-likelihood is used during optimization to find its minimum value, while the structure `fit` is used for evaluating surrogate predictions at prediction points.

From the input variable `xvar`, the first ndim components are hyperparameters $(\theta_1, \theta_2, \dots, \theta_{\text{ndim}})$, and the last component is noise fraction $\tau$ (Line 63). In Line 64, input sample pairs (`xs`, `ys`) are used to build the design matrix `H` by calling the `trend` function. Then, the correlation matrix `R` is constructed in Lines 66~70. It is noted that a small positive value `mu` is added in the diagonal terms in order to make the correlation matrix positive definite (Line 69). Also, the sparse matrix is used to store nonzero terms of the correlation matrix (Line 70).

It is noted that Kriging predictions $\hat{y}(\mathbf{x})$ in Eq. (7.28), the coefficient $\hat{\boldsymbol{\beta}}$ of trend function in Eq. (7.16), and the process variance $\hat{\sigma}^2$ in Eq. (7.16) require the inverse of correlation matrix $\mathbf{R}^{-1}$. However, calculating the inverse of correlation matrix is computationally expensive as the number of samples increases. Therefore, instead of calculating and storing the inverse of correlation matrix, Cholesky factorization is performed in Line 71, where matrix $\mathbf{C}$ is a lower triangular matrix that satisfies the relationship of $\mathbf{R} = \mathbf{C}^T\mathbf{C}$. But, for the convenience of coding, the transpose of $\mathbf{C}$ is stored, such that $\mathbf{R} = \mathbf{C}\mathbf{C}^T$. Therefore, the inverse of the correlation matrix can be decomposed as $\mathbf{R}^{-1} = \mathbf{C}^{-T}\mathbf{C}^{-1}$. Then, the regression equation for the coefficients $\hat{\boldsymbol{\beta}}$ can be rewritten as

$$\begin{aligned}&(\mathbf{X}^T\mathbf{R}^{-1}\mathbf{X})\hat{\boldsymbol{\beta}} = \mathbf{X}^T\mathbf{R}^{-1}\mathbf{y}\\&(\mathbf{X}^T\mathbf{C}^{-T}\mathbf{C}^{-1}\mathbf{X})\hat{\boldsymbol{\beta}} = \mathbf{X}^T\mathbf{C}^{-T}\mathbf{C}^{-1}\mathbf{y}\end{aligned} \tag{7.46}$$

Then, by introducing $\mathbf{F}_t = \mathbf{C}^{-1}\mathbf{X}$ (Line 73) and $\mathbf{Y}_t = \mathbf{C}^{-1}\mathbf{y}$ (Line 74), the above equation can be written as

$$(\mathbf{F}_t^T\mathbf{F}_t)\hat{\boldsymbol{\beta}} = \mathbf{F}_t^T\mathbf{Y}_t \tag{7.47}$$

Then, this equation is nothing but the regression equation assuming $\mathbf{F}_t$ is a modified design matrix. The same regression equation can be solved using qr-decomposition, where the regression equation can be written as

$$\mathbf{F}_t\hat{\boldsymbol{\beta}} = \mathbf{Y}_t \tag{7.48}$$

In the qr-decomposition, the $n_y \times n_b$ matrix $\mathbf{F}_t$ can be decomposed into (Line 75)

$$[\mathbf{Q}, \mathbf{G}] = \mathrm{qr}(\mathbf{F}_t) \tag{7.49}$$

where $\mathbf{Q}$ is a $n_y \times n_b$ orthonormal matrix and $\mathbf{G}$ is a $n_b \times n_b$ upper triangular matrix. In regression, normally $n_y > n_b$. The matrices $\mathbf{Q}$ and $\mathbf{G}$ are related to the modified design matrix such that $\mathbf{F}_t = \mathbf{Q} \cdot \mathbf{G}$. Then the regression equation in Eq. (7.48) can be used to solve for the unknown coefficients as (Line 76)

$$\hat{\boldsymbol{\beta}} = \mathbf{G}^{-1}\mathbf{Q}^T\mathbf{Y}_t \tag{7.50}$$

Since $\mathbf{G}$ is an upper triangular matrix, the coefficients can be quickly calculated using back substitution.

In order to calculate the process variance $\hat{\sigma}^2$, a temporary variable $\boldsymbol{\rho} = \left(\mathbf{Y}_t - \mathbf{F}_t\hat{\boldsymbol{\beta}}\right) = \mathbf{C}^{-1}\left(\mathbf{y} - \mathbf{X}\hat{\boldsymbol{\beta}}\right)$ is defined first (Line 77). Then the vector $\boldsymbol{\alpha}$ in Eq. (7.29) can be calculated as $\boldsymbol{\alpha} = \boldsymbol{\rho}^T\mathbf{C}^{-1} = \left(\mathbf{y} - \mathbf{X}\hat{\boldsymbol{\beta}}\right)^T\mathbf{R}^{-1}$ (Line 78). Then, the process variance can be defined as (Line 80)

$$\hat{\sigma}^2 = \frac{\boldsymbol{\rho}^T\boldsymbol{\rho}}{n_y - n_b} = \frac{\left(\mathbf{y} - \mathbf{X}\hat{\boldsymbol{\beta}}\right)^T\mathbf{R}^{-1}\left(\mathbf{y} - \mathbf{X}\hat{\boldsymbol{\beta}}\right)}{n_y - n_b} \tag{7.51}$$

The determinant of correlation matrix can be calculated using the product of the square of diagonal components of matrix $\mathbf{C}$. In order to prevent numerical overflow, a logarithmic is calculated instead as (Line 79)

$$\log|\mathbf{R}| = \log\prod_{i=1}^{n_y} C_{ii}^2 = \sum_{i=1}^{n_y} \log C_{ii}^2 \tag{7.52}$$

This is another benefit of using Cholesky factorization. Then, the negative log-likelihood in Eq. (7.32) can be calculated as (Line 81)

$$L = \log\left(|\mathbf{R}|\hat{\sigma}^{2(n_y - n_\beta)}\right) = \log|\mathbf{R}| + \left(n_y - n_\beta\right)\log\hat{\sigma}^2 \tag{7.53}$$

Also, variables that are necessary to calculate the Kriging predictions and prediction variance are stored in the structure fit (Line 82).

Function `trend`: Function `trend` calculates the basis functions at locations x for the given order of trend function. If input array x includes sample locations, it calculates design matrix **H**. On the other hand, if input array x includes prediction points, it calculates basis functions at prediction points. The current implementation allows ordinary Kriging (order = 0) and universal Kriging (order = 1 or 2). When order = 0, the design matrix **H** becomes a column vector of ones, i.e., a constant trend function. When order = 1, the design matrix becomes

$$\mathbf{H} = \begin{bmatrix} 1 & x_{1,1} & x_{2,1} & \cdots & x_{\text{ndim},1} \\ 1 & x_{1,2} & x_{2,2} & \cdots & x_{\text{ndim},2} \\ \vdots & \vdots & \vdots & \ddots & \vdots \\ 1 & x_{1,n_y} & x_{2,n_y} & \cdots & x_{\text{ndim},n_y} \end{bmatrix} \tag{7.54}$$

where $x_{i,j}$ is the $j$th sample location of the $i$th dimension. The dimension of the design matrix is $n_y \times n_b$. When order = 2, the basis functions include quadratic polynomials. In this case, the number of coefficients is $n_b = (\text{ndim} + 1)(\text{ndim} + 2)/2$ (Line 92). For example, when ndim = 2, the vector of basis functions becomes

$$\xi(\mathbf{x})^T = \left\{1, x_1, x_2, x_1^2, x_1x_2, x_2^2\right\} \tag{7.55}$$

**Example 7.13 1D Kriging Surrogate for Predictions and Confidence Intervals**
Consider true function $y(x) = 0.5(6x - 2)^2 \sin(12x - 4) + 10(x - 0.5) + 0.5$. Eight samples are generated using the true function at $x_s = [0.1548, 0.7148, 0.4934, 0.9932, 0.5862, 0.0007, 0.3638, 0.8198]$. Use an ordinary Kriging surrogate, plot the approximate function and 95% prediction intervals along with the true function and samples.

**Solution:**
The following Matlab script is used to generate surrogate prediction and 95% prediction intervals.

```
myfun=@(x) 0.5*((6*x-2).^2.*sin(12*x-4))+10*(x-0.5)+5;   % In-line function
xs=[0.1548,0.7148,0.4934,0.9932,0.5862,0.0007,0.3638,0.8198]';    % Samples
ys=myfun(xs);                                  % Samples from true function
trPRS=0;                                                  % Ordinary Kriging
xp=(0:0.01:1)';                                          % Prediction points
[yp, ysig, nb]=Kriging(xs,ys,xp,trPRS,0);          % Build Kriging surrogate
ny=size(ys,1);                                           % Number of samples
cilb = yp + tinv(0.025,ny-nb)*ysig;  % Lower-bounds of confidence intervals
ciub = yp + tinv(0.975,ny-nb)*ysig;  % Upper-bounds of confidence intervals
ytrue=myfun(xp);                  % True function values at prediction points
figure; hold on;                                          % Plotting results
xShaded=[xp;sort(xp,'descend')]; yShaded=[cilb;ciub(end:-1:1)];
patch(xShaded,yShaded,[.24 .73 .89],'EdgeColor','none');
plot(xp,yp,'r-',xs,ys,'k+',xp,ytrue,'k-','Markersize',10,'Linewidth',1.5)
xlabel('x');ylabel('y(x)');
legend('95% PI','Kriging prediction','Samples','True function');
```

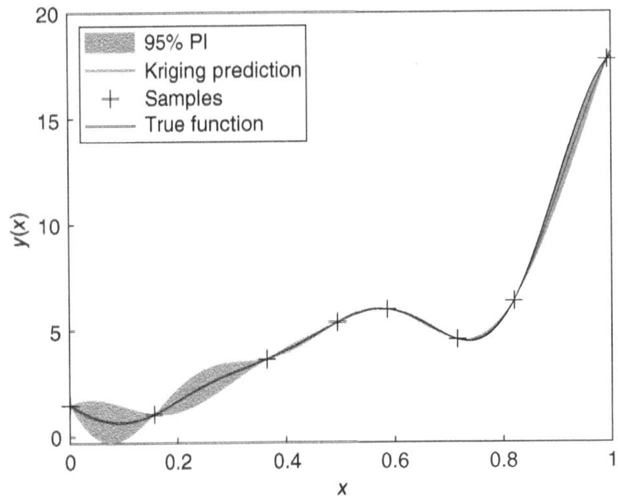


**Figure 7.23** Kriging predictions and 95% prediction intervals.

Since the analytical expression of the true function is available, in-line function is used to define the true function (Line 1). Then, the true function is called with eight sample locations $\mathbf{x}_s$ (Line 2) to calculate samples $\mathbf{y}_s$ (Line 3). For ordinary Kriging, a constant basis function is used (Line 4). Prediction points are defined between $\mathbf{x}_p \in [0, 1]$ with increment of 0.01 (Line 5). Calling Kriging function returns predictions $\mathbf{y}_p$ and prediction standard deviation $\hat{\sigma}(\mathbf{x}_p)$ (Line 6). In order to calculate the upper and lower bounds of prediction intervals, the inverse of t-distribution is used, as shown in Lines 8~9. Figure 7.23 shows Kriging predictions, 95% prediction intervals, eight samples, and the true function. The optimum hyperparameter that maximizes the likelihood is 0.272. It turns out that the Kriging predictions are accurate, and the level of uncertainty is small.

**Example 7.14 Kriging Surrogate for Branin–Hu Function**

The Branin–Hu function is a popular two-dimensional analytical function that is used for testing surrogate models. The functional form is defined in $x_1 \in [-5, 10]$, and $x_2 \in [0, 15]$ as

$$f(x_1, x_2) = \left(x_2 - \frac{5.1}{4\pi^2}x_1^2 + \frac{5}{\pi}x_1 - 6\right)^2 + 10\left(1 - \frac{1}{8\pi}\right)\cos(x_1) + 10$$

Generate 20 samples using Latin hypercube sampling, and fit the samples using an ordinary Kriging surrogate. Evaluate the accuracy of the Kriging surrogate against the true function at $31 \times 31$ uniform grids in terms of RMS error, average error, and maximum error. Plot the Kriging surrogate and true function.

**Solution:**

The following Matlab script is used to generate surrogate predictions, plot the surrogate, and calculate error metrics.

```
BraninHu = @(x) (x(:,2)-1.275*(x(:,1)/pi).^2+5*(x(:,1)/pi)-6).^2 ...
               +10*(1-1/(8*pi))*cos(x(:,1))+10;
rng('default'); N=31; dN=15/(N-1); ny=20;
xs=lhsdesign(ny,2,'iteration',5000);                       % 20 LHS samples
xs=xs.*[15, 15]+[-5, 0];
```

```
6   ys=BraninHu(xs);                                  % Samples from true function
7   trPRS=0;                                          % Ordinary Kriging
8   [XX,YY]=meshgrid(-5:dN:10, 0:dN:15);              % NxN grid of predictions
9   xp=[reshape(XX,N^2,1) reshape(YY,N^2,1)];         % Prediction points
10  [yp, ysig, nb]=Kriging(xs,ys,xp,trPRS,0);         % Build Kriging surrogate
11  YP=reshape(yp,N,N);                               % Predictions at grid
12  YT=reshape(BraninHu(xp),N,N);                     % True function at grid
13  figure(1); scatter(xs(:,1),xs(:,2),'filled');     % Plot sample locations
14  axis([-5 10 0 15]); xlabel('x_1');ylabel('x_2');
15  figure(2); hold on; surf(XX,YY,YP);               % Plot of predictions
16  plot3(xs(:,1),xs(:,2),ys,'ro');                   % Plot of samples
17  axis([-5 10 0 15]);xlabel('x_1');ylabel('x_2');zlabel('y');
18  figure(3); hold on; surf(XX,YY,YT)                % PLot of true function
19  plot3(xs(:,1),xs(:,2),ys,'ro');                   % Plot of samples
20  axis([-5 10 0 15]);xlabel('x_1');ylabel('x_2');zlabel('y');
21  eRMS=sqrt(sum(sum((YT-YP).^2))/N^2)               % RMS error at NxN grid points
22  eav=sum(sum(abs(YT-YP)))/N^2                      % Average error at NxN grid points
23  emax=max(max(abs(YT-YP)))                         % Maximum error at NxN grid points
```

First the analytical expression of the Branin–Hu function is defined as an in-line function (Lines 1~2). Since the Latin hypercube sampling process involves random number generation, `rng('default')` command makes sure that the same random numbers are generated. Samples are generated using `lhsdesign` command in Line 4. In order to improve the space-filling performance, 5,000 iterations are performed (see Section 3.4 for details of Latin hypercube sampling). Since `lhsdesign` generates samples in the range of [0,1], input variables are scaled in Line 5. The `BraninHu` function is called with input samples to generate output samples in Line 6. We will use a constant trend function in Line 7. For the purpose of plotting, $N \times N$ grids are created in the range of input variables. These $N \times N$ grid points are reshaped to meet the format of input prediction points in Line 9. Line 10 fits a Kriging surrogate and returns prediction points `yp`. In order to plot the predictions, they are reshaped into $N \times N$ grid points in Line 11. At the same time, we calculate the true functions at $N \times N$ grid points in Line 12. Line 13 plots sample locations. Lines 15~16 plot Kriging predictions along with samples. Lines 18~19 plot the true function for the purpose of comparison. Lines 21~23 calculate three error metrics.

The optimum hyperparameters that maximize the likelihood are $\theta_1 = 5.207$ and $\theta_2 = 18.15$. Figure 7.24(a) shows the locations of 20 samples. It is noted that the samples cover the entire design domain well, except for the lower left corner and upper right corner. Indeed, the Kriging predictions in Figure 7.24(b) are close to the true function in Figure 7.24(c), except for these two regions. Therefore, Kriging surrogate models share the same challenge common in other surrogate models: good performance in the interpolation region but significant error in the extrapolation region. Adding four corner samples can drastically reduce the error metrics (see Exercise Problem 7.22).

The three error metrics are $e_{\text{RMS}} = 12.0$, $e_{\text{av}} = 4.30$, and $e_{\max} = 119.6$. Considering the range of function $y_{\max} - y_{\min} = 307.7$, the RMS error and average error are reasonable, but the maximum error is large. In fact, the maximum error occurs at the lower left corner of the design space, which is an extrapolation region.

It would be beneficial to look at the behavior of likelihood as a function of hyperparameters. We already plotted the negative log-likelihood as a function of a single hyperparameter in one-dimensional problems in the previous section. It would be difficult to plot the likelihood function with more than two dimensions. Figure 7.25 shows the reduced negative log-likelihood in Eq. (7.32)

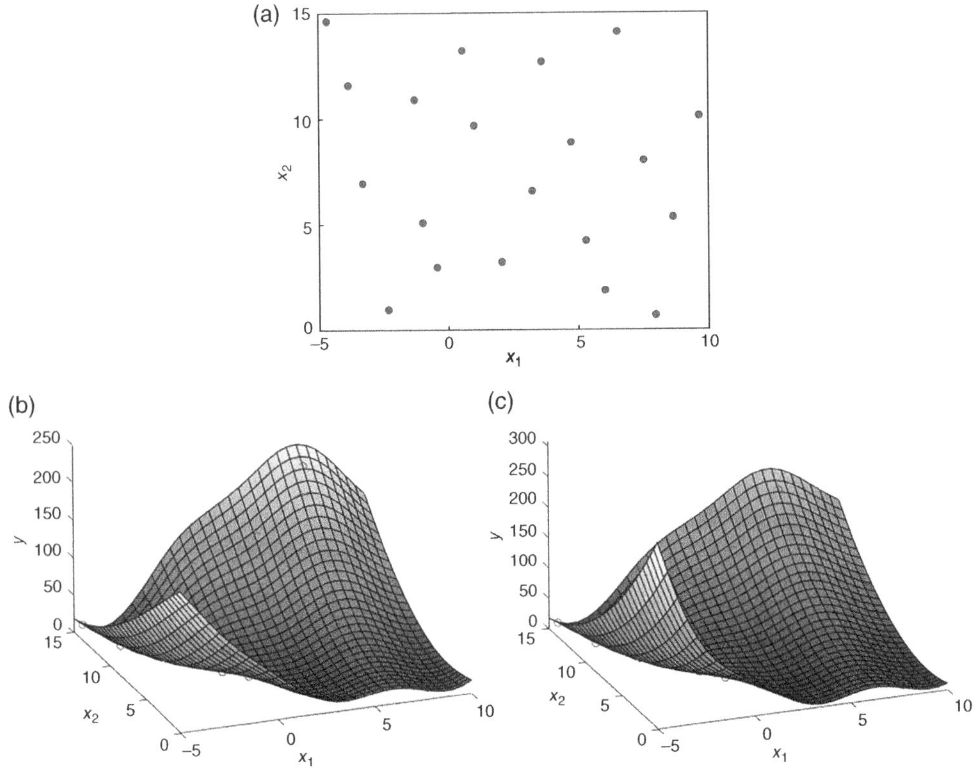


**Figure 7.24** Kriging predictions of Branin–Hu function. (a) Sample locations, (b) Kriging predictions, and (c) true function.

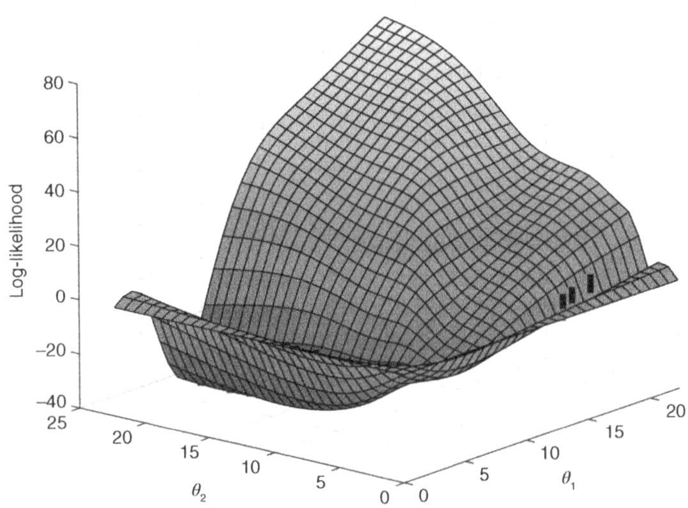


**Figure 7.25** Negative log-likelihood as a function of hyperparameters for Branin–Hu function.

of Branin–Hu function in Example 7.14 (see Exercise Problem 7.25). It is noted that the likelihood function shows a complicated behavior over the range of hyperparameters. It is obvious that the first hyperparameter $\theta_1$ has its minimum value at $\theta_1 = 5.2$ as it shows a narrow valley there. This is because the Branin–Hu function shows a clear period along $x_1$ direction. On the other hand, the function does not show a clear period along $x_2$ direction. Because of this behavior, the likelihood continues to decrease along the $\theta_2$ direction. However, there is a local minimum around $\theta_2 = 18$. However, when a function does not show clear periodic behavior, the contribution of the hyperparameter is not important. Therefore, the Kriging prediction will not be significantly different when $\theta_2 > 18$.

One of the challenges in high-dimensional surrogate modeling is to plot the surrogate predictions and prediction intervals. It is possible to plot surrogate predictions in two-dimensional function as shown in Figure 7.24, but it is difficult to plot them in more than two dimensions. Prediction intervals are difficult to plot even in two dimensions. One possible remedy is to plot a high-dimensional function along a line that goes though one of the diagonals of the design domain. In the case of the Branin–Hu function in Example 7.14, for example, since this function is in two dimensions, we can define a parametric line that connects two points $A(-5, 0)$ and $B(10, 15)$ or that connects $C(-5, 15)$ and $D(10, 0)$. We can define $n_p$ prediction points along the line and calculate Kriging surrogate predictions and intervals. The following Matlab script fits a Kriging surrogate using randomly generated 20 samples from the Branin–Hu function and plot the Kriging surrogate predictions and prediction intervals along the two diagonals.

```
BraninHu = @(x) (x(:,2)-1.275*(x(:,1)/pi).^2+5*(x(:,1)/pi)-6).^2 ...
              +10*(1-1/(8*pi))*cos(x(:,1))+10;
rng('default'); ny=20;
xs=lhsdesign(ny,2,'iteration',5000);                          % 20 LHS samples
xs=xs.*[15, 15]+[-5, 0];                        %xs=[xs;-5 0;10 0;-5 15;10 15];
ys=BraninHu(xs);                                    % Samples from true function
trPRS=0;                                                      % Ordinary Kriging
s=(0:0.01:1)'; N=size(s,1);               % Parameter along a diagonal direction
xp=[-5+15*s 15*s];                                        % xp=[-5+15*s 15-15*s];
[yp,ysig,nb]=Kriging(xs,ys,xp,trPRS,0);                % Build Kriging surrogate
yt=BraninHu(xp);                                % True function for comparison
cilb = yp + tinv(0.025,ny-nb)*ysig;       % Lower-bounds of confidence intervals
ciub = yp + tinv(0.975,ny-nb)*ysig;       % Upper-bounds of confidence intervals
figure; hold on;                                              % Plotting results
xShaded=[s;sort(s,'descend')]; yShaded=[cilb;ciub(end:-1:1)];
patch(xShaded,yShaded,[.24 .73 .89],'EdgeColor','none');
plot(s,yp,'r-',s,yt,'k-','Markersize',10,'Linewidth',1.5)
xlabel('s');ylabel('y(s)');
legend('95% PI','Kriging prediction','True function');
eRMS=sqrt(sum((yt-yp).^2)/N);                      % RMS error at NxN grid points
eav=sum(abs(yt-yp))/N;                         % Average error at NxN grid points
emax=max(abs(yt-yp));                          % Maximum error at NxN grid points
fprintf('RMS error:%8.3g\nAve error:%8.3g\nMax error:%8.3g\n',eRMS,eav,emax);
```

This script is similar to the scripts in Examples Example 7.13 and Example 7.14. Therefore, no detailed explanations are provided. However, variable s is used to connect two points, such that $s = 0$ corresponds to point $A(-5, 0)$, and $s = 1$ corresponds to point $B(10, 15)$. Figure 7.26(a) plots

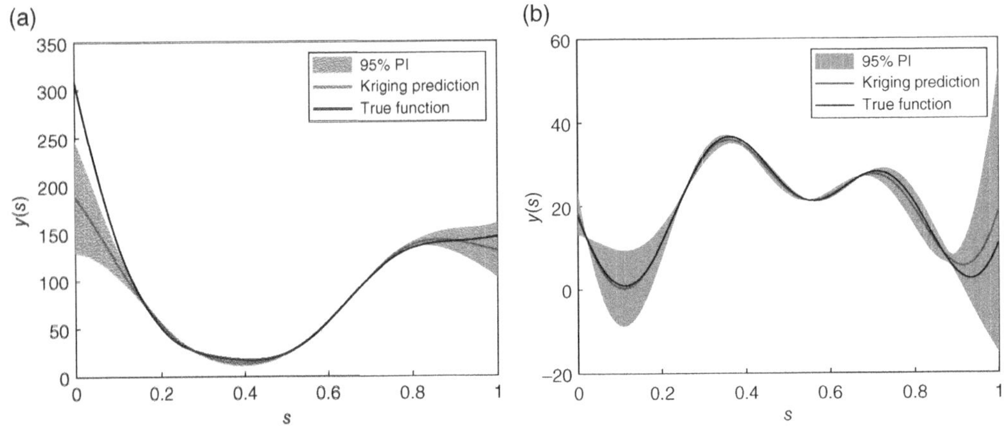


**Figure 7.26** Predictions and prediction intervals of Branin–Hu function along two diagonals. (a) From *A*(−5, 0) to *B*(10, 15) and (b) from *C*(−5, 15) to *D*(10, 0).

the Kriging predictions and prediction intervals along this line and compares against the true function. Along this line, the three error metrics are $e_{RMS} = 23.6$, $e_{av} = 9.32$, and $e_{max} = 119.6$. It is noted that the maximum error is the same, but the RMS error and averaged error are almost double compared to the error metrics in Example 7.14. This is mostly because large errors are present at corner points, where a large fraction of the line belongs to the extrapolation region. The figure also confirms that the prediction errors and prediction variance in the range of $s \in [0, 0.1]$ and $s \in [0.9, 1.0]$ are large compared to those in the range of $s \in [0.1, 0.9]$. Figure 7.26(b) shows Kriging predictions and prediction intervals along the other diagonal line between $C(-5, 15)$ and $D(10, 0)$. Along this line, the three error metrics are $e_{RMS} = 1.76$, $e_{av} = 0.972$, and $e_{max} = 7.69$. Although all error metrics are much better than the other diagonal, the figure confirms that large errors are located near the corner region.

## 7.6 Kriging with Nuggets—Fitting with Noisy Data (Gaussian Process Regression)

So far, we developed the Kriging surrogate model assuming that samples are correct with uncertainty in the model form. This is the strength of the Kriging surrogate but also a drawback. In reality, many samples include random noise even if they are from computer simulation.

Figure 7.27 shows the nature of noise (or error) in computer simulation. A relatively simple geometry as shown in Figure 7.27(a) is modeled using finite-element simulation software Abaqus. The design variable is gradually changed from 25 to 25.5 mm with 50 increments. Since the design variable will remove material, we expect that the area will decrease (Figure 7.27(b)). In addition, since removal of material will make the structure more flexible, we expect that the tip displacement will increase. Since the design is defined in the computer-aided design (CAD) model, when the design variable is changed, the CAD model is updated and finite-element mesh is re-created at the new design, which can cause a meshing error; the mesh in the previous design may not be consistent with the mesh in the current design. It turned out that the area is relatively stable in terms of design and shows a linearly decreasing trend (Figure 7.27(b)). However, displacement at the load application point shows a fluctuating trend, even if globally it increases (Figure 7.27(c)). In optimization, a small perturbation size is often used to calculate sensitivity, i.e., gradient information, so

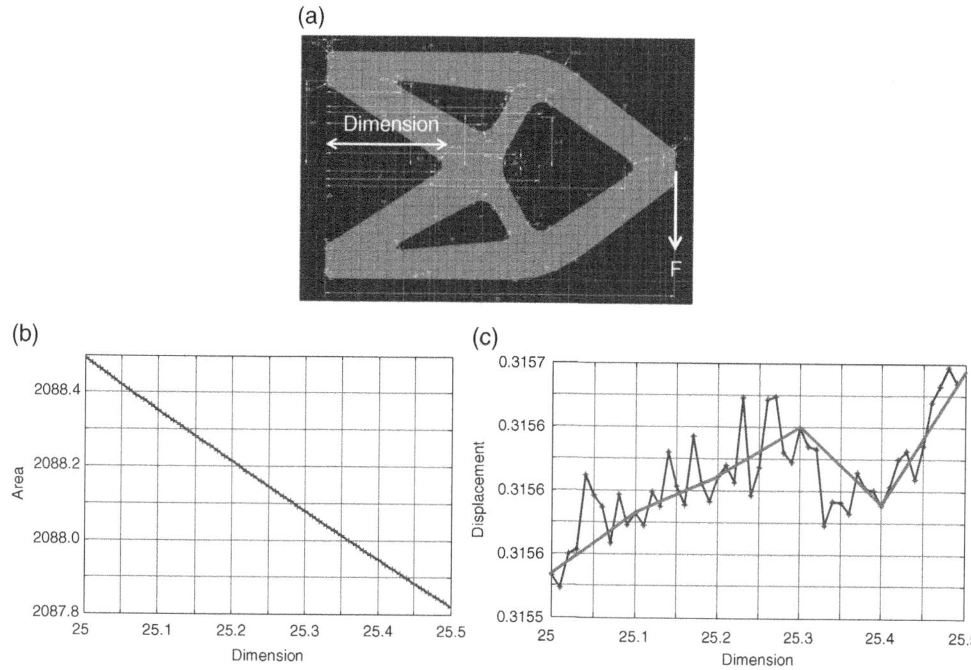


**Figure 7.27** Noise in finite-element simulation due to mesh change. (a) Finite-element model, (b) area as a function of dimension, and (c) tip displacement as a function of dimension.

that the optimization algorithm can decide whether the design variable should be increased or decreased. However, due to the noise, the sensitivity calculated by a small perturbation can be completely wrong. In addition to random noise, the mean trend shows a sudden drop when the number of elements changes. The large drop between dimension = 25.4 and 25.4 is due to the fact that the number of elements is changed. In such a case, interpolating samples can fit the noise and bias rather than fitting the trend. Therefore, it would be necessary to develop a Kriging surrogate model with regression capability.

A version of Kriging surrogate modeling with noisy data is called Kriging with nuggets. The name "nuggets" comes from the fact that additional values are added to the diagonal terms of the correlation matrix, which is referred to as the nugget effect. Matlab's implementation of Kriging with nuggets is called Gaussian process regression (GPR) .

When samples include noise, the Kriging approximation in Eq. (7.3) can be revisited, where samples and prediction have the relationship as $y_i = y(\mathbf{x}_i) = \hat{y}(\mathbf{x}_i)$. Under noise in data, this relationship is not true anymore. Instead, the samples are generated from the following model:

$$y(\mathbf{x}) = \hat{y}(\mathbf{x}) + \epsilon = \xi(\mathbf{x})^T\boldsymbol{\beta} + s(\mathbf{x}) + \epsilon \tag{7.56}$$

where the noise term $\epsilon$ follows a zero-mean Gaussian distribution

$$\epsilon \sim N(0, \boldsymbol{\Sigma}_n) \tag{7.57}$$

where $\boldsymbol{\Sigma}_n$ is the covariance matrix of the noise term. Considering that the local departure term $s(\mathbf{x}) \sim N(0, \sigma^2)$, the Kriging with nuggets needs to consider another covariance from the noise. Therefore, it is expected that the effect of covariance due to correlation between samples and the effect of noise can be combined together.

As we discussed in samples, the correlation between samples is a critically important aspect of Kriging surrogate modeling. For the same reason, depending on the properties of $\boldsymbol{\Sigma}_n$, three classes of noise can be considered.

- $\boldsymbol{\Sigma}_n = \sigma_n^2\mathbf{I}$: The variance of the noise is the same for all samples, and they are not correlated. This is called homogeneous noise, in which the variance $\sigma_n^2$ is constant. The noise follows an independent and identically distributed (iid) Gaussian distribution.
- $\boldsymbol{\Sigma}_n = \text{diag}\left(\sigma_{n,i}^2\right)$: The covariance matrix is diagonal with different diagonal components. This is called independent heterogeneous (heteroscedastic) noise, in which the noise variance $\sigma_{n,i}^2$ differ for each sample but not correlated.
- $\boldsymbol{\Sigma}_n$: This is called general heteroscedastic noise, in which the noise variance can differ for each sample and noise is correlated with other samples.

### 7.6.1 Kriging Surrogate with Correlated Noise

In the case of general heteroscedastic noise, the effect of correlation between samples and that of noise are combined in the form of covariance matrix as

$$\mathbf{K} = \sigma^2\mathbf{R} + \boldsymbol{\Sigma}_n \tag{7.58}$$

In this case, it is not possible to separate the correlation matrix and process variance. With this covariance matrix, the likelihood function given in Eq. (7.13) has the same expression after replacing the $\sigma^2\mathbf{R}$ term with $\mathbf{K}$ as

$$f\left(\mathbf{y} \mid \boldsymbol{\beta}, \sigma^2, \boldsymbol{\theta}\right) = \frac{1}{\sqrt{(2\pi)^{n_y}|\mathbf{K}|}} \exp\left(-(\mathbf{y}-\mathbf{X}\boldsymbol{\beta})^T\mathbf{K}^{-1}(\mathbf{y}-\mathbf{X}\boldsymbol{\beta})\right) \tag{7.59}$$

In the case of noise-free Kriging, the vector of hyperparameters $\boldsymbol{\theta}$ is identified first by maximizing the likelihood function, and then, they are used to estimate other parameters, $\boldsymbol{\beta}$ and $\sigma^2$. In the case of correlated noise, three parameters, $\sigma^2$, $\boldsymbol{\theta}$, $\boldsymbol{\Sigma}_n$, need to be identified by maximizing the likelihood function. This would be challenging unless the noise covariance matrix, $\boldsymbol{\Sigma}_n$, is parameterized. Therefore, a simplified version or parameterized version of noise covariance is often used in practice. Once these parameters are estimated, the coefficients of trend function can be determined similar to Eq. (7.16) as

$$\hat{\boldsymbol{\beta}} = \left(\mathbf{X}^T\mathbf{K}^{-1}\mathbf{X}\right)^{-1}\left\{\mathbf{X}^T\mathbf{K}^{-1}\mathbf{y}\right\} \tag{7.60}$$

Then, the mean Kriging prediction in Eq. (7.28) needs to be modified to

$$\hat{y}(\mathbf{x}) = \boldsymbol{\xi}(\mathbf{x})^T\hat{\boldsymbol{\beta}} + \mathbf{c}(\mathbf{x})^T\mathbf{K}^{-1}\left(\mathbf{y}-\mathbf{X}\hat{\boldsymbol{\beta}}\right) \tag{7.61}$$

where $\mathbf{c}(\mathbf{x}) = \sigma^2\mathbf{r}(\mathbf{x})$ is the cross-covariance vector between the prediction point $\mathbf{x}$ and samples. Unfortunately, the process variance cannot be written explicitly. The prediction variance in Eq. (7.38) can be written as

$$\hat{\sigma}_y^2(\mathbf{x}) = \left[\sigma^2 - \mathbf{c}(\mathbf{x})^T\mathbf{K}^{-1}\mathbf{c}(\mathbf{x}) + \left(\boldsymbol{\xi}(\mathbf{x}) - \mathbf{X}^T\mathbf{K}^{-1}\mathbf{c}(\mathbf{x})\right)^T\left(\mathbf{X}^T\mathbf{K}^{-1}\mathbf{X}\right)^{-1}\left(\boldsymbol{\xi}(\mathbf{x}) - \mathbf{X}^T\mathbf{K}^{-1}\mathbf{c}(\mathbf{x})\right)\right] \tag{7.62}$$

Compared to Eq. (7.38), the prediction variance is equivalent if $\mathbf{r}(\mathbf{x})$ and $\mathbf{R}$ are replaced by $\mathbf{c}(\mathbf{x})$ and $\mathbf{K}$, respectively.

Although the Kriging formulation for correlated noise covariance provides the most general formulation, it is impractical due to the complication of identifying the noise covariance matrix $\boldsymbol{\Sigma}_n$ via maximum likelihood estimate. In addition, it is rare to consider correlated noise distribution in practice. In order to estimate the noise covariance, it would be necessary to parameterize it in terms of handful parameters, which then can be estimated using maximum likelihood estimate. The simplest form of noise covariance is the case of homogeneous noise in the following subsection.

### 7.6.2 Kriging Surrogate with Homogeneous Noise

In the case of homogeneous noise, it is assumed that all samples include noise that is from a normal distribution with zero mean and variance $\sigma_n^2$. In addition, noise samples are uncorrelated to each other, i.e., they follow independent and identically distributed Gaussian distribution. In such a case, the covariance matrix of noise can be written as $\sigma_n^2\mathbf{I}$, which is added to the covariance matrix as

$$\mathbf{K} = \sigma^2\mathbf{R} + \sigma_n^2\mathbf{I} \tag{7.63}$$

Equation (7.63) does not mean that actual noise samples need to be normally distributed. Instead, the Kriging surrogate will estimate the noise distribution based on this assumption.

Since two variances, $\sigma^2$ and $\sigma_n^2$, contribute to the covariance matrix in Eq. (7.63), it would be convenient to define a fraction between them. Let $\sigma_{\text{total}}^2 = \sigma^2 + \sigma_n^2$ be the total variance of the Kriging process. Then, the fraction of noise variance is defined as

$$\tau = \frac{\sigma_n^2}{\sigma_{\text{total}}^2} \tag{7.64}$$

When $\tau = 0$, it becomes noise-free Kriging in Section 7.3. When $\tau = 1$, the effect of local departure disappears, and it becomes polynomial response surface in Chapter 2 with trend function. Using this noise fraction, it is possible to separate the correlation matrix from the covariance matrix as

$$\mathbf{K} = \sigma_{\text{total}}^2[(1-\tau)\mathbf{R} + \tau\mathbf{I}] \equiv \sigma_{\text{total}}^2\tilde{\mathbf{R}} \tag{7.65}$$

Then, the estimation process of Kriging surrogate with noise becomes identical to that of a noise-free case by substituting $\sigma_{\text{total}}^2$ and $\tilde{\mathbf{R}}$ with the process variance and correlation matrix. For example, the likelihood function in Eq. (7.13) can be written as

$$f(\mathbf{y} \mid \boldsymbol{\beta}, \sigma_{\text{total}}^2, \boldsymbol{\theta}, \tau) = \frac{1}{\sqrt{\left(2\pi\sigma_{\text{total}}^2\right)^{n_y}|\tilde{\mathbf{R}}|}}\exp\left(-\frac{(\mathbf{y}-\mathbf{X}\boldsymbol{\beta})^T\tilde{\mathbf{R}}^{-1}(\mathbf{y}-\mathbf{X}\boldsymbol{\beta})}{2\sigma_{\text{total}}^2}\right) \tag{7.66}$$

It is noted that the additional parameter $\tau$ needs to be estimated using the maximum likelihood estimate simultaneously with other parameters. In practice, instead of maximizing the likelihood function, the following reduced negative log-likelihood is minimized

$$(\boldsymbol{\theta}, \tau) = \arg\min\left[\ln\left(\hat{\sigma}_{\text{total}}^{2(n_y - n_p)}|\tilde{\mathbf{R}}|\right)\right] \tag{7.67}$$

In calculating the reduced negative log-likelihood, the coefficients of trend function and total variance are defined as

$$\hat{\boldsymbol{\beta}} = \left(\mathbf{X}^T\tilde{\mathbf{R}}^{-1}\mathbf{X}\right)^{-1}\left\{\mathbf{X}^T\tilde{\mathbf{R}}^{-1}\mathbf{y}\right\} \tag{7.68}$$

$$\hat{\sigma}^2_{\text{total}} = \frac{\left(\mathbf{y}-\mathbf{X}\hat{\boldsymbol{\beta}}\right)^T \tilde{\mathbf{R}}^{-1}\left(\mathbf{y}-\mathbf{X}\hat{\boldsymbol{\beta}}\right)}{n_y - n_p} \tag{7.69}$$

Once all hyperparameters are identified, the mean prediction of Kriging surrogate at point $\mathbf{x}$ can be obtained as

$$\hat{y}(\mathbf{x}) = \boldsymbol{\xi}(\mathbf{x})^T\hat{\boldsymbol{\beta}} + \tilde{\mathbf{r}}(\mathbf{x})^T\tilde{\mathbf{R}}^{-1}\left(\mathbf{y}-\mathbf{X}\hat{\boldsymbol{\beta}}\right) \tag{7.70}$$

where the cross-correlation between samples and prediction point is $\tilde{\mathbf{r}}(\mathbf{x}) = (1-\tau)\mathbf{r}(\mathbf{x})$. Also, the prediction variance at point $\mathbf{x}$ can be obtained as

$$\hat{\sigma}_y^2(\mathbf{x}) = \sigma^2_{\text{total}}\left[1 - \tilde{\mathbf{r}}(\mathbf{x})\tilde{\mathbf{R}}^{-1}\tilde{\mathbf{r}}(\mathbf{x}) + \mathbf{u}(\mathbf{x})^T\left(\mathbf{X}^T\tilde{\mathbf{R}}^{-1}\mathbf{X}\right)^{-1}\mathbf{u}(\mathbf{x}))\right] \tag{7.71}$$

where $\mathbf{u}(\mathbf{x}) = \boldsymbol{\xi}(\mathbf{x}) - \mathbf{X}^T\tilde{\mathbf{R}}^{-1}\tilde{\mathbf{r}}(\mathbf{x})$. It is noted that the prediction variance is not zero at sample locations because of noise variance.

The same Matlab function `Kriging.m` in Section 7.5 can be used to build a Kriging surrogate with noise in samples. When the last argument `noise = 1`, it assumes that samples include noise and estimate the noise fraction $\tau$ during the maximum likelihood estimation.

**Example 7.15 1D Kriging Surrogate for Predictions and Confidence Intervals**
Consider a true function $y(x) = \sin x$ in the range of $x \in [0, 2\pi]$. Generate 10 equally spaced samples from the true function and fit an ordinary Kriging surrogate. Add randomly generated noise $\epsilon \sim N(0, 0.1^2)$ to the samples and fit an ordinary Kriging surrogate with the nuggets effect. Compare these two surrogates in terms of plots, the three error metrics, process noise, and prediction intervals.

**Solution:**
The following Matlab script is used to fit $y(x) = \sin x$ using 10 samples.

```
myfun = @(x) sin(x);                                    % In-line function
rng('default'); ny=10; nstd=0.1;          % No. of samples, and noise std
xs=linspace(0,2*pi,ny)';                                      % Samples
ys=myfun(xs);                                % Samples from true function
noise=lhsnorm(0,nstd^2,ny); noise=nstd*(noise-mean(noise))/std(noise);
fprintf('Sample noiseSTD:%8.3g\n',std(noise));
ys=ys+noise;
trPRS=0;                                               % Ordinary Kriging
xp=(0:0.01:2*pi)';                                   % Prediction points
[yp, ysig, nb]=Kriging(xs,ys,xp,trPRS,1);      % Build Kriging surrogate
cilb = yp + tinv(0.025,ny-nb)*ysig;  % Lower-bounds of confidence intervals
ciub = yp + tinv(0.975,ny-nb)*ysig;  % Upper-bounds of confidence intervals
ytrue=myfun(xp);                  % True function values at predictio points
figure; hold on;                                       % Plotting results
xShaded=[xp;sort(xp,'descend')]; yShaded=[cilb;ciub(end:-1:1)];
patch(xShaded,yShaded,[.24 .73 .89],'EdgeColor','none');
plot(xp,yp,'r-',xs,ys,'k+',xp,ytrue,'k-','Markersize',10,'Linewidth',1.5)
xlabel('x');ylabel('y(x)');
legend('95% PI','Kriging prediction','Samples','True function');
```

Since most parts of the Matlab script are similar to those of previous examples, it would be unnecessary to explain line by line. Instead, only different parts will be explained here. Random noise samples are generated in Line 5. First, the `lhsnorm` function is used to generate normally distributed samples using Latin hypercube sampling. Even if the noise samples are generated from a normal distribution with zero mean and `nstd` standard deviation, the actual samples have different mean and standard deviation. Therefore, the samples are shifted and scaled such that the mean becomes zero and the standard deviation is equal to `nstd`. The random noise samples are added to the samples in Line 7, and `Kriging.m` is called with `noise = 1`. If Line 7 is commented, then the samples do not include noise. If `noise = 0` is used when calling `Kriging.m` in Line 10, samples will be fitted without considering the nuggets effect.

In order to compare the effect of noise in a Kriging surrogate, four different cases are presented by combining with and without noise in samples and with and without considering the nuggets effect in the fitting process. Figure 7.28(a) shows the Kriging surrogate without noise

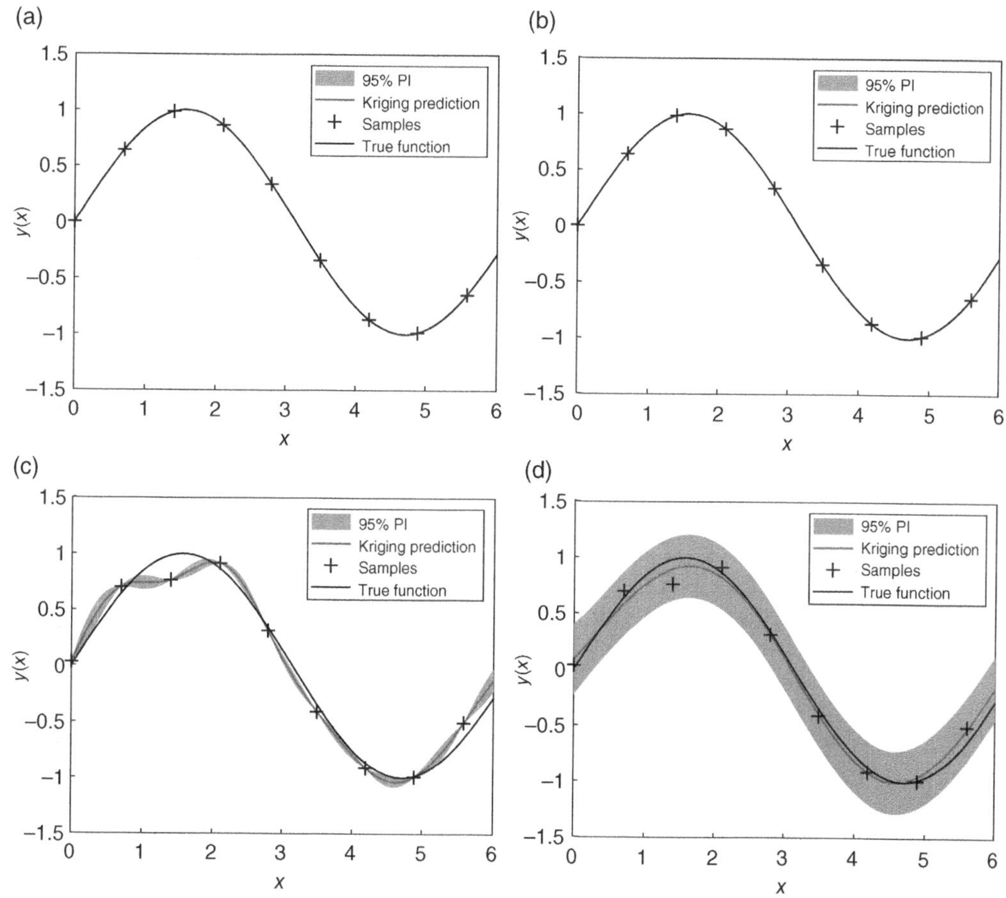


**Figure 7.28** Ordinary Kriging predictions and prediction intervals of $y(x) = \sin x$. (a) No noise in samples and fitting Kriging without the nuggets effect, (b) no noise in samples and fitting Kriging with the nuggets effect, (c) noise in samples but fitting Kriging without the nuggets effect, and (d) noise in samples and fitting Kriging with the nuggets effect.

in samples and without considering the nuggets effect. It identifies hyperparameter $\theta = 4.43$ and process noise $\hat{\sigma} = 2.85$. As shown in the figure, the prediction intervals are ignorable, and prediction is accurate. Figure 7.28(b) shows the Kriging surrogate without noise in samples but considering the nuggets effect in the fitting process. It identifies hyperparameter $\theta = 4.34$, total process noise $\hat{\sigma} = 2.61$, and noise standard deviation $\hat{\sigma}_n = 9.24 \times 10^{-6}$. Basically, even if the fitting process considers the effect of noise, it identifies zero noise. As shown in the figure, the prediction intervals are ignorable, and prediction is accurate. Figure 7.28(c) shows the case where samples include noise, but the fitting process ignores the effect of noise. Since random noise fluctuates, the fitting algorithm considers this effect by identifying a smaller value of hyperparameter $\theta = 1.15$ with a smaller value of process noise $\hat{\sigma} = 0.444$. Due to the interpolation characteristics of Kriging, the predictions pass through all noises, which corresponds to fitting the noise, not the trend. Therefore, it is not a good idea to fit noisy samples without considering the effect of noise. Figure 7.28(d) shows the case where samples include noise, and the fitting process ignores the effect of noise. It identifies hyperparameter $\theta = 2.53$, total process noise $\hat{\sigma} = 0.735$, and noise standard deviation $\hat{\sigma}_n = 0.0704$. As expected, the prediction intervals are not zero at the sample locations. It is interesting to note that the identified noise standard deviation $\hat{\sigma}_n = 0.0704$ is less than the actual noise standard deviation $\sigma_n = 0.1$.

It would be interesting to see how the likelihood function behaves as a function of hyperparameters. Figure 7.29(a) shows the plot of negative log-likelihood of 10 samples used in Example 7.15 when samples do not include noise. It is clear that the negative log-likelihood decreases rapidly as noise fraction $\tau$ approaches zero. Therefore, the optimization algorithm identifies the noise fraction $\tau = 0$ even if the nuggets effect is considered during the fitting process. On the other hand, Figure 7.29(b) shows the plot of negative log-likelihood of 10 samples used in Example 7.15 when samples include random noise. Opposite to the previous case, the negative log-likelihood rapidly increases as noise fraction $\tau$ approaches zero in this case. Therefore, if the $\tau = 0$ condition is used when samples include noise, the hyperparameter will be identified as $\theta = 1.15$ as in Figure 7.28(c).

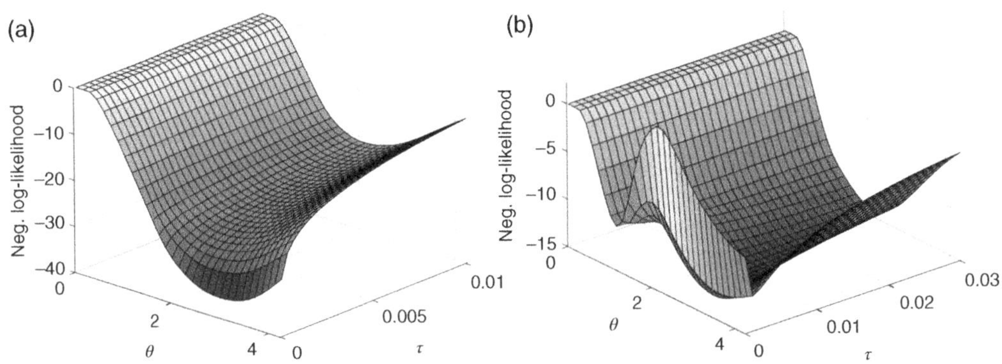


**Figure 7.29** Negative log-likelihood as a function of hyperparameters. (a) No noise in samples and (b) noise in samples.

An important observation in the previous example is that the identified noise standard deviation is underestimated from the actual noise standard deviation in samples. Table 7.1 shows the fitting results of $y = \sin x$ using 20 equally spaced samples with different levels of randomly generated noise. It is interesting to note that the ratio $\hat{\sigma}_n/\sigma_n \sim 0.72$. That is, Kriging consistently underestimates the noise standard deviation. However, this ratio depends on the number of samples and complexity of the function. The estimated hyperparameter gradually decreases as the level of noise increases, which means that the correlation decays faster due to the effect of noise. All error metrics increase as the level of noise increases, which is expected as the surrogate will be less accurate. Table 7.2 shows the fitting results of $y = \sin x$ with random noise $\epsilon \sim N(0,0.1^2)$ with different numbers of samples. When the number of samples is larger than 10, there is no significant trend in Kriging: about 70% of noise standard deviation from the actual one. However, when the number of samples is too small, Kriging significantly underestimates the noise standard deviation. When the number of samples is too small, Kriging tends to fit the noise instead of filtering out.

The reason that Kriging with nuggets underestimates the noise variance is related to the fact that estimation of $\hat{\boldsymbol{\beta}}$, $\hat{\sigma}^2$, and $\hat{\sigma}_n^2$ are correlated. Especially with a small number of samples, Kriging with nuggets tends to overfit the samples, which yields a smaller $\hat{\sigma}^2$ and underestimates $\hat{\sigma}_n^2$.

**Table 7.1** Hyperparameters and error metrics for different levels of noise standard deviation.

| Sample $\sigma_n$ | $\hat{\theta}$ | $\tau$ | $\hat{\sigma}_{total}$ | $\hat{\sigma}_n$ | $e_{RMS}$ | $e_{av}$ | $e_{max}$ |
|---|---|---|---|---|---|---|---|
| 0.00 | 3.36 | 0 | 0.785 | 0 | 1.83E−6 | 1.45E-6 | 5.34E−6 |
| 0.05 | 3.29 | 0.000862 | 1.220 | 0.0357 | 0.0219 | 0.0182 | 0.0425 |
| 0.10 | 2.91 | 0.00503 | 0.994 | 0.0705 | 0.044 | 0.0362 | 0.088 |
| 0.15 | 2.54 | 0.0183 | 0.788 | 0.107 | 0.0665 | 0.0543 | 0.135 |
| 0.20 | 2.31 | 0.0408 | 0.707 | 0.143 | 0.0879 | 0.0714 | 0.179 |
| 0.25 | 2.14 | 0.0747 | 0.661 | 0.181 | 0.109 | 0.0877 | 0.221 |
| 0.30 | 2.0 | 0.118 | 0.642 | 0.221 | 0.129 | 0.104 | 0.261 |

**Table 7.2** Hyperparameters and error metrics for different numbers of samples.

| No. samples | $\hat{\theta}$ | $\tau$ | $\hat{\sigma}_{total}$ | $\hat{\sigma}_n$ | $e_{RMS}$ | $e_{av}$ | $e_{max}$ |
|---|---|---|---|---|---|---|---|
| 5 | 1.41 | 0.00041 | 0.592 | 0.012 | 0.0853 | 0.0747 | 0.143 |
| 7 | 1.62 | 0.00292 | 0.552 | 0.0298 | 0.0942 | 0.0811 | 0.178 |
| 10 | 2.53 | 0.00917 | 0.735 | 0.0704 | 0.0657 | 0.0579 | 0.124 |
| 15 | 3.06 | 0.00558 | 0.993 | 0.0741 | 0.0459 | 0.0398 | 0.0816 |
| 20 | 2.91 | 0.00503 | 0.994 | 0.0705 | 0.044 | 0.0362 | 0.088 |
| 25 | 2.7 | 0.00799 | 0.835 | 0.0747 | 0.0326 | 0.0268 | 0.0683 |
| 30 | 2.66 | 0.00874 | 0.772 | 0.0722 | 0.0328 | 0.0262 | 0.117 |

## Exercise

**7.1** Consider the two functions $f_1 = (x+y)^2$ and $f_2 = x$ in the unit square where $x$ and $y$ are in [0, 1]. Calculate their correlations from a sample of 20 points using LHS and check how accurately you can get the correlation with increasing numbers of iterations using the different options of LHS. These include maximizing the minimum distance between points, minimizing the correlation, and with the "smooth" option that puts the points in the center of the LHS cell.

**7.2** Calculate the correlation coefficient of Problem 7.1 analytically.

**7.3** Calculate the correlation coefficient between (i) $y_1 = x$ and $y_2 = -x$ and (ii) $y_1 = x$ and $y_2 = 2x$ by generating 50 random samples of $x \in [-1, 1]$.

**7.4** Plot the isotropic and anisotropic correlation functions in a two-dimensional design space $x_1$, $x_2 \in [-5, 5]$, as shown in Figure 7.4. For use $\theta = 2$ for isotropic correlation and $\theta_1 = 2$ and $\theta_2 = 1$ for anisotrpic correlation function.

**7.5** Consider two functions $f(X) = X$ and $g(X) = X^2$. When $X$ is a normal random variable, find the correlation coefficient between $f(X)$ and $g(X)$ using an analytical method, and compare it with numerical correlation coefficients obtained using 200 samples. Use (i) $X \sim N(0, 1^2)$, (ii) $X \sim N(1, 1^2)$, and (iii) $X \sim N(2, 1^2)$. Refer to Wikipedia or statistical reference for the mean of monomials of $X$.

**7.6** For Example 7.2, plot the determinant of the correlation matrix as a function of $\theta \in [0.1, 2.0]$.

**7.7** The maximum of the likelihood function in Eq. (7.13) depends on the difference between the samples and global function $(\mathbf{y} - \mathbf{X}\boldsymbol{\beta})$ as well as on the inverse of correlation matrix $\mathbf{R}^{-1}$. In order to find the relationship between these two effects, consider a simple case with two samples. The normalized difference can be expressed as $(\mathbf{y} - \mathbf{X}\boldsymbol{\beta}) = [1, \alpha]^T$, where $0.1 \le \alpha \le 1.0$, and the correlation coefficient can vary between $0 \le \rho \le 0.9999$. Let the likelihood function be simplified by $L = \exp[-(\mathbf{y} - \mathbf{X}\boldsymbol{\beta})^T \mathbf{R}^{-1} (\mathbf{y} - \mathbf{X}\boldsymbol{\beta})]$. Plot $L$ versus $\rho$ curves at different values of $\alpha$. Compare the value of $\alpha$ that maximizes $L$ with the correlation coefficient $\rho$.

**7.8** In Example 7.4, it is shown that the global function depends on the hyperparameter $\theta$. Show that the variance of the local departure term also depends on the hyperparameter and plot it as a function of the hyperparameter.

**7.9** A total of 11 samples are generated in the equal intervals of design space $x \in [0, 1]$. When a linear global function is used, plot the 11 weight functions. Assume that the hyperparameter is $\theta = 0.05$.

**7.10** Repeat Example 7.6 with a linear correlation function. The linear correlation function is defined as

$$\phi(x, s) = \max(0, 1 - |x - s|/\theta)$$

where $\theta = 0.05, 1.5, 2.5$.

**7.11** When the global function is the same as the true function, it is difficult to estimate the hyperparameter and prediction uncertainty because the negative log-likelihood function approaches negative infinity due to $\mathbf{y} - \mathbf{X}\boldsymbol{\beta} = \mathbf{0}$. Generate five equally spaced samples from true function $y = 5x - 10$ and fit the samples using the Kriging surrogate with a linear global function. Plot the negative log-likelihood function versus hyperparameter and discuss the difficulty of finding the optimal hyperparameter. Plot the Kriging predictions and prediction intervals and compare them with the true function.

**7.12** Show the instability of a negative log-likelihood function with a large hyperparameter, as shown in Figure 7.11. First, generate equally spaced 13 samples in the domain $x \in [0, 2]$, and generate function values using $y = \sin \pi x$. Then, gradually increase the hyperparameter $\theta \in [0.02, 2.0]$ and calculate the constant term of the global function and negative log-likelihood. Plot the log-likelihood as a function of the hyperparameter.

**7.13** When the number of samples is too small, the negative log-likelihood function continues to increase after the initial flat region. Plot the negative log-likelihood function versus hyperparameter and compare Kriging predictions at different values of the hyperparameter. Use true function $y = \sin \pi x$ to generate five equally spaced samples in the range of $x \in [0, 2]$ and fit the samples using an ordinary Kriging surrogate. Plot the negative log-likelihood function versus hyperparameter $\theta$. Also plot Kriging predictions with different values of the hyperparameter.

**7.14** When the number of samples increases, the negative log-likelihood function becomes unstable before it reaches a minimum value. This happens when too many samples are present within the wavelength of the function. Plot the negative log-likelihood function versus hyperparameter at different numbers of samples and identify the hyperparameter value that minimizes the negative log-likelihood function. Use true function $y = \sin \pi x$ to generate samples in the range of $x \in [0, 2]$ and fit the samples using an ordinary Kriging surrogate. Plot the negative log-likelihood function versus hyperparameter $\theta$. Also plot the optimum hyperparameter versus the number of samples.

**7.15** Sometimes the negative log-likelihood function may not have a minimum within the range of hyperparameter. Generate equally spaced 20 samples from $y = \sin x + 0.2x$ in the range of $x \in [0, 2\pi]$ and fit the samples using the ordinary Kriging surrogate. Plot the negative log-likelihood function versus hyperparameter $\theta$. Also plot Kriging surrogate predictions at various values of $\theta$ in the range of $x \in [-\pi, 3\pi]$ and compare the prediction accuracy in the interpolation and extrapolation regions.

**7.16** Show the relationship between the negative log-likelihood function and the number of samples by plotting (i) the negative log-likelihood versus hyperparameter $\theta$ and (ii) the optimum hyperparameter $\theta$ versus the number of samples (see Figure 7.14). Generate samples from true function $y = \sin(\pi x)$ and add uniformly distributed random noise$\sim U(-0.005, 0.005)$. Use the number of samples between 6 and 14 and gradually increase the hyperparameter between 0.01 and 1.0. Implement an algorithm to stop increasing the hyperparameter when the correlation matrix becomes ill-conditioned.

**7.17** Six equally spaced samples in the domain of $x \in [0, 1]$ are used to approximate function $y(x) = x^2$ with an ordinary Kriging surrogate. Generate six samples using the true function $y(x) = x^2$. Use a constant global function and Gaussian correlation. (i) Find the hyperparameter that minimizes the negative log-likelihood function by plotting it with different hyperparameter values. (ii) Plot the Kriging approximation along with samples. (iii) Plot the error between the Kriging approximation and the true function.

**7.18** The coefficients of global function are determined from maximum likelihood estimate. In the case of simple Kriging, however, the constant global function is assumed to be known, and often, $\beta = 0$ is used for simple Kriging. Compare the accuracy of simple Kriging with ordinary Kriging, which calculates the constant global function to satisfy the maximum likelihood estimate. Use eight equally spaced samples from $y = \sin x + 0.2x$ in the range of $x \in [0, 2\pi]$. Compare optimal hyperparameter $\theta_{\text{opt}}$ and Kriging predictions in the range of $s \in [-\pi, 3\pi]$.

**7.19** Figure 7.22 shows Kriging predictions of $y = \sin x + 0.2x$ with eight equally spaced samples in the range of $x \in [0, 2\pi]$ using a constant (ordinary Kriging) and a linear (universal Kriging) global function. Reproduce Figure 7.22 and discuss the effect of the global function in the extrapolation region.

**7.20** Consider a true function is given as $y(x) = x\sin(x)$. Generate 10 equally spaced samples in the design space $x \in [0, 14]$, and fit the samples using an ordinary Kriging surrogate. Show the mean prediction and 90% confidence intervals.

**7.21** We are given a function with four pairs of $(x, y)$: $(0, 3)$, $(0.5, 1)$, $(0.7, 3)$, $(1, 17)$, similar to Figure 7.15. (i) Fit a Kriging surrogate and a quadratic polynomial to the data. (ii) Plot and compare their standard error as a function of $x$ (square root of prediction variance).

**7.22** Add four corner points in addition to 20 Latin hypercube samples in Example 7.14 for a Branin–Hu function and fit a Kriging surrogate. Use the three error metrics to compare the accuracy of Kriging surrogate against the one with 20 interior samples. Use a genetic algorithm to find the optimum hyperparameters.

**7.23** Repeat Example 7.14 using different optimization algorithms, such as gradient-based algorithm, genetic algorithm, and simulated annealing algorithm, to find the hyperparameters. Compare the identified hyperparameters with that of Example 7.14.

**7.24** When four corner samples are included, repeat Example 7.14 using different optimization algorithms, such as gradient-based algorithm, genetic algorithm, and simulated annealing algorithm, to find the hyperparameters. Compare the identified hyperparameters with that of Example 7.14.

**7.25** For the Branin–Hu function shown in Example 7.14 with 20 Latin hypercube samples, plot the reduced negative log-likelihood as a function of two hyperparameters, $\theta_1$ and $\theta_2$, as shown in Figure 7.25 and check if the identified hyperparameters are indeed the minimum values of the likelihood.

**7.26** Generate samples from 2D function $y(x_1, x_2) = \sin(x_1)\cos(x_2)$ in the range of $x_1, x_2 \in [0, 2\pi]$ to fit an ordinary Kriging surrogate model. Generate 20 samples from the Latin hypercube sampling method, and use the three error metrics to assess the accuracy of Kriging.

**7.27** Consider Rosenbrock function given as $f(x_1, x_2) = 100\left(x_2 - x_1^2\right)^2 + (a - x_1)^2$. This function has a minimum value 0 at $(x_1, x_2) = (a, a^2)$. When $a = 1$, approximate this function with an ordinary Kriging surrogate in the range of $x_1 \in [-2, 2]$, $x_2 \in [-1, 3]$. Use 24 Latin hypercube samples including four corner points. Assess the accuracy of the Kriging surrogate against the true function using the three error metrics at $31 \times 31$ grid points.

**7.28** Reproduce the negative log-likelihood plots in Figure 7.29 of $y = \sin x$ with 10 samples when (i) samples do not include noise and (ii) samples include random noise $\epsilon \sim N(0, 0.1^2)$.

# 8

# Neural Network Model

## 8.1 Introduction

Due to the recent burst of data science and deep learning technology, neural network (NN) models have been popular in engineering and science. Different names have been used depending on the field of applications, such as machine learning, artificial intelligence, deep learning, etc. In this text, however, the NN is considered as a surrogate model for the prediction of engineering phenomena. It has all the ingredients of being a surrogate model. First, it does not need a physics-based model to predict physical phenomena. Second, it uses samples to fit a mathematical model. In polynomial response surfaces, the mathematical model corresponds to the linear combination of polynomials. In the case of NN, the mathematical model includes weights, biases, and activation functions at different layers. Third, the fitting process determines unknown parameters. In polynomial response surfaces, they are unknown coefficients of polynomials. In the case of NN, the unknown parameters are weights and biases at different layers. In general surrogates, determining unknown parameters is called the fitting process, while in NN, it is called the training process. However, since NNs are developed in a different community from that of surrogates, they use different terminologies. For example, determining unknown parameters is called training in NN, while it is referred to as fitting in surrogate. Also, the samples for fitting a surrogate model are called learning or training data in NN. Since NN is the area that evolves rapidly, it is impractical to cover all the developments in one textbook. Instead, this chapter presents the basics of NN models from the perspective of surrogate models. This includes the effect of the number of samples, the effect of model forms, challenges in the fitting process, and uncertainty associated with prediction.

It is well-known that a brain has about 86 billion neurons, which are the nerve cells that process inputs to lead to action on the basis of these inputs. These neurons are connected with others to form 100 trillion connections to each other. Each neuron is connected to more than 1000 other neurons. When the brain is not satisfied with the results of the action, there is a process of adjusting the connectivity and interaction between neurons in search of improved results. Artificial NNs try to imitate the processing and the training.

Artificial NNs are computing systems vaguely inspired by the biological NNs that constitute brains. The data structures and functionality of networks are designed to simulate associative memory of the brain. As the brain learns the relationship between inputs and action, the computer models also learn by processing inputs and output data, which is also referred to as samples. After learning by enough data, neural nets can predict results from new inputs. The prediction accuracy of NNs depends on the number and diversity of samples processed by NNs. Therefore, even if

*Surrogate Modeling and Optimization: Theories, Applications, and Limitations*, First Edition. Nam-Ho Kim.

Companion website: www.wiley.com/go/kim/surrogate_modeling

artificial NNs have a fancy name to mimic the process of brains, basically they are a kind of surrogate model. That is, unknown parameters are determined (or trained) using samples, and these parameters are used to predict unsampled inputs.

The counterpart of neurons in the brain are nodes in artificial NNs. The collection of connected nodes loosely models the neurons in biological brain. Each connection can transmit a signal to other nodes. A node receives a signal then processes it and sends the processed signal to the nodes connected to it. In computer implementation, the signal at a connection is represented by a real number, and the output of each node is computed by a nonlinear function of the sum of its inputs. Nodes typically have a weight that adjusts as learning proceeds. It is common that the network architecture is divided by three layers: input layer, hidden layer, and output layer. The input layer receives input data from users, while the output layer provides prediction corresponding to the inputs. Although input and output layers are a single layer, the hidden layer can be multiple layers.

The recent development of machine learning has been driven by "deep" learning, which is part of a broader family of machining learning methods based on artificial NNs with representation learning. Learning can be supervised, semi-supervised, or unsupervised. The adjective "deep" comes from the use of multiple layers in the network. Deep learning is a modern variation that is concerned with an unbounded number of layers of bounded size. Due to its flexibility in the number of layers, it allows complicated predictions in many practical applications. Since deep learning significantly increases the number of unknown parameters, it provides computational challenges in the training process, especially optimization algorithms.

The earliest artificial NN was introduced in 1943 by Warren McCulloch, a neurophysiologist, and a mathematician named Walter Pitts [87]. In 1949, Donald Hebb published a book, "The Organization of Behavior," where he formulated the classical Hebbian rule, which is proved to be the basis of almost all neural learning procedures [88]. Perceptron, a simple mathematical representation of neuron, was developed by Frank Rosenblatt in 1958 [89]. Unfortunately, this concept has proved to be limited and can only solve linearly separable problems in Marvin Minsky and Seymour Papert's book Perceptrons, resulting research on NN to an abrupt halt [90].

After long silence in this area, John Hopfield invented associative NN in 1982, which is now more commonly known as Hopfield Network [91]. He applied this technology into a useful device and persuaded many scientists to join this area. At the same time, backpropagation algorithm, a generalization of the Widrow–Hoff learning algorithm, was invented by Rumelhart, Hinton, and Williams [92]. John Hopfield and Geoffrey Hinton received the Nobel Prize in Physics in 2024 due to their contributions for foundational discoveries and inventions that enable machine learning with artificial NNs. These events caused the renaissance of research in NNs. Today, NNs are used to solve a wide variety of problems such as forecasting, classification, and statistical pattern recognition. In this text, however, NNs are used for approximating a function. That is, the main focus of this text is to use NN for the purpose of regression.

The NN algorithm is a representative data-driven prediction method, in which a network model learns a way to produce desired outputs, such as displacement and stress, by reacting to given inputs, such as dimensions and loads. The relationship between inputs and outputs depends on how the NN architecture is constructed and which functions (called the transfer or activation function) are associated with the architecture. Once the network model learns enough about the relationship between inputs and outputs, it can be used for the purpose of prediction.

There are many different types of NN, such as the feedforward NN [93] that conveys the information in one direction, the radial basis function network [94] that usually uses the Gaussian radial basis function, the recurrent NN [95] that has local feedback connections between the input and outputs, and so forth. There are other NNs, such as fuzzy-neural [96], wavelet [97], associative-memory [98], modular [99] and hybrid [100, 101] NNs. Among these types, the feedforward NN is the most common one, and it is introduced in this chapter.

## 8.2 Feedforward Neural Network Model

### 8.2.1 Concept of Feedforward Neural Network

The basic processing units in NNs are neurons or nodes. Based on different functions of nodes, they can be grouped into different layers such as input layer, output layer, and hidden layer. The first layer is an input layer that is used for receiving input data from outside, while the last layer is an output layer used to send processed data out of the NN. For the purpose of surrogate modeling, the number of input nodes is the same as the dimension of input variables. Since surrogate modeling normally predicts one function, a single output node is used. The hidden layer, existing between input and output layers and as the name suggested, is invisible from outside and does not have any interaction with outside. Normally the input and output layers are a single layer, but the hidden layer can be multiple layers in the modern deep learning architecture.

The most popular architecture of NN is feedforward neural network (FFNN), which means that information only moves in the forward direction from the input layer, through the hidden layer and to the output layer. There is no feedback or flow within layers. A two-layer FFNN is illustrated in Figure 8.1, which is a basic form of it. In the figure, circles represent nodes, which can belong to three different layers: input, hidden, and output layers. The input nodes and output nodes represent the input and output variables, respectively. The hidden nodes connect the input and output variables, i.e., they are fed input information and then forward the information to the output node. This is why it is called a FFNN. The number of nodes in the hidden layer should be determined to properly express the mechanism between the input and output layers.

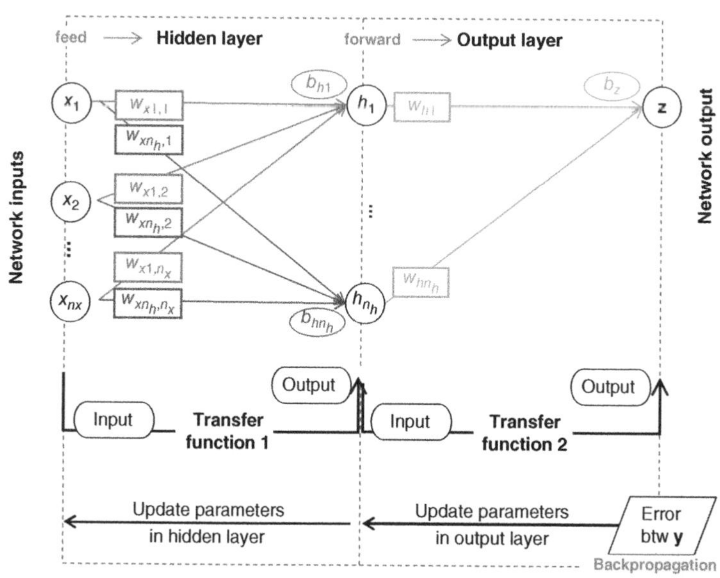


**Figure 8.1** Illustration of a typical network model: feedforward-backpropagation NN.

There are two different ways of defining layers in NN. First, each set of nodes in the same column can be called a layer, i.e., input, hidden, and output layers. Also, a layer can be understood as an interface between different columns of nodes. These two definitions are basically the same, but we follow the second one, as shown in Figure 8.1. That is, the hidden layer is located between input and hidden nodes, while the output layer is between hidden and output nodes. All nodes in a layer are connected with all nodes in the following layer. That is, all nodes in a layer affect all nodes in the following layer.

There are two types of parameters to be estimated in Figure 8.1; the rectangles and the ellipses are called weights and biases, respectively. Weights are interconnection parameters between different nodes, which represent the contribution for each input data when calculating the weighted sum of all inputs. They can be either positive or negative, where a positive weight is considered as excitation and a negative one as inhibition. For each node in hidden nodes and output node, there is a bias attached to it. Biases are usually used as threshold value added after weighted sum of inputs from the previous layer is calculated.

We define the name of weights and biases as follows. Weights in the hidden layer are input weights ($w_x$) since they are associated with the input nodes. On the other hand, weights in the output layer are hidden weights ($w_h$). Also, there are hidden biases ($b_h$) and an output bias ($b_z$), which are biases for function inputs in each layer. These parameters are integrated into some functions called transfer or activation functions to determine the relation between input and output variables. The input variables with input weights and hidden biases become the input for an activation function in the hidden layer, and whose outputs are assigned into the hidden nodes. Then, the hidden nodes (outputs in the hidden layer) with the hidden weights and output bias are used as the input for an activation function in the output layer, and whose output is the final network output. The mechanism of FFNN will be explained in the following section in more detail.

In the perspective of surrogate modeling, the number of hidden layers and hidden nodes and the types of activation functions correspond to the form of the surrogate model, while the weights and biases are unknown model parameters. The model form becomes more complicated when more hidden layers and nodes are used. The complexity of the model form is accompanied by the increased number of model parameters, which make it difficult to fit the model with samples. As we discussed in Chapter 3, it is normally expected that the number of samples is about two or three times more than that of unknown parameters in order to prevent overfitting. Compared to other surrogate models, the number of unknown parameters for a complicated NN model is much more than that of other surrogate models. Therefore, the NN model requires a substantial number of training samples. When not enough samples are available, special techniques are required to prevent overfitting the samples.

The training process is equivalent to finding the optimal parameters, weights, and biases, such that the network model accurately represents the relationship between inputs and outputs. Once the network model is trained enough, the model is functionalized using activation functions, weights, and bias parameters. For this process, the backpropagation is typically employed, which propagates the error between training data and the network outputs backward. That is, updating-related parameters (using gradient or Jacobian) for hidden weights and output bias (output layer) are updated first, and then the parameters for input weights and hidden biases (hidden layer) are updated next based on the updated parameters in the output layer. This is basically an optimization process to find the best weights and biases to minimize the mean squared error (MSE) between network predictions and training data. Since the FFNN is usually trained with backpropagation method, it is often called the backpropagation NN. Once all parameters are determined, they can be used for predicting output at unsampled locations. This process is called simulation. The process of the backpropagation will be explained in the following section.

Different from linear regression in polynomial response surfaces, the training process in NN models may not yield the best combinations of parameters. In linear regression, the square sum of errors is a quadratic function of unknown coefficients. Therefore, the Karush–Kuhn–Tucker necessary condition yields the global optimum parameters. In the case of NNs, however, the MSE is a nonlinear function of parameters with many local optima. Therefore, gradient-based optimization algorithms can only find local optima. If different initial values of parameters are used, it often yields different optimum parameters. Therefore, in the case of NN, there is uncertainty associated with the train process itself. This aspect will be discussed in detail later in this chapter.

### 8.2.2 Feedforward Mechanism

The FFNN model combines the following three principles: (i) The output of a neuron is a function of inputs from the connected neurons, (ii) a neuron favors inputs from some connections over others, and (iii) a neuron saturates if it gets too much input. The basic mathematical relationship in the feedforward network model is a linear combination of input nodes with weights and biases as an input to the activation function in the next layer. The weights are usually multiplied with the value at their associated nodes, and then the biases are added to the sum of the multiplication results to be an input to the activation function. Consider an example of neurons connection shown in Figure 8.2. The current neuron receives input from $n$ connected nodes. Each connection has its own weight $w_k$. The current neuron generates linear combination of all connected inputs as

$$u = w_1x_1 + w_2x_2 + \ldots + w_nx_n + b \tag{8.1}$$

The linear combination in Eq. (8.1) is nothing but linear polynomials, which is not flexible enough to represent nonlinear relationship. NNs incorporate this linear relationship with activation functions to describe nonlinear behaviors of functional relationships. In addition, this linear combination can increase the current neuron significantly when the inputs increase. The activation function prevents such trend by saturating the output. Therefore, the output from the current neuron becomes

$$z = \Phi(u) \tag{8.2}$$

The idea of saturation is that the output cannot exceed a lower or upper limit. A straightforward way to achieve this limit is

$$z(u) = \begin{cases} \Phi_{\text{max}}, & \text{if } u > \Phi_{\text{max}} \\ \Phi_{\text{min}}, & \text{if } u < \Phi_{\text{max}} \end{cases} \tag{8.3}$$

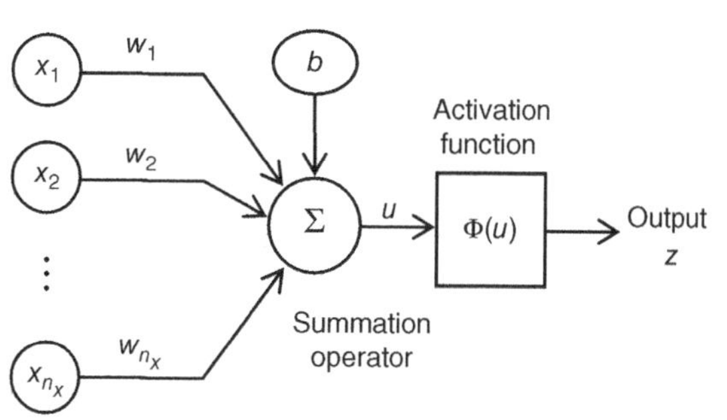


**Figure 8.2** Linear combination of inputs and activation function for NNs.

However, this limit yields a non-smooth function, which is difficult to manage in gradient-based optimization algorithms. Instead, a smooth approximation of Eq. (8.3) can be employed. For that purpose, the following sigmoid (logistic) function is often used:

$$\Phi(u) = \frac{e^u}{1 + e^u} \tag{8.4}$$

This function is monotonic and differentiable, as shown in Figure 8.3. It is desirable to have linear behavior near the origin and to show saturation at large input values. Differentiability is important as we will use gradient-based optimization algorithm for the training process. It is possible to use other functions that have equivalent properties, such as hyperbolic tangent function, arctangent function, and error function.

In order to explain the feedforward mechanism, let us consider a network structure with $n_x$ input variables, $n_h$ hidden nodes, and a single output node, as shown in Figure 8.1. The feedforward mechanism can be explained as

$$\begin{aligned} \mathbf{h} &= \Phi_h(\mathbf{W}_x\mathbf{x} + \mathbf{b}_h) \\ z &= \Phi_z(\mathbf{w}_h\mathbf{h} + b_z) \end{aligned} \tag{8.5}$$

where $\mathbf{W}_x$ is an $n_h \times n_x$ input weight matrix associated with a $n_x \times 1$ vector of input variables, $\mathbf{x}$, and $\mathbf{b}_h$ is a $n_h \times 1$ hidden bias vector, which is the input for an activation function, $\Phi_h$, in the hidden layer, whose function outputs are in $\mathbf{h}$, a $n_h \times 1$ vector of hidden layer outputs. For the relationship between the hidden and output layers, we will consider the case when the network model predicts a scalar output. That is, the output layer has a single node. Then, hidden node $\mathbf{h}$ is combined with $\mathbf{w}_h$, that is a $1 \times n_h$ row vector of hidden weights to be the input for the activation function, $\Phi_z$, in the output layer with output bias $b_z$. Then, $z$ is of the output variable (i.e., prediction of the network model). For the convenience of following derivations, intermediate variables can be introduced. Let $\mathbf{u} = \mathbf{W}_x\mathbf{x} + \mathbf{b}_h$ be the linear combination of inputs before applying the activation functions. Then the hidden node becomes $\mathbf{h} = \Phi_h(\mathbf{u})$. Similarly, let $v = \mathbf{w}_h\mathbf{h} + b_z$ be the linear combination of hidden nodes before applying the activation function. Then the output becomes $z = \Phi_z(v)$.

During the training step, the vector of $n_y$ sample data, $\mathbf{y}$, is available. These samples are measured at $n_y$ different inputs. That is, the following pairs of measurement data are available: $(\mathbf{x}_1, y_1), (\mathbf{x}_2, y_2), \ldots, \left(\mathbf{x}_{n_y}, y_{n_y}\right)$. The training process with pairs of inputs and output samples is

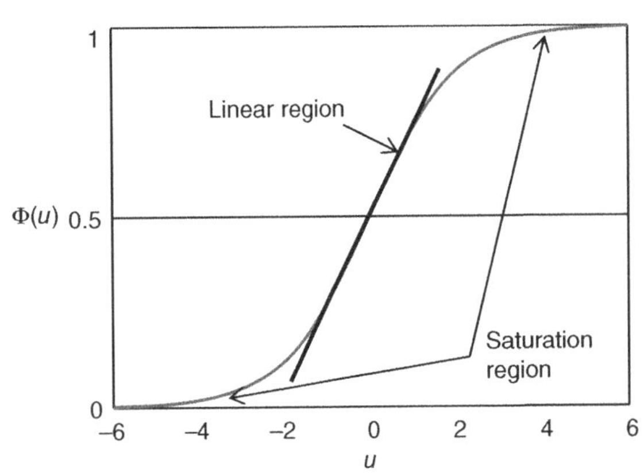


**Figure 8.3** Sigmoid (logistic) function as an activation function.

referred to as supervised learning. Let $\mathbf{X} = [\mathbf{x}_1, \mathbf{x}_2, ..., \mathbf{x}_{n_y}]$ be a $n_x \times n_y$ matrix of input data at all sample locations, and $\mathbf{y} = \{y_1, y_2, ..., y_{n_y}\}^T$ be a column vector of output samples. For given weights and bias parameters, the vector of outputs, $\mathbf{z}$, can be calculated at the same $n_y$ sample locations. Then, the optimization algorithm finds the weights and bias parameters that minimize the difference between the outputs, $\mathbf{z}$, and the sample data, $\mathbf{y}$. The training process utilizes $n_y$ training data in a batch process. Therefore, Eq. (8.5) that is for one pair of data can be rewritten by considering the number of measurement/training data, $n_y$, as follows:

$$\begin{aligned} \mathbf{H} &= \Phi_h(\mathbf{W}_x\mathbf{X} + \mathbf{B}_h) \\ \mathbf{z} &= \Phi_z(\mathbf{w}_h\mathbf{H} + \mathbf{b}_z) \end{aligned} \tag{8.6}$$

where $\mathbf{B}_h = [\mathbf{b}_h \;...\; \mathbf{b}_h]_{n_h \times n_y}$ is a hidden bias matrix, $\mathbf{H}$ is a $n_h \times n_y$ matrix of hidden layer outputs, $\mathbf{b}_z = \{b_z, b_z, ..., b_z\}$ is a $1 \times n_y$ output bias vector, and $\mathbf{z}$ is a $1 \times n_y$ vector of network simulation outputs.

**Example 8.1 Conversion to Two-layer Linear Network Model**

Write the expression of the output of a two-layer FFNN with two input and one hidden nodes as a function of input variables, weights and biases. Use a pure linear function, $z = x$ for the activation functions in both layers.

**Solution:**

This problem can be modeled, as shown in Figure 8.4, which can be formularized with $n_x = 2$ and $n_h = 1$ in Eq. (8.5). The input, weights, and bias matrices/vectors are given as follows:

$$\mathbf{x} = \begin{Bmatrix} x_1 \\ x_2 \end{Bmatrix}, \quad \mathbf{W}_x = [w_{x1}, w_{x2}], \quad \mathbf{b}_h = b_h$$

The input for the activation function in the hidden layer is the same as its output since the function is pure linear. Therefore, the hidden node output can be obtained as

$$h = u = \mathbf{W}_x\mathbf{x} + b_h = w_{x1}x_1 + w_{x2}x_2 + b_h$$

Then, the relationship between inputs and output of the network model can be obtained as

$$\begin{aligned} z = v &= w_h h + b_z \\ &= w_h(w_{x1}x_1 + w_{x2}x_2 + b_h) + b_z \\ &= (w_{x1}w_h x_1 + w_{x2}w_h x_2) + b_h w_h + b_z \end{aligned}$$

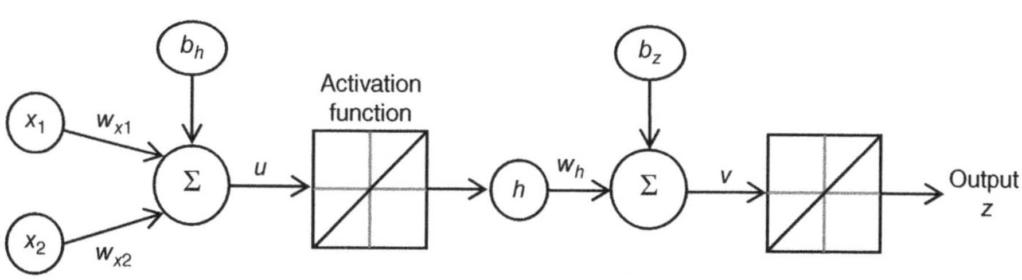


**Figure 8.4** Illustration of two-layer FFNN with two input and one hidden nodes.

Note that the two-layer FFNN with linear activation functions can be reduced to a one-layer network model with a pure linear function, as

$$z = w_{x1}^* x_1 + w_{x2}^* x_2 + b_z^*$$

where $w_{x1}^* = w_{x1} w_h$, $w_{x2}^* = w_{x2} w_h$, and $b_z^* = b_h w_h + b_z$.

The above example can be generalized for the case when the pure linear activation function is used. In such a case, Eq. (8.5) can be rewritten as

$$z = \mathbf{w}_h \mathbf{W}_x \mathbf{x} + \mathbf{w}_h \mathbf{b}_h + b_z \tag{8.7}$$

which is a one-layer network model with $\mathbf{w}_h \mathbf{W}_x$ being input weights and $\mathbf{w}_h \mathbf{b}_h + b_z$ being the output bias. In general, for any multi-layer network model with pure linear activation functions, it is always possible to convert it to a single-layer network model. When the activation function is nonlinear, such as tangent sigmoid function, such a simplification cannot be possible. As shown in Eq. (8.7), no matter how many hidden nodes are used, the network prediction will be a linear function of input variables. The nonlinear behavior appears when nonlinear activation functions are used. Therefore, it is important to use nonlinear activation functions in order to approximate the nonlinear behavior of a function.

### 8.2.3 Activation Functions

The model form of NNs is determined by the number of hidden layers and nodes and activation functions. Activation functions play a critical role in predicting nonlinear behavior of the approximate function. Therefore, it is important to use the right activation functions for training NNs such that they can provide accurate predictions.

The activation functions characterize the relationship between two adjacent layers. For mathematical reasons, the activation functions should be differentiable because the gradient of MSE with respect to parameters is needed in the backpropagation training process. There are numerous activation functions available, such as the sigmoid, logistics, hyperbolic tangent, rectified linear unit (ReLU), inverse, and linear functions [102]. The choice of activation functions mainly depends on the complexity and performance of NN models (see Exercise Problem 8.7). Each activation function has its own unique properties and is suitable for certain use cases. For example, the sigmoid function is ideal for binary classification problems, and the ReLU function helps overcome the vanishing gradient problem. Using the right activation function for the task leads to faster training and better performance. Although there are numerous activation functions available, only a few popular activation functions are discussed in this section.

Pure linear function: The pure linear activation function returns the input value as the output. Mathematically, this can be defined as

$$\Phi(u) = u \tag{8.8}$$

The pure linear activation function is commonly used at the output layer for the purpose of regression. Since the linear function does not distort the input, the actual prediction value is stored in the output node. However, this function is rarely used for the hidden layer. Since the main purpose of the hidden layer is to describe nonlinear behavior, the linear function will deteriorate its performance.

Sigmoid function: The sigmoid activation function is a smooth, continuously differentiable, S-shaped function that is commonly used for an activation function in NNs. The mathematical form of this function is given as

$$\Phi(u) = \frac{e^u}{1 + e^u} \tag{8.9}$$

The input can be in the range of $(-\infty, +\infty)$, while the range of output is (0, 1). Due to the monotonicity and the range of output, this function can be interpreted as probability and, thus, makes it ideal for binary classification problems. For the perspective of backpropagation training, this function is effective because the gradient is high near $u = 0$. On the other hand, this function is difficult to train when the magnitude of input variable is high, where the function is almost flat, and the gradient is close to zero. When the gradient is close to zero, the backpropagation algorithm yields insignificant changes in parameters, which slows down the algorithm.

Hyperbolic tangent function: The tanh (hyperbolic tangent) activation function can be defined as

$$\Phi(u) = \frac{e^u - e^{-u}}{e^u + e^{-u}} \tag{8.10}$$

The input can be in the range of $(-\infty, +\infty)$, while the range of output is $(-1, 1)$. Due to its range, this function works better for negative values of output. This function is symmetric with respect to origin and higher gradient near $u = 0$ than that of the sigmoid function. Because of this reason, this function converges faster. Despite this advantage, the tanh function still suffers from the vanishing gradient problem at large magnitude of input variables. The tanh function is often used for hidden nodes and works well when samples are normalized to have mean zero. Since approximate function can have both positive and negative values, the tanh function is preferred to the sigmoid function.

**Rectified linear unit function**: The ReLU activation function can be defined as

$$\Phi(u) = \max(0, u) \tag{8.11}$$

This function cuts off negative input values and becomes pure linear for positive input values. It does not have the vanishing gradient issue. That is, it does not alter the scale of positive inputs and allows the gradient to pass through unchanged during backpropagation. However, it is not differentiable at $u = 0$, which allows NNs to model complex behavior of approximate functions. The cut-off function makes some input nodes deactivated, leading to more efficient computation. Figure 8.5 shows the four activation functions discussed above. Table 8.1 summarized the characteristics of various activation functions.

In order to show the capability of representing complex behavior using a simple activation function, a two-layer network model is considered that has one input node, one (or five) hidden node, and one output node. The weights and biases are randomly generated from a uniform distribution: $U(-5, 5)$, which is used to calculate the network model output at an arbitrary input based on Eq. (8.5). Then, Figure 8.6 can be obtained for $x \in [-1, 1]$ by repeating the process to calculate the network output five times with randomly generated weights and biases. Figure 8.6(a) and (b) are when the number of hidden nodes is one and five, respectively. The model complexity depends on the number of hidden nodes. For the activation functions, other forms can be used, but the pure linear and the tangent sigmoid functions are usually employed.

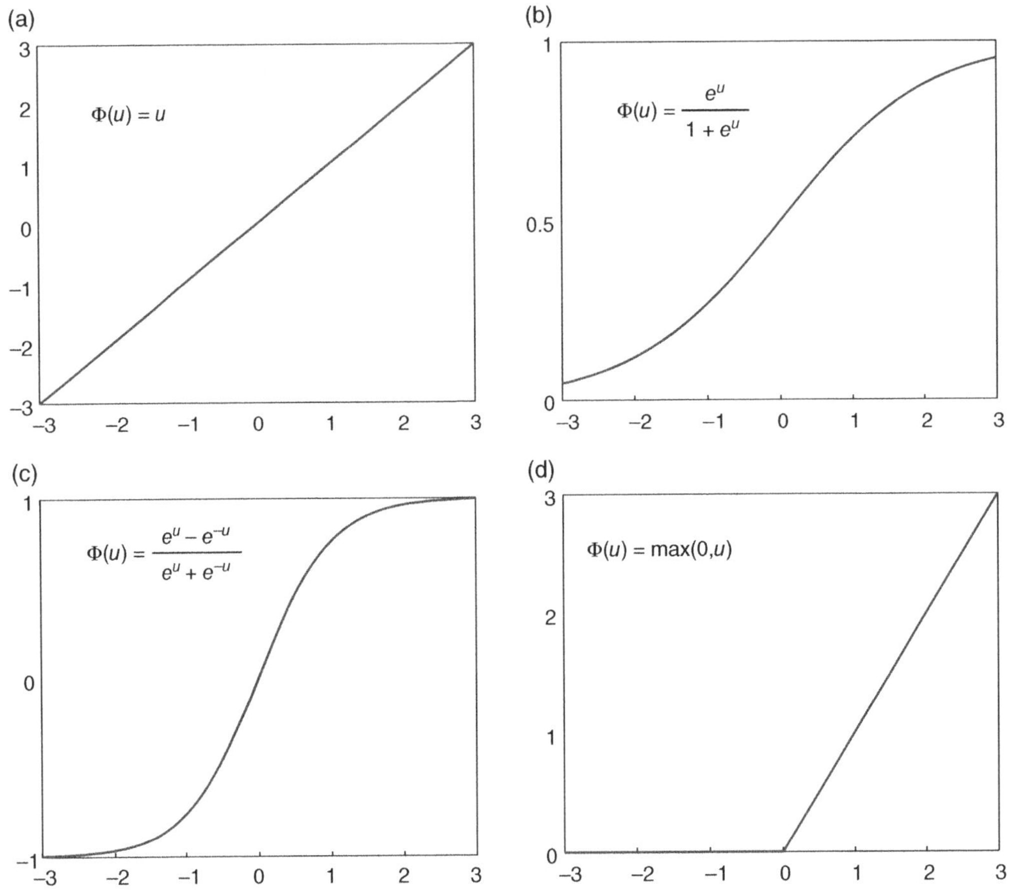


**Figure 8.5** Examples of activation functions. (a) Pure linear, (b) sigmoid, (c) hyperbolic tangent, and (d) rectified linear unit functions.

**Table 8.1** Characteristics of activation functions.

| **Activation function** | **Characteristics** |
|---|---|
| Sigmoid | • Mainly used in one-class classification problems within output layer<br>• Vanishing gradient problem |
| Tanh | • Compensating for shortcomings of sigmoid function (zero-centered; faster convergence)<br>• Still vanishing gradient problem |
| ReLU | • Compensating for vanishing gradient problem<br>• Mainly used for hidden layer<br>• Simple and fast<br>• Dying ReLU problem |
| Leaky ReLU | • Compensating for shortcomings of ReLU function (dying ReLU problem) |
| Softmax | • Mainly used in multiclass classification problems within output layer |
| Linear | • Mainly used in regression analysis problems within output layer<br>• Linear activation function problems |

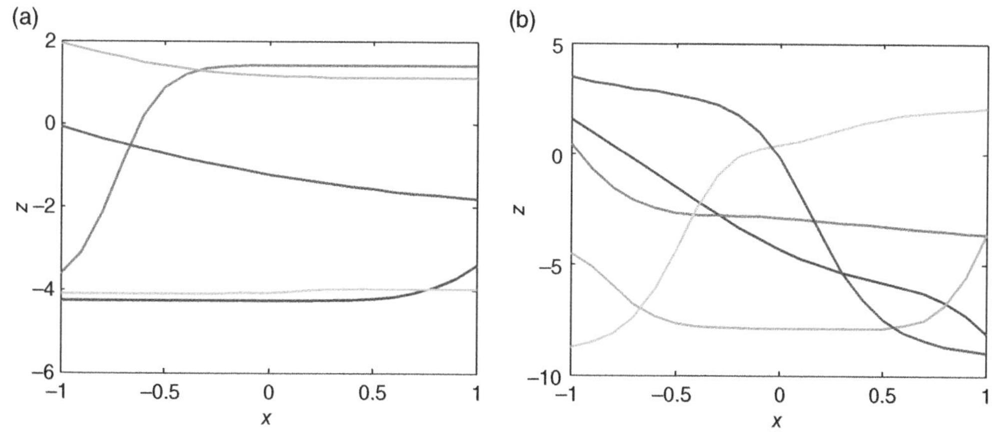


**Figure 8.6** Various network predictions with (a) one hidden node and (b) five hidden nodes.

**Example 8.2 Activation Functions**

Design the FFNN model that can predict quartic polynomial $z(x) = 5x^4 + 2x^3 - x^2 + 3x - 2$ exactly. Determine the number of hidden nodes, activation functions, weights, and biases.

**Solution:**

In order to predict the polynomial function exactly, it would be necessary to design four activation functions, $\Phi_{h1}(u) = u$, $\Phi_{h2}(u) = u^2$, $\Phi_{h3}(u) = u^3$, and $\Phi_{h4}(u) = u^4$. The constant term can be managed by hidden bias term. Therefore, the network model has one input node, four hidden nodes, and one output node, as shown in Figure 8.7. All weights in hidden layers are one, and all hidden biases are zero. This is because we need to apply the activation to the input directly. On the other hand, output weights are the same as the coefficients of polynomials, and the output bias is the constant term. Therefore,

$$W_{xi} = 1, b_{hi} = 0, \quad i = 1, ..., 4$$

$$w_{h1} = 3, \quad w_{h2} = -1, \quad w_{h3} = 2, \quad w_{h4} = 5, \quad b_z = -2$$

Figure 8.7 shows the FFNN model with four hidden nodes.

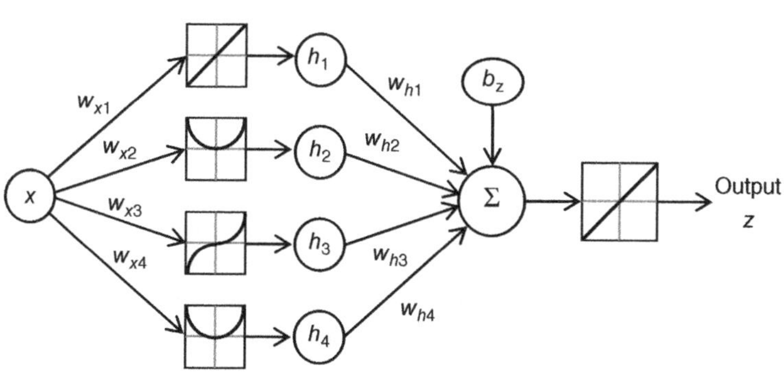


**Figure 8.7** FFNN model with four hidden nodes for quartic polynomials.

### 8.2.4 Backpropagation Process

The backpropagation is a training method to determine weights and biases through a learning/optimization algorithm by backward propagation of the errors between the training data and the network outputs. The backpropagation can be understood as a layer-by-layer updating, which is more effective to find optimum values of the parameters than updating all of them together. In a given network model, the unknown model parameters are weights and biases. For the convenience of notation, the collection of all unknown parameters is defined as $\mathbf{d} = \{\mathbf{W}_x, \mathbf{b}_h, \mathbf{w}_h, b_z\}$. This definition is not mathematically correct because $\mathbf{W}_x$ is a matrix, $\mathbf{b}_h$ and $\mathbf{w}_h$ are vectors, and $b_z$ is a scalr. However, the vector $\mathbf{d}$ can be considered as a collection of all unknown parameters in the network model. The dimension of vector $\mathbf{d}$ is $n_x \times n_h + 2n_h + 1$. The steps for training process with a $n_x \times n_y$ matrix of training input data $\mathbf{X}$ and a $1 \times n_y$ vector of training output data $\mathbf{y}$ as follows:

Step 1. Initial values of the parameters (weights and biases) are set. Selection of initial weight parameters is important as it affects whether NN can find the global optimum and how fast the convergence rate is. Too large initial weights are likely to make them fall in the saturation region easily and the network barely learns, while too small weights lead to a low learning rate. For optimum results, initial weight parameters are usually set as random numbers chosen between −1 and 1.

Step 2. Network simulation outputs $\mathbf{z}$ are calculated with the weights and biases values in Step 1 through the feedforward process (see Figure 8.1 and Eq. (8.6)). This step is performed for all $n_y$ training data. The dimension of $\mathbf{z}$ is $1 \times n_y$.

Step 3. The errors between the training output data $\mathbf{y}$ and the simulation outputs $\mathbf{z}$ are calculated. Different errors can be used, such as MSE, mean absolute error, and sum of squared errors. For example, the following MSE can be used:

$$F(\mathbf{d}) = \frac{1}{n_y}\sum_{i=1}^{n_y}[y_i - z_i(\mathbf{d})]^2 = \frac{1}{n_y}\sum_{i=1}^{n_y}[y_i - z_i(\mathbf{W}_x, \mathbf{b}_h, \mathbf{w}_h, b_z)]^2 \tag{8.12}$$

During the training process, optimization algorithms are used to minimize MSE. Since it represents how much the model differs from data, $F$ is often referred to the loss function. In the optimization perspective, $F$ is the goal of optimization, and because of this reason, it is also referred to as the objective function or the cost function.

Step 4. The changes of weights and biases, $\Delta\mathbf{w}_h$ and $\Delta b_z$, of the output layer are calculated. In the backpropagation stage, the weights and biases at the output layer are updated first. Different optimization algorithms have different methods of calculating $\Delta\mathbf{w}_h$ and $\Delta b_z$. For example, the gradient descent and the Levenberg–Marquardt (LM) algorithms, respectively, calculate a gradient and a Jacobian of MSE with respect to the parameters [103]. The gradient descent is a basic learning algorithm used with the backpropagation, but the LM algorithm is commonly used since it is fast and stable in convergence.

Step 5. The changes in weights and biases, $\Delta\mathbf{W}_x$ and $\Delta\mathbf{b}_h$, of the hidden layer are calculated. Calculation of $\Delta\mathbf{W}_x$ and $\Delta\mathbf{b}_h$ depends on $\Delta\mathbf{w}_h$ and $\Delta b_z$ in Step 4 and also MSE in Step 3.

Step 6. The weights and biases are updated by adding the changes in Step 4 and Step 5.

Step 7. Steps 1∼6 are repeated until the network performance reaches the stopping criterion.

The key ingredient of backpropagation is the gradients of MSE in Eq. (8.12) with respect to parameters. These gradients are used to update parameters iteratively. In FFNN, since the effect of each parameter propagates in the forward direction, the gradients of MSE are calculated in the backward direction, from which the name backpropagation came from. First, consider the gradient of MSE with respect to $j$th hidden weight $w_{hj}$ as

$$\frac{\partial F}{\partial w_{hj}} = \sum_{i=1}^{n_y} \left( \frac{\partial F}{\partial z_i} \times \frac{\partial z_i}{\partial v_i} \times \frac{\partial v_i}{\partial w_{hj}} \right) \tag{8.13}$$

The first term can be calculated by differentiating Eq. (8.12), the second term is the derivative of the activation function $\Phi'_z(v_i)$, and the last term is simply the hidden node $h_j(\mathbf{x}_i)$. Therefore, the gradient can be written as

$$\frac{\partial F}{\partial w_{hj}} = \sum_{i=1}^{n_y} \left( -\frac{2}{n_y}(y_i - z_i) \times \Phi'_z(v_i) \times h_j(\mathbf{x}_i) \right) \tag{8.14}$$

Similarly, the derivative with respect to the output bias can be calculated as

$$\frac{\partial F}{\partial b_z} = \sum_{i=1}^{n_y} \left( -\frac{2}{n_y}(y_i - z_i) \times \Phi'_z(v_i) \right) \tag{8.15}$$

The two terms in Eq. (8.15) appear repeatedly for other gradient calculation; the product of these two terms is often pre-calculated and stored to enhance computational efficiency.

In the backpropagation process, the hidden weights and output bias are updated first. The calculated gradients are used to update parameters during the descent method. The gradient descent method is a representative method in optimization where gradient information is used to find parameters that maximize or minimize the objective function. The gradient descent method calculates changes in weights and biases by multiplying the gradients with a step size. In order to control the step size, a learning rate, $\alpha$, is multiplied with the gradients. The hidden weights and output bias are updated by

$$w_{hj} = w_{hj} - \alpha \frac{\partial F}{\partial w_{hj}}$$

$$b_z = b_z - \alpha \frac{\partial F}{\partial b_z} \tag{8.16}$$

where $\alpha$ is the learning rate. Using this update scheme, the weights and biases that minimize the MSE can be found. In Eq. (8.16), $j = 1, \ldots, n_h$ is the index for hidden nodes.

Once hidden weights and output bias are updated, the input weights and hidden biases are updated using the same gradient descent method. The gradients with respect to input weights are more complicated but basically follow the same chain rule of differentiation. Let $W_{xj,k}$ is the weight that connects $j$th input variable with $k$th hidden node. Then, the gradient can be defined as

$$\frac{\partial F}{\partial W_{xj,k}} = \sum_{i=1}^{n_y} \left( \frac{\partial F}{\partial z_i} \times \frac{\partial z_i}{\partial v_i} \times \frac{\partial v_i}{\partial h_k} \times \frac{\partial h_k}{\partial u_k} \times \frac{\partial u_k}{\partial w_{xj,k}} \right) \tag{8.17}$$

It is noted that even if there are $n_h$ hidden nodes, only $h_k$ is connected with $W_{xj,k}$. Therefore, it is unnecessary to consider all hidden nodes. The first two terms on the right-hand side of Eq. (8.17) are identical to that of Eq. (8.13). The third term is the hidden weight $w_{hk}$, the fourth term is the derivative of hidden activation function $\Phi'_h(u_k)$, and the last term is input $x_j$. Therefore, the gradient can be written as

$$\frac{\partial F}{\partial W_{xk,j}} = \sum_{i=1}^{n_y} \left( -\frac{2}{n_y}(y_i - z_i) \times \Phi'_z(v_i) \times w_{hj} \times \Phi'_h(u_j) \times x_k \right) \tag{8.18}$$

It is noted that variables $x_k$, $u_j$, and $v_i$ should be calculated using $i$th input sample $\mathbf{x}_i$. Also, the updated hidden weight $w_{hj}$ in Eq. (8.16) is used. That is, the information is updated in the backward direction. Similarly, the derivative with respect to the hidden bias can be calculated as

$$\frac{\partial F}{\partial b_{hj}} = \sum_{i=1}^{n_y}\left(-\frac{2}{n_y}(y_i - z_i) \times \Phi_z'(v_i) \times w_{hj} \times \Phi_h'(u_j)\right) \tag{8.19}$$

The calculated gradients are used to update the input weights and hidden biases, as

$$W_{xj,k} = W_{xj,k} - \alpha\frac{\partial F}{\partial W_{xj,k}}$$

$$b_{hj} = b_{hj} - \alpha\frac{\partial F}{\partial b_{hj}} \tag{8.20}$$

where $j = 1, \ldots, n_h$ is the index for hidden nodes, and $k = 1, \ldots, n_x$ is the index for input nodes. Using this update scheme, the weights and biases that minimize the MSE can be found.

When all parameters are updated using Eqs. (8.16) and (8.20), one iteration of learning is finished, which is called an epoch. This updating process is repeated until the maximum allowed number of epochs is reached. The updating process can be stopped before the maximum number of epochs when certain conditions are met, which is referred to as early stopping.

As we discussed in Chapter 5, all gradient-based optimization algorithms can only find local optima unless the optimization problem is convex. Especially with nonlinear activation functions, the MSE is highly nonlinear function of parameters, and many local optima exist. Therefore, the local optimum that the gradient descent method in Eqs. (8.16) and (8.20) found strongly depends on the initial starting values of parameters. Many gradient descent algorithms generate the initial values randomly. Therefore, it is useful to repeat the training algorithm multiple times to increase the chance to find better values of trained parameters. More details will be discussed in the next section.

In Chapter 5, we also discussed that the steepest descent method is not computationally efficient because the algorithm does not use information from the previous iteration. Example 5.4 illustrates the zigzagging behavior of the steepest descent method. In order to improve the convergence rate of an optimization algorithm, it would be necessary to use more advanced methods, such as the conjugate gradient method or those methods using Hessian information. The gradient descent method in Eqs. (8.16) and (8.20) will perform worse than the steepest gradient method because each parameters are updated independently, and there is no one-dimensional line search. Basically, the learning rate $\alpha$ in Eqs. (8.16) and (8.20) plays the role of step size in one-dimensional line search, but it is a fixed value, not determined by function's behavior. However, the gradient descent method is still dominantly used in the machine learning community because of its simplicity. The uniqueness of training optimization in NN is the huge size of parameters. Since there are so many parameters, calculating/updating the Hessian matrix will be infeasible computationally. However, it would be an interesting research topic to explore the possibility of using the conjugate gradient method or including line search algorithm.

**Example 8.3 Backpropagation**

Two samples $(x_i, y_i) = (0, -1), (1, 1)$ are generated from a linear function $y(x) = 2x - 1$. Use the gradient descent method to fit the FFNN model with learning rate $\alpha = 0.2$ and $\alpha = 0.5$. The FFNN model has a single hidden node and pure linear activation function for both hidden and output layers. Assume the initial weights are one and initial biases are zero.

**Solution:**

Since there is one input node and one hidden node, the unknown parameters are $w_x$, $b_h$, $w_h$, and $b_z$. Starting from their initial values, 50 iterations are performed to update these parameters by reducing the mean square sum of errors. The updating cycle is stopped when the mean square sum of error becomes less than $10^{-6}$. In the backpropagation process, first $\Delta w_h$, and $\Delta b_z$ are calculated, followed by $\Delta w_x$, and $\Delta b_h$. The following Matlab script is used.

```
clear;
x=[0 1]; y=[-1 1];
alpha=0.2;    %alpha=0.5;
wx=1; wh=1; bh=0; bz=0;
for i=1:50
 h=wx*x+bh;
 z=wh*h+bz;
 MSE(i)=sum((y-z).^2)/2;
 if MSE(i)
 dwh = -(y(1)-z(1))*h(1)      - (y(2)-z(2))*h(2);
 dbz = -(y(1)-z(1))           - (y(2)-z(2));
 wh = wh -alpha*dwh;
 bz = bz -alpha*dbz;
 dwx = -(y(1)-z(1))*wh*x(1) - (y(2)-z(2))*wh*x(2);
 dbh = -(y(1)-z(1))*wh        - (y(2)-z(2))*wh;
 wx = wx -alpha*dwx;
 bh = bh -alpha*dbh;
end
epoch=1:i; plot(epoch,MSE)
```

In the Matlab script, the learning rate $\alpha = 0.2$ is used first. The convergence history of the mean square sum of errors is shown in Figure 8.8(a). After 20 epochs, the mean square sum of errors

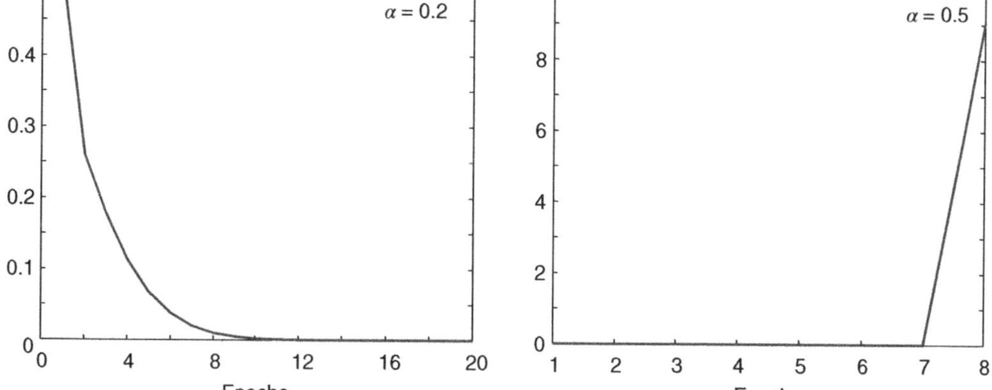


**Figure 8.8** History of MSEs. (a) $\alpha = 0.2$ and (b) $\alpha = 0.5$.

becomes less than the threshold $10^{-6}$. When the learning process is converged, the parameters are determined as

$$W_x = 1.3893, \quad b_h = -0.4240, \quad w_h = 1.4383, \quad b_z = -0.3890$$

At first glance, the network model seems different from the exact model. It is expected that the exact network model should have $W_x = 2$, $b_h = 0$, $w_h = 1$, $b_z = -1$. However, if the equivalent model in Eq. (8.7) is used for pure linear activation functions, the above results can be rewritten as

$$z(x) = w_h W_x x + w_h b_h + b_z = 2x - 1$$

Therefore, the trained network model can produce the exact function. This happens because there are more parameters than samples. Many different combinations of parameters can yield the same model.

The learning rate also plays a significant role in convergence. As shown in Figure 8.8(b), the learning process failed to converge when the learning rate $\alpha = 0.5$. Initially the MSE slowly increases, but after five epochs, it increased dramatically and became not-a-number (NaN) in Matlab. Therefore, it is important to choose an appropriate learning rate to make the optimization algorithm fast and robust.

## 8.3 Matlab Functions for Feedforward Neural Network

Matlab provides Deep Learning Toolbox and Statistical and Machine Learning Toolbox for NNs. A sequence of Matlab functions can be used to define a network model to set up configurations, to train the network model (i.e., to determine weights and biases), and to predict future degradation behavior. The following are four steps of making predictions using FFNN.

In the following steps, the italic letters represent options that can be modified, but the default options explained so far will be mostly adopted in this book. For more options and understanding the NN process, see the Matlab user's guide [104].

1) The function `feedforwardnet(hiddenSize, trainFcn)` creates a feedforward network model with user-defined number of hidden layers and hidden nodes. The input `hiddenSize` is row vector of hidden nodes. For example, `hiddenSize = [5, 8, 9]` means three hidden layers with 5, 8, and 9 hidden nodes, respectively. The input `trainFcn` stores the character string of training function. Supporting training functions are listed in Table 8.2. The Levenberg–Marquardt algorithm is default. Another Matlab function `fitnet` can be used for the same purpose. *Activation functions* have default options; the tangent sigmoid and the pure linear functions are used for the hidden and the output layers, respectively. Once a network model is created, all other required information is automatically assigned with default options, which is explained in the following steps.
2) The function `configure` normalizes input and output data and assigns them to each node. This step is optional. It automatically *normalizes* the input and output data between −1 and 1, and the *initial values* of weights and biases are randomly selected. Configuration is automatically done when the next step (training) is run, but this step can be added to assign the initial values of the parameters instead of random selection.

**Table 8.2** Matlab training algorithms.

| Training function | Algorithm |
|---|---|
| 'trainlm' | Levenberg–Marquardt |
| 'trainbr' | Bayesian regularization |
| 'trainbfg' | BFGS Quasi-Newton |
| 'trainrp' | Resilient backpropagation |
| 'trainscg' | Scaled conjugate gradient |
| 'traincgb' | Conjugate gradient with Powell/Beale restarts |
| 'traincgf' | Fletcher–Powell conjugate gradient |
| 'traincgp' | Polak–Ribiere conjugate gradient |
| 'trainoss' | One step secant |
| 'traingdx' | Variable learning rate gradient descent |
| 'traingdm' | Gradient descent with momentum |
| 'traingd' | Gradient descent |

3) The function `train`[1] trains the network model with the LM backpropagation for *training method* and the MSE for the *error function* (or performance evaluation function). For MATLAB implementation, the early stopping option has several criteria. One of the stopping criteria is based on validation error, which is the most critical criterion of preventing overfitting. In this criterion, training data are divided into three sets based on random *data division method*, 70% for training, 15% for validation, and 15% for test. The training sets are used to update the weights and biases based on their error in the network output. The errors of the training set keep decreasing during the training process, and the errors in the validation set also decrease, which means the training process goes well. On the other hand, errors in the validation set start increasing after some phase of the training process, i.e., satisfactory performance at the training set but unenjoyable performance at the validation set (new points, prediction points), which means that the network model overfits the training data. Therefore, the training process stops when the validation error starts increasing. The test data sets are not used during the training process but used for testing the prediction accuracy with trained parameters.
4) The function `sim` simulates/predicts the network outputs at new inputs based on the identified weights and biases during the training process.

**Example 8.4 Feedforwardnet**

Degradation of battery capacity is measured at five different times: $(t_i, y_i)$ = (0, 1), (5, 0.99), (10, 0.99), (15, 0.94), (20, 0.95). Perform the FFNN by using the Matlab function feedforwardnet with one hidden node and plot predictions at $t \in [-5, 25]$.

1 The function `train` uses all data at once, while `adapt` that is another learning method uses data one by one and is suitable for online learning. For real problems, `adapt` can be utilized since it does not require to store tremendous amount of monitoring data, but `train` is used in this book since it is more efficient for most practical problems.

**Solution:**

Since input is time, we will consider $x = t$. The following Matlab script shows how to use the feed-forwardnet Matlab toolbox to perform FFNN.

```
x=[0    5    10    15    20];                    %[1 x 5] training input data
y=[1 0.99 0.99 0.94 0.95];                      %[1 x 5] training output data
nh=1;                                                   % num. of hidden node
net=feedforwardnet(nh);                     % create two-layer network model
net=configure(net,x,y);                                    % this is optional
[netModel,trainRecor]=train(net,x,y);               % train the model 'net'
xNew=-5:0.1:25;                                            % new input points
z=sim(netModel,xNew);                                    % simulation results
```

The process itself is quite simple. First, the measurement times (cycles) and the degradation data are defined as training input data, `x`, and output data, `y`, respectively. The network model, `net` is created with function `feedforwardnet` after defining the number of hidden nodes in `nh = 1`. The next configuration step, `configure` is optional, but with this, we can extract the initial values of weights and biases by using the following variables: `net.iw` for input weights, `net.lw` for hidden weights, and `net.b` for biases, which are shown in Figure 8.9 (values at the top). Note that the initial values can be different every time when configure is called since they are randomly generated.

Now, the network model is trained based on function `train` and the trained model is named as `netModel`. The optimum values of parameters can be shown in the same way, but using `netModel` instead of `net`, which are also shown in Figure 8.9 ( values at the bottom). The results can also vary every time since the initial values of parameters and the training sets will be different.

The number of training data that is three in this problem is too small, which is even less than the number of parameters to be estimated. In such a case, the estimated network model is not accurate. In general, it is required that the number of training data must be larger than that of network model parameters. Especially since only 70% of training data are used for the purpose of training, it is important to have enough training data to have robust training results.

Training records are stored in `trainRecor`. For example, the indexes for training, validation, and test sets can be shown by using `trainRecor.trainInd`, `trainRecor.valInd`, and `trainRecor.testInd`, respectively, whose results are 2, 3, 4 for training, 5 for validation, 1 for test sets (again, these results can vary). Also, the performance (MSE) of training, validation and test sets are recorded, and it can be plotted with `plotperf(trainRecor)`, which is shown in Figure 8.10. There is a total of eight epochs that can be known with `trainRecor`.

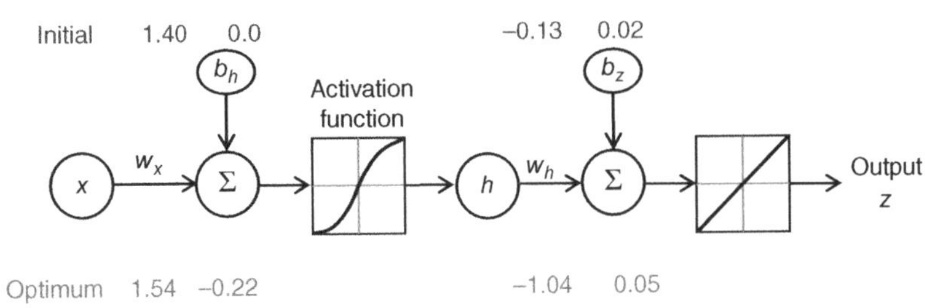


**Figure 8.9** Feedforward network model and initial, optimum values of parameters.

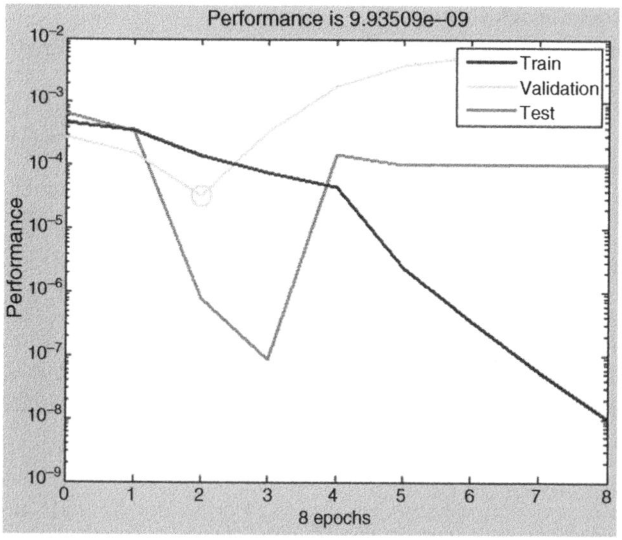


**Figure 8.10** Convergence history plot of training.

`num_epochs`, and the training stops at two epochs that can be known with `trainRecor.best_epoch` based on the early stopping. Since the performance of validation sets are recorded in `trainRecor.vperf`, the circle in the figure can be plotted with the following codes:

```
bestE=trainRecor.best_epoch;
bestP=trainRecor.best_vperf; % or trainRecor.vperf(bestE+1)
hold on; plot(bestE,bestP,'o');
```

Note that the performance of the validation set starts to increase after the best epoch at the circle in the figure, while that of the training set still decreases.

Once training the network model is completed, simulation outputs can be obtained with sim for $t \in [-5, 25]$, which is shown in Figure 8.11.

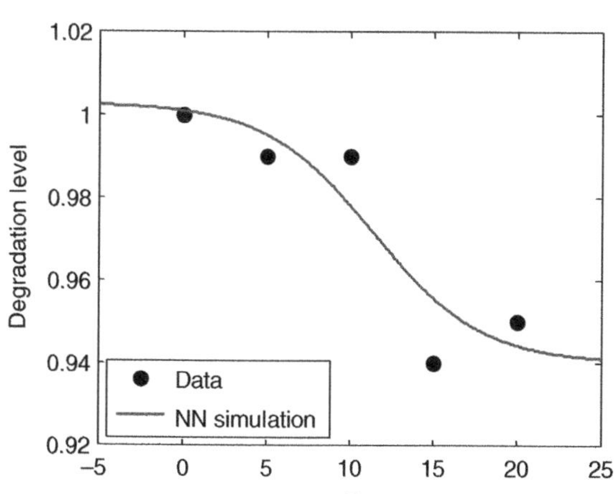


**Figure 8.11** FFNN simulation prediction.

**Example 8.5 Feedforwardnet of Rosenbrock Function**
Approximate the Rosenbrock function using a NN model and calculate the root-mean-squared-error (RMSE). The Rosenbrock function is given as

$$y(x_1, x_2) = 100\left(x_1^2 - x_2\right)^2 + (1 - x_1)^2 \tag{8.21}$$

The range of input data is given as $x_1, x_2 \in [0.5, 1.5]$. Generate 20 samples using Latin hypercube sampling. Train the NN model with five hidden nodes. Calculate the RMSE using $51 \times 51$ equally spaced grid points.

**Solution:**
The purpose of this example is to test the approximation accuracy of NN models. Samples are generated from a known true function, and a network model is trained using the samples. Then it is possible to assess the accuracy of the network model at unsampled locations. The following Matlab script is used to generate samples, train the network model, and plot figures.

```
%Plotting the actual function at grid points
clear; n=51;
x=linspace(0.5,1.5,n); y=linspace(0.5,1.5,n);
[X,Y]=meshgrid(x,y);
Z=zeros(n);
for i=1:n, for j=1:n
 Z(i,j)=Rosenbrock([y(j) x(i)]);
end, end
figure(1); cs=surface(X,Y,Z); hold on;
%
%LHS samples
rng default;
xx=0.5+lhsdesign(20, 2,'iterations',5000,'smooth','off');
yy=Rosenbrock(xx);
plot3(xx(:,1), xx(:,2), yy,'or'); hold off
%
% FFNN
nh=5;                                      % num. of hidden node
net=feedforwardnet(nh);            % create two-layer network model
net=configure(net,xx',yy');                     % this is optional
[netModel,trainRecor]=train(net,xx',yy');      %train the model 'net'
ZNN=zeros(n);
for i=1:n, for j=1:n
 ZNN(i,j)=sim(netModel,[y(j) x(i)]');
end, end
figure(2); cs=surface(X,Y,ZNN); hold on;
plot3(xx(:,1), xx(:,2), yy,'or'); hold off
%
RMSE=sqrt(sum(sum(abs(Z-ZNN)))/(n*n))
```

First, $51 \times 51$ equally space grid points were generated within the range of input data $x_1, x_2 \in [0.5,1.5]$. The purpose of this grid is to plot the function and to calculate the RMSE. Matlab command `lhsdesign` is used to generate 20 random samples. The Matlab command `rng default` keep the sequence of random numbers same, which can remove sampling randomness. In order to distribute the samples evenly, the minimum distance between the samples is maximized with 5000 iterations. The FFNN is generated with five hidden nodes. Once the network model is trained, it is used to simulate the network prediction at 51 × 51 equally space grid points. These points were used to calculate the RMSE. The RMSE of the trained network model turned out to be $e_{\mathrm{RMSE}} = 4.4287$. Considering the range of function [0,306.5], this means that the network model has about 1.5% error on average.

Figure 8.12(a) shows the true function with 20 Latin hypercube sample locations. It is noted that even if the samples are evenly distributed over the range of inputs, there are no samples at the four corners. Since the function shows maximum values at these corners, it would be difficult to approximate the function near the corners. Figure 8.12(b) is the plot of network model. Overall, the function has a reasonable approximation near the center of the range, but it shows significant errors near the four corners. This is a common issue in surrogate modeling. The four corner points belong to the extrapolation region, where all surrogate models experience difficulty. It would be interesting to add additional four samples at the corners and train the network model again. Figure 8.12(c) shows the training history of the FFNN. The network model was trained up to 10 epochs. However, since the validation set started increasing at the fourth epoch, the network model determined to use the fourth epoch as the trained model. Figure 8.12(d) shows training results. It shows that Levenberg–Marquardt algorithm is used for training, and the MSE was used for optimization.

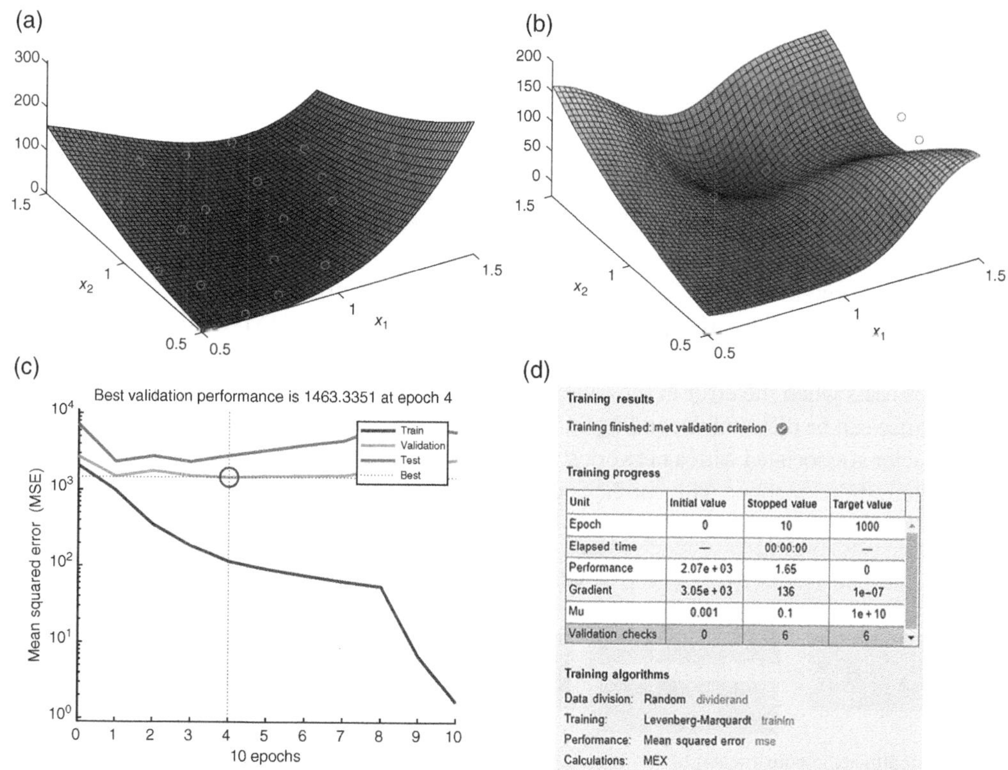


**Figure 8.12** Approximating Rosenbrock function using FFNN. (a) Actual function with 20 samples, (b) FFNN approximation, (c) training history, and (d) training results.

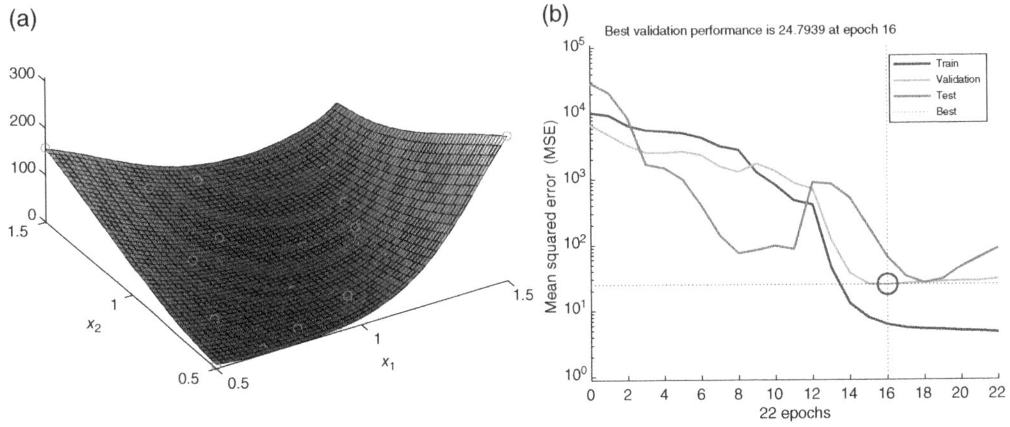


**Figure 8.13** Approximating Rosenbrock function using FFNN with addition four corner samples (a total of 24 samples). (a) Rosenbrock function and (b) training history.

As mentioned earlier, the network model tends to yield a large error in the extrapolation region. In order to show the effect of the extrapolation error, the same network model as in Example 8.5 is used. In addition to 20 samples, additional four samples were added at the four corners of the input range (see Exercise Problem 8.6). Figure 8.13(a) shows the NN model for the Rosenbrock function. With additional four corner samples, the approximate function is much closer to the true function in Figure 8.12(a). In fact, the RMSE turned out to be $e_{\mathrm{RMSE}} = 2.0519$, which corresponds to about 0.67% error on average. Such improved accuracy was achieved by training the model up to 16 epochs, as shown in Figure 8.13(b).

## 8.4 Uncertainty Quantification in Neural Network Models

In the training process of NN, three factors contribute to uncertainty, as shown in Figure 8.14. The first factor is related to the early stopping criterion, which is widely used to prevent overfitting. The early stopping criterion divides training data into training and validation sets and terminates the training process when the error in the validation set begins to increase. In this process, the model performance can be different depending on how the training and validation sets are selected. The second factor is associated with a network structure, including the number of nodes and layers and types of activation functions. This corresponds to a model-form uncertainty of NN. The last factor is

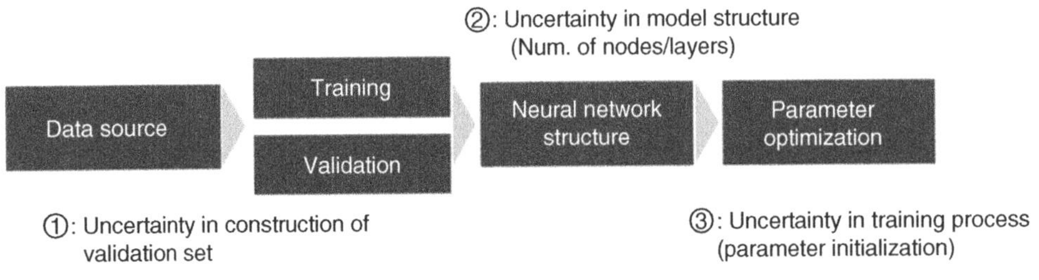


**Figure 8.14** Sources of training uncertainty in NNs.

due to optimization algorithms. Since optimization algorithms can only find local optima based on initial values of parameters, the trained parameters may be different with different starting points. Since the initial values are randomly selected, the trained results may be different even if the same training data and the same network model are used. The uncertainty caused by these three factors is referred to as training uncertainty in this book. The prediction results from the network model must be assessed by taking this into account these sources of uncertainty.

### 8.4.1 Training Uncertainty

It is important to quantify uncertainty in surrogate predictions. Some uncertainties are random by nature (i.e., aleatory), but some uncertainties are related to the lack of knowledge (i.e., epistemic). Noise in data is an example of aleatory uncertainty. This type of uncertainty is considered uncontrollable, and engineers have to live with it. A common method of considering aleatory uncertainty is conservatism. For example, the A-basis of B-basis allowable strength of a material is the outcome of conservative estimate of failure strength, which considers both aleatory uncertainty in material itself and sampling uncertainty. Epistemic uncertainty, on the other hand, may not be random but the accurate value is unknown. A common example of epistemic uncertainty is model error, which is caused by simplification and idealization of physical phenomena in a mathematical form. Another example of epistemic uncertainty is sampling uncertainty, where a finite number of samples are used to predict the statistics of population. Epistemic uncertainty is controllable in the sense that the model error can be reduced by improving the mathematical model or use more samples to reduce sampling uncertainty. In computational solid mechanics, where the system's response is calculated using discretization, the model error can be reduced by using a fine mesh.

Training uncertainty in Figure 8.14 can be considered as a mixture of aleatory and epistemic uncertainties. The random selection of the validation set is aleatory. When different validation sets are selected, the corresponding training set will be changed, which leads to different training results. Unless different early stopping criteria are used, this uncertainty remains. For the given dataset, there is no systematic method to select the optimum validation set. The cross-validation in Chapter 2 can be used instead of selecting the validation set. However, calculation of the PRESS error and utilizing it in training is not straightforward.

The model-form uncertainty related to the model structure is epistemic. It has been an issue since the birth of machine learning. The number of hidden layers, the number of hidden nodes in each layer, and the types of activation functions can significantly contribute to the model predictions, and yet, there is no general rule to select them without understanding the trend of dataset. In polynomial response surfaces, there is a process called backward elimination based on confidence of coefficients. The idea is to start with a complex model with many coefficients and gradually eliminate those coefficients that have the lowest confidence. In NNs, calculating the confidence of parameters is difficult due to nonlinear relationship caused by activation functions.

The last source of uncertainty related to optimization algorithm is epistemic. All the training processes listed in Table 8.2 use gradient-based optimization. As we discussed in Chapter 5, gradient-based optimization algorithms are local optimizers. In other words, they are called hill-descending algorithms. That is, starting from an initial point, the algorithms move to the next point that is better than the current one. Once the algorithms find the bottom of a valley, the algorithms claim that it is the optimum point. However, this is nothing but the closest local optimum point from the starting point. There is no optimization algorithm that guarantees to finding the global optimum. Since optimization algorithms select the initial point randomly, the training process may yield different local optima.

Although it is not easy to remove training uncertainty, it is important to quantify the level of training uncertainty and its contribution to the prediction uncertainty. In general, a common source of aleatory uncertainty is from random noise in data. Since it is assumed that the noise in data is out of user's control, it can be considered as a reference for allowable level of uncertainty. That is, if training uncertainty is less than that of data, the model and training process can be considered acceptable. If training uncertainty is not acceptable, it is necessary to reduce the training uncertainty either by increasing/decreasing hidden layers and nodes, increasing the number of training data, or changing training algorithms.

In order to identify the contributions from different sources of uncertainty, we will use a known true function $z(x) = 0.2x + \sin x$ to generate synthetic samples in the range of $x \in [0, 4\pi]$. Among different sources of uncertainty described in Figure 8.14, we will activate one at a time to check its contribution to the prediction uncertainty. The accuracy of trained network models can be measured using the $e_{\text{RMSE}}$. The prediction uncertainty can be measured by repeating the training process and estimate the standard deviation of $e_{\text{RMSE}}$. The following Matlab script is used for quantifying different sources of uncertainty.

```
 1: %
 2: clear; close all; rng default
 3: a=0; b=4*pi; n=51; ns=15; nh=5;
 4: x=linspace(a,b,n);                              % Test grid points
 5: z=0.2*x+sin(x);                                   % True function
 6: %
 7: xx=linspace(a,b,ns);                       % Equally spaced samples
 8: yy=0.2*xx+sin(xx);
 9: %
10: figure(1); hold on
11: for i=1:20                                     % Repetition of FFNN
12: net=feedforwardnet(nh);           % create two-layer network model
13: net=configure(net,xx,yy);
14: if i==1, wb=getwb(net); end
15: net = setwb(net,wb);                     % Fix initial parameters
16: %
17: net.divideFcn = 'divideind';                      % Fixed datasets
18: net.divideParam.trainInd=[1 2 3 4 6 7 9 11 12 14 15];
19: net.divideParam.valInd=[5 10];
20: net.divideParam.testInd=[8 13];
21: %
22: [netModel,trainRecor]=train(net,xx,yy);      % train the model 'net'
23: zNN=sim(netModel,x);                     % FFNN model predictions
24: plot(x,zNN,'k');
25: %
26: RMSE(i)=sqrt((z-zNN)*(z-zNN)'/n);                % RMSE at test grid
27: end
28: plot(x,z,'r',xx, yy,'or'); hold off
29: m=mean(RMSE), s=std(RMSE) % Mean and standard deviation of RMSE
```

In the Matlab script, $n = 51$ grid points, $n = 15$ samples, and $nh = 5$ hidden nodes are used. Lines 14~15 fix the initial parameter values. In the first cycle, the initial parameters are randomly generated and saved. Then, in the following cycles, the saved initial parameters are used for the training process. Therefore, these lines remove uncertainty related to the initialization of parameters. If users want to make initial parameters random, Line 15 needs to be commented out. Lines 17~20 fix the training, validation, and test datasets. These lines remove uncertainty associated with random selection of training datasets. If users want to make the dataset selection random, these lines should be commented out. The Matlab script repeats 20 times to calculate the RMSE between the true function and network prediction at grid points. The mean and standard deviation are calculated using these 20 repetitions. If the current Matlab script is executed, the 20 repetitions will produce an identical result with the mean of $e_{\mathrm{RMSE}}$ is 0.2798 and standard deviation zero. The value of $e_{\mathrm{RMSE}}$ is the error due to the current selection of training set and the current selection of initial parameters. Figure 8.15(a) shows the true function (a curve with circles on it), 15 samples (circles), and network model prediction (curves without circles).

In order to estimate the uncertainty due to the random initialization of parameters, Line 15 is commented out. Figure 8.15(b) shows 20 network prediction curves. In this case, the mean of $e_{\mathrm{RMSE}}$ is 0.19, which is better than that of no-uncertainty case. This is because of the averaging effect.

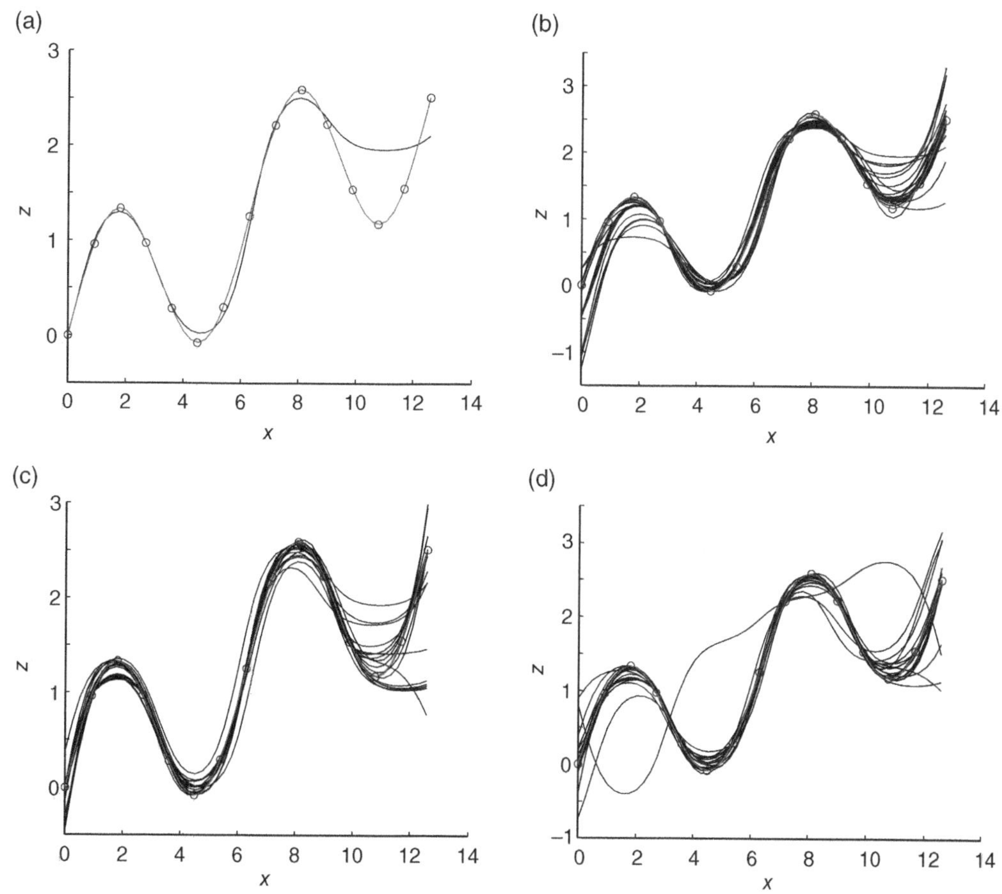


**Figure 8.15** Uncertainty quantification of the training process. (a) No training uncertainty, (b) uncertainty with parameter initialization, (c) uncertainty with selecting training dataset, and (d) uncertainty with both parameter initialization and selecting training dataset.

However, the standard deviation of $e_{\text{RMSE}}$ turned out to be 0.1216, which is about 30% of the mean. Therefore, the effect of initial parameters on the training process is significant.

Next, the initial parameters are fixed, while the training data are randomly selected. In order to estimate the uncertainty due to the random selection of training dataset, Lines 17~20 are commented out. Figure 8.15(c) shows 20 network prediction curves. In this case, the mean of $e_{\text{RMSE}}$ is 0.1888, which is better than the previous two cases. The standard deviation of $e_{\text{RMSE}}$ turned out to be 0.1106, which is smaller than the uncertainty due to random initialization.

Lastly, both uncertainties are activated by commenting out Linear 15, 17~20. Figure 8.15(d) shows 20 network prediction curves when both sources of uncertainty are activated. Out of 20 predictions, it is obvious that one prediction completely loses predicting the trend. This often happens when the error in the validation set starts increasing at an early epoch, and the training process stops prematurely. When the two sources of uncertainty are present, the mean of $e_{\text{RMSE}}$ is 0.1930, which is in the same level with the previous case. Therefore, having more uncertainty may not affect the mean of $e_{\text{RMSE}}$. On the other hand, the standard deviation of $e_{\text{RMSE}}$ turned out to be 0.2258, which is the combined effect of both sources of uncertainty. Considering $\sqrt{0.1216^2 + 0.1106^2} = 0.1644 < 0.2258$, the two sources of uncertainty show a high level of correlation. More specifically, if the distributions of the two uncertainties are assumed to be Gaussian, the correlation coefficient $\rho = 0.9$. Table 8.3 compares the statistics of the RMSE.

The effect of random initialization and random selection of training set was quantified by the standard deviation of the RMSE. The last puzzle of training uncertainty is model-form uncertainty, which is caused by selecting the number of hidden layers and nodes and the types of activation functions. The effect of this model-form uncertainty can be measured by the mean of RMSE, not standard deviation. This is because this uncertainty is not aleatory but epistemic. A good model approximates the true function better by having smaller errors.

In order to test the effect of model-form uncertainty, the number of hidden nodes is changed from 1 to 10. For such a simple example, it would not make sense to change the number of hidden layers. Therefore, we keep the single hidden layer. The Matlab script shown previously is modified to add an additional loop for hidden nodes. More specifically, the following lines were added before Line 11:

```
for j=1:10
 nh=j;

  % Lines 11 ~ 27 should be here

 m(j)=mean(RMSE), s(j)=std(RMSE)  % Mean and standard deviation of RMSE
end
```

**Table 8.3** Comparison of statistics of RMSE (no noise in samples).

| Source of uncertainty | Mean of $e_{\text{RMSE}}$ | Standard deviation of $e_{\text{RMSE}}$ |
|---|---|---|
| No uncertainty | 0.2798 | 0.0 |
| Initialization uncertainty | 0.1900 | 0.1216 |
| Selecting training set uncertainty | 0.1888 | 0.1106 |
| Both uncertainties | 0.1930 | 0.2258 |

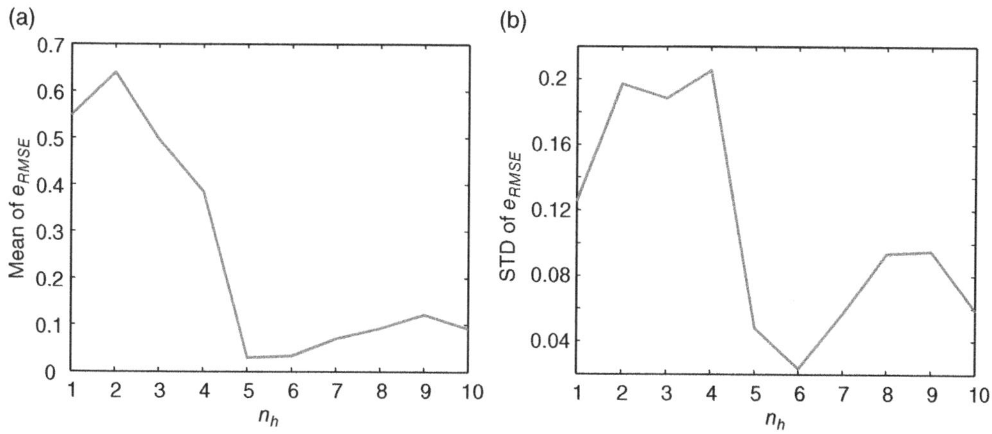


**Figure 8.16** Effect of the number of hidden nodes on network model accuracy. (a) Mean of RMSE and (b) standard deviation of RMSE.

For this test, a total of 30 samples were used. The initial parameters and the training set were randomly selected. Figure 8.16(a) shows the mean of $e_{\mathrm{RMSE}}$ with different number of hidden nodes. Although it does not show a clear trend, the $e_{\mathrm{RMSE}}$ decreases significantly after $n_h = 5$. Therefore, the function needs at least five hidden nodes to approximate the complex behavior of the function. In addition, the trend of standard deviation of $e_{\mathrm{RMSE}}$ also decreases significantly after $n_h = 5$. This test also reveals that having more hidden nodes than necessary may not help improve prediction accuracy. In fact, the mean and standard deviation of RMSE slightly increased with more hidden nodes. This is because increasing the number of hidden nodes may increase the possibility of overfitting with a fixed number of samples. Therefore, it seems there is an optimum number of hidden nodes for a given complexity of functional behavior and a given number of samples. However, it would not be easy to determine the optimum number of hidden nodes. In general, the optimum number of hidden nodes increases along with the complexity of functional behavior and the number of samples. Producing the results in Figure 8.16 is left to users (see Exercise Problem 8.11).

### 8.4.2 Sampling Uncertainty

Sampling uncertainty represents uncertainty associated with the input samples. There are two types of uncertainty related to the input samples. The first uncertainty is related to the number of samples, and the second is related to random noise in the samples. The first uncertainty can be considered as epistemic uncertainty as the accuracy of model prediction can be improved with more samples. On the other hand, the second uncertainty is aleatory. In conventional regression, the second type of uncertainty is referred to as sampling uncertainty. In the polynomial response surface in Chapter 2, for example, it is assumed that the model form is exact, while samples include random noise. Then, regression is the process of eliminating the effect of noise in samples. In fact, the name regression comes from the phrase "regression toward the mean."

In the case of polynomial response surfaces, however, there is no training uncertainty. The linear regression minimizes the square sum of errors, which is a quadratic function of unknown coefficients. Therefore, the optimization problem is convex. Therefore, the global minimum can be found by differentiating the square sum of errors with respect to unknown coefficients, which yields a linear system of equations.

In the case of NN model, the same square sum of errors (i.e., MSE) is used to train model parameters. It is also possible to use monomial functions as activation functions, as shown in Example 8.2. Then NN models can reproduce polynomial response surfaces. However, since NN models are designed to model much broader applications than simple regression, it would not be recommended to use monomial functions as activation functions. In fact, the main strength of NNs is to predict complicated functional behavior by making complicated networks. Therefore, general activation functions listed in Table 8.1 are commonly used, instead of using special activation functions customized for specific applications.

Due to nonlinearity introduced by activation functions, the optimization problem is not convex anymore, and optimization algorithms often find local optima. Therefore, even if the samples do not have any random noise, the model predictions are different due to training uncertainty. In addition, these predictions depend on the number of samples used. Since there are many parameters and early stopping criterion is used, it is difficult to determine how many samples are good enough. In the case of polynomial response surfaces, it has been suggested that the number of samples should be two or three times more than that of the unknown coefficients.

In order to understand the effect of the number of samples on network predictions, the same example as shown in Figure 8.15 is considered. All configurations of the network model are the same, except that the number of samples is gradually increased from 15 to 30. Figure 8.17 shows the mean and standard deviation of RMSE with respect to the number of samples. The uncertainty in network predictions is due to the training uncertainty. Although the trend of graph is not monotonic, the figure shows that the mean of $e_{\mathrm{RMSE}}$ decreases until the number of samples becomes 22. The standard deviation also shows a similar trend. Modifying the Matlab script to produce this figure is left as an exercise to users (see Exercise Problem 8.12). This figure shows that prediction accuracy of a network model can be improved with more samples. Adding samples in training is equivalent to improving knowledge of the function. Therefore, this sampling uncertainty is epistemic. However, there is no more improvement in accuracy after a certain number of samples. This is because sampling uncertainty decreases as the number of samples increases. However, the network model cannot improve the prediction accuracy because of model form uncertainty. That is, with five hidden nodes, the model cannot predict better even if more samples are used. The network model has 16 parameters (five input weights, five hidden biases, five hidden weights, and one output bias). This happened to be the same as the number of training samples (16 training samples, 3 validation

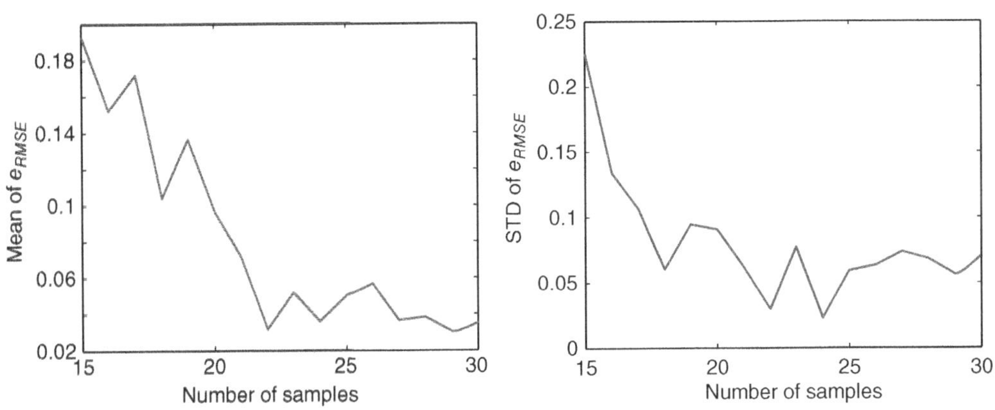


**Figure 8.17** Effect of the number of samples on network model accuracy.

samples, and 3 test samples). Therefore, it seems that the NN models do not require two or three times more samples than the number of parameters. However, this might be related to the fact that NNs use separate validation and test samples.

The number of samples when the model-form uncertainty becomes dominant depends on the complexity of functional behavior and the number of hidden nodes. For example, when three hidden nodes are used, prediction accuracy is worse than that of five hidden nodes. Therefore, the model-form uncertainty is greater than that of five hidden nodes, and the model-form uncertainty becomes dominant with a smaller number of samples (see Exercise Problem 8.13).

The second type of sampling uncertainty is related to the level of noise in samples. So far, all samples have been generated from a true function, and the accuracy of the trained model was measured using the error between the trained model prediction and the true function. In reality, the true function is often unknown, and only samples were available. In addition, these samples normally include random noise, which represents an uncertain environment of measurement. In such a case, it is difficult to assess the accuracy of network models. Only the discrepancy between samples and network predictions is available information. In order to overcome this limitation, synthetic samples are often used to assess the prediction accuracy of network models, which simulates the uncertain environment of measurement by adding randomly generated noise to the true function. Synthetic samples have many benefits for uncertainty study, in particular, for quantifying sampling uncertainty.

For this study, consider the same true function $z(x) = 0.2x + \sin x$. The uncertain environment of measurement is modeled by a Gaussian distribution with zero mean and known standard deviation. That is, the random noise follows statistical distribution $\sim N(0, \sigma^2)$. The following Matlab script can be used to generate $n_s$ numbers of samples that include random noise with $\sim N(0, 0.3^2)$.

```
v=randn(1,ns); v=0.3*(v-mean(v))/std(v);
yy=0.2*xx+sin(xx)+v;                          % Add random noise
```

The script needs some explanations. First, $n_s$ numbers of samples were generated from $\sim N(0, 1^2)$ using Matlab function `randn`. First, even if the samples were generated from population distribution $\sim N(0, 1^2)$, the samples may show different mean and standard deviation because of sampling uncertainty. In order to make the samples have zero mean and standard deviation of 0.3, the samples were shifted and scaled. Then, the scale random noise samples `v` follow $\sim N(0, 0.3^2)$. Then, these scaled random noise samples were added to the true function.

The network model is trained with randomly generated 15 samples, and the network predictions at 51 grid points are compared with that of the true function in terms of the RMSE. This process is repeated 20 times to obtain the distribution of RMSE. Figure 8.18 shows the 20 network predictions (solid curves) and the true function (dashed curve). Circular markers are noisy samples at the last repetition. Producing this plot is left as an exercise problem to users (see Exercise Problem 8.14). Overall, the predictions show a larger error than that of the no-noise case in Figure 8.15, which is expected.

Figure 8.18(a) shows 20 predictions only when sampling uncertainty is present. That is, the initial values of parameters and training set are fixed. Table 8.4 summarizes the statistics of $e_{\text{RMSE}}$. It turns out that the significant effect of sampling uncertainty is on prediction accuracy (i.e., the mean of $e_{\text{RMSE}}$) but not on prediction uncertainty (i.e., the standard deviation of $e_{\text{RMSE}}$). The RMSE 0.3312 is close to the standard deviation of random noise is 0.3. Figure 8.18(b) shows 20 predictions when both sampling uncertainty and the random selection of initial parameters are present. It is reminded that the mean of $e_{\text{RMSE}}$ in Table 8.4 is reduced when initial parameters are randomly selected. This is because the fixed initial parameters are not the best one. However, when samples

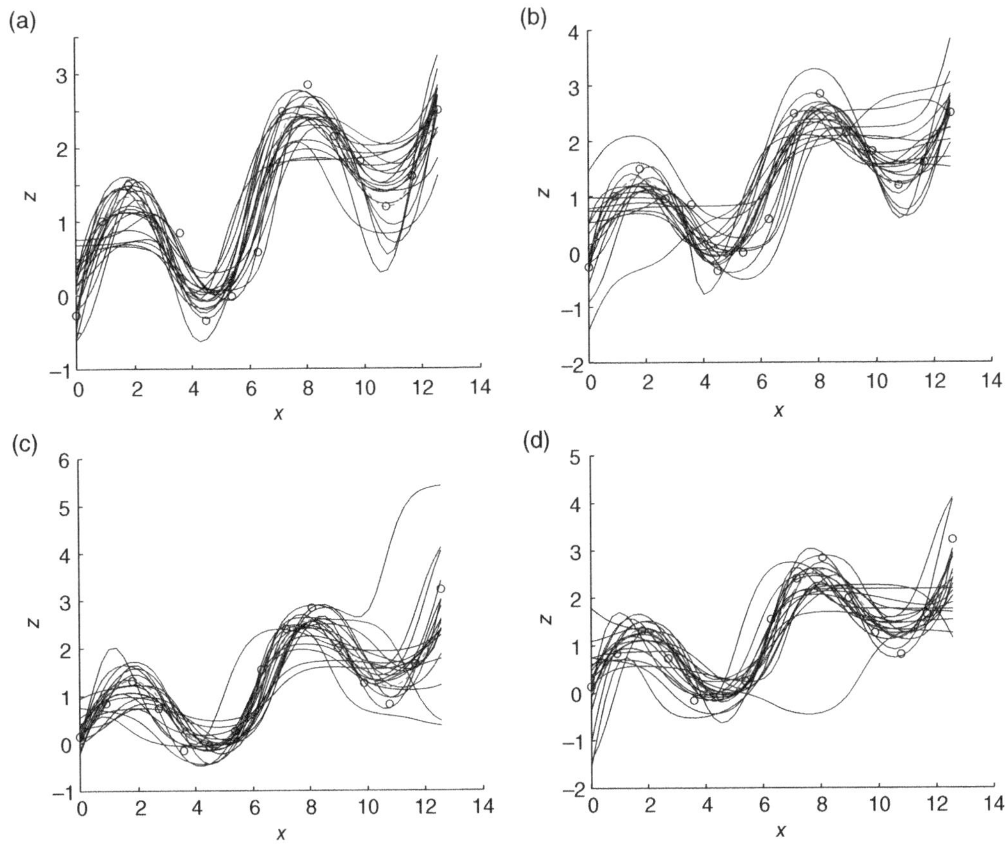


**Figure 8.18** Neural network model prediction with noise in samples. (a) Only sampling uncertainty, (b) sampling uncertainty + random initialization uncertainty, (c) sampling uncertainty + random training set, and (d) all three sources of uncertainty.

**Table 8.4** Comparison of statistics of RMSE (with noise in samples).

| Source of uncertainty | Mean of $e_{RMSE}$ | Standard deviation of $e_{RMSE}$ |
|---|---|---|
| Sampling uncertainty only | 0.3312 | 0.0684 |
| Sampling uncertainty + initialization uncertainty | 0.4374 | 0.1909 |
| Sampling uncertainty + selecting training set uncertainty | 0.4600 | 0.2723 |
| All three uncertainties | 0.4413 | 0.2588 |

include random noise, selecting random initial parameters increases the mean of $e_{RMSE}$. On the other hand, the increase in standard deviation is at a similar level. The same observation can be made for the case of when the training set is randomly selected in Figure 8.18(c). When all sources of uncertainty are considered, the mean of $e_{RMSE}$ is increased from 0.1930 to 0.4413, while the standard deviation remains at the same level. Therefore, it can be concluded that the sampling uncertainty mostly affects prediction accuracy rather than prediction uncertainty.

The results in Figure 8.18 and Table 8.4 are based on 15 samples. As shown before, training uncertainty is relatively high with 15 samples. If the number of samples is increased to 30, the level of training uncertainty will be ignorable compared to sampling uncertainty. In such a case, the mean of $e_{\text{RMSE}}$ is mostly caused by sampling uncertainty, while the standard deviation of $e_{\text{RMSE}}$ may not change much, which is left for users' exercise (see Exercise Problem 8.15).

### 8.4.3 Confidence Intervals and Prediction Intervals

There are several ways to represent uncertainty. The simplest method is that the FFNN process is repeated several times to check the uncertainty as a prediction interval for the training results. At a given value of input variable $\mathbf{x}$, the FFNN model predicts different values at different training repetitions. Let the training process is repeated $n$ times with the same set of training data. Then, the predictions at given $\mathbf{x}$ are denoted by $\mathbf{z} = \{z_1, z_2, \ldots, z_n\}$. Let the average and standard deviation of these predictions be $\bar{z}$ and $s$, respectively. In the statistical viewpoint, predictions at a point are randomly distributed with population mean $\mu_z$ and population standard deviation $\sigma_z$.

If the training process is infinitely repeated, it is possible to determine the mean $\mu_z$ and standard deviation $\sigma_z$ of predictions. However, with a finite number of samples, $\bar{z}$ and $s$ are not accurate, which is referred to as sampling uncertainty. It is obvious that $\bar{z}$ and $s$ will approach $\mu_z$ and $\sigma_z$ as the number of samples increases. The confidence interval estimates the range of the population mean of these predictions for a given point $\mathbf{x}$. First, it is assumed that the samples of prediction are independent each other and normally distributed. Then, the two-sided confidence intervals can be determined as

$$\text{CI}_z = \left[\bar{z} - t_{\alpha/2,n-1} \times \frac{s}{\sqrt{n}}, \bar{z} + t_{\alpha/2,n-1} \times \frac{s}{\sqrt{n}}\right] \tag{8.22}$$

where $t_{\alpha/2,n-1}$ is the $100(1-\alpha/2)$ percentile of the student's $t$ distribution with $n-1$ degrees of freedom. The confidence level of the above confidence interval is $1-\alpha$. For example, when $\alpha = 0.05$, it becomes 95% confidence intervals. It is confident that 95 out of 100 times the prediction mean will fall between the upper and lower values of the confidence interval. As expected, the confidence intervals are inversely proportional to $\sqrt{n}$. That is, the interval is reduced as more samples are used. In the conventional notion, the confidence intervals are caused by uncertainty in samples. However, the confidence intervals in Eq. (8.22) are caused by the training process because we used the same set of training samples, and we repeated the training process.

The prediction intervals is different from the confidence intervals. It asks the following question: knowing the uncertainty in the mean and standard deviation, what is the confident range of additional sample of prediction? One can think of this as taking a sample out of a distribution, whose mean is uncertain as well. Therefore, the uncertainty of the mean and the uncertainty of the sample need to be considered together. More precisely, the variance of the mean and that of the sample need to be added. Therefore, the prediction intervals can be defined as

$$\text{PI}_z = \left[\bar{z} - t_{\alpha/2,n-1} \times \sqrt{s^2 + s^2/n}, \bar{z} + t_{\alpha/2,n-1} \times \sqrt{s^2 + s^2/n}\right] \tag{8.23}$$

where the term $\sqrt{s^2 + s^2/n}$ represents the standard error of the prediction, which accounts for a single observation with respect to the mean. The uncertainty expressed as a prediction intervals includes not only the uncertainty $s^2/n$ due to the population mean estimate, but also the uncertainty $s^2$ associated with new observation. Since the uncertainty associated with the population mean estimate is independent of new observations, the individual uncertainty estimates must be combined using the root-sum-of-squares to yield total uncertainty, i.e., the standard error of the prediction.

**Example 8.6 Uncertainty in NN Simulation**

Plot all simulation results by repeating the process in Example 8.4 five times. Record the initial and optimum values of weights and biases, and the indexes of train, validation, and test sets.

**Solution:**

Whenever the network model is configured or initialized with `net=init(net)`, the initial values of parameters are assigned differently. Also, the train, validation, and test sets change when the training is re-started. These two are the sources of uncertainty in the NN process and make different results. Figure 8.19 can be obtained by repeating the process given in the solution in Example 8.4, and the initial/optimum values of parameters and the indexes of data sets are listed in Table 8.5.

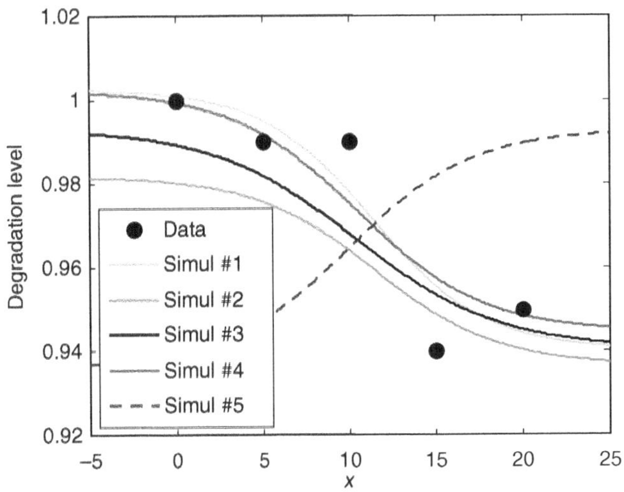


**Figure 8.19** FFNN simulation predictions with five repetitions.

**Table 8.5** Prediction intervals in FFNN simulation.

| **Repetition** | **1** | **2** | **3** | **4** | **5** |
|---|---|---|---|---|---|
| $w_{1,\text{init}}$ | 1.4000 | −1.4000 | 1.4000 | −1.4000 | 1.4000 |
| $b_{1,\text{init}}$ | 0.0000 | 0.0000 | 0.0000 | 0.0000 | 0.0000 |
| $w_{2,\text{init}}$ | −0.1332 | 0.6697 | 0.7417 | −0.3831 | 0.9438 |
| $b_{2,\text{init}}$ | 0.0179 | −0.9266 | −0.8994 | 0.6767 | −0.1827 |
| $w_{1,\text{opti}}$ | 1.5431 | −1.4141 | 1.1967 | −1.3651 | 1.4000 |
| $b_{1,\text{opti}}$ | −0.2173 | 0.2205 | −0.0467 | 0.0674 | 0.0000 |
| $w_{2,\text{opti}}$ | −1.0414 | 0.7558 | −0.8837 | 0.9658 | 0.9438 |
| $b_{2,\text{opti}}$ | 0.0480 | −0.3632 | −0.1038 | 0.1140 | −0.1827 |
| Train set | 2, 3, 4 | 1, 3, 5 | 1, 4, 5 | 3, 4, 5 | 1, 2, 3 |
| Validation set | 5 | 4 | 3 | 1 | 4 |
| Test set | 1 | 2 | 2 | 2 | 5 |

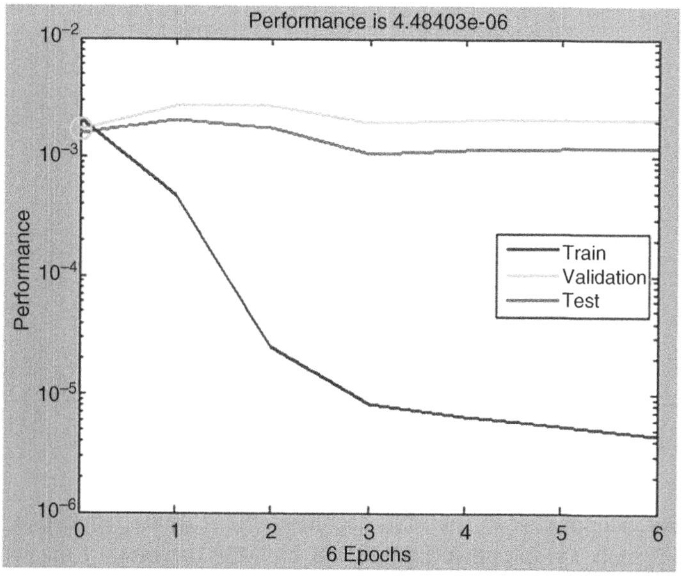


**Figure 8.20** Training history plot of simulation 5 in Example 8.6.

In the figure, most results follow the trend of data except for the fifth simulation result (magenta dashed curve). This is because of the early stopping that sometimes hinders obtaining proper optimization results. As shown in Figure 8.20 and the last column in Table 8.5, it practically stops at the initial stage. The optimization algorithm can be modified to resolve this problem, e.g., modifying the parameters associated with the optimization algorithm or using other optimization algorithms. It would be necessary to apply a constraint based on the property that degradation should increase or decrease along with the cycles.

## 8.5 Issues in Feedforward Neural Network

### 8.5.1 Adaptive Learning Rate

As shown in Example 8.3, the learning rate is important in the training performance. The learning rate used in the gradient descent algorithm of the backpropagation process is a constant for how much change will be made in one learning. One cycle of updating parameters is called epoch. These hyperparameters, such as learning rate, are values that users set directly outside of the machine learning algorithm, and it may take considerable time to find appropriate values. As shown in Figure 8.21, if the learning rate is too small, training is slow, and if it is too large, the optimal value cannot be found. Therefore, instead of using a fixed learning rate for training speed and convergence, it is necessary to change the learning rate through the learning process.

The way the learning rate changes over time (or epoch) is called learning rate decay. The simplest learning rate reduction algorithm linearly or exponentially reduces the learning rate from a large initial value to a small value. This allows large changes in weights at the beginning of the learning process and small changes or fine-tuning near the end of training. An adaptive learning rate algorithm changes depending on the behavior of MSE. It monitors the training performance and adjusts

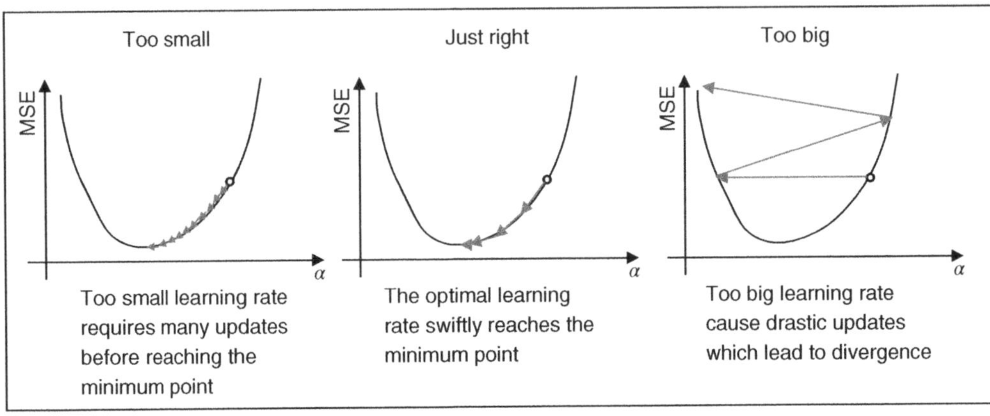


**Figure 8.21** Effect of the learning rate on gradient descent updates.

the learning rate accordingly. For example, when the MSE decreases, the learning rate can be increased to move fast in the direction. On the other hand, when the MSE increases, it means the network has reached a difficult region. Therefore, the learning rate can be decreased to slow down progress. In the case of bold driver algorithm [105], if the MSE decreases compared to that in the previous epoch, the learning rate is slightly increased (1∼5%) to increase the training speed. If the MSE increases, the learning rate is greatly decreased (50%) to slow down the speed.

### 8.5.2 Scaling Input Data

Even if an appropriate learning rate is used, the learning process may slow down frequently, especially when multiple input nodes are used. This happens when some parameters have much larger gradients than others. As shown in Eq. (8.20), the learning process depends on the learning rate and the gradient of MSE with respect to parameters. Therefore, an efficient learning process is expected if the magnitudes of all gradients are more or less the same. However, due to nonlinear activation functions, the gradients are large near the origin and diminish quickly as the input to the activation function increases or decreases. As shown in Figure 8.22, for example, the sigmoid function, which

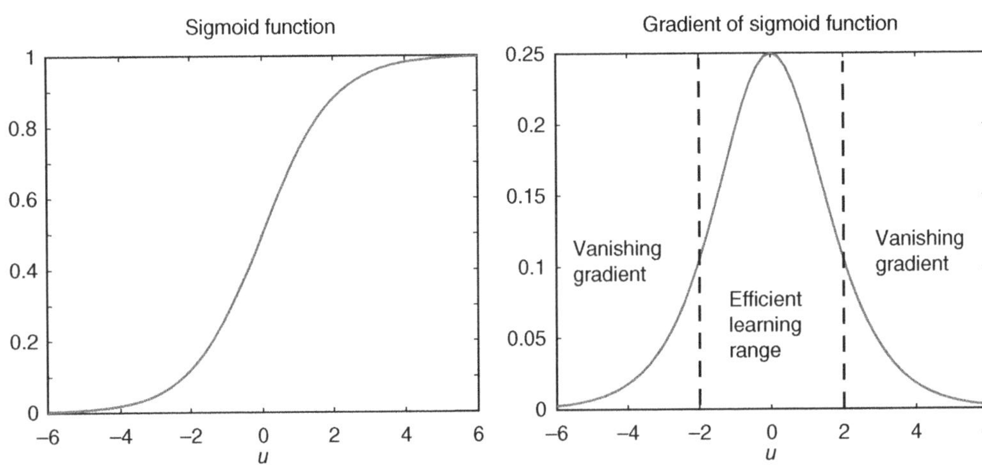


**Figure 8.22** Sigmoid function and its derivative.

is often used as an activation function of the input layer, causes the vanishing gradient problem, in which training slows down rapidly when the input to the activation function exceeds the range of $[-2, 2]$. It is well known that the training speed is faster when the mean of the input variable is close to zero. For the purpose of code generalization and increasing the training speed, it would be good to normalize input data in the range of $[-1, 1]$, which is called scaling input data.

There are two common normalization methods: min-max normalization and Z-score normalization. The former scales input data based on minimum and maximum values, while the latter scales input data such that the mean of input data is zero, and the standard deviation one. Min-max normalization is more suitable for regression problems, where the input data are normalized by

$$\tilde{X}_i = \frac{X_i - X_{\min}}{X_{\max} - X_{\min}} \quad (i = 1, 2, \ldots, n_x) \tag{8.24}$$

where $\tilde{X}_i$ is the normalized $i$th input data, $X_i$ is the original $i$th training data, and $X_{\min}$ and $X_{\max}$ are the minimum and maximum values among the training data, respectively. However, this normalization is not efficient because the range of data is $[0, 1]$, while the most efficient region is around zero. In fact, different activation functions have different input ranges to be efficient. For example, in the case of sigmoid function, the range is $[-2, 2]$. Therefore, the normalization formula is modified as

$$\tilde{X}_i = \frac{X_i - X_{\min}}{X_{\max} - X_{\min}} \times (t_{\max} - t_{\min}) + t_{\min} \quad (i = 1, 2, \ldots, n_x) \tag{8.25}$$

where $t_{\min}$ and $t_{\max}$ are the minimum and maximum target scaling values, respectively. Then, the range of normalized input data is $[t_{\min}, t_{\max}]$. It is important to remember that not only the training data but also prediction inputs need to be normalized because the network is trained in the normalized inputs. The normalized input data should only be used internally. Whenever the normalized input data is reported to the users, it should be denormalized using the same scale factors used in the normalization process as

$$X_i = \frac{\tilde{X}_i - t_{\min}}{t_{\max} - t_{\min}} \times (X_{\max} - X_{\min}) + X_{\min} \quad (i = 1, 2, \ldots, n_x) \tag{8.26}$$

Although it is not related to training, it is also recommended to normalize outputs. The normalization of outputs does not require target scaling. The following normalization will scale the outputs between zero and one:

$$\tilde{y}_i = \frac{y_i - y_{\min}}{y_{\max} - y_{\min}} \quad (i = 1, 2, \ldots, n_y) \tag{8.27}$$

where $y_{\max}$ and $y_{\min}$ are, respectively, the maximum and minimum of output samples. The normalization of outputs may affect the overfitting performance, which will be discussed in the following subsection.

### 8.5.3 Overfitting

Underfitting and overfitting, which are frequently encountered in machine learning, are representative problems that reduce the performance of training. As shown in Figure 8.23, underfitting occurs when the training data is not sufficiently learned, so the resulting model is too simple, and overfitting occurs when the training data is learned too much, so the resulting model is too complex. From a mathematical perspective, an underfitted model has high bias and low variance and is not trained for the problem regardless of the specific sample of the training data. An

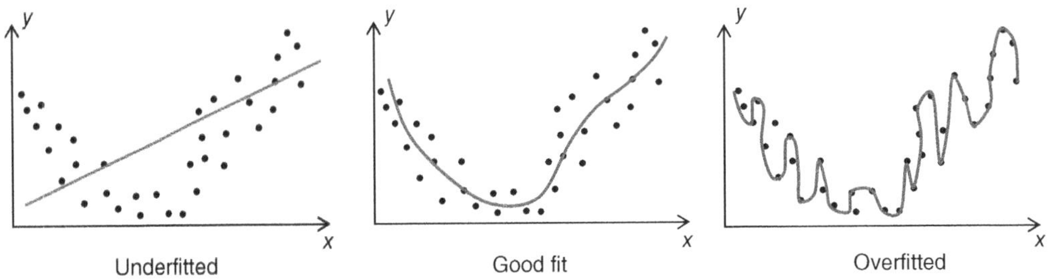


**Figure 8.23** An illustration of underfitting, good fit, and overfitting.

overfitted model has low bias and high variance and is highly accurate for the training data, but does not work properly for new data. In many cases, training data include noise, and overfitting fits noise rather than the trend of a function. A good fit model is a well-generalized model that appropriately learns from the training data and provides meaningful predictions for new inputs.

Underfitting can be resolved easily by increasing the amount of training data or changing the structure of the model, such as adding more layers and/or more nodes to the layers, to increase the capacity of the model. Underfitting often occurs due to insufficient learning due to early termination of learning, and to resolve this, the learning process is initialized and restarted if the MSE is greater than a threshold when learning is complete. However, overfitting can be more difficult to resolve than underfitting, and much research has been conducted to address this problem. In this section, four remedies are introduced: (i) the number of nodes in each layer, (ii) $L_2$ regularization, (iii) early termination of learning, and (iv) penalty coefficients.

1) The effect of the number of nodes in each layer: One way to resolve overfitting is to reduce the complexity of the model. The easiest way to control the complexity of the model is to change the number of hidden nodes. When more nodes are introduced to the model, there are more weights and biases to fit. Since overfitting is related to the number of parameters, introducing more nodes increases the probability of having overfitting.

   In order to understand the relationship between the number of nodes and the overfitting phenomenon, the battery degradation model is used to generate training data with noise. Since this process simulates actual measurement conditions, the data generated by this process is referred to as synthetic data. Synthetic data is useful in order to assess the accuracy of learning parameters and prediction. A simple form of the empirical battery degradation model is given as

$$z = \exp(-bt) \tag{8.28}$$

   where $z$ is the degradation level of battery's capacity, $t$ is the time or cycles, and $b$ is the model parameter, which was assumed to be 0.003 to produce real data. In order to simulate the actual measurement environment, random Gaussian noise with a mean zero and a variance $0.02^2$ was added to the generated data. In addition, it is well known that the capacity of secondary batteries such as lithium-ion batteries deteriorates with the usage cycle, and the failure threshold is defined when the capacity decreases by 30%. Ten training data are generated by substituting time $t = 0, 5, \ldots, 45$ in Eq. (8.28) and adding Gaussian noise.

   In the following comparison, the number of input and output nodes is fixed to be one. The first case uses a single hidden node, while the second case uses five hidden nodes. With 10 training data, the FFNN network model is trained 10 times. Because of various sources of uncertainty in

FFNN training, the trained results are usually different even if the same set of data is used. Figure 8.24 shows 10 training data (plus markers). Figure 8.24(a) and (b) shows 10 trained FFNN models using one hidden node and five hidden nodes, respectively. It is clear that the trained models with five hidden nodes are scattered more than that of one hidden node. This is because the model with one hidden node is simpler than that of five nodes. The effect of overfitting is magnified at the extrapolation region, which is far away from sample locations. In a simple model like the current problem, the number of hidden nodes should be reduced, and if the model is expected to be complex, the number of nodes should be increased.

Next, in order to see effect of input nodes, the number of hidden nodes is fixed to three, and the number of input nodes is set to one and five. In the current model, the input variable is time, and it is difficult to change the number of input nodes. Instead of time, it is possible to use the three degradations at previous times as inputs for predicting the degradation at the current time. For example, the pairs of input and output data can be arranged, as shown in Table 8.6. Similar to the previous example, the training is repeated 10 times to build 10 FFNN models, and their predictions are compared.

As shown in Figure 8.24(c) and (d), the overfitting phenomenon in the training data section is greatly affected by the number of input nodes. All 10 trained models with a single input node show reasonable predictions, while the trained models with five input nodes scatter much larger.

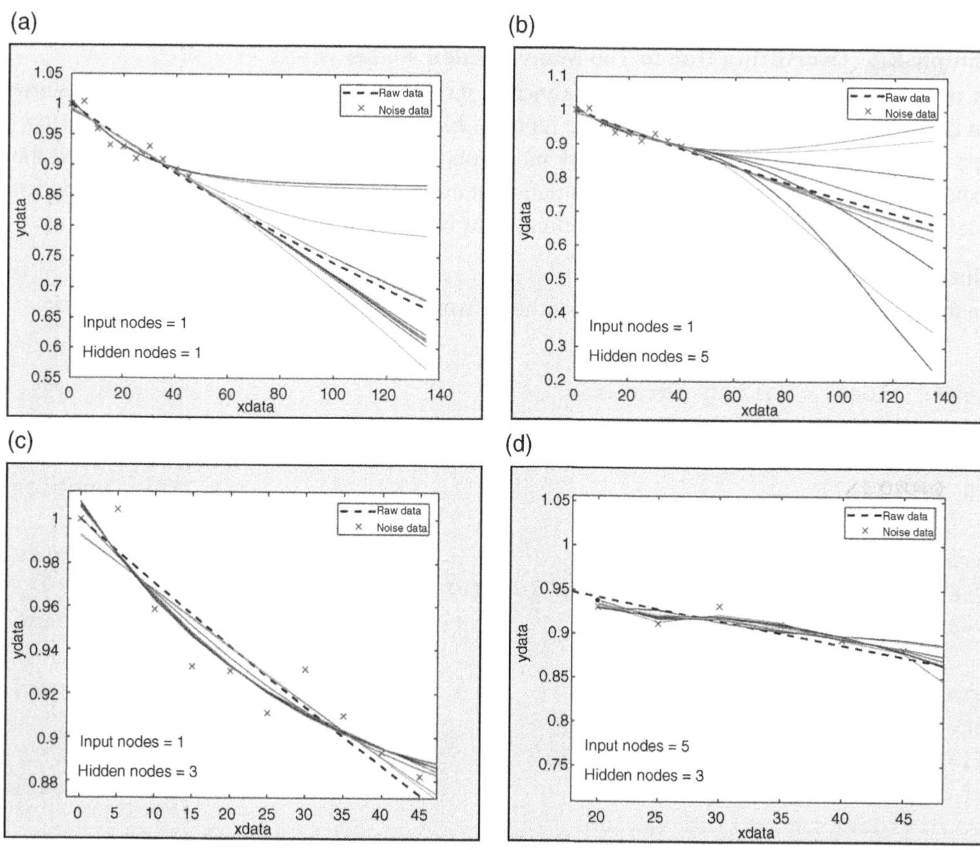


**Figure 8.24** The effect of the number of input and hidden nodes on training results. (a) One input and one hidden, (b) one input and five hidden, (c) one input and three hidden, and (d) five input and three hidden.

**Table 8.6** Training data set for sequential predictions.

| Input samples | Output samples | Predictions |
|---|---|---|
| $(y_1, y_2, y_3, y_4, y_5)$ | $y_6$ | $z_6$ |
| $(y_2, y_3, y_4, y_5, y_6)$ | $y_7$ | $z_7$ |
| $(y_3, y_4, y_5, y_6, y_7)$ | $y_8$ | $z_8$ |
| $(y_4, y_5, y_6, y_7, y_8)$ | $y_9$ | $z_9$ |
| $(y_5, y_6, y_7, y_8, y_9)$ | $y_{10}$ | $z_{10}$ |

As the number of input nodes increases, the chance of overfitting increases as well. First, as shown in Table 8.6, only five training data are available when five input nodes are used. Second, the number of parameters increases as the number of input nodes increases. In the case of single input node and three hidden nodes, there are three input weights. However, in the case of 5 input nodes, 15 input weights need to be determined. Therefore, the complexity of the model increases. In addition, as shown in Figure 8.24(b) and (d), it seems that the number of hidden nodes is related to the complexity of the prediction interval, while the number of input nodes is related to the complexity of the training data interval.

**Example 8.7 Overfitting Due to Too Many Hidden Nodes**
The network models in Figure 8.18 of true function $z(x) = 0.2x + \sin x$ are reasonable in the sense that the models follow the trend of the true function even if samples include random noise. However, when overfitting happens, the network model may pass all noisy samples accurately, but the prediction accuracy is not good. Test the situation of overfitting by making the number of hidden nodes to be 10 and use all samples for training. That is, the validation and test sets are empty.

**Solution:**
The following Matlab script is used to make the network model overfits the noise in samples:

```
%
clear; close all; rng default
a=0; b=4*pi; n=51; ns=15; nh=10;
x=linspace(a,b,n);                              % Test grid points
z=0.2*x+sin(x);                                   % True function
%
xx=linspace(a,b,ns);                      % Equally spaced samples
v=randn(1,ns); v=0.3*(v-mean(v))/std(v);
yy=0.2*xx+sin(xx)+v;                            % Add random noise
%
net=feedforwardnet(nh);             % create two-layer network model
net=configure(net,xx,yy);
all=1:ns; valInd=[]; testInd=[];
net.divideFcn = 'divideind';                       % Fixed datasets
net.divideParam.valInd=valInd;
net.divideParam.testInd=testInd;
net.divideParam.trainInd=setdiff(setdiff(all,valInd),testInd);
```

```
[netModel,trainRecor]=train(net,xx,yy);          % train the model 'net'
zNN=sim(netModel,x);                             % FFNN model predictions
%
plot(x,zNN,'k',x,z,'r',xx, yy,'or'); hold off
RMSE=sqrt((z-zNN)*(z-zNN)'/n)                    % RMSE at test grid
```

In order to prevent early stopping criterion, all 15 samples were used for training, and the validation and test sets were empty. Then, the backpropagation process will repeat as long as the MSE of the training set decreases. After the 67th epoch, the training process stopped with MSE being zero ($9.97 \times 10^{-20}$). However, using 51 equally spaced grid points, the RMSE turned out to be 0.6292. Therefore, the training performance was great, but due to overfitting, the prediction performance is low. Figure 8.25 shows the network model predictions at the grid points. In fact, the network predictions pass through all samples exactly, which is referred to as "fitting noise, not the trend." However, there are significant errors at unsampled locations.

2) $L_2$ regularization: The goal of a NN is to have a trained model that works well on both the data used for training and the new data that the model will be used for prediction. The ability to predict well on previously unobserved inputs is called generalization. Since overfitting is the situation where the model fits well for training data but predicts poorly for unobserved points, generalization is equivalent to preventing overfitting. In general, regularization is the process to modify the learning algorithm to reduce the generalization error, i.e., overfitting.

   As shown in Figure 8.23, overfitted models show fluctuation when input variable varies. Therefore, reducing overfitting means that predictions vary slowly when input variable changes. Among two types of parameters, weights affect generalization more than biases. Weights are multiplied with input variables, while biases are added independently. When a weight is large, the changes in input lead to a large change in output. An FFNN model with large magnitudes of weights tends to overfitting data. If the weight values of the final model after training are large, it means that the variance of the results for the input data is large, and large variance means that it

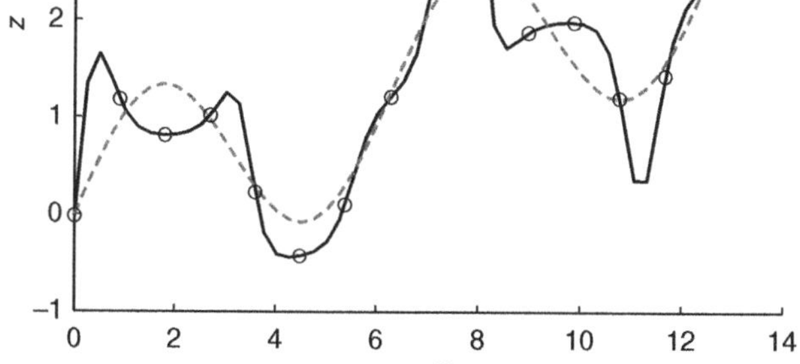


**Figure 8.25** Overfitting of noise in samples.

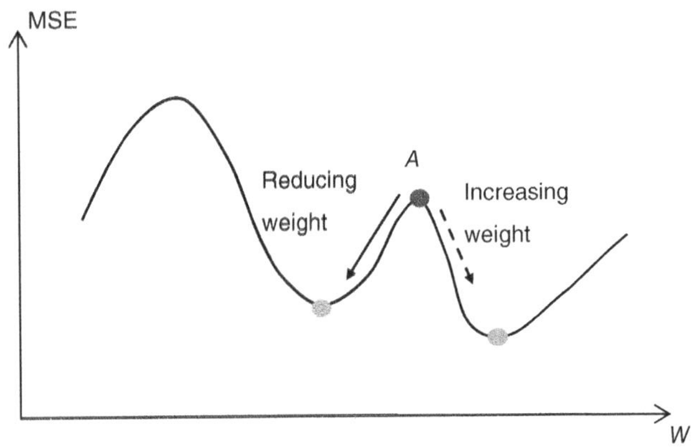


**Figure 8.26** Nonlinear behavior of the MSE with respect to a weight.

is more sensitive to small fluctuations and noise in the new input data. Therefore, when small fluctuations or noises are propagated through a network model with large weights, it generates values in the output layer that are much different than those of a network model with small weights.

In order to prevent overfitting, therefore, it would be helpful to reduce the magnitudes of weights in the trained model. As shown in Figure 8.26, the MSE is a nonlinear function of weights and may have many local minima. Let the current value of a weight be point A. The weight can be increased (dashed arrow) or decreased (solid arrow) because both directions can reduce the MSE. In this particular illustration, increasing the weight can yield a smaller value of MSE. However, this may cause overfitting as the weight is increased. Therefore, it would be better to guide the backpropagation algorithm to move in the direction to reduce the weight. Small weights can resolve the overfitting issue by obtaining a model that is less sensitive to fluctuations and noise in the input data, less complex, and more stable.

The core idea of regularization is to reduce the size of weights while reducing the MSE. The simplest and most common way is to modify the cost function (i.e., MSE) to penalize large weights. A simple penalty function is added to the MSE for the purpose of train. The following two forms are often used:

$$\tilde{F} = (1-\gamma)F + \gamma L_1 = \frac{1-\gamma}{n_y}\sum_{i=1}^{n_y}[y_i - z_i(\mathbf{d})]^2 + \frac{\gamma}{n_w}\sum_{j=1}^{n_w}|w_j| \tag{8.29}$$

$$\tilde{F} = (1-\gamma)F + \gamma L_2 = \frac{1-\gamma}{n_y}\sum_{i=1}^{n_y}[y_i - z_i(\mathbf{d})]^2 + \frac{\gamma}{n_w}\sum_{j=1}^{n_w}w_j^2 \tag{8.30}$$

where $\mathbf{d} = \{\mathbf{W}_x, \mathbf{b}_h, \mathbf{w}_h, b_z\}$ is the vector of all unknown parameters, $\gamma$ is that performance ratio that determines the influence of the regularization term and is called the regularization parameter, and $n_w = n_h \times n_x + n_h$ is the number of weights used in the model, including both the hidden and output layers. The original MSE is recovered when $\gamma = 0$. The effect of regularization increases along with $\gamma$. The $L_1$-norm in Eq. (8.29) penalizes all weights equally because the gradients of all weights will have the same magnitude. It is useful when trying to reduce the magnitude of all weights. The $L_2$-norm in Eq. (8.30) penalizes larger weights more than small ones. In addition, the $L_2$-norm provides smooth gradients, while the $L_1$-norm is non-differentiable when the weight is zero. Because of smoothness, regularization using the $L_2$-norm is commonly used.

For the purpose of backpropagation, the derivatives of the regularization term with respect to weights need to be calculated, and weights are updated using the gradients. This calculation is straightforward as the expression is simple. For example, the derivative of the $L_2$-norm regularization term with respect to the output weight $w_h$ can be written as

$$\frac{\partial L_2}{\partial w_h} = \frac{\partial}{\partial w_h}\left(\frac{1}{n_w}\sum_{j=1}^{n_w} w_j^2\right) = \frac{2}{n_w} w_h \tag{8.31}$$

Using these gradients, the weight update formulas for the gradient descent method in Eq. (8.20) is modified as follows:

$$w_{hj} = w_{hj} - \alpha\left((1-\gamma)\frac{\partial F}{\partial w_{hj}} + \gamma\frac{\partial L_2}{\partial w_{hj}}\right)$$

$$W_{xj,k} = W_{xj,k} - \alpha\left((1-\gamma)\frac{\partial F}{\partial W_{xj,k}} + \gamma\frac{\partial L_2}{\partial W_{xj,k}}\right) \tag{8.32}$$

The regularization parameter $\gamma$ in Eq. (8.30) plays a critical role in the performance of regularization. When it is too small, the effect of regularization disappears, and overfitting may occur. Figure 8.27(a) shows the case when the regularization parameter is too small. The effect of overfitting is minuscule in the interpolation region ($0 \le \text{xdata} \le 45$ in the figure), but it becomes significant in the extrapolation region ($45 \le \text{xdata}$ in the figure). When the regularization parameter is too large, the effect of regularization dominates that of the MSE. Then, the trained model loses its prediction capability. Figure 8.27(b) shows the case when the regularization parameter is too large. Even if the training data shows a clear trend, due to the dominant contribution of $L_2$-norm, the prediction becomes almost constant.

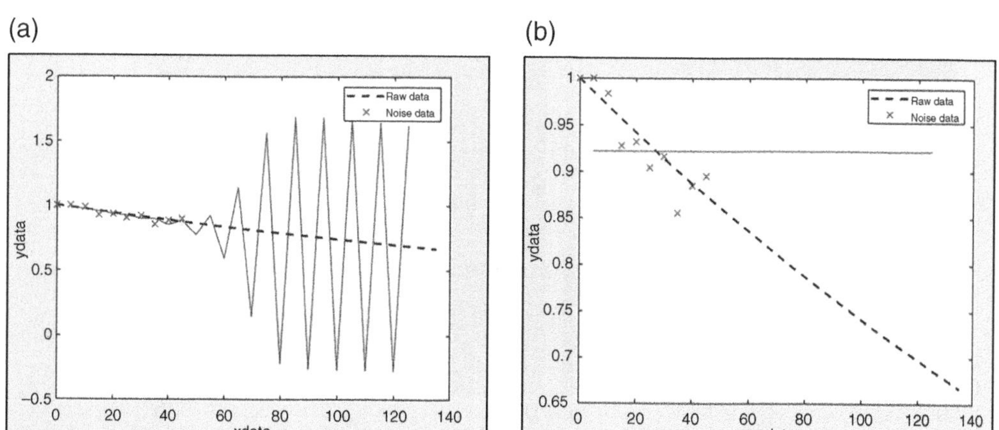


**Figure 8.27** The effect of the regularization parameter. (a) Too small $\gamma = 10^{-5}$ and (b) too large $\gamma = 1$.

It is difficult to provide an appropriate regularization parameter without knowing the behavior of underlying function. This is because it depends on the complexity of the network model, nonlinearity of output function, the level of noise in training data, and the number of training data. In addition, the parameter also depends on the range of output data **y**. This is because it will propagate to MSE. The contribution of the regularization term in Eq. (8.30) depends on the magnitude of MSE. It would be necessary to normalize the output data **y** in addition to scaling the input data. The users need to try different values of regularization parameters for training and find appropriate values of the regularization parameter.

With the adaptive learning rate, input data normalization, and $L_2$-norm regularization, the learning speed can be improved drastically. The adaptive learning rate will improve convergence near the optimal parameters. The input data normalization will help the vanishing gradient issue in activation functions. $L_2$-norm regularization will help to prevent large weights, which causes instability in predictions. Although it is not always the case, Figure 8.28 shows the contour of the cost function with respect to two weights, $w_1$ and $w_2$, along with convergence history of the training process. The left figure shows the learning history when MSE is used as a cost function, while the right figure is when the $L_2$-norm regularized cost function is used. The unregularized cost function converged after 100 epochs, while the regularized one converged in 17 epochs. In addition, the unregularized history shows zigzagging oscillation. It is clear that adding $L_2$-norm regularization can lead to faster progress toward the minimum in the gradient descent method, which not only prevents overfitting but can also increase convergence speed.

The main issue in the early stopping criterion is that it does not use all available samples for training the network model. In the case of Matlab implementation, 30% of samples are used for validation and testing. However, it is obvious that a better training outcome can be achieved by utilizing all available samples. The problem is that it is difficult to prevent overfitting when all samples are used for training. Instead of using separate samples for validation and testing, it is possible to utilize $L_2$-norm regularization to prevent overfitting, while using all samples for training. The following example demonstrates how $L_2$-norm regularization is used instead of separate validation and test datasets.

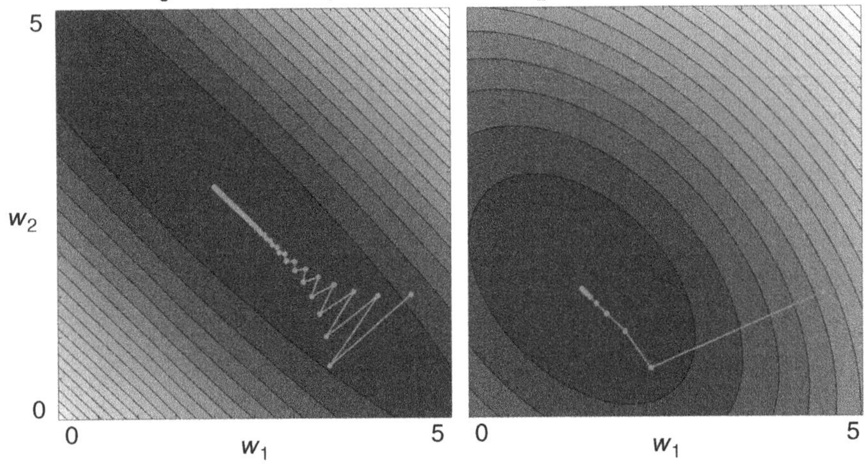


**Figure 8.28** Improved convergence speed pf gradience descent due to $L_2$ regularization.

**Example 8.8 Regularization Parameter**

Using the same network model in Example 8.7, apply $L_2$-norm regularization to remove overfitting phenomenon. Use all 15 samples for training and 10 hidden nodes. Study the effect of regularization parameter $\gamma = 0$, 0.01, 0.1, 0.5 and discuss its effect on prediction accuracy and prediction uncertainty.

**Solution:**

By default, Matlab uses Levenberg–Marquardt algorithm (Matlab command `'trainlm'`), which updates the Hessian information with only first-order derivative information. In general, the LM algorithm is fast and stable, but the current algorithm will stop training at the second epoch because the validation set is empty, and the penalty parameter used in the Jacobian matrix becomes infinity. Therefore, instead of the LM algorithm, the Fletcher–Powell conjugate gradient method (Matlab command `'traincgf'`) is used instead. The train algorithm can be specified in `feedforwardnet(nh,'traincgf')`. In order to activate the regularization parameter, `net.performParam.regularization` is set to be $\gamma$. The following Matlab script is used to generate 20 different sets of random noise and fit the network model with all samples.

```
%
clear; close all; rng default
a=0; b=4*pi; n=51; ns=15; nh=10;
x=linspace(a,b,n);                                   % Test grid points
z=0.2*x+sin(x);                                      % True function
%
xx=linspace(a,b,ns);                          % Equally spaced samples
yy=0.2*xx+sin(xx);
%
figure(1); hold on
for i=1:20% Repetition of FFNN
  v=randn(1,ns); v=0.3*(v-mean(v))/std(v);
  yy=0.2*xx+sin(xx)+v;                               % Add random noise
  %
  net=feedforwardnet(nh,'traincgf');       % create two-layer network model
  net=configure(net,xx,yy);
  net.performParam.regularization=0.01;
  % all=1:ns; valInd=[]; testInd=[];
  net.divideFcn = 'divideind';                       % Fixed datasets
  net.divideParam.valInd=valInd;
  net.divideParam.testInd=testInd;
  net.divideParam.trainInd=setdiff(setdiff(all,valInd),testInd);
  %
  [netModel,trainRecor]=train(net,xx,yy);        % train the model 'net'
  zNN=sim(netModel,x);                         % FFNN model predictions
  plot(x,zNN,'k');
  %
  RMSE(i)=sqrt((z-zNN)*(z-zNN)'/n);                  % RMSE at test grid
 end
 plot(x,z,'r',xx, yy,'or'); hold off
 m=mean(RMSE), s=std(RMSE) % Mean and standard deviation of RMSE
```

Figure 8.29 shows the network predictions with different regularization parameters. With 10 hidden nodes, there are 31 parameters to train, while the number of training samples is 15. Therefore, it is expected to have overfitting without regularization. Figure 8.29(a) shows the network predictions with $\gamma = 0$, which is identical to the case with no regularization, as shown in Figure 8.25. All predictions are not smooth and do not follow the trend of the true function. When an appropriate size of regularization parameter is used as in Figure 8.29(b), the predictions follow the trend of the true function and the RMSE shown in Table 8.7 has a low value of mean and standard deviation. However, it is unfortunate that it is not trivial to find the appropriate regularization parameter without trial and error. When the regularization parameter increases, as shown in Figure 8.29(c), the predictions become smoother, but the prediction errors increase. Lastly, when the regularization parameter becomes $\gamma = 0.5$, which means that the regularization effect is equally weighted with the MSE, the predictions become almost straight, as shown in Figure 8.29(d), with significantly increased error. At this point, regularization dominates the fitting error. In general, it is suggested to start with a small regularization parameter, such as $\gamma = 0.01$, and to monitor the smoothness of network predictions. If the predictions are not smooth, it indicates that overfitting occurs. The smallest regularization parameter that provides smooth predictions is probably the best parameter value.

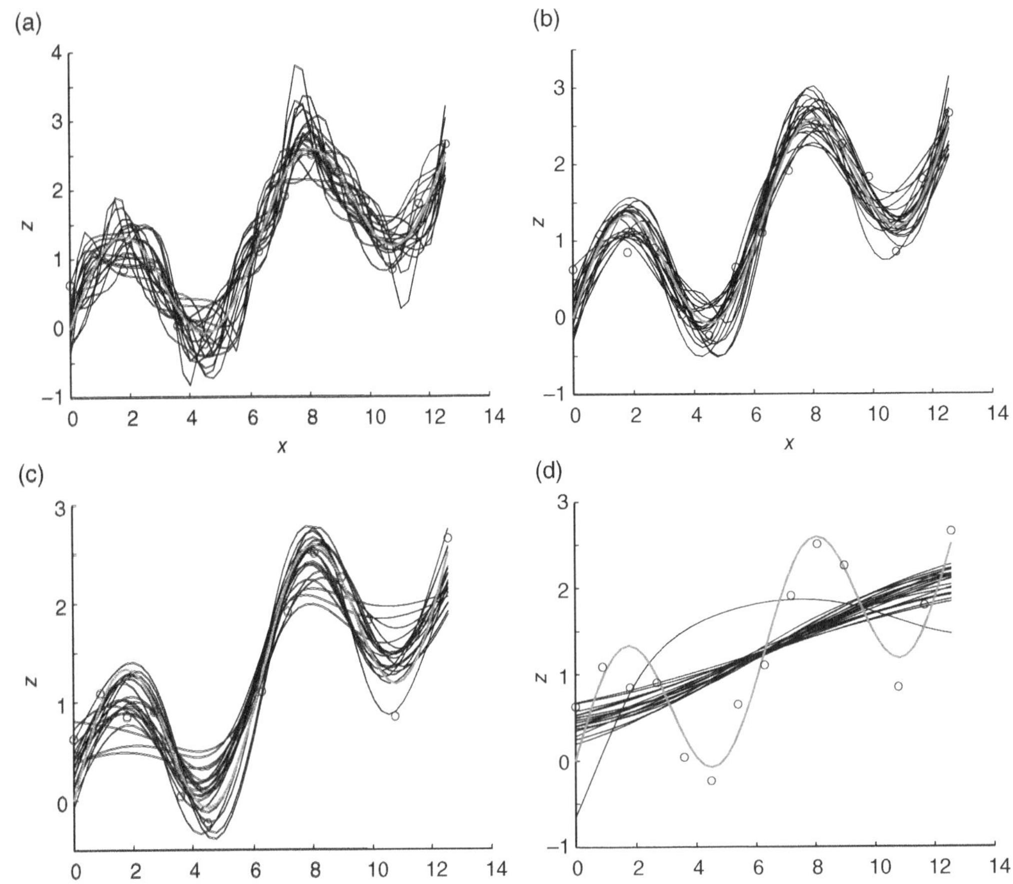


**Figure 8.29** $L_2$-norm regularization. (a) $\gamma = 0$, (b) $\gamma = 0.01$, (c) $\gamma = 0.1$, and (d) $\gamma = 0.5$.

**Table 8.7** Network prediction accuracy and uncertainty for different regularization parameters.

| Regularization parameter $\gamma$ | Mean of $e_{RMSE}$ | STD of $e_{RMSE}$ |
|---|---|---|
| 0.0 | 0.3052 | 0.0373 |
| 0.01 | 0.2132 | 0.0245 |
| 0.1 | 0.2503 | 0.0884 |
| 0.5 | 0.6549 | 0.0437 |

3) Early termination of learning: One of the important questions in machine learning is how many times to update NN training (epochs). If there are insufficient training epochs, the model will underfit data and fail to follow the trend of data. If there are too many training epochs, the model may overfit the training data, resulting in inferior performance on prediction. Normally, most training algorithms have the maximum number of epochs as a user-defined parameter and repeat training until it reaches the maximum number of epochs. However, this can be considered a waste of resources if the training algorithm converges before the maximum allowed number of epochs, in addition to the danger of overfitting. Therefore, it would be a good strategy to stop the learning process when it satisfies some stopping criteria.

Since the backpropagation process updates parameters using their gradients, the learning (updating) process slows down when the magnitudes of gradients become small. This can happen when actually gradients are close to zero near the optimum or due to the vanishing gradient problem in activation function. Either way, when the gradients are smaller than a user-defined threshold, additional training iterations may not improve the loss function significantly. Therefore, the magnitude of gradients can be used as an early stopping criterion. This is equivalent to checking the change of parameters (weight and biases) between two consecutive iterations

$$
\begin{aligned}
|\Delta w_h| &= \left|\alpha \frac{\partial F}{\partial w_h}\right| \le \beta \\
|\Delta b_z| &= \left|\alpha \frac{\partial F}{\partial b_z}\right| \le \beta \\
\left|\Delta W_{xj,k}\right| &= \left|\alpha \frac{\partial F}{\partial W_{xj,k}}\right| \le \beta \\
\left|\Delta b_{hj}\right| &= \left|\alpha \frac{\partial F}{\partial b_{hj}}\right| \le \beta
\end{aligned}
\tag{8.33}
$$

where $\beta$ is a user-defined parameter. If the changes of all parameters are less than the predefined threshold $\beta$ for continuous several iterations, the training can be stopped.

The above stopping criterion based on the changes of parameters can only reduce the computational cost by stopping unnecessary repetitive calculations. It does not have a significant effect on solving overfitting. It would be beneficial to introduce an early stopping criterion to prevent overfitting. For this purpose, it would be necessary to separate some data not to use training but to use for monitoring overfitting. This separated data set is referred to as validation set in machine learning. For example, 80% of data are used for the training set and 20% of them are used for the validation set. Then, the backpropagation process is performed only using

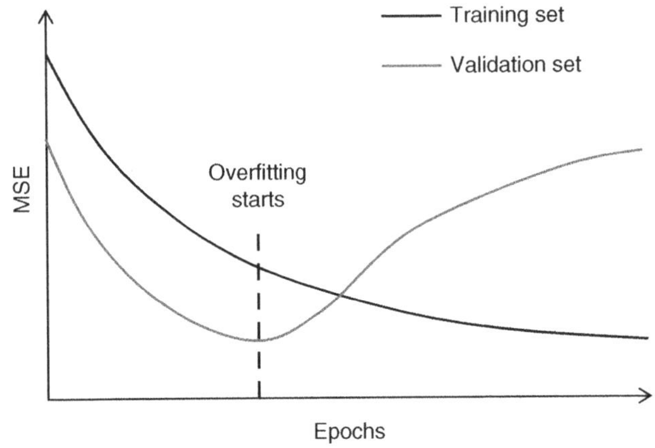


**Figure 8.30** MSEs of training set and validation set.

training data set. At each epoch, the MSEs are calculated separately for the two sets. The MSE of the training set indicates the performance of training, while that of the validation set shows the performance of generalization, i.e., the performance of prediction capability. When the loss function is reduced for both training and validation sets, the training process continues because the training fits data well and predicts unobserved data well. However, when the MSE of the training set decreases but that of the validation set increases, it indicates that the training process starts overfitting the training data. Then the training process stopped. As shown in Figure 8.30, the training process is stopped when MSE of the validation set starts increasing.

The early stopping criterion using the validation set is widely used in NN training because it is simple and can avoid overfitting effectively. The only disadvantage of this criterion is that not all data are used for training. This is the strategy Matlab implemented in the Machine Learning Toolbox. In Matlab, the entire data is divided into three sets: training data, validation data, and test data sets. By default, 70% of data are used for the training set, 15% for the validation set, and remaining 15% for the test set. The test set is used to assess the prediction accuracy. However, in many applications, especially when the number of data is not enough, it may not be necessary to have the test set. The selection of these three sets is random, which also causes training uncertainty in NNs. Since the training of NNs is stochastic and noisy, the performance of the model on the validation set shown in the graph can change (increase or decrease) several times. In fact, the validation error curve almost always has more than one local minimum. Therefore, a slight delay or "patience" in early stopping training is often used. That is, the training process is stopped if MSE of validation set increases for several iterations. When the training process is stopped by satisfying this condition, then the final trained model parameters are the values when MSE of the validation set start increasing.

## 8.6 Neural Networks with Constraints

As emphasized previously, the key point in NN modeling is not the accuracy of fitting the current data but the accuracy at unobserved points. In the conventional NN community, this is referred to as generalization. The surrogate community calls this prediction accuracy. In the previous chapters, we also discussed the difference between the predictions in the interpolation region and that of the

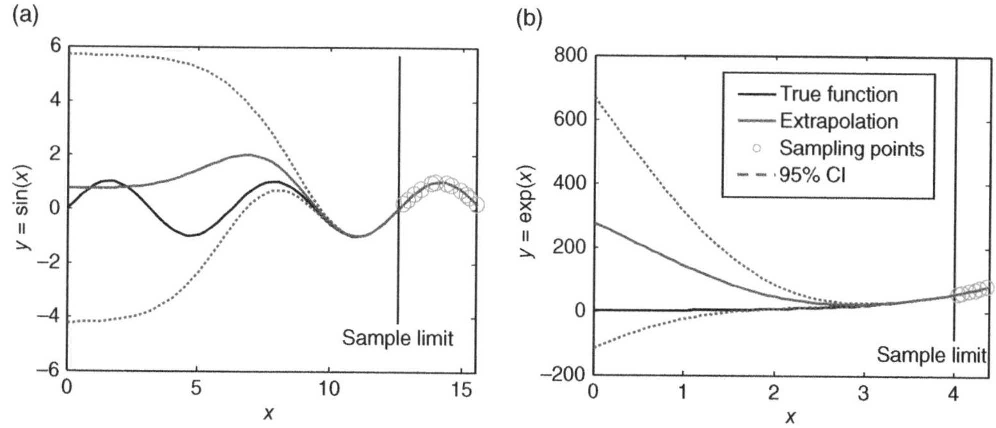


**Figure 8.31** Extrapolation of (a) $y = \sin(x)$ and (b) $y = \exp(x)$.

extrapolation region. The weakest point of NNs is the prediction accuracy at the extrapolation region. This is because network models are trained by samples, and without utilizing physical knowledge, it is difficult to guarantee accuracy at unobserved points. For example, Figure 8.31 shows two examples of extrapolating simple functions, $y = \sin(x)$ and $y = \exp(x)$. It is assumed that samples are given only at 20% of the entire domain, and the remaining 80% is the extrapolation region. The surrogate approximations were accurate in the interpolation region, but they deviate from the true one significantly as the distance increases from the sample limit. This is a common behavior of data-driven surrogate models because they do not consider any information about the functional behavior at the unsampled domain.

In order to improve the prediction accuracy at unsampled locations, physical constraints are often imposed on the NN model. For example, if users know the function has a monotonically increasing or decreasing behavior, this knowledge can be imposed in the form of constraint in the training process. For example, if users know that the exponential function in Figure 8.31(b) is monotonic, the confidence interval in the figure can significantly be reduced. In addition, if the positive condition of the exponential function is imposed, then the network model may yield quite accurate predictions at the extrapolation region. Therefore, providing additional information (physical or non-physical) can enhance the accuracy of predictions and reduce uncertainty in the extrapolation region.

Imposing additional constraints in the training process is similar to reducing overfitting using $L_2$-norm regularization in the previous section. The normal training process solves an optimization problem that minimizes the MSE. This corresponds to unconstrained optimization in Section 5.4. With additional constraints, however, the training process becomes a constrained optimization problem, which was discussed in Section 5.6. Solving a constrained optimization problem is quite different from unconstrained one because optimization results are determined by constraints. The basic concept in the gradient descent method is that the optimum results are when all gradients become zero, which is called the first-order optimality criterion (Karush–Kuhn–Tucker condition). Therefore, a simple algorithm is developed to move parameters until their gradients become zero. However, in the case of constrained optimization, the optimum results are not located in the zero-gradient locations, but on the constraint boundary, where the gradients are not zero.

### 8.6.1 Penalty Method for Constraints

Although there are many different kinds of constraints, a simple constraint is used here for the purpose of illustration. We utilize simple physical and engineering information that we already know to prevent meaningless training results and obtain a more robust and reliable trained model. As shown in battery degradation problem in Eq. (8.28), the input is time, and the output is the degradation level of battery. Discrete time intervals are given as $\{t_1, t_2, ..., t_{n_y}, t_{n_y+1}, ..., t_N\}$. The degradation data at the first $n_y$ intervals are used for training: $\mathbf{y} = \{y_1, y_2, ..., y_{n_y}\}$. Therefore, time intervals $\{t_1, t_2, ..., t_{n_y}\}$ are the interpolation region, while time intervals $\{t_{n_y+1}, ..., t_N\}$ are the extrapolation region. The network model predictions at the discrete times are denoted as $\mathbf{z} = \{z_1, z_2, ..., z_{n_y}, z_{n_y+1}, ..., z_N\}$. Since degradation is a monotonically decreasing process, the slopes of the predictions at discrete times are used as constraints. Then the constrained optimization problem is to find $\mathbf{d} = \{\mathbf{W}_x, \mathbf{b}_h, \mathbf{w}_h, b_z\}$ such that

$$\begin{aligned} &\text{Minimize} \quad F = \frac{1}{n_y}\sum_{i=1}^{n_y}[y_i - z_i(\mathbf{d})]^2 \\ &\text{subject to} \quad z_i' \le 0, \quad i = 1, ..., t_N \end{aligned} \tag{8.34}$$

where $z_i'$ is the time derivative of degradation.

It is noted that Eq. (8.34) has inequality constraints. In general, it is possible to have equality constraints as well. However, equality constraints are much more difficult to satisfy than inequality ones. It is possible that an equality constraint can be replaced by two inequality constraints. That is, equality constraint $g(x) = 0$ can be replaced by $g(x) \le 0$ and $g(x) \ge 0$. Greater-than-or-equal-to constraint, $g(x) \ge 0$, can be converted into less-than-or-equal-to $-g(x) \le 0$. Therefore, the inequality constraints in Eq. (8.34) can cover a broad range of engineering constraints.

Additional constraints make the optimization problem difficult. In the optimization algorithms in Section 5.5, it has been shown that the constrained optimization problem in Eq. (8.34) can be converted into a unconstrained optimization problem using either Lagrange multiplier method or the penalty method. Here we utilize the penalty method to convert Eq. (8.34) into a unconstrained optimization problem. Before we convert it, first it is noted that there are $N$ constraints. Different from the cost function, the magnitudes of constraints are not important as long as they satisfy the inequality. Therefore, instead of considering all $N$ constraints, it is possible to consider the maximum constraint, which is called the maximum violation. Therefore, the constraints in Eq. (8.34) are equivalent to $P(\mathbf{d}) = \max(0, \mathbf{z}'(\mathbf{d})) \le 0$, where $P(\mathbf{d})$ is called the penalty function. This is also called the positive part function. Penalty $P(\mathbf{d})$ is equal to zero when all constraints are satisfied. When some constraints are violated (i.e., positive), it has the value of the maximum violation. This penalty function can serve the purpose, but it causes a non-differentiability issue at the constraint boundary. In order to make a smooth transition, the penalty term is squared, which is often called a quadratic loss function. $P(\mathbf{d}) = \max(0, \mathbf{z}'(\mathbf{d}))^2 \le 0$. Adding a penalty function can cause a serious change in slope at the boundary of the graph, which can interfere with the general minimization program. To solve this, the quadratic loss function is multiplied by penalty parameter $\lambda$. Finally, the modified cost function with penalty term is defined as

$$\tilde{F} = F(\mathbf{d}) + \lambda P(\mathbf{d}) = \frac{1}{n_y}\sum_{i=1}^{n_y}[y_i - z_i(\mathbf{d})]^2 + \lambda \cdot \max(0, \mathbf{z}'(\mathbf{d}))^2 \tag{8.35}$$

Minimizing the modified cost function does not guarantee that all constraints are satisfied. This is because the penalty method allows a small constraint violation. But when the constraint is violated, the cost function increases, such that the optimization algorithm tries to reduce the constraint violation. The constraint violation will be reduced if the penalty parameter is increased. If $\lambda$ is increased too much, however, the optimization algorithm will have difficulty in convergence due to ill-conditioning. It is noted that the modified cost function in Eq. (8.35) is similar to that in Eq. (8.30) by considering the regularization parameter as the penalty parameter. Therefore, a similar approach can be used to minimize the modified cost function.

### 8.6.2 Regularization Using Soft-maximum

The quadratic penalty loss function in Eq. (8.35) makes a single constraint $P(\mathbf{d}) = \max\left(0, z_1'\right)^2 \leq 0$ smooth. However, the maximum of multiple constraints can still show non-smooth behavior as parameters vary. In order to illustrate this issue clearly, Figure 8.32 shows hypothetical behaviors of multiple constraints. In the figure, each constraint has a different maximum point. In such a case, the penalty function shows ups and downs with many peaks. This behavior of a function makes the optimization algorithm difficult to satisfy constraint because when the parameter increases to reduce $z_1'$, other constraint $z_2'$ starts increasing. Therefore, it is highly likely the optimization algorithm can be stuck in worthless local optima.

In order to resolve the oscillating behavior of the maximum function, the softplus function can be used. The softplus function is a smooth version of the rectified linear unit (ReLU) function. Let us consider a positive part function $f(x) = \max(0, x)$. This function is non-smooth at $x = 0$. In order to approximate this function, the softplus function can be defined as

$$g(x) = \log(1 + e^x) \tag{8.36}$$

As shown in Figure 8.33, this function yields the same value as the positive part function when $x \leq -3$ or $x \geq 3$. When $-3 < x < 3$, this function smoothly changes its slope. In fact, the gradient of the softplus function can be obtained as

$$g'(x) = \frac{e^x}{1 + e^x} \tag{8.37}$$

which is the logistic sigmoid function.

The multivariate generalization of the softplus function is the logarithm of the sum of the exponents of multiple variables (LogSumExp), which is a smooth approximation of the maximum, i.e., soft maximum. The LogSumExp function is defined as

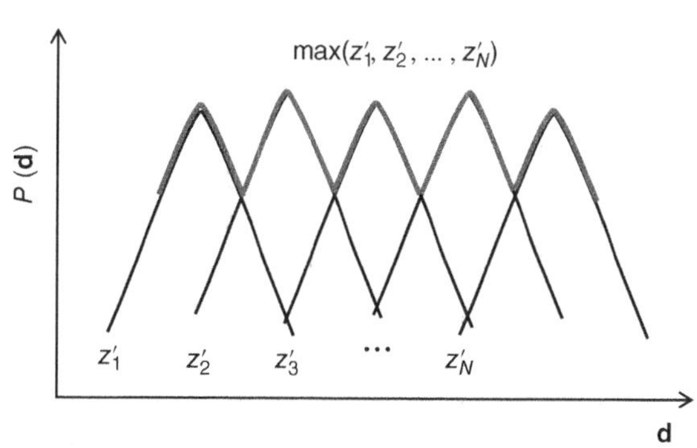


**Figure 8.32** Non-smoothness of maximum function.

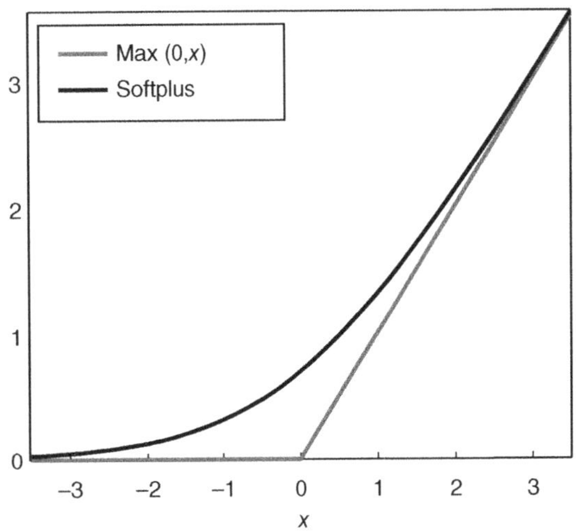


**Figure 8.33** Rectified linear unit function versus softplus function.

$$g(x_1, x_2, ..., x_n; k) = \frac{1}{k}\log\left[1 + \sum_{i=1}^{n}\exp(kx_i)\right] \tag{8.38}$$

where $k$ is a parameter to control the accuracy of LogSumExp function to the actual multi-variate maximum function. When variables $x$'s are statistically distributed such that only a single $x_i$ is the maximum, a relatively small value of $k$ will approximate the actual maximum well. However, if multiple $x$'s have the same maximum, a large $k$ is required to be accurate.

**Example 8.9 Accuracy of the LogSumExp function**

Test the accuracy of the LogSumExp function by comparing it with the actual maximum. (i) Generate 10 random numbers in the range of $x \in [0, 10]$ and find the maximum using LogSumExp. (ii) Generate 10 copies of $x_i = 10$, $i = 1, ..., 10$ and find the maximum using LogSumExp. Consider two cases of $k = 10$ and $k = 50$.

**Solution:**

a) Ten random numbers are generated using Matlab function `x=10*rand(1,10)` as

$$\mathbf{x} = [2.7692 \quad 0.4617 \quad 0.9713 \quad 8.2346 \quad 6.9483 \quad 3.1710 \quad 9.5022 \quad 0.3445 \quad 4.3874 \quad 3.8156]$$

In these samples, the actual maximum value is

$$f(\mathbf{x}) = \max(x_1, x_2, ..., x_{10}) = 9.5022$$

The LogSumExp function calculates

$$g(\mathbf{x}) = \frac{1}{k}\log\left[1 + \sum_{i=1}^{10}\exp(kx_i)\right] = 9.5022$$

This result is the same for both $k = 10$ and $k = 50$.

b) Ten samples of maximum are generated using Matlab function `x=10*ones(1,10)`. It is obvious that the actual maximum is 10. The LogSumExp function yields $g(\mathbf{x}) = 10.2303$ with $k = 10$, while g(**x**) = 10.0461 with $k = 50$. Therefore, with $k = 10$, the LogSumExp function has about 2.3% error, while the error is 0.46% when $k = 50$. However, these errors are conservative in the sense that the soft maximum is greater than the actual maximum.

**Example 8.10 Soft-maximum function**

Consider five functions $z_i(d) = -(2d - 3i)^2 + 2i$, $i = 1, ..., 5$, in the range of $d \in [0, 10]$. Find the soft maximum of $f(\mathbf{z}) = \max(z_1, z_2, ..., z_5)$ using $k = 0.5$ and $k = 1.0$. Compare the true maximum with the soft maximum for the range of parameter $d$.

**Solution:**

The following Matlab code is used to plot the actual maximum and the soft maximum.

```
k=1.0; %k=0.5;
d=0:0.1:10;
for i=1:5
 z(i,:)=-(2*d-3*i).^2+2*i;
end
tmax=max(0,z);
smax=log(1+sum(exp(k*z)))/k;
plot(d,z(1,:),'k-',d,z(2,:),'k-',d,z(3,:),'k-',d,z(4,:),'k-'...
    ,d,z(5,:),'k-',d,tmax,'b',d,smax,'r')
axis([0 10 -1 10])
```

Figure 8.34 shows the plots of five functions (dashed curves), the actual maximum (dark curve), and the soft maximum (light curve). When the control parameter is large, $k = 1.0$, the soft maximum closely follows the actual maximum function, except for the sharp valleys. When the control parameter is small, $k = 0.5$, the smoothness of the soft maximum function is improved, while the error between the soft maximum and the actual maximum increases as well. It would be unnecessary to use a large value of the control parameter because the soft maximum overestimates the actual maximum and smoothness is better than the case with a large control parameter.

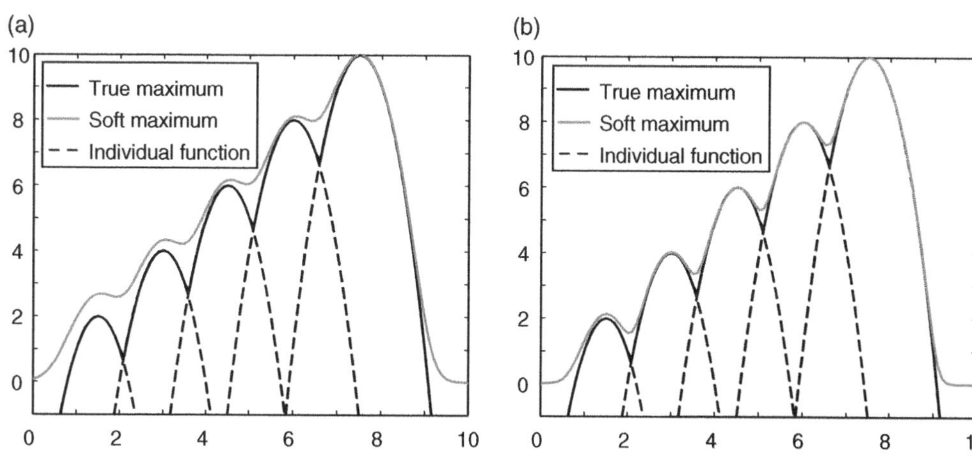


**Figure 8.34** Soft maximum of five functions. (a) $k = 0.5$ and (b) $k = 1.0$.

It is also noted that the soft maximum function approaches zero when the actual maximum is negative. This is because of the added one in the logarithmic function in Eq. (8.38). The soft maximum function follows the negative side of the actual maximum function without this term. Since the penalty constraint should be zero when the maximum constraint is negative, this term is required for this purpose.

### 8.6.3 Backpropagation of Penalty Constraints

In order to implement the constraint that the gradient of the trained model must be less than or equal to zero, the gradients $\mathbf{z}'$ of the simulation output were calculated for each data point. The soft maximum function in Eq. (8.38) is used to impose the slope constraints in the original constrained optimization problem in Eq. (8.34). With the penalty parameter $\lambda$ and control parameter $k$, the unconstrainted optimization problem minimizes the following modified loss function:

$$\tilde{F} = F(\mathbf{d}) + \lambda P(\mathbf{d}) = \frac{1}{n_y}\sum_{i=1}^{n_y} [y_i - z_i(\mathbf{d})]^2 + \lambda \cdot \frac{1}{k} \log\left[1 + \sum_{i=1}^{N} \exp\left(kz_i'(\mathbf{d})\right)\right] \tag{8.39}$$

It is noted that the slope constraints are imposed not only on the training data but also on future prediction points. This will prevent unrealistic models that violate physical behavior of degradation. Since the soft maximum function is a single constraint, minimizing the modified loss function in Eq. (8.39) is considered computationally inexpensive than considering $N$ individual constraints.

In order to use the backpropagation algorithm, it is necessary to calculate the gradients of the modified loss function with respect to model parameters, $\mathbf{d} = \{\mathbf{W}_x, \mathbf{b}_h, \mathbf{w}_h, b_z\}$. Since the gradients of the MSE were already derived in Section 8.2, the gradients of the penalty function will be discussed here. First, time derivatives of the output can be approximate by the difference between the two consecutive times: $z_i' \approx \Delta z_i = z_i - z_{i-1}$. First, consider the gradient of the penalty function with respect to $j$th hidden weight $w_{hj}$ as

$$\frac{\partial P}{\partial w_{hj}} = \sum_{i=2}^{N}\left(\frac{\partial P}{\partial z_i} \times \frac{\partial z_i}{\partial v_i} \times \frac{\partial v_i}{\partial w_{hj}}\right) \tag{8.40}$$

Note that the summation is over the entire time intervals except for the first interval. The first term can be calculated by differentiating $P(\mathbf{d})$ in Eq. (8.39), the second term is the derivative of the activation function $\Phi_z'(v_i)$, and the last term is simply the hidden node $h_j(\mathbf{x}_i)$. The first term can be calculated by

$$\frac{\partial P}{\partial z_i} = \frac{1}{kP(\mathbf{d})}[\exp(k\Delta z_i) - \exp(k\Delta z_{i-1})] \tag{8.41}$$

The other two terms were already calculated in the backpropagation section. Therefore, the gradient can be written as

$$\frac{\partial P}{\partial w_{hj}} = \sum_{i=2}^{N}\left(\frac{\partial P}{\partial z_i} \times \Phi_z'(v_i) \times h_j(\mathbf{x}_i)\right) \tag{8.42}$$

Similarly, the derivative with respect to the output bias can be calculated as

$$\frac{\partial P}{\partial b_z} = \sum_{i=2}^{N_y}\left(\frac{\partial P}{\partial z_i} \times \Phi_z'(v_i)\right) \tag{8.43}$$

The gradients with respect to input weights are more complicated but basically follow the same chain rule of differentiation. Let $W_{xj,\,k}$ is the weight that connects $j$th input variable with $k$th hidden node. Then, the gradient can be defined as

$$\frac{\partial P}{\partial W_{xj,k}} = \sum_{i=2}^{N}\left(\frac{\partial P}{\partial z_i} \times \frac{\partial z_i}{\partial v_i} \times \frac{\partial v_i}{\partial h_k} \times \frac{\partial h_k}{\partial u_k} \times \frac{\partial u_k}{\partial w_{xj,k}}\right) \tag{8.44}$$

It is noted that even if there are $n_h$ hidden nodes, only $h_k$ is connected with $W_{xj,k}$. Therefore, it is unnecessary to consider all hidden nodes. The first two terms on the right-hand side of Eq. (8.44) are identical to Eq. (8.41). The third term is the hidden weight $w_{hk}$, the fourth term is the derivative of hidden activation function $\Phi_h'(u_k)$, and the last term is input $x_j$. Therefore, the gradient can be written as

$$\frac{\partial P}{\partial W_{xk,j}} = \sum_{i=2}^{N}\left(\frac{\partial P}{\partial z_i} \times \Phi_z'(v_i) \times w_{hj} \times \Phi_h'\left(u_j\right) \times x_k\right) \tag{8.45}$$

It is noted that variables $x_k$,$u_j$, and $v_i$ should be calculated using $i$th input sample $\mathbf{x}_i$. Similarly, the derivative with respect to the hidden bias can be calculated as

$$\frac{\partial P}{\partial b_{hj}} = \sum_{i=2}^{N}\left(\frac{\partial P}{\partial z_i} \times \Phi_z'(v_i) \times w_{hj} \times \Phi_h'\left(u_j\right)\right) \tag{8.46}$$

In the same way, $\partial P/\partial b_z$, $\partial P/\partial W_{xj,k}$, and $\partial P/\partial b_{hj}$ can be calculated. The gradients of the calculated penalty term are reflected in the existing parameter update equation in Eqs. (8.16) and (8.20) and modified as

$$\begin{aligned}
w_{hj} &= w_{hj} - \alpha\left(\frac{\partial F}{\partial w_{hj}} + \lambda\frac{\partial P}{\partial w_{hj}}\right)\\
b_z &= b_z - \alpha\left(\frac{\partial F}{\partial b_z} + \lambda\frac{\partial P}{\partial b_z}\right)\\
W_{xj,k} &= W_{xj,k} - \alpha\left(\frac{\partial F}{\partial W_{xj,k}} + \lambda\frac{\partial P}{\partial W_{xj,k}}\right)\\
b_{hj} &= b_{hj} - \alpha\left(\frac{\partial F}{\partial b_{hj}} + \lambda\frac{\partial P}{\partial b_{hj}}\right)
\end{aligned} \tag{8.47}$$

Based on the modified equation, we can find the weights and biases that minimize the MSE term and penalty term in the training process.

### 8.6.4 Updating Penalty Parameter

From the perspective of optimization problems, adding a penalty function to the loss function can cause a serious change in the gradient at the constraint boundary, which makes the algorithm difficult to converge. In fact, it has been confirmed that the training results with penalty terms frequently experience overfitting compared to the results without them. In particular, the larger the value of the penalty parameter $\lambda$ is, which controls how much the constraint violation is allowed, the more severe the overfitting phenomenon becomes. If the size of the parameter is reduced, the chance of overfitting can be reduced, but the influence of the penalty term is reduced, which can lead to training results that do not satisfy the constraints. To resolve this issue, an algorithm can be applied that gradually increases the penalty parameter $\lambda$ during the training process.

First, the training process starts with small penalty parameter $\lambda$, and the optimal parameters are found when no serious change in the gradient appears. However, since the influence of the penalty term is not sufficiently large, it may not satisfy the constraints. Then, the optimal parameters found are used as the initial parameters for the next training with the increased value of $\lambda$. By repeating the training process with gradual increase of the penalty parameter, the optimal training result that satisfies the constraints can be obtained.

Figure 8.35(a) shows the training result with fixed $\lambda = 1$. The training process is repeated five times. Although the trained model yields a smooth prediction, some of them failed to satisfy monotonically decreasing constraint. Figure 8.35(b) shows the training result with fixed $\lambda = 100$. Due to the large penalty parameter, the trained model shows spurious oscillation and failed to follow the trend of data. Figure 8.35(c) is the training result with gradually increasing penalty parameter. Starting from initial value $\lambda = 0.1$, the parameter is increased by 10 times every repetition of training. The training process is repeated five times. Finally, the optimal weights and biases were obtained when $\lambda = 10{,}000$. If the training process failed to converge, the training process started again with the initial parameter value. As shown in Figure 8.35(c), all predictions satisfy the constraint, and the overfitting phenomenon that frequently occurred when $\lambda = 100$ can be resolved.

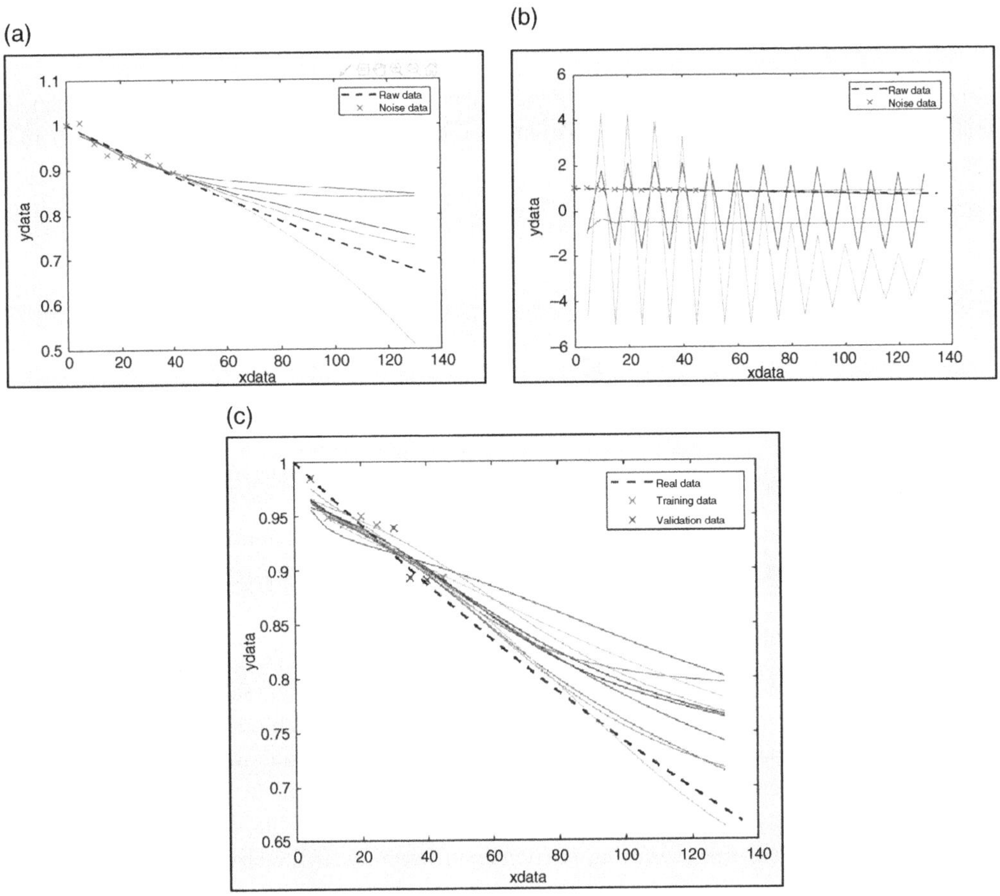


**Figure 8.35** The effect of penalty parameter on overfitting. (a) Fixed $\lambda = 1$, (b) fixed $\lambda = 100$, and (c) gradually increased $\lambda = 10{,}000$.

### 8.6.5 Numerical Examples

As mentioned before, the training results will be different even if the same dataset is used. This is because of various sources of uncertainty in the training process. Therefore, it would be necessary to repeat the training process multiple times to estimate the prediction intervals. In the following examples, the training process is repeated 30 times and estimate prediction intervals. Figure 8.36(a) shows the prediction results for the battery degradation problem. Data are generated from the degradation model in Eq. (8.28) with added random noise. Without imposing monotonicity constraint, some predictions violate the physical phenomenon. Figure 8.36(b) shows the prediction results with monotonicity constraint. Overall, the predictions with the constraint show consistent behavior and narrower prediction intervals.

Next example is the fatigue crack growth example. When cyclic fatigue loads are applied to a flat panel, the pre-existing initial crack gradually increases over the loading cycles. The Paris model [106] describes the relationship between the crack growth rate $da/dN$ with the stress intensity factor $\Delta K = \Delta\sigma\sqrt{\pi a}$:

$$\frac{da}{dN} = C(\Delta K)^m \tag{8.48}$$

where $C$ and $m$ are model parameters. By integrating this equation, the following crack growth model can be obtained:

$$a(N) = \left[N \cdot C\left(1 - \frac{m}{2}\right)(\Delta\sigma\sqrt{\pi})^m + a_0^{1-\frac{m}{2}}\right]^{\frac{2}{2-m}} \tag{8.49}$$

where $a_0$ is the initial crack size, $a(N)$ is the crack size at cycle $N$, and $\Delta\sigma$ is the range of fatigue stress. Simulated data was generated assuming $\Delta\sigma = 75$ MPa, $m = 3.8$, $C = 1.5\times10^{-10}$ and $a_0 = 0.01$ m, to which Gaussian noise with mean zero and variance $0.001^2$ was added. It was used as training data. The model in Eq. (8.49) can calculate a complex number when the crack size becomes too large. When that happens, it is replaced with an arbitrary large crack size value of 1.0 m.

The prediction results of 30 training processes are shown in Figure 8.37. Different from the battery degradation example, since the crack size must monotonically increase, a constraint is required that the slope calculated at each point must be positive ($-z' \le 0$), which is imposed using the

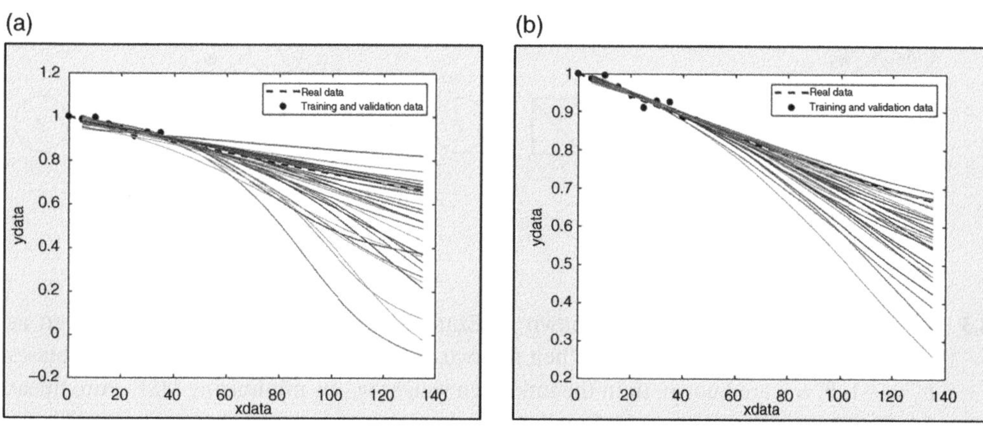


**Figure 8.36** Battery degradation model training results. (a) 30 training results without penalty and (b) 30 training results with penalty.

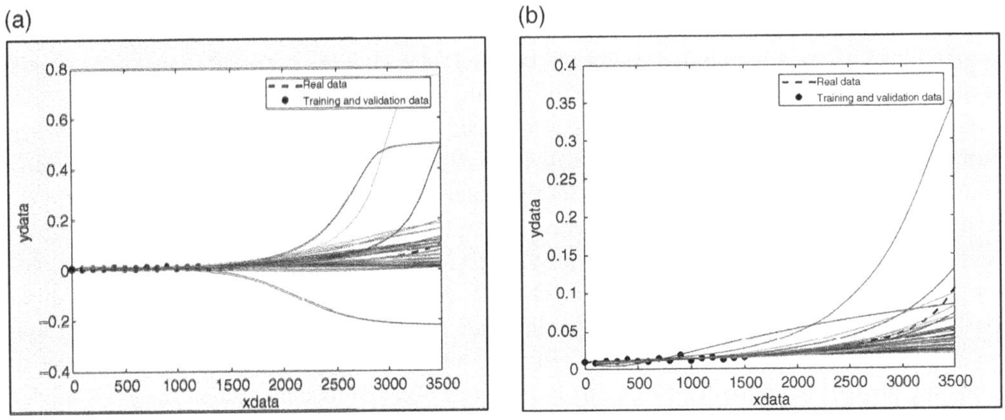


**Figure 8.37** Crack growth training results. (a) 30 training results without penalty, (b) 30 training results with penalty.

penalty method. Without the constraint in Figure 8.37(a), the predicted crack size can decrease or even negative crack sizes, which is physically impossible. On the other hand, with the slope constrain in Figure 8.37(b), all predictions show positive slopes. In addition, the prediction intervals are smaller compared to the case without the constraint.

## Exercise

**8.1** A NN model has two input nodes, two hidden nodes, and one output node. When all activation functions are pure linear, convert the network model without the hidden layer by calculating equivalent input weights and output bias.

**8.2** A network model with one hidden node is converted into a network model without hidden node, as shown in the figure. Calculate the weights $(w_1^*, w_2^*, w_3^*)$ and bias $(b_1^*)$ in terms of the original weights $(w_1, w_2, w_3, w_4)$ and biases $(b_1, b_2)$.

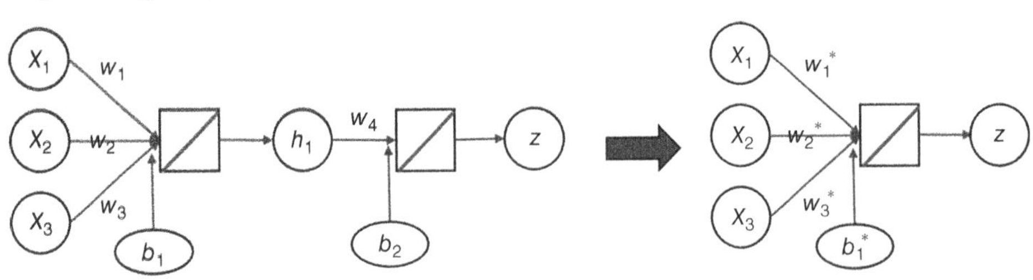


**8.3** A NN model is constructed, as shown in Example 8.1. Two training data are given as $\{x_1, x_2, z\} = \{1, 2, 3\}$ and $\{2, 3, 4\}$. When the two weights $w_{x2} = 2$, $w_h = -1$ and two biases $b_h = -1$, $b_z = 1$ are known, train the unknown weight $w_{x1}$ by minimizing MSE. Pure linear function is used for the transfer functions.

**8.4** Design a FFNN model that can express the Rosenbrock function in Eq. (8.21) exactly.

**8.5** Repeat the backpropagation algorithm in Example 8.3 when the hyperbolic tangent function is used for the hidden layer. Use the learning rate $\alpha = 0.2$. Test the accuracy of the trained network model by plotting the network prediction with the true function.

**8.6** The prediction accuracy of the NN model of the Rosenbrock function in Example 8.5 was not good at the four corner regions because they are extrapolation regions. Add additional four samples at the four corners and train the network model. Compare the RMSE with and without corner samples.

**8.7** Activation functions play an important role in modeling nonlinear behavior of a function. By default, Matlab uses hyperbolic tangent function (`tansig`) for hidden layers and pure linear function (`purelin`) for the output layer. Train the Rosenbrock function in Example 8.5 with different activation functions (pure linear, sigmoid, ReLU, and tanh) and compare them in terms of plots and RMSE.

**8.8** The Branin–Hoo function is a popular analytical function that is used for testing surrogate models. The functional form is defined in $x_1 \in [-5, 10]$ and $x_2 \in [0, 15]$ as

$$f(x_1, x_2) = \left(x_2 - bx_1^2 + cx_1 - 6\right)^2 + 10(1 - t)\cos(x_1) + 10$$

where $b = 5.1/(4\pi^2)$, $c = 5/\pi$, and $t = 1/(8\pi)$. Trained the Branin–Hoo function with 30 Latin hypercube samples and estimate the RMSE at $51 \times 51$ grid points. Repeat the process with additional four corner samples. The Branin–Hoo function has three global optima of $f(\mathbf{x}^*) = 0.397887$ at $\mathbf{x}^* = (-\pi, 12.275), (\pi, 2.275), (9.42478, 2.475)$. Evaluate the accuracy of the NN model at these optimum points.

**8.9** Generate 30 samples from a linear function $y = x$ and add random noise $\sim N(0, 1^2)$. Using the samples, train the NN model with five hidden nodes. Repeat the training process 30 times to estimate uncertainty and compare it with that of PRS in Chapter 2.

**8.10** The mean and standard deviation of $e_{\text{RMSE}}$ of the one-dimensional example shown in Figure 8.15 depends on the number of training samples. Repeat the same example with a total of 30 samples and compare the statistics of $e_{\text{RMSE}}$. Use four samples for the validation and test sets.

**8.11** Figure 8.16 estimates the model-form uncertainty by gradually changing the number of hidden nodes from 1 to 10 with 30 samples. Reproduce the results in Figure 8.16.

**8.12** Figure 8.17 estimates the prediction accuracy of a network model by gradually changing the number of samples from 15 to 30. Reproduce the results in Figure 8.17.

**8.13** The NN model in Figure 8.17 shows that the minimum number of samples is 22 when 5 hidden nodes are used. The minimum number of samples varies depending on the complexity of functional behavior and the number of hidden nodes. Find the minimum number of samples when three hidden nodes are used.

**8.14** Figure 8.18 shows the NN model prediction when samples include random noise. Reproduce the results in Figure 8.18.

**8.15** In Figure 8.18, the level of training uncertainty is similar to that of sampling uncertainty. Therefore, it would be hard to clearly separate the effect of different sources of uncertainty. Repeat the same process as with that in Figure 8.18 with 30 samples. The training uncertainty will be much smaller with 30 samples.

**8.16** In Example 8.3, the backpropagation process diverged when the learning rate $\alpha = 0.5$. Using the adaptive learning rate in Section 8.5, perform the backpropagation start with $\alpha = 0.5$ and show that the learning process converges. Increase the learning rate by 5% when the MSE decreases, while decrease the learning rate by 50% when the MSE increases.

# 9

# Multi-fidelity Surrogate Models

## 9.1 Introduction

As mentioned throughout this book, surrogate models are used to significantly reduce the computational cost of evaluating quantities-of-interest (QoIs) without actually performing expensive computer simulations. In order to build a surrogate model, a set of samples needs to be generated first via experiments or computer simulations, which is referred to as design of experiments, as discussed in Chapter 3. The performance (i.e., prediction accuracy) of a surrogate model depends on (i) the functional form of the surrogate model, (ii) the number and locations of samples, and (iii) the accuracy of samples. So far, we have discussed the first two contributions to the accuracy of a surrogate model. The accuracy or fidelity of samples can obviously contribute to the accuracy of a surrogate model, which is the main topic of this chapter. Since surrogate models are built based on given samples, their accuracy depends on that of samples no matter how good the fit is.

When building a surrogate model, samples can be generated with different fidelities. This is particularly the case when computer simulations are used to generate samples [107]. The first difference is in the fidelity of mathematical models (Figure 9.1(a)). In computational fluid dynamics, for example, inviscid Euler equations or viscous Reynolds-averaged Navier–Stokes equations can be solved to calculate fluid velocity fields. Since the former has more simplification than the latter, the former can be considered a low-fidelity physical model, while the latter can be a high-fidelity one. The difference in fidelity can also come from dimensions (Figure 9.1(b)). Computer simulation in a one-dimensional domain can be a low-fidelity model as it assumes the response is identical in other two dimensions. On the other hand, computer simulation in a three-dimensional domain can be a high-fidelity one as spatial variation of response can be modeled. The level of discretization in mesh or grid can be an important difference in fidelity as well (Figure 9.1(c)). It is well known that numerical solutions converge to the accurate ones when a fine grid is used. Therefore, the simulation results with a coarse grid can be considered low-fidelity ones, while those from a fine grid can be high-fidelity ones. When data come from different fidelities, it would not be trivial to determine how to use them to build a better surrogate.

Computing power has increased dramatically over the past few decades, with modern smartphones having more computing power than the most advanced supercomputers from just a few decades ago. Moore's law has held true for more than half a century, which indicates that exponential growth is at the heart of the rapid increase of computing capabilities. Nevertheless, the progress that we make in algorithms and computer speed seems to always be utilized for raising the bar in terms of the acceptable accuracy of simulations. This is because the additional modeling fidelity can

*Surrogate Modeling and Optimization: Theories, Applications, and Limitations*, First Edition. Nam-Ho Kim.

Companion website: www.wiley.com/go/kim/surrogate_modeling

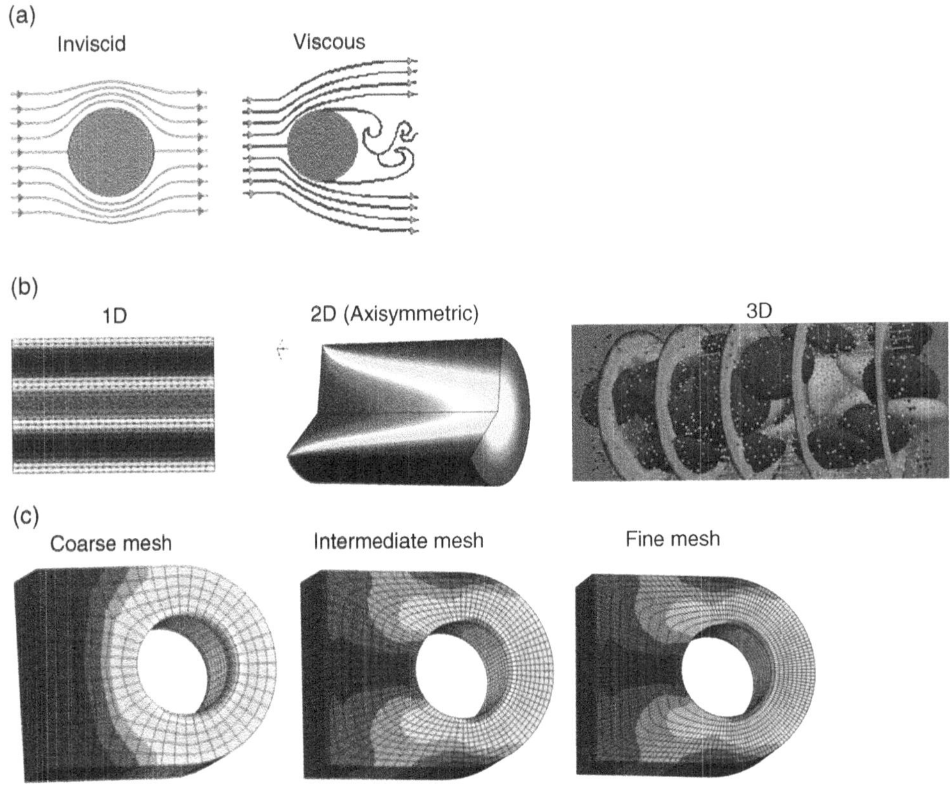


**Figure 9.1** Different types of fidelities in computer simulations. (a) Fidelity in a physical model, (b) fidelity in numerical models, and (c) fidelity in discretization.

be translated to better engineering design. For optimization and uncertainty quantification where we need thousands of simulations, however, this creates a problem even if we want to use surrogates. When a single simulation takes weeks to complete, we cannot afford to do enough simulations to create an accurate surrogate.

The fundamental premise of multi-fidelity modeling is that high-fidelity (HF) models satisfy the requirement of accuracy but are expensive to generate samples. On the other hand, low-fidelity (LF) models are cheap to generate samples but do not meet the requirement of accuracy. In general, it is assumed that computational cost is prohibitive to generate enough numbers of HF samples to build a reliable surrogate model. Otherwise, multi-fidelity modeling cannot be justified. It is also assumed that for a given computational budget, we can afford many LF simulations but only a small number of HF simulations. With a small number of HF samples, we cannot predict QoI accurately in the entire domain. Especially in high dimensions, due to a small number of HF samples, most predictions are extrapolation, where most data-driven approaches experience difficulty. On the other hand, when a surrogate model is built only using LF samples, the prediction accuracy may be compromised or even mislead.

Multi-fidelity (MF) surrogate models try to address this dilemma by integrating both LF and HF samples and making trade-off between prediction accuracy and computational cost. As shown in Figure 9.2, a multi-fidelity surrogate tries to take advantage of accuracy in HF models and

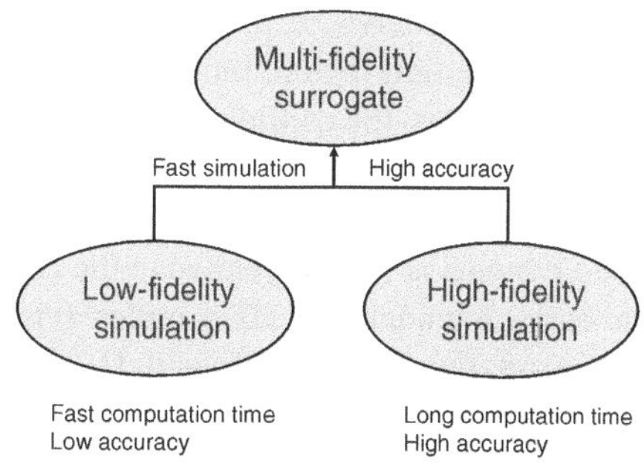


**Figure 9.2** Multi-fidelity surrogate from low- and high-fidelity simulations.

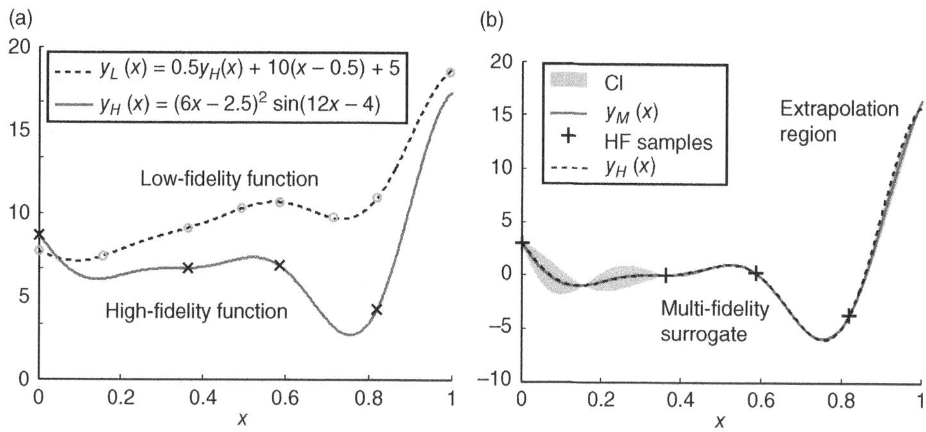


**Figure 9.3** Multi-fidelity surrogate model in one dimension. (a) Low- and high-fidelity functions and samples and (b) multi-fidelity surrogate with prediction uncertainty.

computational efficiency in LF models. In the construction of MF surrogate models, many LF samples are used to reduce computational cost, and a few HF samples are used to improve the prediction accuracy. MF surrogates usually combine a large number of LF simulations with a small number of HF simulations. This works well usually when the correlations between the two are high. An important assumption determining the successful use of the MF surrogate models is that the LF model should capture the overall trend of the responses from the HF model.

In order to explain the key ingredient of MF surrogate models, Figure 9.3(a) shows one-dimensional low- and high-fidelity functions, where samples are generated from these functions. First, the HF function is defined by $y_H(x) = (6x - 2.5)^2 \sin(12x - 4)$. Then, the LF function is obtained by scaling the HF function by half and adding an additional linear function as $y_L(x) = 0.5y_H(x) + 10(x - 0.5) + 5$. Although the LF function has a large error against the HF function, it is well correlated with the HF function. For example, when 21 equally spaced samples are used, the correlation coefficient between the LF and HF samples is 0.71. Eight samples are generated from the LF function at $x_L = \{0.0,\ 0.15, 0.36, 0.49, 0.59, 0.71, 0.82, 0.99\}$, while only four samples are generated from the HF function at $x_H = \{0.0, 0.36, 0.59, 0.82\}$. These 12 samples are used to build a multi-fidelity surrogate in Figure 9.3(b). In general, an MF surrogate (solid curve) matches well with the true HF function (dashed curve) with only four HF samples. It is noted that LF samples cover the entire domain $x \in [0, 1]$, while HF samples only cover a partial domain

$x \in [0, 0.82]$. Therefore, there is no sample in the extrapolation region $x \in [0.82, 1]$. Also, within the interpolation region, there is a relatively large unsampled region in $x \in [0, 0.36]$. Due to the lack of samples, the confidence intervals in these regions are large. However, it is impressive that MF surrogate predictions are accurate in the extrapolation region even if there is no HF sample. This is because the LF surrogate model is able to follow the trend of HF function. In other words, the LF model is well correlated to the HF model. Therefore, the success of an MF surrogate relies on both LF models and HF samples. It is emphasized that it is not the accuracy of LF models but the correlation of LF models with HF models that can contribute to the prediction accuracy of MF surrogate models [108].

Natural questions related to MF surrogate models are as follows: (i) How much cost can be saved by using MF surrogate models for the same level of accuracy? (ii) How much accuracy can be improved by using MF surrogate models for the same computational cost? (iii) When should we use MF surrogate models rather than HF surrogate models based on sparser samples? (iv) When MF surrogate models perform worse than HF surrogate models? We will address these questions throughout this chapter.

## 9.2 Multi-fidelity Surrogate Models

There are different approaches to combine models with different fidelities. The first approach is the space mapping method, which was originally developed to find an optimum design using models in two different fidelities [109]. However, the main difference between the two fidelities is in the input variable not output QoI. More specifically, the relationship between input variables in two fidelities can be approximated by $x_L \cong p(x_H)$ where $p(\cdot)$ is a mapping function. Then the multi-fidelity function can be obtained using this mapping function as $y_M(x_H) = y_L(p(x_H))$. This method is useful when two fidelities are defined in different input spaces. For example, one is defined in a coarse model space, while the other is in the fine model space. The main idea of the space mapping technique is to use the coarse model to gain information about the fine model and to apply this in the search for an optimal solution of the latter. Thus, the technique iteratively establishes a mapping between the parameters of the two models, which relate similar model responses. Having this mapping, most of the model evaluations can be directed to the fast coarse model. Although this method is useful to find an optimum design mostly using a cheap LF model, it may not be useful to build an MF surrogate model, where the main purpose is approximating QoIs not finding an optimum one. Therefore, this method will not be discussed further.

The second approach is to use a correction function to match the two fidelity models [110]. This approach can further be divided into three different forms: (i) multiplicative correction function, (ii) additive correction function, and (iii) comprehensive correction function. The first one multiplies a correction function to LF function as

$$y_M(x) = \rho(x) y_L(x) \tag{9.1}$$

where $\rho(x)$ is the correction function, $y_L(x)$ represents the LF function, and $y_M(x)$ is the MF function. For the purpose of generating samples, $y_H(x)$ represents the HF function. The purpose is to make the MF function $y_M(x)$ close to the HF function $y_H(x)$ by building a correction surrogate model $\rho(x)$. Since the correction function is the ratio between the two fidelity functions, we fit a surrogate of the ratio. This form is useful when the difference is proportional to the value of function. The second one uses additive correction function as

$$y_M(x) = y_L(x) + \delta(x) \tag{9.2}$$

where $\delta(x)$ is called a discrepancy function as it represents the difference between LF and HF functions. This form is useful when the LF function has a bias error, which is modeled by the discrepancy function. The last one uses both multiplicative and additive correction function and is referred to as a comprehensive form, defined as

$$y_M(x) = \rho y_L(x) + \delta(x) \tag{9.3}$$

In the comprehensive form, the multiplicative correction is not a function but a constant, which is referred to as an LF scale factor. This form of the MF surrogate is considered the most advanced one.

Even if a single MF function is used in the definition, it is possible that multiple LF functions can be used simultaneously. For example, a linear combination of multiple LF functions can be used as $y_L(x) = b_1 y_{L1}(x) + b_2 y_{L2}(x) + \dots + b_m y_{Lm}(x)$. The coefficient $b_i$ of the LF function can be considered as a weight of each LF function. It is also possible that more than two fidelities can be used. In such a case, the above formulas can be used in a sequence from the lowest fidelity to the highest one. However, we will only discuss the MF surrogate with two fidelity levels in the chapter.

In MF surrogate modeling, it is assumed that the form of the HF function $y_H(x)$ is unknown, but we can evaluate the HF function at a given input location $x$, albeit it is expensive. This is a common case of expensive numerical simulations, such as finite-element models or computational fluid dynamics with a fine grid. It is assumed that for the given computational budget, only a small number of HF samples, $y_{Hi} = y_H(x_i)$, are affordable, which is not enough to build a reliable HF surrogate. On the other hand, the LF function is used for two purposes. The first is to generate samples to fit the MF surrogate, and the second is to evaluate the fitted MF surrogate at prediction points. It is possible that the explicit form of the LF function $y_L(x)$ is available or it can be evaluated using numerical simulation. If an explicit functional form is available, it can be used to generate samples $y_{Li} = y_L(x_i)$. Otherwise, LF numerical simulations are used to generate samples. It is assumed that the computational cost of evaluating the LF model is ignorable compared to that of evaluating the HF model. Also, it is possible that LF samples are used to build an LF surrogate model, and then, it is used for MF surrogate.

An important step of MF surrogate modeling is to fit the correction function $\rho(x)$ in Eq. (9.1) for the multiplicative form, the discrepancy function $\delta(x)$ in Eq. (9.2) for the additive form, or the scale factor and discrepancy function in Eq. (9.3) for the comprehensive form. In the case of the multiplicative form, the correction function can be defined as $\rho(x) = y_H(x)/y_L(x)$. It is noted that HF function $y_H(x)$ is used instead of $y_M(x)$ because the goal is to make $y_M(x)$ close to $y_H(x)$. Since the functional form of the correction function is unknown, a correction surrogate is built using samples. The samples that we need to build a correction surrogate are $\rho_i = \rho(x_i) = y_H(x_i)/y_L(x_i) = y_{Hi}/y_{Li}$, $i = 1, \dots, n_H$. Therefore, we need both HF and LF samples. The multiplicative form has inherently difficulty when the function value is zero.

In the case of the additive form in Eq. (9.2), MF surrogate modeling is to fit the discrepancy function $\delta(x)$, which is defined as $\delta(x) = y_H(x) - y_L(x)$. In this case, a discrepancy surrogate is built using discrepancy samples $\delta_i = \delta(x_i) = y_H(x_i) - y_L(x_i) = y_{Hi} - y_{Li}$, $i = 1, \dots, n_H$. Again, both HF and LF samples are required at the same locations.

The comprehensive form in Eq. (9.3) not only fits the discrepancy function but also determines the scale factor. Two approaches can be used to determine the scale factor and discrepancy function: sequential or simultaneous approaches. The first approach determines the scale factor $\rho$ first by minimizing the error between HF samples and LF models. Then, MF surrogate modeling fits the discrepancy function $\delta(x)$, which is defined as $\delta(x) = y_H(x) - \rho y_L(x)$. In this case, a discrepancy surrogate is built using discrepancy samples $\delta_i = \delta(x_i) = y_H(x_i) - \rho y_L(x_i) = y_{Hi} - \rho y_{Li}$, $i = 1, \ldots, n_H$. Again, both HF and LF samples are required at the same locations. The second approach is to determine the scale factor and discrepancy function simultaneously. The challenge in this approach is that the discrepancy samples $\delta_i$ cannot be determined in advance as the scale factor $\rho$ is unknown. Therefore, the discrepancy samples need to be defined during the fitting process. Since the second approach is more flexible than the first one, it is expected that the second approach yields a better scale factor and discrepancy function. In fact, the scale factor in the second approach will not minimize the error between HF samples and the LF model. Instead, the scale factor will be determined to make the discrepancy function a simple functional form.

Before discussing how to fit the correction or discrepancy function, it would be necessary to discuss how to prepare samples. Let $y_{Hi} = y_H(x_{Hi})$ be HF samples such that the pairs of HF samples are given as $(x_{Hi}, y_{Hi})$, $i = 1, \ldots, n_H$, where $n_H$ is the number of HF samples. When the LF model is available, LF samples are generated at the same HF locations as $(x_{Hi}, y_{Li})$, $i = 1, \ldots, n_H$. Note that we evaluate the LF function at HF sample locations. This is necessary in order to fit the discrepancy surrogate $\delta(x) = y_M(x) - y_L(x)$ in Eq. (9.2) using discrepancy samples $\delta_i = y_{Hi} - y_{Li}$, $i = 1, \ldots, n_H$. When an LF model is not available, LF samples are generated first $(x_{Li}, y_{Li})$, $i = 1, \ldots, n_L$, where $n_L$ is the number of LF samples. Then an LF surrogate $\hat{y}_L(x)$ is fitted first. Then, this LF surrogate is used to generate LF samples corresponding to HF samples $(x_{Hi}, \hat{y}_{Li})$, $i = 1, \ldots, n_H$. However, these samples may include prediction errors of the LF surrogate. Therefore, it would be better for LF sample locations to include HF sample locations.

Once HF and LF samples are prepared, the next step is to fit the MF surrogate, which comes with two methods: deterministic and probabilistic. The deterministic method minimizes the errors at HF sample points [111]. This is equivalent to regression in Chapter 2, where the square sum of errors is minimized to find unknown coefficients. This method is often used when a discrepancy function is modeled using polynomial response surfaces (PRSs). The probabilistic method maximizes the likelihood function, which is used for fitting a Kriging surrogate in Chapter 7. This method is mostly used when the discrepancy function is modeled using a Kriging surrogate [110]. When both the LF function and discrepancy function are fitted using Kriging, it is referred to as co-Kriging surrogate.

**Example 9.1 Multi-fidelity Surrogate Using Discrepancy Function**

Consider the following LF and HF functions: $y_L(x) = 0.5\sin(3x)$ and $y_H(x) = \sin(3x) + 0.25\sin(0.3x)$ in the domain $x \in [0, 1]$. For the LF function, we have data at seven equally spaced points [0.2, 0.3, 0.4, 0.5, 0.6, 0.7, 0.8]. For the HF function, samples are available only at three points [0.2, 0.4, 0.6]. Fit LF samples with quadratic PRS and HF samples with a linear one and compare their accuracy against the true functions. Using the additive correction form in Eq. (9.2), fit the difference between the two sets of samples using linear PRS and compare the accuracy against the true HF function.

**Solution:**

The HF function is the sum of a high-frequency sine function and a low-frequency sine function. The LF function has only the high-frequency component, and its amplitude is off by a factor of two. We have seven LF samples and three HF samples. Since HF samples are only three, we use linear polynomials to fit the samples, while quadratic polynomials are used to fit seven LF samples. The following Matlab script defines LF and HF samples and fit LF and HF surrogates.

```
fLF= @(x) 0.5*sin(3*x); fHF= @(x) sin(3*x) + 0.25*sin(0.3*x);
xtrainLF=[0.2,0.3,0.4,0.5,0.6,0.7,0.8];
ytrainLF=fLF(xtrainLF);                        % LF training samples
xtrainHF=[0.2,0.4,0.6];
ytrainHF=fHF(xtrainHF);                        % HF training samples
xplot=0:0.01:1; yplotLF=fLF(xplot); yplotHF=fHF(xplot);
% Low-fidelity surrogate (quadratic PRS)
X_LF=[ones(length(xtrainLF),1) xtrainLF' xtrainLF.^2'];
[B_LF,BINT_LF,R_LF,RINT_LF,STATS_LF]=regress(ytrainLF',X_LF);
y_LF=[ones(length(xplot),1) xplot' xplot.^2']*B_LF;
% High-fidelity surrogate (linear PRS)
X_HF=[ones(length(xtrainHF),1) xtrainHF'];
[B_HF,BINT_HF,R_HF,RINT_HF,STATS_HF]=regress(ytrainHF',X_HF);
y_HF=[ones(length(xplot),1) xplot']*B_HF;
```

In the above Matlab script, `fLF` and `fHF` are inline LF and HF functions, respectively. LF sample pairs are given as (`xtrainFL`, `ytrainLF`), and HF sample pairs are given as (`xtrainHF`, `ytrainHF`). `X_LF` and `X_HF` are design matrices to fit LF and HF surrogates, respectively. The Matlab function `regress` is used to determine coefficients `B_LF` and `B_HF` of polynomial response surfaces. The constructed LF quadratic PRS is $\hat{y}_L(x) = -0.0725 + 2.1829x - 2.0876x^2$, while the HF linear PRS is $\hat{y}_H(x) = 0.4144 + 1.0974x$.

For the additive correction, we fit the difference between the two functions at the three common points by a linear polynomial. The samples of a discrepancy surrogate are from the difference between LF and HF samples. Since LF sample locations include HF sample locations, the three samples of difference can be calculated first. Then, these three samples are fitted for a linear PRS using the `regress` function. The constructed discrepancy surrogate is $\hat{\delta}(x) = 0.2073 + 0.5859x$. The MF surrogate is obtained by adding the quadratic LF surrogate with the linear discrepancy surrogate as $\hat{y}_M(x) = 0.1348 + 2.7688x - 2.0876x^2$. The following Matlab script is used to calculate an MF surrogate.

```
% Multi-fidelity surrogate (linear discrepancy PRS)
ydiff=fHF(xtrainHF)-fLF(xtrainHF);
X_diff=[ones(length(xtrainHF),1) xtrainHF'];
[B_diff,BINT_diff,R_diff,RINT_diff,STATS_diff]=regress(ydiff',X_diff);
y_MF=[ones(length(xplot),1) xplot' xplot.^2']*B_LF ...
    +[ones(length(xplot),1) xplot']*B_diff;
```

The following Matlab script is used to plot the two true functions, LF and HF surrogates, and MF surrogate.

```
plot(xplot,yplotLF,'b',xplot,yplotHF,'r',xplot,y_LF,'b-',...
     xplot,y_HF,'r-',xplot,y_MF,'k-',xtrainLF,ytrainLF,'bo',...
     xtrainHF,ytrainHF,'r*');
legend('LF function','HF function','LF prediction',...
       'HF prediction', 'MF prediction');
```

Figure 9.4 plots two true LF and HF functions, LF and HF surrogates, and MF surrogate along with LF and HF samples (circles and asterisks). It turns out that the HF linear PRS cannot follow

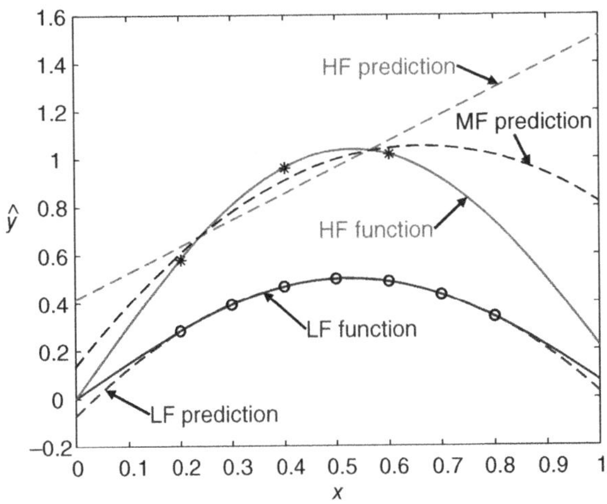


**Figure 9.4** Multi-fidelity surrogate with additive form.

the trend of HF function. In addition, although the LF quadratic PRS fits well for the LF function, it shows significant errors compared to the HF function. It is also observed that both LF and HF surrogates show a significant error in the extrapolation region. The MF surrogate (black dashed curve) shows substantial improvement compared to the HF surrogate. However, the fitting quality is still poor in the extrapolation region.

**Example 9.2 Multi-fidelity Surrogate Using Comprehensive Form**
Build an MF surrogate model using the comprehensive form in Eq. (9.3) with the same sets of samples in Example 9.1.

**Solution:**
In the previous example, we calculate discrepancy samples $\delta_i = y_{Hi} - y_{Li}$, $i = 1, \ldots, n_H$ and then fit them for the discrepancy surrogate. In the case of a comprehensive approach, discrepancy samples cannot be used because the scale factor in $\delta_i = y_{Hi} - \rho y_{Li}$ is still unknown. In order to determine scale factor $\rho$ and linear discrepancy function $\delta(x) = a + bx$ simultaneously, it is required to use the HF samples directly. First, the MF surrogate has the following form:

$$y_M(x) = \rho y_L(x) + a + bx \tag{9.4}$$

where the vector of unknown parameters is defined as $\mathbf{b}_M = \{\rho, a, b\}$. It is noted that the MF surrogate in Eq. (9.4) is linear in terms of $\mathbf{b}_M$. Therefore, linear regression in Chapter 2 can be used to determine the unknown coefficients. The only difference is that instead of monomial bases in PRS, LF function $y_L(x)$ is used as an additional basis function. Since there are three HF samples, the vector of basis functions and design matrix can be defined as

$$\xi(x) = \{y_L(x) \;\; 1 \;\; x\}^T, \quad \mathbf{X} = \begin{bmatrix} y_L(x_1) & 1 & x_1 \\ y_L(x_2) & 1 & x_2 \\ y_L(x_3) & 1 & x_3 \end{bmatrix}$$

and the vector of HF samples is defined as $\mathbf{y}_H = \{y_{H1} \;\; y_{H2} \;\; y_{H3}\}^T$. Then the standard regression equation in Chapter 2 is used to determine the vector of unknown coefficients as

$$\mathbf{b}_M = \left(\mathbf{X}^T\mathbf{X}\right)^{-1}\mathbf{X}^T\mathbf{y}_H$$

and the MF surrogate prediction can be obtained as

$$y_M(x) = \xi(x)^T \mathbf{b}_M = \rho \hat{y}_L(x) + a + bx$$

where LF surrogate $\hat{y}_L(x) = -0.0725 + 2.1829x - 2.0876x^2$ is used as the first basis function.

As a continuation of the Matlab script in Example 9.1, the following Matlab script first builds the design matrix **X**, determines the vector of coefficients **b** using the regress function, and predicts the MF surrogate.

```
X_comp=[[ones(length(xtrainHF),1) xtrainHF' xtrainHF.^2']*B_LF ...
          ones(length(xtrainHF),1) xtrainHF'];
[B_comp,BINT_comp,R_comp,RINT_comp,STATS_comp]=regress(ytrainHF',X_comp);
y_MF_comp=[[ones(length(xplot),1) xplot' xplot.^2']*B_LF ...
            ones(length(xplot),1) xplot']*B_comp;
```

It is noted that $\hat{y}_L(x_i)$ is calculated using the LF surrogate. But it is possible to use LF samples directly as LF samples are available at the same locations.

To help understand the details of the process, we show here the design matrix **X**, which has the shape functions at the sample points.

$$\mathbf{X} = \begin{bmatrix} 0.281 & 1 & 0.2 \\ 0.467 & 1 & 0.4 \\ 0.486 & 1 & 0.6 \end{bmatrix}$$

The second and third columns correspond to the constant and linear basis functions. The first column corresponds to the LF surrogate $\hat{y}_L(x_i)$ at sample locations.

When we look at the vector of coefficients found by the fit, $\mathbf{b}_M = \{1.95 \quad 0.013 \quad 0.097\}^T$, the first coefficient $\rho$ is close to the ratio of 2 between the main components, and the linear coefficient 0.097 is of the order of a linear approximation to 0.25 sin(0.3$x$), which would be 0.075$x$. Figure 9.5

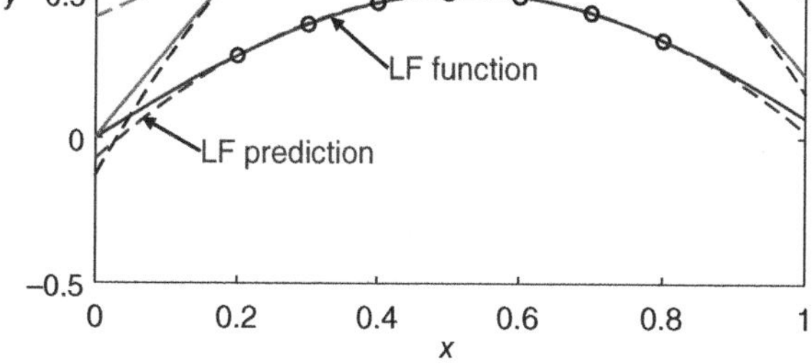


**Figure 9.5** Multi-fidelity surrogate with a comprehensive form.

shows the MF surrogate, which is much improved from the discrepancy form in Figure 9.4. Significant improvement is due to the scale factor $\rho \approx 1.95$. This is because the LF function has $0.5\sin(3x)$, while the HF function has a $\sin(3x)$ term. The regression process determines the scale factor to make the discrepancy function simple.

In the previous example, it is possible that the HF samples are used to fit a quadratic PRS. However, this is dangerous as the number of samples is the same as the number of unknown coefficients. In this particular example, a quadratic PRS with three samples turns out to be a great fit but without any possibility for an error estimate. This highlights the problem that it is often difficult to decide whether to fit the HF samples alone or to use an MF surrogate.

In Example 9.1, it is assumed that the LF function is not available in MF surrogate predictions. Instead, LF samples are used to fit an LF surrogate, which is used in MF surrogate predictions. Therefore, the error in an LF surrogate propagates to the error in an MF surrogate. On the other hand, since the discrepancy samples are obtained based on the exact samples from the LF function, the identified coefficients of discrepancy surrogate will not change. In Example 9.2, the LF surrogate is used in fitting both the discrepancy surrogate and MF predictions. Therefore, the error in the LF surrogate affects significantly in MF predictions. It would be a good exercise to use the exact LF function and compare the prediction accuracy of the MF surrogate (see Exercise Problem 9.1).

## 9.3 Regression-based Multi-fidelity Surrogate

Regression-based MF surrogate is a simple and yet powerful model based on regression or least-squares fit. In particular, when all unknown parameters are linearly combined, it is referred to as linear regression MF surrogate. The advantage of this surrogate is that all theoretical developments in Chapter 2 can directly be applicable, including standard errors and prediction variance. In this surrogate, MF prediction is approximated by a linear combination of LF prediction and a discrepancy function. In this section, the discrepancy function is modeled by a linear combination of monomial basis functions, i.e., polynomial response surface (PRS). However, the approach is also applicable to other basis functions such as radial basis functions. The key idea is to consider the LF model as an additional basis function in the MF surrogate with the scale factor as a regression coefficient. Since the LF model is considered as a basis, including multiple LF models is straightforward in this approach. The design matrix in the regress equation consists of both LF models and discrepancy basis functions. Then the scale factors and coefficients of the basis functions are obtained simultaneously by solving the regression set of linear equations. The most attractive characteristic of a linear regression MF surrogate is that various tools presented in Chapters 2 and 3, such as prediction variance and D-optimal designs, are directly applicable to MF surrogate. In addition, the fitting process of linear regression solves a convex problem, where the global optimum parameters can be found by solving a linear system of equations. Therefore, there is no uncertainty associated with the fitting process. This is a great advantage of a linear regression MF surrogate compared to the Bayesian framework in the following section, which relies on expensive global optimization to find the surrogate parameters and the cost increases exponentially with the number of samples and dimensionality.

The linear regression MF surrogate uses the comprehensive form, where the discrepancy can be captured by the scale factor and the polynomial form of additive discrepancy [111]. With a small number of high-fidelity samples, it would make sense to use a simple functional form for the discrepancy. Model validation schemes, such as cross-validation, could be used to evaluate the effectiveness of the MFS predictions and avoid potential misusage. In the comprehensive form, the MF surrogate is defined as

$$\hat{y}_M(\mathbf{x}) = \rho y_L(\mathbf{x}) + \hat{\delta}(\mathbf{x}) \tag{9.5}$$

where $\rho$ is the scale factor for the LF function $y_L(\mathbf{x})$ and $\hat{\delta}$ is the discrepancy surrogate. It is noted that we do not use any approximation on the LF function. It will be shown later that it is unnecessary to have an explicit expression of the LF function. It is enough that $y_L(\mathbf{x})$ can be evaluated at a given sample location using computer simulation.

Without loss of generality, we assume that the discrepancy function is represented in the form of polynomial response surface as

$$\hat{\delta}(\mathbf{x}) = \sum_{j=1}^{p} \xi_j(\mathbf{x}) b_j = \boldsymbol{\xi}(\mathbf{x})^T \mathbf{b} \tag{9.6}$$

where $\xi_j(\mathbf{x})$ denotes the $j$th monomial basis and $b_j$ is the unknown coefficient of $\xi_j(\mathbf{x})$. Although we explain our approach using a polynomial response surface, the proposed approach will work for any surrogates that uses linear regression. Since the number of HF samples is often limited, the number of terms in Eq. (9.6) is limited. We often use constant or low-order polynomials for approximating the discrepancy function.

Since the model forms in Eqs. (9.5) and (9.6) are fixed, the only remaining task is to determine the unknown coefficients. The main observation is that we view the LF function as another basis function with the scale factor $\rho$ as an unknown coefficient. Then, the augmented vector $\tilde{\mathbf{b}} = \{\rho, b_1, \ldots, b_p\}^T$ of all unknown coefficients can be determined using linear regression. Such seemingly natural observation turns out to be an extremely simple MF surrogate that can be easily applicable with single linear regression. Since we view the LF function as a basis, it is also possible to include multiple LF functions together with different scale factors.

As with the standard least-squares method, the errors between MF surrogate predictions and HF samples are defined as

$$\begin{aligned} e_i &= y_{Hi} - \hat{y}_M(\mathbf{x}_{Hi}) \\ &= y_{Hi} - \rho y_L(\mathbf{x}_{Hi}) - \sum_{j=1}^{p} \xi_j(\mathbf{x}_{Hi}) b_j, \quad i = 1, \ldots, n_H \end{aligned} \tag{9.7}$$

Or, in the vector-matrix form, the vector of errors at HF samples can be written as

$$\mathbf{e} = \mathbf{y}_H - \mathbf{X}\tilde{\mathbf{b}} \tag{9.8}$$

where

$$\mathbf{e} = \begin{Bmatrix} e_1 \\ \vdots \\ e_{n_H} \end{Bmatrix}, \quad \mathbf{y}_H = \begin{Bmatrix} y_{H1} \\ \vdots \\ y_{Hn_H} \end{Bmatrix}, \quad \mathbf{X} = \begin{bmatrix} y_L(\mathbf{x}_1) & \xi_1(\mathbf{x}_1) & \ldots & \xi_p(\mathbf{x}_1) \\ \vdots & \vdots & \ddots & \vdots \\ y_L(\mathbf{x}_{n_H}) & \xi_1(\mathbf{x}_{n_H}) & \ldots & \xi_p(\mathbf{x}_{n_H}) \end{bmatrix}, \quad \tilde{\mathbf{b}} = \begin{Bmatrix} \rho \\ b_1 \\ \vdots \\ b_p \end{Bmatrix} \tag{9.9}$$

In the above equation, $\mathbf{e}$ is the vector of residual errors, $\mathbf{y}_H$ is the vector of HF samples, $\mathbf{X}$ is the augmented design matrix, and $\tilde{\mathbf{b}}$ is the vector of unknown coefficients. By augmenting the design matrix, the scale parameter and unknown coefficients of the discrepancy functions are estimated simultaneously. It is clear from Eq. (9.9) that the LF function is included in the design matrix with other basis functions of the discrepancy function.

The unknown coefficients in the regression-based MF surrogate are obtained by minimizing the square sum of errors as

$$\underset{\tilde{\mathbf{b}}}{\text{minimize}}\ \mathbf{e}^T\mathbf{e} = \left(\mathbf{y}_H - \mathbf{X}\tilde{\mathbf{b}}\right)^T\left(\mathbf{y}_H - \mathbf{X}\tilde{\mathbf{b}}\right) \tag{9.10}$$

where the standard regression technique can yield the following form of unknown coefficients:

$$\tilde{\mathbf{b}} = \left(\mathbf{X}^T\mathbf{X}\right)^{-1}\mathbf{X}^T\mathbf{y}_H \tag{9.11}$$

Linear regression assumes that the residual errors follow independent and identically distributed Gaussian distributions. The standard error (i.e., the estimate standard deviation of noise) is estimated by

$$\hat{\sigma} = \sqrt{\frac{\mathbf{e}^T\mathbf{e}}{n_H - (p+1)}} \tag{9.12}$$

and the standard error of prediction (i.e., standard deviation of prediction) at a point $\mathbf{x}$ is obtained as

$$\sigma_{y_M}(\mathbf{x}) = \hat{\sigma}\sqrt{\tilde{\xi}(\mathbf{x})^T\left(\mathbf{X}^T\mathbf{X}\right)^{-1}\tilde{\xi}(\mathbf{x})} \tag{9.13}$$

where $\tilde{\xi}(\mathbf{x}) = \left\{y_L(\mathbf{x}) \quad \xi(\mathbf{x})^T\right\}^T$ is the augmented vector of basis functions.

The scale factor $\rho$ implies the level of trend similarity between $\hat{y}_M(\mathbf{x})$ and $y_L(\mathbf{x})$ and plays a critical role in approximating MF data. Negative values or extremely large values of $\rho$ indicates a risky prediction, which is likely to be associated with an undesirable LF function, inappropriate surrogate forms, or inadequate samples. The discrepancy function $\hat{\delta}(\mathbf{x})$ is likely to be modeled well by a low-order PRS when $y_L(\mathbf{x})$ had a similar trend as $\hat{y}_M(\mathbf{x})$.

**Example 9.3 Accuracy of MF Surrogate**

Consider an HF function $y_H(x) = \sin(x) + 0.2x + (x-5)^2/16 + 0.6$ and two LF functions $y_{L1}(x) = (x-5)(x-4)(x-9)/20 + 2$ and $y_{L2}(x) = \sin(x) + 0.2x + 0.5$ in the domain of $x \in [0, 10]$. Five HF sample locations are given as $\mathbf{x}_H = \{1\ 3\ 5\ 7\ 9\}^T$. Build an MF surrogate using (i) $y_{L1}(x)$ and linear discrepancy function and (ii) $y_{L2}(x)$ and linear discrepancy function and compare the accuracy against the HF function. Discuss the accuracy of MF surrogate using the correlation between HF and LF functions.

**Solution:**

a) First, five HF samples are calculated using HF function as $\mathbf{y}_H = \{2.54, 1.49, 0.54, 2.81, 3.71\}^T$. With $y_{L1}(x)$, the MF surrogate can be defined as

$$\hat{y}_M(x) = \rho y_{L1}(x) + \hat{\delta}(x) = \rho y_{L1}(x) + b_1 + b_2 x$$

The design matrix $\mathbf{X}$ at these five sample locations with $y_{L1}(\mathbf{x}_H)$ and the vector of unknown coefficients $\tilde{\mathbf{b}} = \{\rho\ b_1\ b_2\}^T$ can be calculated as

$$\mathbf{X} = \begin{bmatrix} 2.60 & 1 & 1 \\ 2.75 & 1 & 3 \\ 1.10 & 1 & 5 \\ 0.05 & 1 & 7 \\ 2.00 & 1 & 9 \end{bmatrix}, \quad \tilde{\mathbf{b}} = \left(\mathbf{X}^T\mathbf{X}\right)^{-1}\mathbf{X}^T\mathbf{y}_H = \begin{Bmatrix} 0.4031 \\ 0.2259 \\ 0.2615 \end{Bmatrix}$$

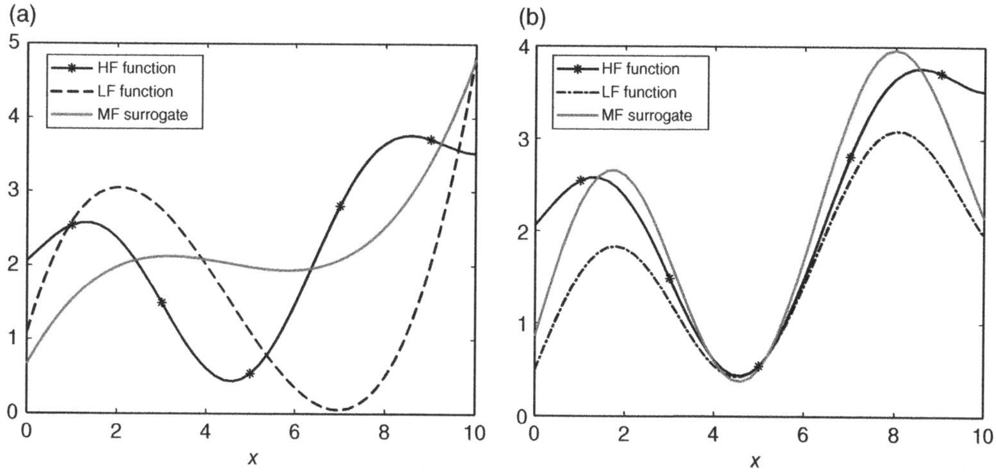


**Figure 9.6** Multi-fidelity surrogates with (a) $y_{L1}(x)$ with $e_{RMSE}$ = 0.9696 and (b) $y_{L2}(x)$ with $e_{RMSE}$ = 0.4665.

Therefore, the identified MF surrogate becomes

$$\hat{y}_M(x) = 0.4031y_{L1}(x) + 0.2259 + 0.2615x$$

Figure 9.6(a) shows the HF function (curve with asterisks) with HF samples (asterisks), LF function (dashed curve), and MF surrogate (solid curve). The performance of the MF surrogate is not satisfactory as the root mean squared error (RMSE) using 51 equally spaced points is 0.9696. This is mostly because of the low level of correlation between the LF and HF functions. The correlation coefficient between the HF and $y_{L1}(x)$ functions is 0.0833. The RMSE of cubic PRS using HF samples is 0.6644, which is better than that of the MF surrogate. Therefore, when the LF function is not correlated with the HF function, it can deteriorate the prediction accuracy of the MF surrogate.

b) With $y_{L2}(x)$, the MF surrogate can be defined as

$$\hat{y}_M(x) = \rho y_{L2}(x) + \hat{\delta}(x) = \rho y_{L2}(x) + b_1 + b_2 x$$

The design matrix $\mathbf{X}$ at these five sample locations with $y_{L2}(\mathbf{x}_H)$ and the vector of unknown coefficients $\tilde{\mathbf{b}} = \{\rho\ b_1\ b_2\}^T$ can be calculated as

$$\mathbf{X} = \begin{bmatrix} 1.54 & 1 & 1 \\ 1.24 & 1 & 3 \\ 0.54 & 1 & 5 \\ 2.56 & 1 & 7 \\ 2.71 & 1 & 9 \end{bmatrix}, \quad \tilde{\mathbf{b}} = (\mathbf{X}^T\mathbf{X})^{-1}\mathbf{X}^T\mathbf{y}_H = \begin{Bmatrix} 1.4537 \\ 0.1351 \\ -0.0830 \end{Bmatrix}$$

Therefore, the identified MF surrogate becomes

$$\hat{y}_M(x) = 1.4537y_{L2}(x) + 0.1351 - 0.083x$$

Figure 9.6(b) shows the HF function (curve with asterisks) with HF samples (asterisks), LF function (dash-dotted curve), and MF surrogate (solid curve). The performance of the MF surrogate is satisfactory as the RMSE using 51 equally spaced points is 0.4665. MF predictions are much better in the interpolation region $x \in [1, 9]$, and most errors come from the extrapolation region. In fact, the RMSE in the interpolation region is 0.2450. The MF surrogate with $y_{L2}(x)$ performs better than that with $y_{L1}(x)$ because the former has better correlation than the latter. In fact, the correlation coefficient between $y_H(x)$ and $y_{L2}(x)$ is 0.9067.

**Example 9.4 Prediction Intervals of MF Surrogate**

Plot 90% prediction intervals of an MF surrogate using $y_{L2}(x)$ in Example 9.3.

**Solution:**

In order to calculate the prediction intervals, the standard error in Eq. (9.12) is calculated first:

$$\hat{\sigma} = \sqrt{\frac{\mathbf{e}^T\mathbf{e}}{n_H - (p+1)}} = 0.4815$$

where $n_H = 5$ and $p = 2$. Then, with the augmented basis vector $\tilde{\xi}(x) = \{y_{L2}(x) \quad 1 \quad x\}^T$, the standard error of prediction $\sigma_{y_M}(x)$ in Eq. (9.13) is calculated at 51 equally spaced points. The 90% prediction interval can be calculated from Student's $t$-distribution $[t_{0.05,4}, \ t_{0.95,4}] = [-2.13, 2.13]$ as $PI = [-2.13\sigma_{y_M}(x), 2.13\sigma_{y_M}(x)]$. Figure 9.7 shows the HF function (curve with asterisks) with HF samples, MF surrogate (dashed curve), and 90% prediction intervals (gray area). Even if the mean prediction is reasonable, predictions come with a large level of uncertainty due to a small number of HF samples.

In the previous example, it is noted that the difference between the HF function and the second LF function is a quadratic function, while a linear polynomial is used for the discrepancy function. If a quadratic polynomial is used for the discrepancy function, it is expected that the MF surrogate can be accurate compared to the HF function (see Exercise Problem 9.2).

There are nice properties of a linear regression MF surrogate. First, it could be easily applied to more than two-fidelity models by augmenting the design matrix with multiple low-fidelity models with multiple scale factors as

$$\hat{y}_M(\mathbf{x}) = \sum_{i=1}^{q} \rho_i y_{Li}(\mathbf{x}) + \sum_{j=1}^{p} \xi_j(\mathbf{x}) b_j \tag{9.14}$$

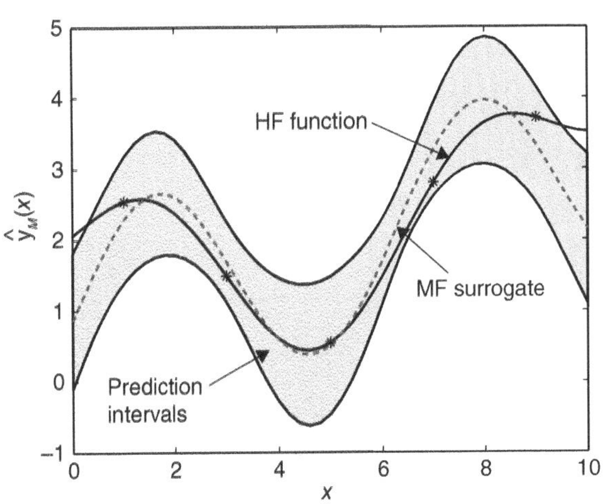


**Figure 9.7** 90% prediction intervals of MF surrogate with $y_{L2}(x)$.

where $q$ is the number of LF functions. Second, it is possible to use other basis functions beyond monomials to approximate the discrepancy function. Third, the established linear regression MF surrogate can have several advantages, such as (i) handling noise from different distributions; (ii) estimating confidence intervals, prediction intervals, and tolerance limits; (iii) reducing model-form uncertainty via stepwise regression; and (iv) obtaining optimal design of experiments, e.g., D-optimal designs.

An important question when multiple LF functions are used is to distinguish important LF functions from unimportant ones. As discussed in Park et al. [110], some LF models can deteriorate the performance of MF surrogate, while some LF models enhance it. Therefore, it is important to identify the importance of individual LF functions toward the prediction accuracy of MF surrogate. An important observation is that coefficients $\rho_i$ and $b_j$ in Eq. (9.14) can be identified using linear regression, where LF functions are considered as basis functions. Therefore, the same theoretical study in Chapter 2 can be applied to estimate the confidence of each coefficient and to identify unimportant basis functions. In Section 2.5, in particular, we derived the uncertainty of coefficients in the form of coefficient of variation and eliminated the basis function that has the largest coefficient of variation. The process of backward elimination stops when the adjusted coefficient of multiple determination $R_a^2$ starts increasing. The same backward elimination can be applied here to eliminate unimportant LF functions.

**Example 9.5 Backward Elimination of LF Functions**

Use the two MF functions in Example 9.3 to build an MF surrogate. Using $R_a^2$, eliminate LF functions until $R_a^2$ decreases.

**Solution:**

With $y_{L1}(x)$ and $y_{L2}(x)$, the MF surrogate can be defined as

$$\hat{y}_M(x) = \rho_1 y_{L1}(x) + \rho_2 y_{L2}(x) + b_1 + b_2 x$$

The design matrix $\mathbf{X}$ at these five sample locations with $y_{L2}(\mathbf{x}_H)$ and the vector of unknown coefficients $\tilde{\mathbf{b}} = \{\rho_1\ \rho_2\ b_1\ b_2\}^T$ can be calculated as

$$\mathbf{X} = \begin{bmatrix} 2.60 & 1.54 & 1 & 1 \\ 2.75 & 1.24 & 1 & 3 \\ 1.10 & 0.54 & 1 & 5 \\ 0.05 & 2.56 & 1 & 7 \\ 2.00 & 2.71 & 1 & 9 \end{bmatrix}, \quad \tilde{\mathbf{b}} = (\mathbf{X}^T\mathbf{X})^{-1}\mathbf{X}^T\mathbf{y}_H = \begin{Bmatrix} 0.2506 \\ 1.4052 \\ -0.4964 \\ -0.0252 \end{Bmatrix}$$

Therefore, the identified MF surrogate becomes

$$\hat{y}_M(x) = 0.2506 y_{L1}(x) + 1.4052 y_{L2}(x) - 0.4964 - 0.0252x$$

Figure 9.8(a) shows the HF function with HF samples (asterisks), LF functions, and MF surrogate. The performance of the MF surrogate is not better than that with $y_{L2}(x)$ in Example 9.3. The RMSE $e_{\text{RMSE}} = 0.3915$, which is improved, but the standard error $\hat{\sigma} = 0.4944$ is worse. Accordingly, the 90% prediction intervals in Figure 9.8(b) are slightly larger than that of Example 9.4. This indicates that having more LF functions does not always improve the prediction accuracy.

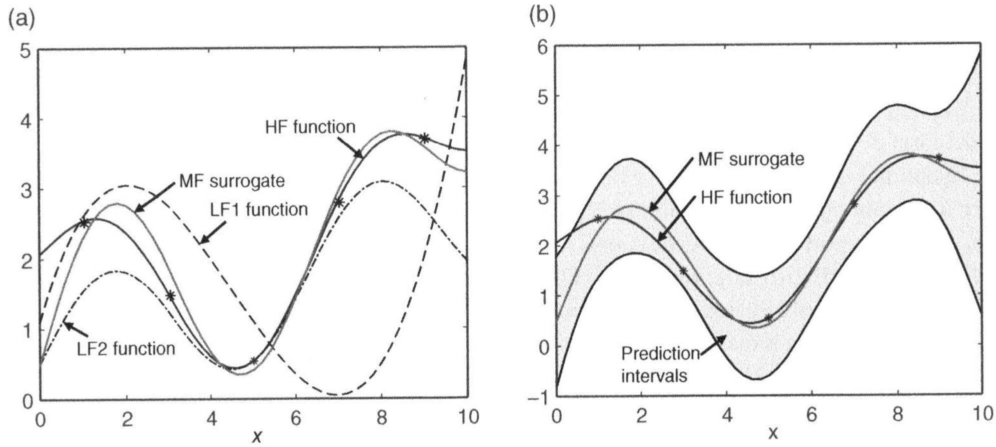


**Figure 9.8** Multi-fidelity surrogates with two LF functions. (a) MF surrogate and (b) 90% prediction intervals.

In order to assess the quality of the MF surrogate, the adjusted coefficient of multiple determination $R_a^2$ is calculated using Eqs. (2.21)–(2.27) as

$$\bar{y}_H = 2.2186, \quad SS_y = 6.0244, \quad SS_r = 5.78, \quad R^2 = 0.9594, \quad R_a^2 = \infty$$

Since the total number of samples $n_H = 5$ and denominator $n_H - n_\beta - 1 = 5$ in Eq. (2.27), the adjusted coefficient of multiple determination $R_a^2$ becomes infinity. That is, the number of samples is too small to evaluate the quality of the surrogate.

Using the standard error $\hat{\sigma} = 0.4944$, the uncertainty associated with the regression parameter can be calculated using the standard deviation of coefficients as

$$\sigma_{bi} = \hat{\sigma}\sqrt{\left[\left(\mathbf{X}^T\mathbf{X}\right)^{-1}\right]_{ii}} = \{0.2647 \quad 0.3534 \quad 0.8517 \quad 0.1180\}$$

The following coefficient of variation can measure the level of uncertainty in individual coefficients compared to their nominal values:

$$c_i = \frac{\sigma_{bi}}{|b_i|} = \{1.0560 \quad 0.2515 \quad 1.7158 \quad 4.6772\}$$

Based on the backward elimination in Chapter 2, it seems that the linear coefficient of the discrepancy function is the most uncertain and therefore needs to be eliminated first. However, the goal in this example is to eliminate the LF function that does not contribute to the prediction of an MF surrogate. Therefore, the elimination process focuses on two terms, $\rho_1 y_{L1}(x)$ and $\rho_2 y_{L2}(x)$. Even if the standard deviation of $\rho_2$ is greater than that of $\rho_1$, the coefficient of variation of $\rho_1$ is greater because the coefficient value is small. This is expected as in Example 9.3, it is shown that the correlation between $y_{L2}(x)$ and HF function is much higher than that of $y_{L1}(x)$ and HF function. Therefore, the first LF function $\rho_1 y_{L1}(x)$ will be eliminated.

With $y_{L2}(x)$ alone, the MF surrogate can be written as

$$\hat{y}_M(x) = \rho y_{L2}(x) + b_1 + b_2 x$$

It is noted that this MF surrogate is identical to part (b) of Example 9.3. Therefore, MF surrogate becomes

$$\hat{y}_M(x) = 1.4537 y_{L2}(x) + 0.1351 - 0.083x$$

The adjusted coefficient of multiple determination $R_a^2$ is calculated as

$$\bar{y}_H = 2.2186, \quad SS_y = 6.0244, \quad SS_r = 5.5608, \quad R^2 = 0.9230, \quad R_a^2 = 0.6922$$

Although it is impossible to compare $R_a^2$ because it was infinity previously, $R^2$ is reduced slightly. In addition, the RMSE has also increased from 0.3915 to 0.4665. Therefore, the MF surrogate with both $y_{L1}(x)$ and $y_{L2}(x)$ is slightly better than the MF surrogate with $y_{L2}(x)$. Therefore, it is better to use two LF functions.

## 9.4 Kriging-based Multi-fidelity Surrogate

In the previous section, PRS surrogate models were used to model LF and discrepancy functions. In this section, Kriging surrogate models that were introduced in Chapter 7 are used for the same purpose. Kriging or Gaussian process is one of the most popular surrogate models. Kriging naturally provides the prediction variance at an unsampled point. The Bayesian multi-fidelity surrogate developed by Kennedy and O'Hagan [113] introduces the correlation between multi-fidelity datasets. The main advantage of PRS and Kriging surrogates is that they provide prediction uncertainty at unsampled points. Although different forms of MF surrogate are shown in Eqs. (9.1)~(9.3), we will only consider the comprehensive form in Eq. (9.3) in this section because it is the most advanced form.

There are two possible scenarios in applying MF surrogate models. First, it is possible that an analytical expression of the LF function is available. Even if an analytical expression is not available, it is enough if we can evaluate the LF function at prediction points using numerical simulation. This corresponds to the case when the computational cost of evaluating the LF function is ignorable. In this case, a discrepancy surrogate needs to be built using HF samples. The main advantage of this scenario is that the LF function does not cause any error and uncertainty in MF predictions.

The second scenario is that the LF function is not available, but a number of samples from the LF function is available. This often happens when the LF function is relatively expensive to evaluate it. The LF function is used to obtain samples, but an LF surrogate is used to evaluate prediction points. In this case, LF samples are used to fit an LF surrogate first, and then it is used to generate a discrepancy surrogate. Therefore, both LF and discrepancy surrogates can cause prediction uncertainty. This scenario can be further separated into two conditions. First, HF sample locations are a subset of LF sample locations. In this case, the discrepancy samples do not have any prediction uncertainty because they are the difference of two samples at the same location. Therefore, this situation is preferred. The second condition is when HF samples are placed in different locations than the LF sample locations. In this case, the LF surrogate is used to evaluate LF function values at HF sample locations, which involves prediction uncertainty from the LF surrogate.

We will consider multi-fidelity surrogate modeling when the LF function is available first, followed by multi-fidelity surrogate modeling when LF samples are available.

### 9.4.1 Multi-fidelity Surrogate with Low-fidelity Function

When an LF function $y_L(x)$ is available or the LF function can be used for predictions, the comprehensive form in Eq. (9.3) requires to identify the LF scale factor $\rho$ and to fit the discrepancy function $\delta(x)$. The discrepancy function is defined as $\delta(x) = y_H(x) - \rho y_L(x)$. Then, the Kriging surrogate for the discrepancy function is modeled as

$$\hat{\delta}(x) = \boldsymbol{\xi}(x)^T \boldsymbol{\beta} + s(x) \tag{9.15}$$

The discrepancy surrogate is fitted using discrepancy samples $\delta_i = \delta(x_i) = y_H(x_i) - \rho y_L(x_i) = y_{Hi} - \rho y_{Li}$, $i = 1, \ldots, n_H$. Therefore, it is required to evaluate the LF function at HF sample locations. If the LF scale factor $\rho$ is known, this process is identical to fitting a Kriging surrogate using $n_H$ numbers of discrepancy samples. However, due to an unknown scale factor, the fitting process is slightly different from the regular Kriging surrogate. The scale factor can be identified by maximizing the log-likelihood function, which is defined as

$$\ln\left[f\left(y \mid \boldsymbol{\beta}, \sigma^2, \theta, \rho\right)\right] = -\frac{n_y}{2}\ln(2\pi) - \frac{n_y}{2}\ln\left(\sigma^2\right) - \frac{1}{2}\ln|\mathbf{R}| - \frac{(\boldsymbol{\delta} - \mathbf{X}\boldsymbol{\beta})^T\mathbf{R}^{-1}(\boldsymbol{\delta} - \mathbf{X}\boldsymbol{\beta})}{2\sigma^2} \tag{9.16}$$

where $\boldsymbol{\delta} = \{\delta_1, \delta_2, \ldots, \delta_{n_H}\}^T$ is the vector of discrepancy samples and it depends on the LF scale factor. Although the likelihood function depends on hyperparameter $\theta$, it is assumed to be given at this point. It will be determined using maximum likelihood estimation at the end. However, we will express other parameters, $\boldsymbol{\beta}$, $\sigma^2$, and $\rho$, in terms of hyperparameter $\theta$, and then the hyperparameter will be determined through optimization. The maximum likelihood estimate can be obtained by differentiating the log-likelihood function with respect to $\boldsymbol{\beta}$, $\sigma^2$, and $\rho$ and using the stationary conditions as

$$\frac{\partial \ln f}{\partial \boldsymbol{\beta}} = 0 \Rightarrow \hat{\boldsymbol{\beta}} = \left(\mathbf{X}^T\mathbf{R}^{-1}\mathbf{X}\right)^{-1}\left\{\mathbf{X}^T\mathbf{R}^{-1}(\mathbf{y}_H - \rho\mathbf{y}_L)\right\} \tag{9.17}$$

$$\frac{\partial \ln f}{\partial \sigma^2} = 0 \Rightarrow \hat{\sigma}^2 = \frac{\left(\boldsymbol{\delta} - \mathbf{X}\hat{\boldsymbol{\beta}}\right)^T\mathbf{R}^{-1}\left(\boldsymbol{\delta} - \mathbf{X}\hat{\boldsymbol{\beta}}\right)}{n_H - n_p} \tag{9.18}$$

$$\frac{\partial \ln f}{\partial \rho} = 0 \Rightarrow \mathbf{y}_L^T\mathbf{R}^{-1}\left(\mathbf{y}_H - \rho\mathbf{y}_L - \mathbf{X}\hat{\boldsymbol{\beta}}\right) = 0 \tag{9.19}$$

Different from the conventional Kriging formulation, these parameters are interrelated with each other. Therefore, we substitute $\hat{\boldsymbol{\beta}}$ in Eq. (9.17) into Eq. (9.19) to solve for the scale factor, as

$$\rho = \frac{\mathbf{y}_L^T\mathbf{R}^{-1}\mathbf{y}_H - \mathbf{y}_L^T\mathbf{R}^{-1}\mathbf{X}\left(\mathbf{X}^T\mathbf{R}^{-1}\mathbf{X}\right)^{-1}\mathbf{X}^T\mathbf{R}^{-1}\mathbf{y}_H}{\mathbf{y}_L^T\mathbf{R}^{-1}\mathbf{y}_L - \mathbf{y}_L^T\mathbf{R}^{-1}\mathbf{X}\left(\mathbf{X}^T\mathbf{R}^{-1}\mathbf{X}\right)^{-1}\mathbf{X}^T\mathbf{R}^{-1}\mathbf{y}_L} \tag{9.20}$$

Although the expression is rather complicated, the scale factor is determined based on the correlation between LF and HF samples. In order to simplify the expression, we utilize the numerical method that we discussed in Section 7.5. By utilizing Cholesky factorization, $\mathbf{R} = \mathbf{C}\mathbf{C}^T$, and qr-decomposition, $\text{qr}(\mathbf{C}^{-1}\mathbf{X}) = \text{qr}(\mathbf{F}_t) = \mathbf{Q}\mathbf{G}$, the scale factor can be simplified as

$$\rho = \frac{\mathbf{y}_L^T\mathbf{C}^{-T}\left(\mathbf{I} - \mathbf{Q}\mathbf{Q}^T\right)\mathbf{C}^{-1}\mathbf{y}_H}{\mathbf{y}_L^T\mathbf{C}^{-T}\left(\mathbf{I} - \mathbf{Q}\mathbf{Q}^T\right)\mathbf{C}^{-1}\mathbf{y}_L} \tag{9.21}$$

Once the LF scale factor is determined, the discrepancy samples can be determined as $\delta_i = y_{Hi} - \rho y_{Li}$, $i = 1, \ldots, n_H$. Then, the two parameters, $\hat{\boldsymbol{\beta}}$ and $\hat{\sigma}^2$, can be determined using Eqs. (9.17) and (9.18).

As we mentioned before, all three parameters depend on hyperparameter $\theta$, which can be determined by maximizing the likelihood function or minimizing the negative log-likelihood function. This process is identical to the Kriging surrogate in Chapter 7 as

$$\theta = \arg\min\left[\ln\left(\hat{\sigma}^{2(n_H - n_p)}|\mathbf{R}|\right)\right] \tag{9.22}$$

The only difference is that the process variance $\hat{\sigma}^2$ is calculated from Eq. (9.18).

Once the optimal hyperparameter $\theta$ that minimizes the negative log-likelihood function is determined, it is used to calculate the three parameters in Eqs. (9.17)~(9.19). Then, the Kriging surrogate for the discrepancy can be obtained as

$$\begin{aligned}\hat{\delta}(\mathbf{x}) &= \boldsymbol{\xi}(\mathbf{x})^T\hat{\boldsymbol{\beta}} + \mathbf{r}(\mathbf{x})^T\mathbf{R}^{-1}\left(\boldsymbol{\delta} - \mathbf{X}\hat{\boldsymbol{\beta}}\right) \\ &= \boldsymbol{\xi}(\mathbf{x})^T\hat{\boldsymbol{\beta}} + \mathbf{r}(\mathbf{x})^T\boldsymbol{\alpha}\end{aligned} \tag{9.23}$$

where the term $\boldsymbol{\alpha} = \mathbf{R}^{-1}\left(\boldsymbol{\delta} - \mathbf{X}\hat{\boldsymbol{\beta}}\right)$ only depends on the discrepancy samples.

Similar to the Kriging surrogate in Chapter 7, the prediction variance of the discrepancy surrogate can be rewritten as

$$\hat{\sigma}_{\delta}^2(\mathbf{x}) = \hat{\sigma}^2\left[1 - \mathbf{r}(\mathbf{x})^T\mathbf{R}^{-1}\mathbf{r}(\mathbf{x}) + \mathbf{u}(\mathbf{x})^T\left(\mathbf{X}^T\mathbf{R}^{-1}\mathbf{X}\right)^{-1}\mathbf{u}(\mathbf{x})\right] \tag{9.24}$$

where $\mathbf{u}(\mathbf{x}) = \boldsymbol{\xi}(\mathbf{x}) - \mathbf{X}^T\mathbf{R}^{-1}\mathbf{r}(\mathbf{x})$.

When the global function is constant (ordinary Kriging), the design matrix $\mathbf{X}$ becomes a column vector of ones, $\mathbf{X} \equiv \mathbf{1} = [1, 1, ..., 1]^T$. Also, the basis vector $\boldsymbol{\xi}(\mathbf{x}) = 1$. Therefore, the prediction variance can be simplified as

$$\hat{\sigma}_{\delta}^2(\mathbf{x}) = \hat{\sigma}^2\left[1 - \mathbf{r}(\mathbf{x})^T\mathbf{R}^{-1}\mathbf{r}(\mathbf{x}) + \frac{u(\mathbf{x})^2}{\mathbf{1}^T\mathbf{R}^{-1}\mathbf{1}}\right] \tag{9.25}$$

where $u(\mathbf{x}) = 1 - \mathbf{1}^T\mathbf{R}^{-1}\mathbf{r}(\mathbf{x})$.

The following 100-line Matlab code can build a multi-fidelity surrogate when the LF function is available. The code includes Matlab function `MFSLF.m` and `fitDIS.m`. In order to run the code, an additional Matlab function `trend.m` is required from Chapter 7. All these three functions can be combined in a single file `MFSLF.m`. The three files can also be saved in separate files: `MFSLF.m`, `fitDIS.m`, and `trend.m`. Since this code is similar to `Kriging.m` in Chapter 7, more detailed explanations can be found there.

```
function [yDp,yDsig2,rho,nbD]=MFSLF(xs,ys,yLs,xp,trD,noise)
% Fit Bayesian multi-fidelity surrogate with LF function
% INPITS: xs: High-fidelity sample locations
%         ys: High-fidelity samples
%         yLs: Low-fidelity samples
%         xp: Prediction points
%         trD: Order of discrepancy function (0, 1, 2)
%         noise: 0~no noise samples, 1~noisy samples
% OUTPUTS: yDp: Discrepancy predictions
%          yDsig2: Discrepancy prediction variance
%          rho: LF scale factor
%          nbD: # of coefficients in discrepancy function
%
[ny,ndim]=size(xs);nvar=ndim+1;np=size(xp,1);              % Scaling samples
meanx=mean(xs);stdx=std(xs);                 % Mean and STD of sample locations
xs=(xs-meanx)./stdx;                                  % Scaling sample locations
meany=mean(ys);stdy=std(ys);                          % Mean and STD of samples
ys=(ys-meany)/stdy;yLs=(yLs-meany)/stdy;                       % Scaling samples
xp=(xp-meanx)./stdx;                               % Scaling of prediction points
                                                                               %
nij=ny*(ny-1)/2; l=0;               % # of upper-triangular components of RD
dH=zeros(nij,ndim);                              % Distance between HF samples
```

```
ijH=zeros(nij,2);                             % Index of correlation matrix
for i=1:ny-1
 l=l(end)+(1:ny-i);
 ijH(l,:)=[repmat(i,ny-i,1),(i+1:ny)'];
 dH(l,:)=repmat(xs(i,:),ny-i,1)-xs(i+1:ny,:);
end
%
                                  % Estimating discrepancy hyper-parameters
parD=struct('xs',xs,'ys',ys,'yLs',yLs,'d',dH,'ij',ijH,'trPRS',min(trD,2));
hyp0 = [ny^(-1/4)*ones(ndim,1); min(0.1,noise)];
lb=[0.01*ones(ndim,1);0];ub=[6*ones(ndim,1);noise];% Lower- & upper-bounds
options=optimset('Display','off','TolFun',1e-6,'DiffMinChange',0.001);
 %hypOpt=fmincon(@(hyp) fitDIS(hyp,parD),hyp0,[],[],[],[],lb,ub,[],options);
hypOpt = ga(@(hyp) fitDIS(hyp,parD),nvar,[],[],[],[],lb,ub,[],options)';
    %hypOpt = simulannealbnd(@(hyp) fitDIS(hyp,parD),hyp0,lb,ub,options);
hypD=hypOpt(1:size(xs,2)); tau=max(0,hypOpt(end));
fprintf('Discrepancy hyperparam:                  %8.3g\n',hypD.*stdx');
if noise ~= 0, fprintf('Discrepancy noise fraction:    %8.3g\n',tau); end
%
[~,fitD]=fitDIS(hypOpt,parD);                     % Fitting Kriging surrogate
rho=fitD.rho; fprintf('LF scale factor:                        %8.3g\n',rho);
%
                       % Evaluating discrepancy surrogate at prediction points
xi = trend(xp,trD); nbD=size(xi,2);      % Trend basis vector at np points
xrep=repmat(xp',ny,1); xrep=transpose(reshape(xrep,ndim,np*ny));
d=xrep-repmat(xs,np,1);                   % Distance b/w predictions & samples
r = (1-tau)*exp(sum(-(d./hypD').^2,2)); % Correlation b/w preds & samples
r = reshape(r,ny,np);
yDp=xi*fitD.beta + (fitD.alpha*r)';                    % Kriging predictions
yDp=yDp*stdy+(1-rho)*meany;                           % Unscaling predictions
%
                              % Prediction variance of discreepancy surrogate
rt=fitD.C\r;
q=fitD.G\(xi'-fitD.Ft'*rt);
sigma2=fitD.sigma2.*stdy.^2;                               % Process variance
yDsig2=sigma2.*(1-sum(rt.^2,1)+sum(q.^2,1))';           % Prediction variance
fprintf('Disc. process std :                %8.3g\n',sqrt(sigma2)*stdy);
if noise~=0, fprintf('Disc. Noise std :%8.3g\n',sqrt(tau*sigma2)*stdy);end
end
%
%
function [negLogLK,fit]=fitDIS(xvar,par)
% Negative log-likelihood estimate for given hyper-parameter
% INPUTS  xvar         : Hyper-parameter (hyp) & noise (tau)
%         par          : Model parameters structure
% OUTPUTS negLogLK     : Negative log-likelihood function
%         fit          : Necessary parameters for function evaluation
%
hyp=xvar(1:size(par.xs,2)); tau=xvar(end);
H = trend(par.xs,par.trPRS);                                  % Design matrix
ny=size(par.ys,1); nb=size(H,2);        % ny: # of samples, nb: # of coeff.
r = (1-tau)*exp(sum(-(par.d./hyp(:)').^2, 2));    % Correlation b/w samples
idx=find(r>0);                % Index of nonzero components in sparse matrix
```

```
76  ii=(1:ny)';                                               % Diagonal index
77  mu=ones(ny,1)+(10+ny)*eps;           % Small diagonal to prevent singularity
78  R=sparse([par.ij(idx,1);ii],[par.ij(idx,2);ii],[r(idx);mu]); % Correlation
79  [C,~]=chol(R);                              % Cholesky decomposition R = C'*C
80  C=C';                                                % Transpose: R = C*C'
81  Ft=C\H;                                                    % Ft=[C^1][X]
82  [Q,G]=qr(Ft,0);                               % qr-factorization, Ft = Q*G
83  %
84  Ylt=C\par.yLs;          % Maximum likelihood estimate of LF scale factor
85  Yht=C\par.ys;
86  taul=(eye(ny)-Q*Q')*Ylt;
87  tauh=(eye(ny)-Q*Q')*Yht;
88  rhol=max(0.1,min(3,taul'*tauh/(taul'*taul)));              % LF scale factor
89  %
90  dis=par.ys-rhol*par.yLs;                               % Discrepancy samples
91  Yt=C\dis;
92  b=G\(Q'*Yt);                              % Coefficients of trend function
93  rho=(Yt-Ft*b);                                       % rho = Cinv*(y-Xb)
94  alpha=rho'/C;                                     % alpha = (y-Xb)'*Rinv
95  detR = sum(log(full(diag(C)).^2));      % Determinant of correlation matrix
96  sigma2=sum(rho.^2)/(ny-nb);                              % Process variance
97  negLogLK=detR + (ny-nb)*log(sigma2);                % Negative log-likelihood
98  fit = struct('sigma2',sigma2,'beta',b,'alpha',alpha,'C',C,'Ft',Ft,...
99                'G',G,'rho',rhol);
100 end
```

Function MFSLF: The Matlab function takes HF sample pairs `(xs,ys)`, LF sample pairs `(xs, yLs)`, prediction points `xp`, the order to discrepancy trend function `trD`, and the indicator of noise in data `noise` as inputs. The function produces discrepancy predictions `yDp` corresponding points `xp`, discrepancy variance `yDsig2`, LF scale factor `rho`, and the number of coefficients `nbD` of discrepancy trend function as outputs. The matrix `xs` of sample locations is `nHy×ndim` matrix, with `nHy` being the number of HF samples and `ndim` being the dimensions of input variables. The vector `ys` of HF samples is a `nHy×1` vector with sample values. The matrix `xp` of prediction points is `np×ndim` matrix, with `np` being the number of prediction points. In order to build a discrepancy Kriging surrogate, the user also needs to specify the order of trend function. The current implementation allows up to quadratic polynomials. Therefore, `trD` should be 0, 1, or 2. When samples include noise, `noise = 1`, otherwise use `noise = 0`. The discrepancy Kriging function returns with the vector `yDp` of predictions at `np` points and the vector `yDsigs` of prediction standard deviations at `np` points. The function also returns the number of coefficients `nbD` in the trend function. In the case of $\text{trD} = 0$, $\text{nbD} = 1$, when $\text{trD} = 1$, $\text{nbD} = 1 + \text{ndim}$, and when $\text{trD} = 2$, $\text{nbD} = (\text{ndim} + 1)(\text{ndim} + 2)/2$.

For the purpose of a stable and robust process, it would be good practice to normalize all samples. Lines 14~19 normalize sample locations and predictions using the mean and standard deviation of the samples. It is important to note that it is necessary to denormalize the predictions and prediction variances before they are returned (Lines 52 and 57). In Kriging, correlation between samples is determined using distances. Even if correlation depends on hyperparameters, the distances remain the same. Therefore, it would be efficient to store distances between samples, which are calculated in Lines 21~28. Since distances between two samples are symmetric, only the upper triangular portion is stored in the sparse matrix format. The array `ijH` stores the index, while the array `dH` stores distances. The distances are measured in different dimensions.

Hyperparameters are determined in Lines 30~40. First, all necessary variables are stored in the parameter structure `parD` so that they can be transferred to the `fitDIS` function (Line 31). In current implementation, hyperparameters are $\boldsymbol{\theta} = \{\theta_1, \theta_2, \ldots, \theta_{\text{ndim}}, \tau\}^T$, where $\tau$ is the fraction of noise variance. For the moment, it is enough to set `noise = 0`, such that $\tau = 0$ throughout optimization. The initial values of hyperparameters `hyp0` are set to be $n_y^{-1/4}$ in Line 32. For noise fraction, the initial value is 0.1 when samples include noise. Otherwise, the initial value is zero. The lower and upper bounds of hyperparameters are established in Line 33. Optimum hyperparameters were sought in the range of [0.01,6]. It means hyperparameters can increase up to six times the standard deviation of input variables. On the other hand, the noise fraction is sought in the range of [0,1] when samples include noise or [0,0] when samples do not include noise. The `fmincon` function is used to minimize the negative log-likelihood calculated in the `fitDIS` function by changing hyperparameter `hyp` (Line 35). The `fmincon` function uses a gradient-based constrained optimization algorithm. The optimum values of hyperparameters are stored in `hypmle` for hyperparameter and `tau` for noise fraction (Line 38). When printing the optimum hyperparameters, they need to be denormalized to the original variables (Line 39).

Matlab provides different optimization algorithms. Although the `fmincon` function is robust, different algorithms can be used. For example, a genetic algorithm can be used by uncommenting Line 36, or a simulated annealing algorithm can be used by uncommenting Line 37.

Once the hyperparameters are optimized, the `fitDIS` function is used again to calculate the Kriging structure `fit` for Kriging predictions and prediction variance (Line 42). It is noted that during optimization, the negative log-likelihood from `fitDIS` is used, while for the purpose of Kriging predictions, the Kriging structure `fit` is required from `fitDIS`. The identified LF scale factor is retrieved in Line 43. Prediction points are stored in the `xp` matrix, whose dimension is `np×ndim`. Then, the `trend` function is used to calculate the basis functions at prediction points (Line 46). Then, the next task is to calculate correlation between np prediction points and `nHy` sample points. In order to do this, the `xp` matrix is copied `nHy` times, and the copied matrix is rearranged to `ndim×np*nHy` dimensions (Line 47). Then the distances between `np` prediction points and `nHy` sample points are stored in the matrix `d` (Line 48), which is used to calculate correlation `r` (Line 49). The correlation is reshaped to a `nHy×np` matrix in Line 50. The discrepancy predictions are calculated using Line 51. Since these predictions are normalized, they need to be denormalized in Line 52. The process of denormalization of discrepancy function needs some explanation. Since both LF and HF samples are normalized, let us define a normalized multi-fidelity function $y'_H = \rho y'_L + \delta'$, where $y'_H = (y_H - m_y)/s_y$ and $y'_L = (y_L - m_y)/s_y$. Then, the `MFSLF.m` fits the following relationship:

$$\frac{y_H - m_y}{s_y} = \rho\frac{y_L - m_y}{s_y} + \delta'$$

Therefore, we can recover the original multi-fidelity form as

$$y_H = \rho y_L + s_y\delta' + (1-\rho)m_y$$

Therefore, the denormalization process of the discrepancy surrogate is

$$\hat{\delta}(x) = s_y\delta'(x) + (1-\rho)m_y \tag{9.26}$$

This denormalization is shown in Line 52. The prediction variance of the discrepancy surrogate is calculated in Lines 55~58, which follows the same procedure as `Kriging.m`.

Function `fitDIS`: The function `fitDIS` is similar to `fitKRG` in Chapter 7 except that it also calculates the LF scale factor $\rho$ from maximum likelihood estimate. After determining the LF scale factor, it calculates discrepancy samples, from which the negative log-likelihood is calculated for the given values of hyperparameters. The main purpose of this function is to calculate the likelihood function for given values of hyperparameters, such that the hyperparameters that can maximize the likelihood can be estimated, i.e., maximum likelihood estimate. This function also calculates other variables that are necessary to calculate Kriging prediction and prediction variance. The function `fitDIS` takes the vector `xvar` of optimization variables and the parameter structure `par` of the model database. Then, the function `fitDIS` returns with the negative log-likelihood `negLogLK` and the Kriging structure `fit` of surrogate parameters. The negative log-likelihood is used during optimization to find its minimum value, while the Kriging structure `fit` is used for evaluating surrogate predictions at prediction points.

From the input variable `xvar`, the first ndim components are hyperparameters ($\theta_1, \theta_2, \ldots, \theta_{\text{ndim}}$), and the last component is noise fraction $\tau$ (Line 71). In Line 72, input sample locations `xs` are used to build the design matrix `H` by calling the `trend` function. Then, the correlation matrix `R` is constructed in Lines 74~78. It is noted that a small positive value `mu` is added in the diagonal terms in order to make the correlation matrix positive definite (Line 77). Also, the sparse matrix is used to store non-zero terms of the correlation matrix (Line 78). The main difference from `fitKRG` is that this correlation matrix and LF and HF samples are used to determine the LF scale factor in Lines 84~88, which uses Eq. (9.21). Once the LF scale factor is determined, it is used to calculate the discrepancy sample in Line 90. Using these discrepancy samples, the modified negative log-likelihood is calculated using the same procedure with `fitKRG`.

It is noted that the Matlab function `MFSLF` does not calculate multi-fidelity predictions. Instead, it calculates discrepancy surrogate predictions and the LF scale factor. The multi-fidelity predictions can be calculated by combining them with LF predictions. The following example explains how to use `MFSLF` to calculate multi-fidelity predictions.

**Example 9.6 Multi-fidelity Surrogate When LF Function Is Available.**
Generate three samples from the following HF function at samples locations $\mathbf{x}_s = \{0.5862, 0.3638, 0.8198\}^T$ and fit a comprehensive form of the multi-fidelity surrogate using Kriging:

$$y_H(x) = (6x-2)^2 \sin(12x-4)$$

The following form of LF function is available:

$$y_L(x) = 0.5\left[(6x-2)^2 \sin(12x-4)\right] + 10(x-0.5) + 5$$

Plot predictions and 95% prediction intervals for the multi-fidelity surrogate and discrepancy surrogate in the range of $x_p \in [0, 1]$. Also, assess the accuracy of the multi-fidelity surrogate against the true HF function using the three error metrics.

**Solution:**
The following Matlab script is used to build a multi-fidelity surrogate using `MFSLF.m`. In Lines 1 and 2, LF and HF functions are defined. The HF function is used only for generating samples and assessing the accuracy of the multi-fidelity surrogate. On the other hand, the LF function is used at the prediction stage as well. Since the LF function is available, both the LF and HF samples are calculated at the same sample locations in Line 4. Using these samples, `MFSLF` is used to calculate discrepancy prediction `yDp` and prediction variance `yDsig2` in Line 6. `MFSLF` also calculates the LF scale factor rho and the number of coefficients in the trend function.

By combining discrepancy predictions with LF predictions along with the LF scale factor, multi-fidelity predictions are calculated in Line 9. Also, the prediction variances are calculated in Line 10. It is noted that since the LF function is available, only the prediction variances of the discrepancy function contribute to those of a multi-fidelity surrogate.

```
1   lowfun = @(x) 0.5*((6*x-2).^2.*sin(12*x-4))+10*(x-0.5)+5;
2   highfun = @(x) (6*x-2).^2.*sin(12*x-4);
3   xs=[0.5862,0.3638,0.8198]';nHy=size(xs,1);      % HF sample locations & size
4   yLs=lowfun(xs);yHs=highfun(xs);                             % LF & HF samples
5   x=(0:0.01:1)';N=size(x,1);                                % Prediction points
6   yLtrue=lowfun(x);yHtrue=highfun(x);                  % True LF & HF functions
7   [yDp,yDsig2,rho,nbD] = MFSLF(xs,yHs,yLs,x,0,0);        % fitting Bayesian MFS
8   %
9   yHp=rho*yLtrue+yDp;                                % Multi-fidelity predictions
10  yHsig=rho*sqrt(yDsig2);                                      % Prediction STD
11  PI = yHp + tinv([0.025 0.975],nHy-nbD).*yHsig;         % Prediction intervals
12  figure(1); hold on;                               % Plotting result MF surrogate
13  xShaded=[x;sort(x,'descend')]; yShaded=[PI(:,1);PI(end:-1:1,2)];
14  patch(xShaded,yShaded,[.24 .73 .89],'EdgeColor','none');
15  plot(x,yHp,'r-',xs,yHs,'k+',x,yHtrue,'k-','LineWidth',2,'Markersize',10);
16  legend('95% PI','y_M(x) prediction','y_H(x) samples','y_H(x) true');
17  xlabel('x'); set(gca,'Fontsize',14)
18  %
19  yDsig=sqrt(yDsig2); yDtrue=yHtrue-rho*yLtrue;% Discrepancy STD & prediction
20  PI = yDp + tinv([0.025 0.975],nHy-nbD).*yDsig;         % Prediction intervals
21  figure(2); hold on;                              % Plotting discrepancy surrogate
22  xShaded=[x;sort(x,'descend')]; yShaded=[PI(:,1);PI(end:-1:1,2)];
23  patch(xShaded,yShaded,[.24 .73 .89],'EdgeColor','none');
24  plot(x,yDp,'r-',xs,yHs-rho*yLs,'k+',x,yDtrue,'k-','LineWidth',2,'Markersize',10)
25  legend('95% PI','y_D(x) prediction','y_D(x) samples','y_D(x) true');
26  xlabel('x'); set(gca,'Fontsize',14)
27  %
28  eRMS=sqrt(sum((yHtrue-yHp).^2)/N);                  % RMS error at N grid points
29  eav=sum(abs(yHtrue-yHp))/N;                    % Average error at NxN grid points
30  emax=max(abs(yHtrue-yHp));                     % Maximum error at NxN grid points
31  fprintf('RMS error:%8.3g\nAve error:%8.3g\nMax error:%8.3g\n',eRMS,eav,emax);
```

Figure 9.9(a) shows the plots of LF and HF functions along with the three HF samples. The correlation between LF and HF functions is medium, where the correlation coefficient is 0.74. When the genetic algorithm is used for optimization, the hyperparameter that minimizes the negative log-likelihood turned out to be $\theta = 1.37$. Considering the range of input variable is $x \in [0, 1]$, the hyperparameter is relatively large, which means that the discrepancy function does not fluctuate much over the design space. In fact, the discrepancy surrogate shown in Figure 9.9(c) shows almost a straight line. This is desirable because with only three HF samples, it is difficult to capture complex behavior. The processing standard deviation of the discrepancy surrogate is 29.8. However, the prediction variance is almost zero between the three sample locations. High prediction variance is observed in the extrapolation range.

The optimization algorithm identifies the LF scale factor $\rho = 1.76$. As shown in Figure 9.9(a), since the LF function is above the HF function, it is expected that the scale factor should be less than one to fit the HF function. However, the optimization algorithm moves the LF function in

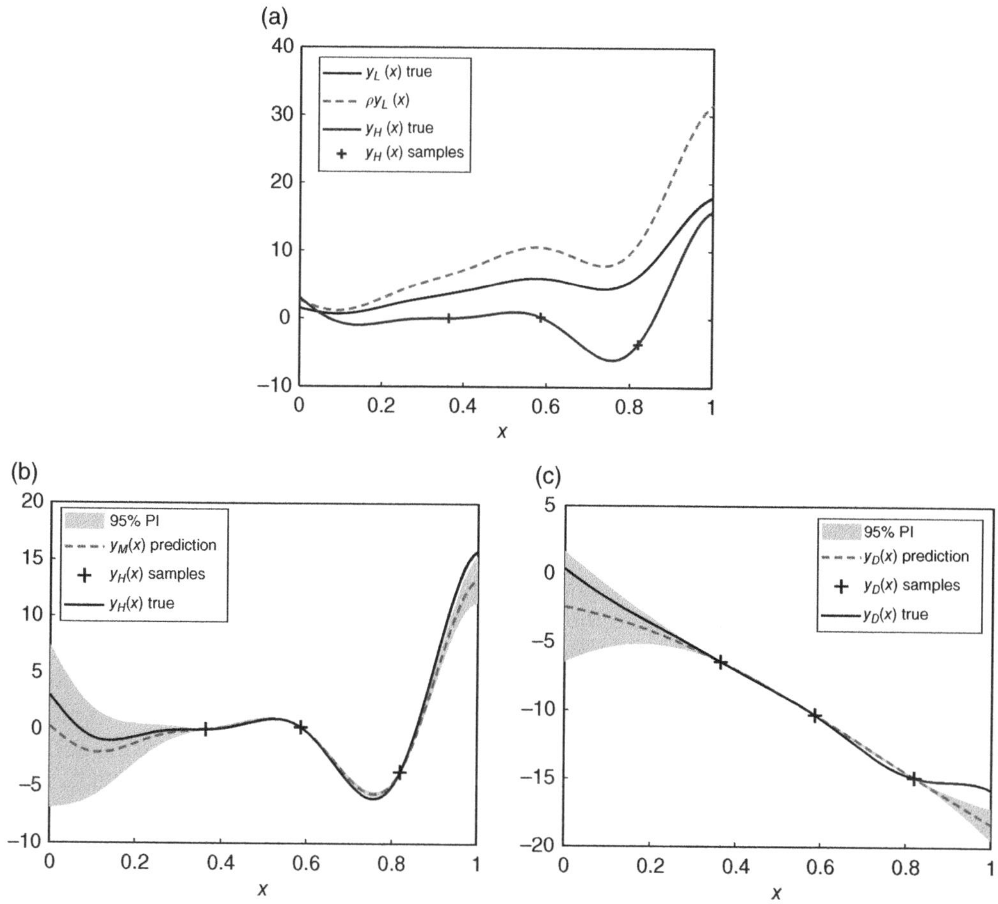


**Figure 9.9** Multi-fidelity surrogates when LF function is available. (a) LF and HF functions with three HF samples, (b) multi-fidelity surrogate with 95% prediction intervals, and (c) discrepancy surrogate with 95% prediction intervals.

the opposite direction to separate the LF function from HF function further. By scaling the LF function, the discrepancy function can have a simpler form with less fluctuation.

The predictions of a multi-fidelity surrogate are shown in Figure 9.9(b) along with 95% prediction intervals. Since the LF function is available, most errors in predictions and prediction variance follow the same trend with the discrepancy function. The three error metrics are $e_{\text{RMS}} = 1.03$, $e_{\text{av}} = 0.664$, and $e_{\max} = 2.77$. Overall, the accuracy of the surrogate is good, and most errors occur in the extrapolation region. It would be good exercise to include an additional sample and compare the performance (see Exercise Problem 9.6).

In the previous example, it was shown that the LF scale factor was identified to be $\rho = 1.76$. It would be interesting to assess the quality of the multi-fidelity surrogate when the scale factor is not used. This would correspond to the discrepancy form in Eq. (9.2). This can be achieved by fixing `rho = 1` in Line 88 of `MFSLF.m`. Figure 9.10(a) shows multi-fidelity predictions and 95% prediction intervals. Compared to the comprehensive case, predictions show clear difference from the true HF function even within the interpolation region. Also, the prediction intervals are larger than before.

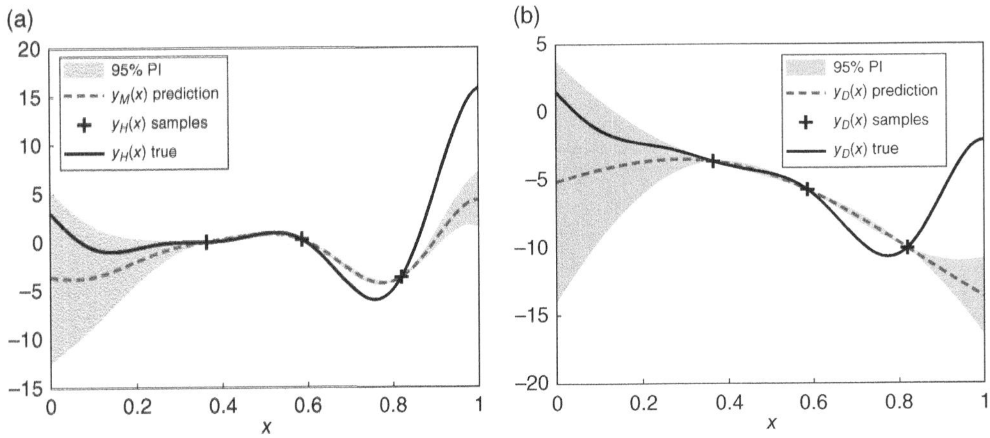


**Figure 9.10** Multi-fidelity surrogates when $\rho = 1$. (a) Multi-fidelity surrogate with 95% prediction intervals and (b) discrepancy surrogate with 95% prediction intervals.

The three error metrics are $e_{\mathrm{RMS}} = 3.63$, $e_{\mathrm{av}} = 2.23$, and $e_{\mathrm{max}} = 11.6$. They are more than three times larger than the previous case. Therefore, the quality of the multi-fidelity surrogate with a fixed scale factor $\rho = 1$ deteriorates significantly. The hyperparameter that minimizes the negative log-likelihood turned out to be $\theta = 0.865$, which is about 40% smaller than the comprehensive form. This indicates that the discrepancy surrogate varies rapidly, which is confirmed from the plot of discrepancy surrogate in Figure 9.10(b). The true discrepancy shows fluctuation, while the discrepancy surrogate failed to capture this complex behavior only with three samples. This concludes that including the LF scale factor is important in the fitting performance of a multi-fidelity surrogate.

### 9.4.2 Multi-fidelity Surrogate with Low-fidelity Samples

When LF samples are available but not LF function, it is necessary to build an LF surrogate as well before building the discrepancy surrogate. This situation is considered as the most complicated case as it involves fitting two Kriging surrogates and estimating the LF scale factor. Even if the LF function is not available, it is still assumed that LF samples can be obtained at a much lower cost than that of HF samples. This situation commonly occurs when LF samples are obtained from simplified simulations, while HF samples are from complex simulations. As shown in Figure 9.1, for example, LF samples are from inviscid Euler simulations, while HF samples are from viscous Reynolds-averaged Navier-Stokes simulations. It is still considered that the cost of generating an LF sample is much lower than that of an HF sample. Therefore, it is expected that the number of LF samples is much larger than that of HF samples. Even if LF samples are cheap to obtain, it is still considered too expensive to use them for the purpose of predictions. Therefore, it would be necessary to build an LF surrogate first and use it for predictions.

As discussed earlier, it is recommended that the HF samples are co-located with LF samples such that the discrepancy samples can be calculated at the same locations. If two sets of samples are not located at the same locations, the LF surrogate is used to predict LF samples at HF sample locations. This can cause additional prediction uncertainty in HF samples. In the implementation stage, however, an LF surrogate is built first, and then, it is used to evaluate LF samples at HF sample locations. This should not cause additional uncertainty when LF samples are co-located with HF samples because Kriging has no prediction uncertainty in sample locations.

The following Matlab code can build a multi-fidelity surrogate using LF and HF samples. The code includes the Matlab function `MFS.m`. In order to run the code, additional Matlab functions `fitKRG.m`, `fitDIS.m`, and `trend.m` are required.

```
function [yLp,yLsig2,yDp,yDsig2,rho,nbL,nbD]=MFS(xLs,yLs,xs,ys,xp,trL,trD,noise)
% Fit Bayesian multi-fidelity surrogate
% INPITS: xLs: Low-fidelity sample locations
%         yLs: Low-fidelity samples
%         xs: High-fidelity sample locations
%         ys: High-fidelity samples
%         xp: Prediction points
%         trL: Order of LF trend function (0, 1, 2)
%         trD: Order of discrepancy function (0, 1, 2)
%         noise: 0~no noise samples, 1~noisy samples
% OUTPUTS: yLp: LF predictions
%          yLsig2: LF prediction variance
%          yDp: Discrepancy predictions
%          yDsig2: Discrepancy prediction variance
%          rho: LF scale factor
%          nbL: # of coefficients in LF trend function
%          nbD: # of coefficients in discrepancy function
%
% Scaling samples
[nLy,ndim]=size(xLs);nvar=ndim+1;ny=size(xs,1);np=size(xp,1);
meanx=mean(xLs);stdx=std(xLs);              % Mean and STD of sample locations
xLs=(xLs-meanx)./stdx;xs=(xs-meanx)./stdx;          % Scaling sample locations
meany=mean(yLs);stdy=std(yLs);                       % Mean and STD of samples
yLs=(yLs-meany)/stdy;ys=(ys-meany)/stdy;                     % Scaling samples
xp=(xp-meanx)./stdx;                               % Scaling of prediction points
%
nij=nLy*(nLy-1)/2; l=0;              % # of upper-triangular components of RL
dL=zeros(nij,ndim);                             % Distance between LF samples
ijL=zeros(nij,2);                               % Index of correlation matrix
for i=1:nLy-1
l=l(end)+(1:nLy-i);
ijL(l,:)=[repmat(i,nLy-i,1),(i+1:nLy)'];
dL(l,:)=repmat(xLs(i,:),nLy-i,1)-xLs(i+1:nLy,:);
end
%
nij=ny*(ny-1)/2; l=0;                % # of upper-triangular components of RD
dH=zeros(nij,ndim);                             % Distance between HF samples
ijH=zeros(nij,2);                               % Index of correlation matrix
for i=1:ny-1
l=l(end)+(1:ny-i);
ijH(l,:)=[repmat(i,ny-i,1),(i+1:ny)'];
dH(l,:)=repmat(xs(i,:),ny-i,1)-xs(i+1:ny,:);
end
%
% Estimating LF hyper-parameters
parL=struct('xs',xLs,'ys',yLs,'d',dL,'ij',ijL,'trPRS',min(trL,2));
hyp0 = [ny^(-1/4)*ones(ndim,1); min(0.1,noise)];                % Initial values
lb=[0.01*ones(ndim,1);0];ub=[6*ones(ndim,1);noise];     % Lower- & upper-bounds
options=optimset('Display','off','TolFun',1e-6,'DiffMinChange',0.001);
```

```
%hypOpt=fmincon(@(hyp) fitKRG(hyp,parL),hyp0,[],[],[],[],lb,ub,[],options);
hypOpt = ga(@(hyp) fitKRG(hyp,parL),nvar,[],[],[],[],lb,ub,[],options)';
%hypOpt = simulannealbnd(@(hyp) fitKRG(hyp,parL),hyp0,lb,ub,options);
hypL=hypOpt(1:size(xLs,2)); tau=max(0,hypOpt(end));
fprintf('LF Hyperparameters:%8.3g\n',hypL.*stdx');
if noise ~= 0, fprintf('LF Noise fraction:%8.3g\n',tau); end
%
[~,fitL]=fitKRG(hypOpt,parL);                          % Fitting LF Kriging surrogate
%
% Evaluating LF surrogate at prediction points
 xi = trend(xp,trL); nbL=size(xi,2);             % Trend basis vector at np points
xrep=repmat(xp',nLy,1); xrep=transpose(reshape(xrep,ndim,np*nLy));
d=xrep-repmat(xLs,np,1);                       % Distance b/w predictions & samples
r = (1-tau)*exp(sum(-(d./hypL').^2,2));           % Correlation b/w preds & samples
r = reshape(r,nLy,np);
yLp=meany+stdy*(xi*fitL.beta + (fitL.alpha*r)');        % LF Kriging predictions
%
% Prediction variance of LF surrogate
rt=fitL.C\r;
q=fitL.G\(xi'-fitL.Ft'*rt);
sigma2=fitL.sigma2.*stdy.^2;                                    % Process variance
yLsig2=sigma2.*(1-sum(rt.^2,1)+sum(q.^2,1))';                % Prediction variace
fprintf('LF process std :%8.3g\n',sqrt(sigma2)*stdy);
if noise~=0, fprintf('LF Noise std :%8.3g\n',sqrt(tau*sigma2)*stdy); end
%
% Generating LF samples at HF sample locations
xi = trend(xs,trL);                             % Trend basis vector at ny points
xrep=repmat(xs',nLy,1); xrep=transpose(reshape(xrep,ndim,ny*nLy));
d=xrep-repmat(xLs,ny,1);                       % Distance b/w predictions & samples
r = (1-tau)*exp(sum(-(d./hypL').^2,2));           % Correlation b/w preds & samples
r = reshape(r,nLy,ny);
yLs=xi*fitL.beta + (fitL.alpha*r)';                            % Kriging predictions
%
% Estimating discrepancy hyper-parameters
parD=struct('xs',xs,'ys',ys,'yLs',yLs,'d',dH,'ij',ijH,'trPRS',min(trD,2));
hyp0 = [ny^(-1/4)*ones(ndim,1); min(0.1,noise)];
lb=[0.01*ones(ndim,1);0];ub=[6*ones(ndim,1);noise];      % Lower- & upper-bounds
options=optimset('Display','off','TolFun',1e-6,'DiffMinChange',0.001);
%hypOpt=fmincon(@(hyp) fitDIS(hyp,parD),hyp0,[],[],[],[],lb,ub,[],options);
hypOpt = ga(@(hyp) fitDIS(hyp,parD),nvar,[],[],[],[],lb,ub,[],options)';
%hypOpt = simulannealbnd(@(hyp) fitDIS(hyp,parD),hyp0,lb,ub,options);
hypD=hypOpt(1:size(xs,2)); tau=max(0,hypOpt(end));
fprintf('Discrepancy hyperparam:%8.3g\n',hypD.*stdx');
if noise ~= 0, fprintf('Discrepancy noise fraction:%8.3g\n',tau); end
%
[~,fitD]=fitDIS(hypOpt,parD);                          % Fitting Kriging surrogate
rho=fitD.rho;
fprintf('LF scale factor:%8.3g\n',rho);
%
% Evaluating discrepancy surrogate at prediction points
xi = trend(xp,trD); nbD=size(xi,2);              % Trend basis vector at np points
xrep=repmat(xp',ny,1); xrep=transpose(reshape(xrep,ndim,np*ny));
d=xrep-repmat(xs,np,1);                        % Distance b/w predictions & samples
```

```
103 r = (1-tau)*exp(sum(-(d./hypD').^2,2));        % Correlation b/w preds & samples
104 r = reshape(r,ny,np);
105 yDp=xi*fitD.beta + (fitD.alpha*r)';                       % Kriging predictions
106 yDp=yDp*stdy+(1-rho)*meany;                               % Unscaling predictions
107 %
108 % Prediction variance of discreepancy surrogate
109 rt=fitD.C\r;
110 q=fitD.G\(xi'-fitD.Ft'*rt);
111 sigma2=fitD.sigma2.*stdy.^2;                                   % Process variance
112 yDsig2=sigma2.*(1-sum(rt.^2,1)+sum(q.^2,1))';               % Prediction variance
113 fprintf('Disc. process std :%8.3g\n',sqrt(sigma2)*stdy);
114 if noise~=0, fprintf('Disc. Noise std :%8.3g\n',sqrt(tau*sigma2)*stdy);end
115 end
```

The Matlab function takes LF sample pairs (xLs, yLs), HF sample pairs (xs, ys), prediction points xp, order to LF trend function trL, order of discrepancy trend function trD, and indicator of noise in data noise as inputs. The function produces LF predictions yLp, LF prediction variance yLsig2, discrepancy predictions yDp, discrepancy variance yDsig2, LF scale factor rho, number of coefficients nbL for LF LF trend function, and number of coefficients nbD of discrepancy trend function as outputs. The matrix xLs of sample locations is nLy×ndim matrix, with nLy being the number of LF samples and ndim being the dimensions of input variables. The vector yLs of LF samples is nLy×1 vector with sample values. The matrix xs of sample locations is nHy×ndim matrix, with nHy being the number of HF samples. The vector ys of HF samples is nHy×1 vector with sample values. The matrix xp of prediction points is np×ndim matrix with np being the number of prediction points. In order to build a Kriging surrogate, the user also needs to specify the order of trend function. The current implementation allows up to quadratic polynomials. Therefore, trL and trD should be 0, 1, or 2. When samples include noise, noise = 1, otherwise use noise = 0. The LF Kriging surrogate returns with the vector yLp of predictions at np points and the vector yLsigs2 of prediction variances at np points. The function also returns the number of coefficients nbL in the trend function. The discrepancy Kriging surrogate returns with the vector yDp of predictions and the vector yDsigs2 of prediction variances at np points. The function also returns the number of coefficients nbD in the trend function.

For the purpose of a stable and robust process, it would be good practice to normalize all samples. Lines 21~25 normalize sample locations and predictions using the mean and standard deviation of the samples. It is important to note that it is necessary to denormalize the predictions and prediction variances before they are returned (Lines 65, 70, 106, and 111). In Kriging, correlation between samples is determined using distances. Even if correlation depends on hyperparameters, the distances remain the same. Therefore, it would be efficient to store distances between samples, which are calculated in Lines 27~34 for LF samples and Lines 36~43 for HF samples.

First, LF hyperparameters are determined in Lines 46~55. First, all necessary variables are stored in the parameter structure parL so that they can be transferred to the fitKRG function (Line 46). In current implementation, hyperparameters are $\boldsymbol{\theta} = \{\theta_1, \theta_2, ..., \theta_{\text{ndim}}, \tau\}^T$, where $\tau$ is the fraction of noise variance. For the moment, it is enough to set noise = 0, such that $\tau = 0$ throughout optimization. The initial values of hyperparameters hyp0 are set to be $n_y^{-1/4}$ in Line 47. For noise fraction, the initial value is 0.1 when samples include noise. Otherwise, the initial value is zero. The lower and upper bounds of hyperparameters are established in Line 48. Optimum hyperparameters were sought in the range of [0.01, 6]. It means hyperparameters can increase up to six times the standard deviation of input variables. On the other hand, the noise fraction is sought in the range of [0, 1] when samples include noise or [0, 0] when samples do not include noise. The `fmincon` function is used to minimize the negative log-likelihood calculated in the `fitKRG` function by changing hyperparameter `hyp` (Line 50). The `fmincon` function uses a gradient-based constrained optimization algorithm. The optimum values of hyperparameters are stored in `hypL` for hyperparameter and `tau` for noise fraction (Line 53). When printing the optimum hyperparameters, they need to be denormalized to the original variables (Line 54).

Matlab provides different optimization algorithms. Although the `fmincon` function is robust, different algorithms can be used. For example, a genetic algorithm can be used by uncommenting Line 51, or a simulated annealing algorithm can be used by uncommenting Line 52.

Once the hyperparameters are optimized, the `fitKRG` function is used again to calculate the structure `fitL` for Kriging predictions and prediction variance (Line 57). It is noted that during optimization, the negative log-likelihood from `fitKRG` is used, while for the purpose of Kriging predictions, the Kriging structure `fitL` is required from `fitKRG`.

Prediction points are stored in the `xp` matrix, whose dimension is `np×ndim`. Then, the `trend` function is used to calculate the basis functions at prediction points (Line 60). Then, the next task is to calculate correlation between np prediction points and `nLy` sample points. In order to do this, the `xp` matrix is copied `nLy` times, and the copied matrix is rearranged to `ndim×np*nLy` dimensions (Line 61). Then the distances between `np` prediction points and `nLy` sample points are stored in the matrix `d` (Line 62), which is used to calculate correlation `r` (Line 63). The correlation is reshaped to the `nLy×np` matrix in Line 64. The LF predictions are calculated (Line 65). Since these predictions are normalized, they need to be denormalized. The same procedure with `Kriging` is used to calculate the prediction variances of the LF surrogate in Lines 68∼73.

Once the LF surrogate is built and LF predictions and prediction variances are calculated, the next step is to calculate LF samples at the same location as HF samples. Similar to calculating LF predictions, the same procedure is applied using `xs` as prediction points (Lines 76∼81). It is noted that it is unnecessary to denormalize the LF predictions because they are used for fitting a discrepancy surrogate.

Discrepancy hyperparameters are determined in Lines 84∼93. First, all necessary variables are stored in the parameter structure `parD` so that they can be transferred to the `fitDIS` function (Line 84). The optimization process is identical with that of LF hyperparameters, except that the `fitDIS` function is used instead of the `fitKRG` function.

Once the hyperparameters are optimized, the `fitDIS` function is used again to calculate the Kriging structure `fitD` for Kriging predictions and prediction variance (Line 95). It is noted that during optimization, the negative log-likelihood from `fitDIS` is used, while for the purpose of Kriging predictions, the Kriging structure `fitD` is required from `fitDIS`. The identified LF scale factor is retrieved in Line 96.

Prediction points are stored in the `xp` matrix, whose dimension is `np×ndim`. Then, the `trend` function is used to calculate the basis functions at prediction points (Line 100). Then, the next task is to calculate correlation between `np` prediction points and `nHy` sample points. In order to do this, the `xp` matrix is copied `nHy` times, and the copied matrix is rearranged to `ndim×np*nHy` dimensions (Line 101). Then the distances between `np` prediction points and `nHy` sample points are stored in the matrix `d` (Line 102), which is used to calculate correlation `r` (Line 103). The correlation is reshaped to a `nHy×np` matrix in Line 104. The discrepancy predictions are calculated (Line 105). Since these predictions are normalized, they need to be denormalized in Line 106. The same procedure with `MFSLF` is used to calculate the prediction variances of discrepancy surrogate in Lines 109∼114.

It is noted that Matlab function `MFS` does not calculate multi-fidelity predictions. Instead, it calculates LF and discrepancy surrogate predictions and the LF scale factor. The multi-fidelity predictions can be calculated by combining them. The following example explains how to use `MFS` to calculate multi-fidelity predictions.

**Example 9.7 Multi-fidelity Surrogate When LF Samples Are Available**
Four HF samples are located at $\mathbf{x}_H = \{0.5862, 0.0007, 0.3638, 0.8198\}^T$, while eight LF samples are located at $\mathbf{x}_L = \{0.1548, 0.7148, 0.4934, 0.9932, 0.5862, 0.0007, 0.3638, 0.8198\}^T$. Generate HF and LF samples using the following true functions:

$$y_H(x) = (6x-2)^2 \sin(12x-4)$$
$$y_L(x) = 0.5\left[(6x-2)^2 \sin(12x-4)\right] + 10(x-0.5) + 5$$

With these LF and HF samples, build a multi-fidelity surrogate using the comprehensive form. Plot predictions and 95% prediction intervals for LF, discrepancy and multi-fidelity surrogates in the range of $x_p \in [0, 1]$. Also, assess the accuracy of the multi-fidelity surrogate against the true HF function using the three error metrics.

**Solution:**
The following Matlab script is used to build a multi-fidelity surrogate using the function `MFS`. In Lines 1 and 2, LF and HF functions are defined. These functions are used only for generating samples and assessing the accuracy of the multi-fidelity surrogate. The LF and HF samples are calculated at the same sample locations in Lines 3 and 4. It is noted that the four HF samples are co-located with those of LF samples. Using these samples, `MFS` is used to calculate LF prediction `yDp`, LF prediction variance `yLsig2`, discrepancy prediction `yDp`, and prediction variance `yDsig2` in Line 12. `MFS` also calculates the LF scale factor `rho` and the number of coefficients in the LF and discrepancy trend functions, `nbL` and `nbH`.

The multi-fidelity predictions are obtained by combining LF prediction, LF scale factor, and discrepancy predictions in Line 13. Also, the prediction standard deviation is calculated in Line 14. Calculating 95% prediction intervals and plotting surrogate predictions are the same as previous examples.

```
1   lowfun = @(x) 0.5*((6*x-2).^2.*sin(12*x-4))+10*(x-0.5)+5;
2   highfun = @(x) (6*x-2).^2.*sin(12*x-4);
3   xLs=[0.1548,0.7148,0.4934,0.9932,0.5862,0.0007,0.3638,0.8198]';      % LF inputs
4   xHs=[0.5862,0.0007,0.3638,0.8198]';                                  % HF inputs
5   yLs=lowfun(xLs);                                           % low-fidelity outputs
6   yHs=highfun(xHs);                                         % high-fidelity outputs
7   trL=0;                                          % LF trend function order (0, 1, 2)
8   trD=0;                                 % Discrepancy trend function order (0, 1, 2)
9   x=(0:0.01:1)';
10  nHy=size(yHs,1);nLy=size(yLs,1);
11  % fitting Bayesian MFS
12  [yLp,yLsig2,yDp,yDsig2,rho,nbL,nbH] = MFS(xLs,yLs,xHs,yHs,x,trL,trD,0);
13  yHp=rho*yLp+yDp;
14  yHsig=sqrt(rho^2*yLsig2+yDsig2);
15  cilb = yHp + tinv(0.025,nHy-nbH)*yHsig;                     % Lower-bounds of PI
16  ciub = yHp + tinv(0.975,nHy-nbH)*yHsig;                     % Upper-bounds of PI
17  yHtrue=highfun(x);
18  % plotting result
19  figure(1); hold on;
```

```
xShaded=[x;sort(x,'descend')]; yShaded=[cilb;ciub(end:-1:1)];
patch(xShaded,yShaded,[.24 .73 .89],'EdgeColor','none');
plot(x,yHp,'r-','LineWidth',2)
plot(xHs,yHs,'k+','LineWidth',2,'Markersize',10)
plot(x,yHtrue,'k-','lineWidth',2)
legend('95% CI','y_M(x) prediction','y_H(x) samples','y_H(x) true');
xlabel('x'); set(gca,'Fontsize',14)
%
yLtrue=lowfun(x); yLsig=sqrt(yLsig2);
cilb = yLp + tinv(0.025,nLy-nbL)*yLsig;                    % Lower-bounds of PI
ciub = yLp + tinv(0.975,nLy-nbL)*yLsig;                    % Upper-bounds of PI
figure(2); hold on;
xShaded=[x;sort(x,'descend')]; yShaded=[cilb;ciub(end:-1:1)];
patch(xShaded,yShaded,[.24 .73 .89],'EdgeColor','none');
plot(x,yLp,'r-','LineWidth',2)
plot(xLs,yLs,'k+','LineWidth',2,'Markersize',10)
plot(x,yLtrue,'k-','lineWidth',2)
legend('95% CI','y_L(x) prediction','y_L(x) samples','y_L(x) true');
xlabel('x'); set(gca,'Fontsize',14)
%
figure(3); hold on;
yDsig=sqrt(yDsig2); yDtrue=yHtrue-rho*yLtrue;
cilb = yDp + tinv(0.025,nHy-nbH)*yDsig;                    % Lower-bounds of PI
ciub = yDp + tinv(0.975,nHy-nbH)*yDsig;                    % Upper-bounds of PI
xShaded=[x;sort(x,'descend')]; yShaded=[cilb;ciub(end:-1:1)];
patch(xShaded,yShaded,[.24 .73 .89],'EdgeColor','none');
plot(x,yDp,'r-','LineWidth',2)
plot(xHs,yHs-rho*lowfun(xHs),'k+','LineWidth',2,'Markersize',10)
plot(x,yDtrue,'k-','lineWidth',2)
legend('95% CI','y_D(x) prediction','y_D(x) samples','y_D(x) true');
xlabel('x'); set(gca,'Fontsize',14)
```

Figure 9.11(a) shows LF and HF true functions with samples. As mentioned before, we will use the true functions for the purpose of generating samples and assessing the accuracy of the surrogate. It is noted that the correlation coefficient between the LF and HF functions is 0.74, which is considered a medium level of correlation. Figure 9.11(b) is the prediction and 95% prediction intervals of LF surrogate. The LF hyperparameter turns out to be $\theta = 0.272$, which indicates that the LF function fluctuates within the range of the input variable. Since we used eight samples and since there is no extrapolation region, the LF surrogate is quite accurate, but there is some level of prediction uncertainty when $x \leq 0.4$. The three error metrics for LF surrogate against the true LF function are $e_{\text{RMS}} = 0.171$, $e_{\text{av}} = 0.078$, and $e_{\max} = 0.613$. Figure 9.11(c) shows the prediction and 95% prediction intervals of the discrepancy surrogate. It is interesting to note that the discrepancy function is almost a straight line with minimal prediction intervals. The discrepancy hyperparameter turns out to be $\theta = 2.01$, which indicates that there is no fluctuation with the range of the input variable. Such a simple form of discrepancy surrogate is achieved by determining the LF scale factor $\rho = 2.15$. Even if $2.15 \times \hat{y}_L(x)$ has more errors, it will make the form of the discrepancy function simpler, which leads to a better approximation in the multi-fidelity surrogate. The three error metrics for discrepancy surrogate against the true discrepancy function are $e_{\text{RMS}} = 0.557$, $e_{\text{av}} = 0.266$, and $e_{\max} = 2.09$. Therefore, even if the prediction uncertainty is small for the discrepancy surrogate, the errors are larger than that of the LF surrogate. Figure 9.11(d) shows the prediction and 95% prediction intervals of the multi-fidelity surrogate. It turns out that the prediction accuracy is

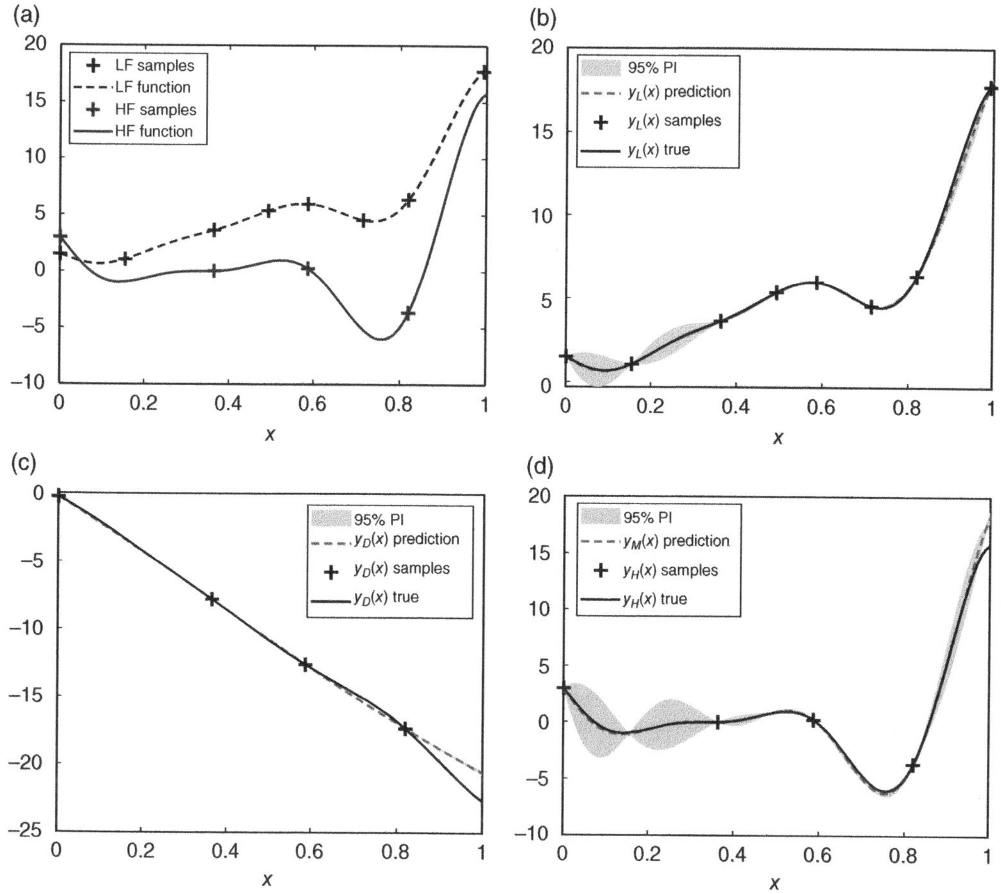


**Figure 9.11** Multi-fidelity surrogates when LF samples are available. (a) LF and HF functions with 4 HF samples, (b) LF surrogate with 95% prediction intervals, (c) discrepancy surrogate with 95% prediction intervals, and (d) multi-fidelity surrogate with 95% prediction intervals.

a combination of LF and discrepancy surrogates. The three error metrics for multi-fidelity surrogate against the true HF function are $e_{\mathrm{RMS}} = 0.375$, $e_{\mathrm{av}} = 0.159$, and $e_{\mathrm{max}} = 2.49$. We can conclude that most prediction errors come from the discrepancy surrogate, while most prediction uncertainty comes from the LF surrogate. Considering the range of the HF function is about 28, $e_{\mathrm{RMS}} = 0.375$ can be assessed as about 1.3% error. In addition, the maximum error occurs at the extrapolation region.

In the previous example, it was shown that the LF scale factor was identified to be $\rho = 2.15$. It would be interesting to assess the quality of multi-fidelity surrogate when the scale factor is not used. This would correspond to the discrepancy form in Eq. (9.2). This can be achieved by fixing `rho = 1` in the `fitDIS` function. Figure 9.12(a) shows multi-fidelity predictions and 95% prediction intervals. Compared to the comprehensive case, predictions show a clear difference from the true HF function even within the interpolation region. Also, the prediction intervals are larger than before. The three error metrics are $e_{\mathrm{RMS}} = 3.16$, $e_{\mathrm{av}} = 1.63$, and $e_{\mathrm{max}} = 10.4$. They are more than five times larger than the previous case. Therefore, the quality of the multi-fidelity surrogate

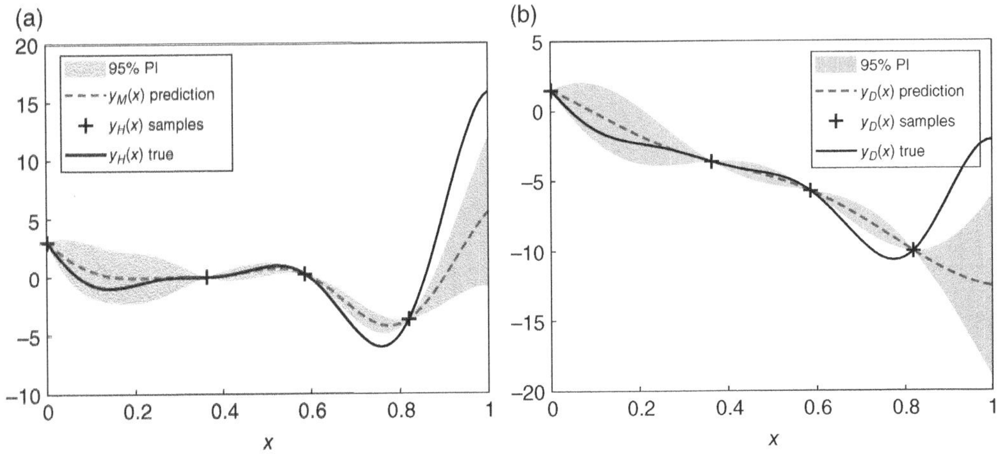


**Figure 9.12** Multi-fidelity surrogates when $\rho = 1$: (a) Multi-fidelity surrogate with 95% prediction intervals and (b) discrepancy surrogate with 95% prediction intervals.

with a fixed scale factor $\rho = 1$ deteriorates significantly. The discrepancy hyperparameter that minimizes the negative log-likelihood turned out to be $\theta = 0.495$, which is about 25% of the comprehensive form case. This indicates that the discrepancy surrogate varies rapidly, which is confirmed from the plot of the discrepancy surrogate in Figure 9.10(b). The true discrepancy shows fluctuation, while the discrepancy surrogate failed to capture this complex behavior only with three samples. This concludes that including the LF scale factor is important in the fitting performance of multi-fidelity surrogate.

## 9.5 Sampling Strategy for Multi-fidelity Surrogate Modeling

The sampling strategy for surrogate modeling is important as the quality of approximation depends on the number and location of samples. It is well known that a space-filling sampling strategy is good for general purpose surrogate models. The most popular space-filling sampling method is Latin hypercube sampling in Section 3.4. In the case of multi-fidelity surrogate modeling, however, the sampling strategy becomes complicated because multi-levels of samples need to be generated. Therefore, how many samples should be for LF and HF surrogates and how they should be located in the design space are important topics to consider in achieving good performance of multi-fidelity surrogate modeling. Of course, it would always be better to have many samples for both LF and HF surrogates. However, each sample means computer simulation or experiment. Therefore, obtaining each sample involves a sampling cost. Therefore, the key ingredient of surrogate modeling is to obtain the required level of accuracy in the surrogate model with the least number of samples. In a multi-fidelity surrogate, an additional issue in the number of samples occurs, which is how to allocate the number of samples between LF and HF surrogates. Even if LF samples are cheaper than HF samples, they are not free either. If the cost ratio between HF and LF samples is high, it would be better to have a large number of LF samples. If the cost ratio is low, then a tradeoff between LF and HF samples would be important. It makes sense to assess the cost of obtaining

LF and HF samples and allocate them appropriately to achieve the best performance. We will investigate the issue of allocating samples at the end of this section. Once the numbers of LF and HF samples are determined, the next question is how to distribute them in the design space. This issue is particularly important in the discrepancy-based multi-fidelity surrogate because discrepancy is defined using the difference between two sets of samples. In this section, two sampling strategies for multi-scale surrogate modeling will be considered: nearest neighbor sampling and nested design sampling.

### 9.5.1 Selecting Locations for LF and HF Samples

For discrepancy-based multi-fidelity surrogates, the sampling strategy that the HF sample set is a subset of the LF sample set is commonly adopted because we need discrepancy samples at common points (Balabanov et al. [114] and Kennedy and O'Hagan [113]). There are two ways to make the HF sample set a subset of the LF sample set and at the same time to make each sample set to satisfy the space-filling requirement: the nearest neighbor sampling and nested design sampling.

The nearest neighbor sampling strategy takes two steps. First, it generates independent initial LF and HF sampling points using Latin hypercube sampling. In the second step, each HF sample point is moved to the nearest LF sampling point. Figure 9.13 shows an example of 40 LF and 10 HF sample points based on the nearest neighbor design. In Figure 9.13(a), 40 LF samples are shown with cross markers, and 10 HF samples are shown with circles. Among LF samples, those samples that were closest to HF samples were shown with light-colored cross markers. Then, these nearest sample locations were moved to HF sample locations as shown in Figure 9.13(b). Now, 10 LF samples are co-located with 10 HF samples, from which the discrepancy samples are calculated.

Another sampling strategy for achieving co-located samples is the nested design sampling, which was initially developed as a space-filling technique for adding samples to an existing sample set in a way to ensure optimal coverage of the union of two sample sets (Jin et al. [115]). Let us assume that we want to generate $n_L$ numbers of LF samples and $n_H$ numbers of HF samples with the condition of $n_L > n_H$. Then, the Latin hypercube sampling method is used to generate $n_L - n_H$ numbers of

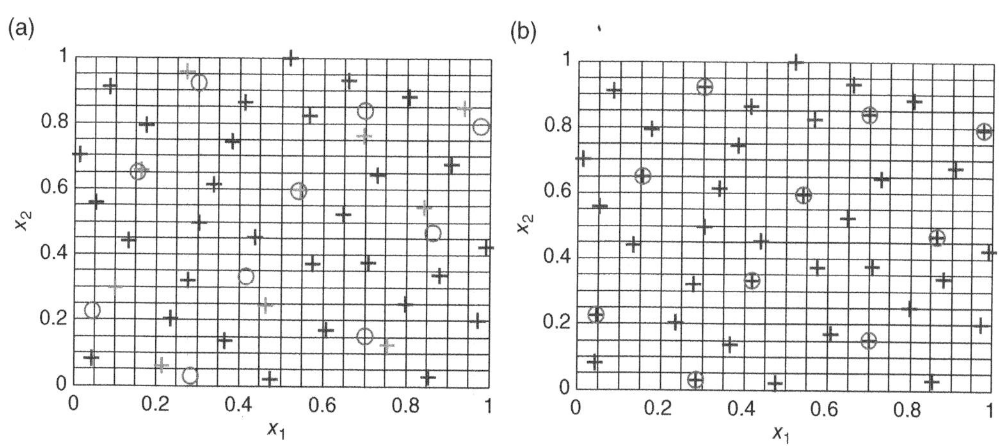


**Figure 9.13** The nearest neighbor sampling for the discrepancy function-based frameworks. (a) 40 LF (blue and green crosses) and 10 HF (red circles) samples and (b) updated 40 LF samples (blue crosses) and 10 HF samples (red circles).

samples first. Then, additional $n_H$ samples are generated by maximizing the minimum distance between existing and new sample locations. These additional sample locations are also used for HF sample locations. These two sampling strategies provide LF and HF sample sets that satisfy the sampling condition of the high-fidelity sampling points being a subset of the low-fidelity sampling points.

### 9.5.2 Allocating LF and HF Samples

Multi-fidelity surrogates are often employed to reduce computational costs and/or to improve accuracy. In this section, we examine multi-fidelity surrogates with three factors that affect the performance: (i) total computational budget, (ii) high-to-low sample evaluation cost ratio, and (iii) high-to-low sample size ratio. We statistically evaluate the performances of multi-fidelity surrogates based on randomly generated samples. The Hartmann-6 and borehole functions were used for the statistical study of frameworks. The Hartmann-6 function is an algebraic function with six input variables. The borehole function is a physical function with three input variables, initially developed to calculate the flow of water through a borehole drilled from the ground surface through two aquifers.

Table 9.1 shows combinations of LF and HF samples for a given total budget and cost ratio. The total computational budget is expressed in terms of the number of HF samples. For example, 56H denotes that we have a computational budget for evaluating 56 HF samples. The sample cost ratio tells how many low-fidelity samples can be evaluated for the cost for a single HF sample. For example, a cost ratio of 4 means that 4 LF samples can be evaluated with the budget for evaluating a single HF sample. For that ratio, a total budget of 56H can be used for either 56 HF samples, or 224 LF samples, or mixes of the two such as the 36 HF and 80 LF samples shown in Table 9.1. These combinations are expressed with the numbers of HF ($n_H$) and LF ($n_L$) samples, such as 36/80. We selected various ratios, ranging from spending most of the budget on HF simulations (36/80), to spending most of the budget on LF simulation (6/200).

We use the Hartmann-6 function over [0.1,1] for all input dimensions as an HF function and an approximated function as an LF function. For each case, we randomly generate 100 different DoEs using nearest neighbor sampling from Latin hypercube samples. The RMSE ($e_{\text{RMS}}$) is calculated based on 10,000 test points, which give an accurate estimate. Among 100 $e_{\text{RMS}}$, the median is chosen as a representative value. For robust comparison, the same test points are used to calculate $e_{\text{RMS}}$ for different DoEs.

**Table 9.1** Cases of sample cost and size ratio combinations for two total computational budgets for Hartmann-6 function.

| Total budget | Sample cost ratio | Sample size ratio $n_H/n_L$ |
|---|---|---|
| 56H | 4 | 36/80, 26/120, 16/160, 6/200 |
| | 10 | 49/70, 46/100, 42/140, 35/210, 28/280, 21/350, 14/420, 7/490 |
| | 30 | 48/240, 46/300, 44/360, 42/420, 40/480, 38/540, 28/840, 18/1140 |
| 28H | 4 | 18/40, 13/60, 8/80, 3/100 |
| | 10 | 22/60, 19/90, 16/120, 13/150, 10/180, 7/210, 4/240 |
| | 30 | 24/120, 22/180, 20/240, 18/300, 16/360, 14/420, 10/540, 6/660, 4/720 |

The HF Hartmann-6 function is defined as

$$y_H(\mathbf{x}) = -\frac{1}{1.94}\left(2.58 + \sum_{i=1}^{4} \alpha_i \exp(\nu_i)\right), \quad \nu_i = -\sum_{j=1}^{6} A_{ij}\left(x_j - P_{ij}\right)^2 \tag{9.27}$$

where $\boldsymbol{\alpha} = \{1, 1.2, 3, 3.2\}^T$ is the vector of model parameters, and matrices $A_{ij}$ and $P_{ij}$ are defined as

$$\mathbf{A} = \begin{bmatrix} 10 & 3 & 17 & 3.5 & 1.7 & 8 \\ 0.05 & 10 & 17 & 0.1 & 8 & 14 \\ 3 & 3.5 & 1.7 & 10 & 17 & 8 \\ 17 & 8 & 0.05 & 10 & 0.1 & 14 \end{bmatrix},$$

$$\mathbf{P} = 10^{-4}\begin{bmatrix} 1312 & 1696 & 5569 & 124 & 8283 & 5886 \\ 2329 & 4135 & 8307 & 3736 & 1004 & 9991 \\ 2348 & 1415 & 3522 & 2883 & 3047 & 6650 \\ 4047 & 8828 & 8732 & 5743 & 1091 & 381 \end{bmatrix}$$

The LF function is defined as

$$y_L(\mathbf{x}) = -\frac{1}{1.94}\left(2.58 + \sum_{i=1}^{4} \alpha_i' f_{\exp}(\nu_i)\right) \tag{9.28}$$

where $\boldsymbol{\alpha}' = \{0.5, 0.5, 2.0, 4.0\}^T$ and $f_{\exp}(x)$ is the approximation function of the exponential function, which is expressed as

$$f_{\exp}(x) = \left[\exp\left(-\frac{4}{9}\right) + \exp\left(-\frac{4}{9}\right)\frac{(x+4)}{9}\right]^9$$

Note that the total function variation of the Hartmann-6 function is 0.33 and the $e_{\mathrm{RMS}}$ of the LF function against the HF function is 0.11. That means the maximum $e_{\mathrm{RMS}}$ of a surrogate based solely on LF samples is 0.11. The LF scale factor will be found within the range of [0.5, 1.5].

Figure 9.14(a) shows the effect of a sample ratio on the median of $e_{\mathrm{RMS}}$ for the cases of total computational budgets 56H with a cost ratio of 30. Legend LF stands for the case when LF

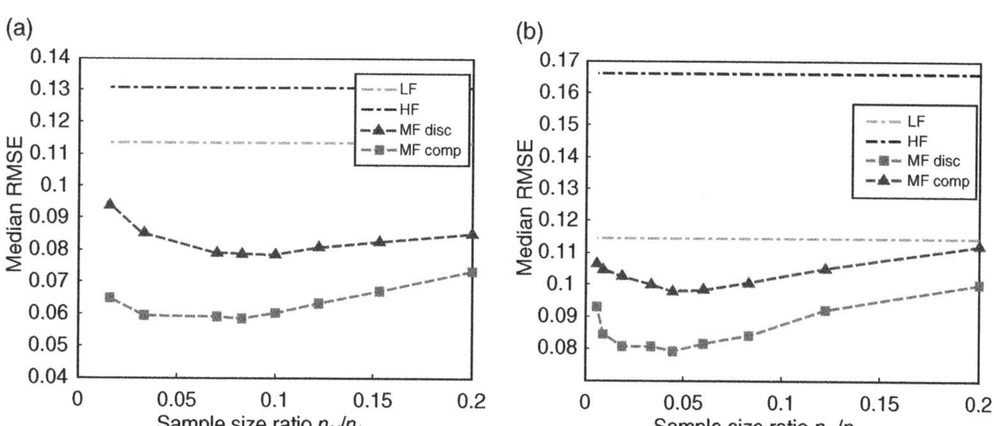


**Figure 9.14** Median RMSEs for a cost ratio of 30. (a) 56H total budget and (b) 28H total budget.

surrogate is built using 1,680 LF samples, while legend HF is the case when 56 HF samples are used to build a surrogate. Since all samples are used to build a single fidelity surrogate, there is no effect of a sample size ratio. The $e_{\text{RMS}}$ of an LF surrogate with 1,680 samples is close to 0.11, which is the $e_{\text{RMS}}$ of the exact LF function. On the other hand, 56 HF samples yield $e_{\text{RMS}} = 0.13$. Therefore, with such a high cost ratio, LF surrogates yield better accuracy than HF surrogates. The reason that an HF surrogate has a larger error is that 56 HF samples are not enough to accurately model a 6-dimensional design space. As we discussed in Chapter 3, if only two samples were generated in each design dimension, a total of 64 HF samples would be required. In the case of total budget 28H as shown in Figure 9.14(b), a similar trend is observed with larger values of $e_{\text{RMS}}$.

Two kinds of multi-fidelity surrogates are considered. The first (MF disc) uses the additive correction function in Eq. (9.2), while the second (MF comp) uses the comprehensive form in Eq. (9.3). The difference between these two forms is the presence of the LF scale factor $\rho$. The multi-fidelity surrogates perform better in terms of median $e_{\text{RMS}}$ than the Kriging surrogates using only LF and HF samples for most sample size ratios. It is interesting to note that multi-fidelity surrogates seem to have an optimal sample size ratio. In the case of 56H, the optimum ratio is around 0.07, while it is around 0.05 in the case of 28H. This indicates that when the cost ratio is high, it is better to have a large number of LF samples and a small number of HF samples. Note that, for the sample size ratio 0, the median $e_{\text{RMS}}$ converged to that of LF surrogates. For this example, the LF scale factor $\rho$ improves the accuracy of a multi-fidelity surrogate significantly.

The next example is a borehole function, which models the flow of water through a borehole penetrating two aquifers. We considered various combinations of LF and HF samples for the given total budget and cost ratio as shown in Table 9.2. The same notation is used for the total budget, cost ratio, and sample size ratio as with the previous example.

The model for the flow was obtained based on assumptions of steady-state flow from the upper aquifer into the borehole and from the borehole into the lower aquifer, no groundwater gradient, and laminar, isothermal flow through the borehole (Morris et al. [116]). The flow rate through the borehole in $\text{m}^3/\text{yr}$ for given conditions was obtained as

$$y_H(R_w, L, K_w) = \frac{2\pi T_u(H_u - H_l)}{\ln\left(\dfrac{R}{R_w}\right)\left(1 + \dfrac{2LT_u}{\ln(R/R_w)R_w^2 K_w} + \dfrac{T_u}{T_l}\right)} \tag{9.29}$$

We assume that the flow rate $y_H(R_w, L, K_w)$ is calculated for input variables $R_w$, $L$, and $K_w$ defining the dimensions and conductivity of a borehole. The input variables and parameters of the function

**Table 9.2** Cases of sample cost and size ratio combinations for two total computational budgets for a borehole example.

| Total budget | Sample cost ratio | Sample size ratio |
|---|---|---|
| 5H | 10 | 4/10, 3/20, 2/30 |
| | 30 | 4/30, 3/60, 2/90 |
| 10H | 10 | 8/20, 7/30, 6/40, 5/50, 4/60, 3/70, 2/80 |
| | 30 | 9/30, 8/60, 7/90, 6/120, 5/150, 4/180, 3/210, 2/240 |

**Table 9.3** Input variables and environmental parameters for a borehole example.

| Input | Description |
|---|---|
| $R_w = [0.05, 0.15]$ m | Radius of a borehole |
| $L = [800, 2800]$ m | Length of a borehole |
| $K_w = [9855, 12045]$ m/yr | Hydraulic conductivity of a borehole |
| Parameters | |
| $R = 25050$ m | Radius of influence |
| $T_u = 89335$ m$^2$/yr | Transmissivity of the upper aquifer |
| $H_u = 1050$ m | Potentiometric head of the upper aquifer |
| $T_l = 89.55$ m$^2$/yr | Transmissivity of the lower aquifer |
| $H_l = 760$ m | Potentiometric head of the lower aquifer |

are presented in Table 9.3. We consider the borehole flow function as an HF function, while an LF function is selected from the literature (Xiong et al. [117]), as

$$y_L(R_w, L, K_w) = \frac{5T_u(H_u - H_l)}{\ln\left(\frac{R}{R_w}\right)\left(1.5 + \frac{2LT_u}{\ln(R/R_w)R_w^2 K_w} + \frac{T_u}{T_l}\right)} \tag{9.30}$$

The RMSE ($e_{\text{RMS}}$) of the LF function with respect to the HF function is 15.0. For the comprehensive form, the LF scale factor ($\rho$) is bounded by the range of [0.5, 1.5].

Same with the Hartmann-6 function, we randomly generate 100 different DoEs using the nearest neighbor sampling from Latin hypercube samples. The RMSE ($e_{\text{RMS}}$) is calculated based on 10,000 test points, which give an accurate estimate. Among 100 $e_{\text{RMS}}$, the median is chosen as a representative value. For robust comparison, the same test points are used to calculate $e_{\text{RMS}}$ for different DoEs.

Figure 9.15(a) shows the effect of the sample ratio on the median of $e_{\text{RMS}}$ for the case with the total computational budgets 10H with a cost ratio of 30. The median $e_{\text{RMS}}$ of LF surrogates is almost the

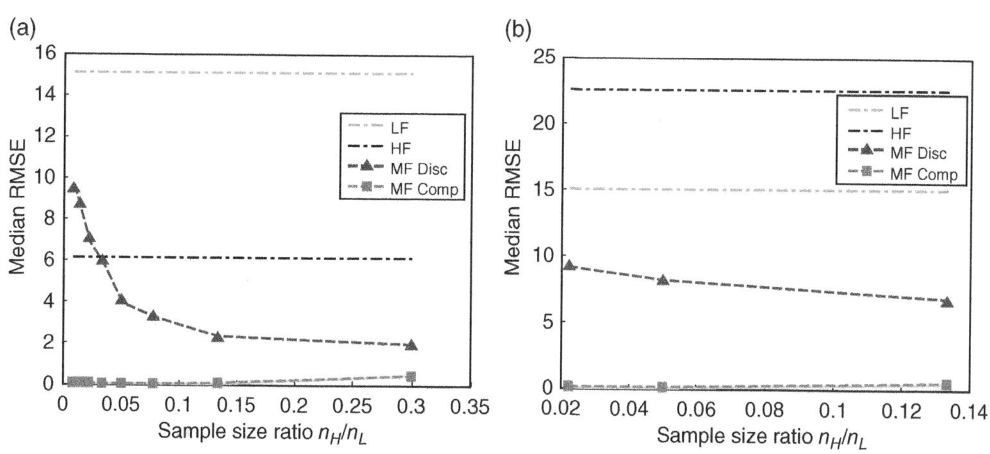


**Figure 9.15** Median RMSEs for a cost ratio of 30. (a) 10H total budget and (b) 5H total budget.

same with the $e_{RMS}$ of the LF function of 15.0. That means the LF surrogates have almost the same error as the exact LF function. This is because $n_L = 10 \times 30 = 300$ samples are enough to approximate the 3D surrogate accurately. On the other hand, an HF surrogate with 10 HF samples turns out to be better than an LF surrogate with 300 LF samples.

In this example, the difference between the discrepancy form and comprehensive form is significant. In particular, when the sample size ratio is low, such as 4/180, 3/210, and 2/240, the discrepancy-based multi-fidelity surrogate is less accurate than a single HF surrogate. On the other hand, the comprehensive form consistently outperforms other surrogates. Therefore, the effect of the LF scale factor $\rho$ is significant on the accuracy of the multi-fidelity surrogate. In the case of 5H, the HF surrogate with 5 HF samples is poorly performed. This is because five samples are too small to approximate 3D surrogate. Other than that, the trend of a 5H case is similar to that of a 10H case.

## 9.6 Challenges and Recommendations

When only a handful of HF samples are available, the HF surrogate with these samples may not be reliable, and therefore, multi-fidelity surrogate is the only viable option. In many cases, however, engineers need to decide whether to use multi-fidelity surrogates or single-fidelity surrogates. Sometimes, engineers also need to choose an LF surrogate from many available LF models or an LF dataset from many datasets. In this section, some challenges and recommendations associated with multi-fidelity surrogate modeling are discussed.

### 9.6.1 Deciding Whether to Use LF Samples

It is not true that multi-fidelity surrogates will always lead to better accuracy than single-fidelity surrogates. As recently reported by Zhang et al. [118], in a strength prediction problem, the surrogate built using only three or more HF samples was found to have better accuracy than a multi-fidelity surrogate with the same three samples, aided by 12 LF samples in a 2D problem. This means that given HF and LF samples, we need a criterion to decide whether to use LF samples to build a multi-fidelity surrogate or to fit a single-fidelity surrogate with HF samples only.

One way of deciding the utility of LF surrogate is to use the comprehensive form in Eq. (9.3) and identify the LF scale factor $\rho$. If the scale factor turns out to be zero or very small, it means that the LF surrogate is not important, and it is better to use HF samples to build single-fidelity surrogates. However, Guo et al. [119] tested the maximum-likelihood estimation for a turbine problem with a single HF model and two alternative LF models (say, LF1 and LF2). They varied the number of HF samples ranging from 4 to 12 with 20 different DoEs for each even number of samples. Therefore, a total of 100 sets of DoEs were conducted. A total of 36 LF samples were used defined in the $6 \times 6$ grid. Their numerical tests showed that multi-fidelity surrogates were more accurate than HF surrogates for 59 out of 100 for LF1 and 18 out of 100 for LF2. But among those cases where HF surrogates were more accurate, the maximum-likelihood estimate correctly selected $\rho = 0$ for only 3 of the 123 cases. This indicates that the maximum likelihood estimate tends to give a preferrable choice to multi-fidelity surrogates. Therefore, the maximum-likelihood estimate is not reliable to determine whether LF samples are useful or not.

Thus, it appears that there is still a lack of a dependable criterion to tell whether we gain additional accuracy by using LF samples. Of course, there are many cases where MF surrogates are clearly more accurate, especially when we have a small number of HF samples. However, more research into the choice between an HF-alone surrogate and multi-fidelity surrogate is clearly called for.

### 9.6.2 Choosing Between Multiple LF Datasets

When there are multiple LF models available, several options are possible: using all LF models, using only a subset of LF models, using the best LF model, or not using any LF model. This is also related to the previous challenge of deciding whether to use LF samples. Guo et al. [119] discussed choosing one between two LF surrogates. In their study, LF1 was more useful than LF2 because it had a better correlation with the HF samples so that the discrepancy function had a significantly lower range of variation. Out of 100 sets of DoEs, the multi-fidelity surrogates with LF1 were 85 times more accurate than those with LF2. Cross-validation identified 71 of these cases, while maximum likelihood identified 81 of them. In addition, the maximum-likelihood estimation could not identify a single case of the 15 opposite cases, and the cross-validation estimation only six out of the 15. This study indicated that the performance of multi-fidelity surrogate depends on DoEs in addition to the quality of LF samples. Moreover, maximum-likelihood and cross-validation estimations were not good at choosing a better LF model for a given DoE. Although this is a limited study, it shows that the choice between multiple LF models may be a challenge, deserving of more research. Another option is to use all available LF models in a single multi-fidelity surrogate as in Section 9.3. However, this was not rigorously exercised in Kriging-based multi-fidelity surrogates.

### 9.6.3 Selecting $\rho$ for Other Surrogates

In Section 9.4, it was shown that introducing the LF scaling factor $\rho$ significantly improved the performance of multi-fidelity surrogates, by reducing fluctuation of the discrepancy surrogate. The constant scaling factor $\rho$ can be found by minimizing the fluctuation of the discrepancy surrogate or by minimizing the error between the scaled LF surrogate predictions with that of HF samples. Park et al. [109] showed that the former performs better than the latter because it is efficient to fit a simple discrepancy with a small number of HF samples. They also observed that maximizing the likelihood function in Kriging surrogates is similar to reducing fluctuation in the discrepancy. However, this approach requires an uncertainty model for prediction; not all surrogates provide one.

While Kriging surrogates are often the most accurate or at least close to the most accurate, there are cases when other surrogates are more accurate. In particular, Kriging does not work well with noisy samples or with a very large number of samples, especially when some of them are tightly clustered. When non-Kriging surrogates are used for constructing a comprehensive multi-fidelity surrogate, it is unclear how fluctuation can be reduced effectively. One may use a Kriging model for obtaining $\rho$, and then, switch to a different surrogate for fitting a multi-fidelity surrogate. This approach may deserve further study. In addition, it is possible to treat the LF surrogate as a basis with the scaling factor $\rho$ as an unknown coefficient in linear regression, as discussed in Section 9.3.

### 9.6.4 Recommendations on Using MF Surrogates

Although there are many different ways of using multi-fidelity surrogates, it is possible to collect some common recommendations to help engineers who start using multi-fidelity surrogates. Below is a short list of recommendations that can be considered for many multi-fidelity surrogate applications:

- Creating an MF surrogate usually involves a substantial investment of time and effort. Therefore, it should be undertaken mainly in cases when using a single-fidelity surrogate is not an option because one cannot afford enough samples for an adequately accurate single-fidelity surrogate.
- Adding the scaling parameter $\rho$ that transforms the additive correction to the comprehensive one often results in substantial improvement in accuracy.
- When the number of HF simulations is severely limited, there is an advantage of using the LF simulations for identifying the most important variables and having the discrepancy function $\delta$ include only these variables. The dependence on the other variables may be still adequately captured through the scaled LF function or surrogate.
- When there are multiple candidates for LF simulations, those simulations that were considered adequate in the past have been the most promising candidates for our applications.
- In choosing between candidate LF simulations, high correlation between LF and HF samples is the most important consideration.
- Selecting an LF simulation that is cheap enough to avoid the need for LF surrogate has two important advantages: (i) It often substantially reduces the implementation effort; (ii) it often avoids a substantial loss of accuracy.

## Exercise

**9.1** In Examples 9.1 and 9.2, the LF surrogate is used for fitting and prediction purposes. This is the case when only LF samples are available. However, when the LF function is available or LF simulation can be used, it is possible that the LF function and LF simulation can directly be used instead of using an LF surrogate. Repeat Examples 9.1 and 9.2 using the LF function directly and compare the prediction accuracy with that of Examples 9.1 and 9.2.

**9.2** In Example 9.3, a linear polynomial is used for discrepancy function. Use a quadratic polynomial for discrepancy function and construct an MF surrogate with $y_{L2}(x)$. Compare the accuracy and prediction uncertainty of the MF surrogate with that of Example 9.3 and explain the differences.

**9.3** In Example 9.5, the first LF function is eliminated even if the linear term of discrepancy function has a larger value of the coefficient of variation. Eliminate the linear term of discrepancy function and evaluate the accuracy of the MF surrogate and compare it with the case of a linear discrepancy function.

**9.4** Repeat Example 9.5 with equally distributed 7 HF samples.

**9.5** Four HF samples are given as $(x_i, y_i)$ = =(0.05, 0.7385), (0.45, 0.4829), (0.7, −4.6058), (1.0, 15.8297). These samples were generated from the true HF function $y_H(x) = (6x-2)^2 \sin(12x-4)$. Build an MF surrogate with a linear discrepancy function $\hat{\delta}(x) = b_1 + b_2x$. Consider the following three LF functions: (a) $y_{L1}(x) = x^2$, (b) $y_{L2}(x) = 5\sin(2\pi - x) - 3$, and (c) $y_{L3}(x) = 0.4y_H(x) + 10(0.4x^4 + 0.1x^2 + 0.2x + 0.2) - 10$. Compare the accuracy of the three MF surrogates against the true HF function using the RMSE of 100 equally spaced points.

**9.6** In Example 9.6, add an additional HF sample at $x = 0.0007$ and build a multi-fidelity surrogate. Compare the accuracy of multi-fidelity surrogate with four samples against that with three samples.

**9.7** In Example 9.7, four HF samples and eight LF samples are used to build an MF surrogate. When both LF and HF have eight samples at the same locations, build an MF surrogate and compare its accuracy against a single fidelity surrogate with HF samples only.

**9.8** An HF function is given as $y_H(x) = \sin(x) + 0.2x + (x-5)^2/16 + 0.5$, and two LF functions are given as $y_{L1}(x) = (x-0.5)(x-4)(x-9)/20 + 2$ and $y_{L2}(x) = \sin(x) + 0.2x + 0.5$ in the range of $x \in [0, 10]$. Generate five HF samples using LHS. (i) Build an MF surrogate model using linear regression with two LF models and compare its accuracy against the true HF function. (ii) Eliminate less important LF models and build an MF surrogate with a single LF model and compare its accuracy against the true HF function. Use a quadratic polynomial for the discrepancy function.

**9.9** Repeat Problem 9.8 when an LF surrogate is used instead of LF functions. Use 20 samples to generate LF surrogates. Use a quadratic polynomial for discrepancy function and a cubic polynomial for LF surrogate.

**9.10** For the LF2 and HF functions in Problem 9.8, build a comprehensive form of the MF surrogate using Kriging.

**9.11** Repeat Problem 9.10 when an LF surrogate is used instead of LF functions. Use 20 samples to generate LF surrogates.

**9.12** An HF function is given as $y_H(x) = (6x-2)^2 \sin(12x-4)$, and two LF functions are given as $y_{L1}(x) = (x-0.5)(x-4)(x-9)/20 + 2$ and $y_{L2}(x) = \cos(x) + (x^2 + x + 20)/25$ in the range of $x \in [0, 1]$. Generate five HF samples using LHS. (i) Build an MF surrogate model using linear regression with two LF models and compare its accuracy against the true HF function. (ii) Eliminate less important LF models and build an MF surrogate with a single LF model and compare its accuracy against the true HF function. Use a quadratic polynomial for the discrepancy function.

**9.13** Repeat Problem 9.12 when an LF surrogate is used instead of LF functions. Use 20 samples to generate LF surrogates. Use a quadratic polynomial for a discrepancy function and a cubic polynomial for an LF surrogate.

**9.14** For the LF2 and HF functions in Problem 9.12, build a comprehensive form of the MF surrogate using Kriging.

**9.15** Repeat Problem 9.14 when an LF surrogate is used instead of LF functions. Use 20 samples to generate LF surrogates.

**9.16** An HF function is given as $y_H(\mathbf{x}) = 4x_1^2 - 2.1x_1^4 + x_1^6/3 + x_1x_2 - 4x_2^2 + 4x_2^4$, and two LF functions are given as $y_{L1}(x) = y_H(0.7x_1, 0.8x_2) + x_1x_2 - 65$ and $y_{L2}(\mathbf{x}) =$

$y_H(0.8x_1, 0.6x_2) - x_1^4 + 32$ in the range of $x \in [-2, 2]$. Generate 15 HF samples using LHS. (i) Build an MF surrogate model using linear regression with two LF models and compare its accuracy against the true HF function. (ii) Eliminate less important LF models and build an MF surrogate with a single LF model and compare its accuracy against the true HF function. Use a quadratic polynomial for the discrepancy function.

**9.17** Repeat Problem 9.16 when an LF surrogate is used instead of LF functions. Use 40 samples to generate LF surrogates. Use a quadratic polynomial for a discrepancy function and a cubic polynomial for the LF surrogate.

**9.18** For the LF2 and HF functions in Problem 9.16, build a comprehensive form of the MF surrogate using Kriging.

**9.19** Repeat Problem 9.18 when an LF surrogate is used instead of LF functions. Use 20 samples to generate LF surrogates.

# 10

# Efficient Global Optimization

## 10.1 Introduction

In many engineering optimization problems, the objective function depends on several design variables, and we want to find the optimum values of the design variables that yield the minimum value of the objective function through iterative processes. When the evaluation of the objective function is computationally expensive, surrogate modeling is often used to approximate this dependence of the objective function on design variables using a simple mathematical form. In the case of a polynomial response surface (PRS), this relationship is expressed in terms of polynomials, while it is expressed in terms of statistical correlation between samples in the case of Kriging. As we discussed in the previous chapters, the quality of approximation depends on the number of samples. Generally, the more the number of samples is, the better the approximation accuracy is. However, it is difficult to decide how many samples are good enough in advance.

So far, optimization using surrogate models means that we build a surrogate model using a number of samples, and then, optimization is performed using the surrogate model instead of the actual function. Therefore, the number of samples is fixed, and the accuracy of the optimum design depends on that of the surrogate model. It is often suggested that once an optimum design is obtained using a surrogate model, the actual function is evaluated at the optimum design to check its accuracy. If the surrogate prediction is different from the actual function, then it would be difficult to assess the quality of the optimum design. Due to the prediction error of the surrogate, the true optimum design might be in a different location.

Another important aspect of optimization using a surrogate model is that the accuracy of the surrogate model over the entire design domain is not important. What matters is the accuracy of the surrogate model near the optimum design. Since the accuracy of a surrogate model is strongly correlated with the density of samples, it makes sense to place more samples near the optimum design if we know where it is.

Considering the abovementioned challenges associated with samples and accuracy in surrogate-based optimization, a new optimization strategy is proposed by Jones et al. [120] in order to make the surrogate model as accurate as possible at optimum design from as few samples as feasible by carefully adding additional samples during optimization process. The key idea of this strategy is to add samples adaptively near the possible optimum design such that the prediction accuracy is improved at the optimum design, not over the entire design space.

The iterative approach called efficient global optimization (EGO) starts with an initial design-of-experiments (DoE). Instead of standard DoEs in Chapter 3, space-filling DoEs such as Latin hypercube sampling are preferred because additional samples will be added at unknown locations. Then, these initial samples are used to build a surrogate model. This step is the same as typical surrogate modeling. At this point, it is assumed that the surrogate is not accurate enough. Therefore, an additional sample is added at the location that can improve the prediction accuracy at the optimum design. The simplest choice would be the location where the current surrogate model predicts the minimum value. It is also possible to choose a location based on the probability that a point can have a chance to be lower than the current minimum value. Different methods are available based on how to choose the new sample location. Selecting a new sample location itself is an optimization problem. Once the new sample location is determined, a computer simulation is performed at this location to generate an additional sample. Then, a new surrogate is fitted with this additional sample, and optimization is performed again. This process is repeated until the stopping criteria are satisfied. How to choose a new sample location is considered as the key ingredient of EGO.

In EGO, the global optimum was sought by "balancing the need to exploit the approximating surface (by sampling where it is minimized) with the need to improve the approximation (by sampling where prediction error may be high)." This algorithm and sequential Kriging optimization [121] use Kriging predictions and prediction variances to seek the point of maximum expected improvement (EI) as the next simulation for the optimization process. The purpose of this chapter is to introduce the EGO algorithm and its variants.

As we discussed in Chapter 4, the Weierstrass theorem or the extreme value theorem [42] says that a global optimum exists when the objective function $f(\mathbf{x})$ is continuous, and the feasible set $S$ is closed and bounded. In Chapter 6, however, we also discussed that there is no algorithm that can guarantee to find the global optimum unless the optimization problem is convex. Instead, most global optimization algorithms in Chapter 6 are in fact global search algorithms, which increase the probability of finding the global optimum. Therefore, the name "efficient global optimization" may mislead the readers that this algorithm may find the global optimum efficiently. In short, EGO is a sequential sampling algorithm to find the best design in the domain using a surrogate model. A surrogate model is built with a given set of samples, and then, the optimum location is found using the surrogate. Due to surrogate error, the optimum point may have prediction error. Therefore, a new sample is added to the optimum design to check if the location is indeed the optimum design. Also, due to prediction uncertainty, a non-optimum point may have the probability of being an optimum. Therefore, EGO also adds a sample when a non-optimum point has a high probability of being an optimum due to prediction uncertainty. Therefore, EGO is a trade-off between exploring the design domain and refining the current optimum design by exploiting near the location.

In Chapter 6, we presented several global search optimization algorithms. However, global search optimization using a surrogate has several advantages. Different from random search algorithms, surrogate-based optimization can be efficient as it requires the fewest function evaluations. This is because engineering functions do not vary randomly. Instead, they have some trends or patterns governed by physics, which can easily be captured by surrogate models. When a surrogate model provides prediction intervals, it provides credible stopping criteria based on the EI for continuing the search. As we discussed in Chapters 2 and 7, both polynomial response surface and Kriging surrogates provide confidence intervals. Therefore, we will use these two surrogates to explain the EGO process.

Consider a 1D example of approximating a true function $y(x) = \sin(x) + 2\sin(2x) + \sin(3x)$ using equally spaced 10 samples as shown in Figure 10.1(a). When a Kriging surrogate is used, the function has a minimum when $x^* = 5.27$ with the optimum function value of $y^* = -3.37$. In order to

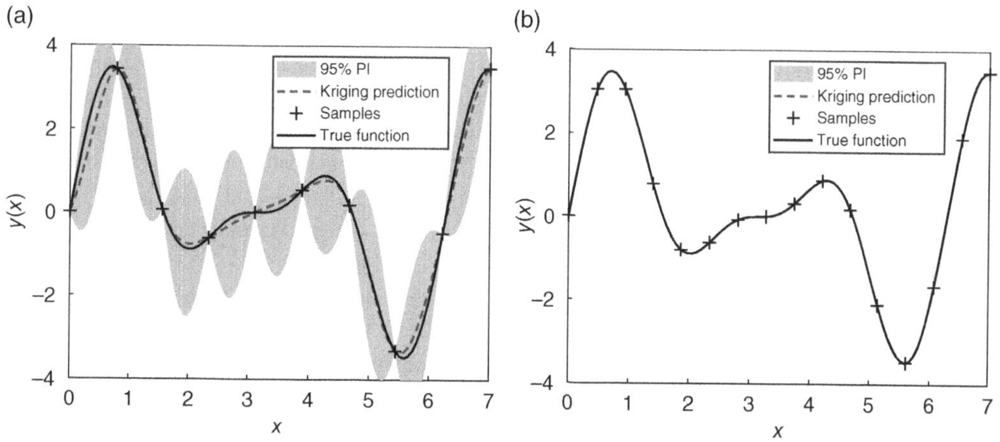


**Figure 10.1** Kriging predictions and uncertainty with different numbers of samples. (a) 10 samples and (b) 16 samples.

compare its accuracy, the true optimal design of the function is when $x_{\text{opt}} = 5.59$ with the optimum function value of $y_{\text{opt}} = -3.48$. Therefore, when the number of samples is small, the optimum design obtained using a surrogate model may not be accurate. This includes both the location of the optimum design and the function value at the optimum design. On top of the inaccuracy of the surrogate predictions, the figure shows a significant level of uncertainty in predictions. Figure 10.1(b) shows an approximation of the same true function using 16 samples. Different from the case with 10 samples, the Kriging predictions are close to the true function, and the level of uncertainty is ignorable. Therefore, the accuracy of the optimum design using surrogate strongly depends on the number of samples. The idea of EGO is to achieve this level of accuracy with a minimum number of samples and place them effectively to find the global optimum point.

This chapter is organized as follows. In Section 10.2, the general concept of EGO is explained based on maximizing EI and the probability of target improvement. In order to apply EGO, it is necessary that the surrogate model provides prediction uncertainty. Therefore, PRS and Kriging surrogate will be considered in Sections 10.3 and 10.4, respectively.

## 10.2 Efficient Global Optimization

EGO algorithms require surrogate predictions and uncertainty associated with predictions. Not all surrogate models provide prediction uncertainty. So far, PRS and Kriging surrogate provide prediction uncertainty. Therefore, it is natural to develop EGO algorithms using these two types of surrogate models. Once a surrogate model is selected, the most important step in EGO is how to select the next sample location. Different algorithms are available based on the selection criterion of the next sample locations. In this section, we will discuss EI, probability of improvement (PI) with adaptive target improvement, and expected feasibility (EF). Of course, there are other criteria available, such as Bayesian expected losses, lower confidence bounds, and choosing multiple sample locations simultaneously, but we refer to the literature for interested readers.

### 10.2.1 Expected Improvement

Although EGO uses surrogate predictions for optimization, it determines an optimum design using a sample. The key idea is to add an additional sample at the most probable location of the global optimum design. For the given set of samples, the sample that has the lowest value is called the present best sample and is denoted by $y_{PBS} = \min(y_i)$. Therefore, the most probable optimum design is predicted using a surrogate model, but the optimum design is determined by an actual sample. This is an additional advantage of EGO because the optimum design does not include any prediction uncertainty of the surrogate model.

After constructing the Kriging model, the algorithm iteratively adds points to the sample set in an effort to improve upon the present best sample $y_{PBS}$. Since Kriging is a stochastic process, the improvement at a point $\mathbf{x}$ is measured in the form of probability as

$$I(\mathbf{x}) = \max(y_{PBS} - Y(\mathbf{x}), 0) \tag{10.1}$$

where $Y(\mathbf{x})$ is an uncertain variable of Kriging predictions. As discussed in Chapter 7, Kriging prediction is normally distributed as $Y(\mathbf{x}) \sim N\left(\hat{y}(\mathbf{x}), s_y^2(\mathbf{x})\right)$, where $\hat{y}(\mathbf{x})$ is the mean prediction and $s_y^2(\mathbf{x})$ is the prediction variance. In the following, we will use the notation that $Y(\mathbf{x})$ is an uncertain variable and $y(\mathbf{x})$ is a realization of the uncertain variable.

Consider Kriging predictions of a true function $y(x) = (6x - 2)^2 \sin(12x - 4)$ using four samples at $\mathbf{x}_s = \{0, 0.5, 0.68, 1.0\}$, as shown in Figure 10.2(a). The present best sample is $y_{PBS} = y(0.68) = -3.68$. As shown in the figure, Kriging predictions show a large level of uncertainty. The optimum of the true function occurs at $x_{opt} = 0.757$ with $y_{opt} = -6.02$. Figure 10.2(b) shows the probability density function (PDF) of Kriging prediction at $x = 0.2$, i.e., $Y(0.2) \sim N\left(\hat{y}(0.2), s_y^2(0.2)\right)$. The area of this PDF below $y_{PBS}$ is the PI.

EI is the expectation of $I(\mathbf{x})$. It is the first moment of $Y(\mathbf{x})$ from the negative infinity to $y_{PBS}$. Mathematically, EI can be defined as

$$\text{EI}(\mathbf{x}) = \mathbb{E}[I(\mathbf{x})] = \int_{-\infty}^{y_{PBS}} (y_{PBS} - y(\mathbf{x})) f_Y(y) dy \tag{10.2}$$

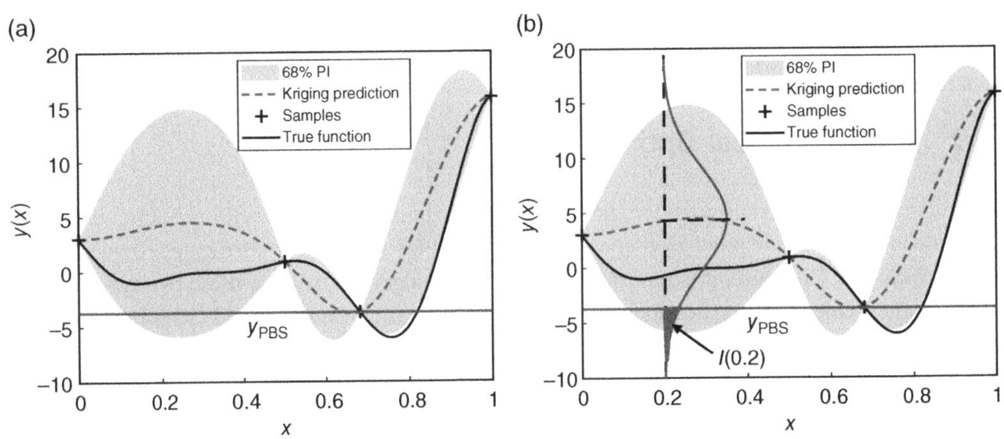


**Figure 10.2** Kriging surrogate of function $y(x) = (6x - 2)^2 \sin(12x - 4)$ using four samples. (a) Predictions and uncertainty and (b) probability of improvement at $x = 0.2$.

where $\mathbb{E}[\cdot]$ is the expectation operator in statistics. In this equation, we add and subtract the mean prediction $\hat{y}(\mathbf{x})$ and split the integral into two as

$$\mathrm{EI}(\mathbf{x}) = \int_{-\infty}^{y_{\mathrm{PBS}}} (y_{\mathrm{PBS}} - \hat{y}(\mathbf{x}))f_Y(y)dy + \int_{-\infty}^{y_{\mathrm{PBS}}} (\hat{y}(\mathbf{x}) - y(\mathbf{x}))f_Y(y)dy$$

In order to make the calculation convenient, the uncertain variable is normalized using $Z = (Y - \hat{y}(\mathbf{x}))/s_y(\mathbf{x})$. Then, the uncertain variable $Z$ follows the standard normal distribution, whose PDF is $\phi(z)$ and the cumulative distribution function (CDF) is $\Phi(z)$. After normalization, the EI can be written as

$$\mathrm{EI}(\mathbf{x}) = (y_{\mathrm{PBS}} - \hat{y}(\mathbf{x}))\Phi\left(\frac{y_{\mathrm{PBS}} - \hat{y}(\mathbf{x})}{s_y(\mathbf{x})}\right) - s_y(\mathbf{x}) \int_{-\infty}^{(y_{\mathrm{PBS}} - \hat{y}(\mathbf{x}))/s_y(\mathbf{x})} z\phi(z)dz$$

The term $y_{\mathrm{PBS}} - \hat{y}(\mathbf{x})$ is the predicted difference between the current minimum and the prediction at $\mathbf{x}$. This term is penalized by the area under the distribution function. Therefore, this term becomes large when $\hat{y}(\mathbf{x})$ is small with respect to $y_{\mathrm{PBS}}$, which promotes exploitation. On the other hand, the negative of the second term becomes large when the prediction uncertainty $s_y(\mathbf{x})$ is large, which promotes exploration. For a standard normal random variable, we have $\phi'(z) = -z\phi(z)$. Therefore, the last integral becomes $-\phi\left((y_{\mathrm{PBS}} - \hat{y}(\mathbf{x}))/s_y(\mathbf{x})\right)$. By introducing the normalized $y_{\mathrm{PBS}}$, the EI can be simplified as

$$\mathrm{EI}(\mathbf{x}) = s_y(\mathbf{x})[z_{\mathrm{PBS}}\Phi(z_{\mathrm{PBS}}) + \phi(z_{\mathrm{PBS}})], \quad z_{\mathrm{PBS}} = \frac{y_{\mathrm{PBS}} - \hat{y}(\mathbf{x})}{s(\mathbf{x})} \tag{10.3}$$

Then, the new sample location is determined by maximizing the EI as

$$\mathbf{x}_{\mathrm{new}} = \arg\max_{\mathbf{x}} \mathrm{EI}(\mathbf{x}) \tag{10.4}$$

In general, the prediction standard deviation $s_y(\mathbf{x})$ is zero at a sample location and, thus, EI($\mathbf{x}$). Therefore, there is no chance that the new sample is co-located with an old sample. It is noted that the normalized variable $z_{\mathrm{PBS}} \geq 0$. Therefore, there are two scenarios where the EI can be large. The first scenario is when $s_y(\mathbf{x})$ is large. This corresponds to the point $x = 0.2$ in Figure 10.2. Although the function value $y(0.2) = -0.6397$ is far away from $y_{\mathrm{PBS}} = -3.6829$, there is a large probability that the actual function can be lower than $y_{\mathrm{PBS}}$ due to the large $s_y(0.2) = 8.4527$. Since prediction uncertainty becomes large when the distance between two adjacent samples is large, this tends to place an additional sample in the region where samples are sparse. In EGO, this is called the *exploration* of the design domain. The second scenario is when a better optimum value is located near the $y_{\mathrm{PBS}}$. Since the $y_{\mathrm{PBS}}$ is a sample, prediction uncertainty $s_y(\mathbf{x})$ near the point is small. However, the EI can be large because of a large $\Phi(z_{\mathrm{PBS}})$ and $\phi(z_{\mathrm{PBS}})$. In such a case, an additional sample tends to be located near the $y_{\mathrm{PBS}}$ for fine-tuning the optimum design. In EGO, this is called *exploitation*. Therefore, the EI in Eq. (10.3) is to find a new sample location for balancing exploration and exploitation in searching for the global optimum design.

In order to show how EGO can determine a new sample location, the same example in Figure 10.2 is used. Figure 10.3(a) and (c) showed that the four-sample Kriging surrogate has the maximum of EI = 0.8271 at $x = 0.246$. This is obviously for the purpose of exploration because there is a big gap between the two adjacent samples $x = 0$ and $x = 0.5$. Therefore, an additional

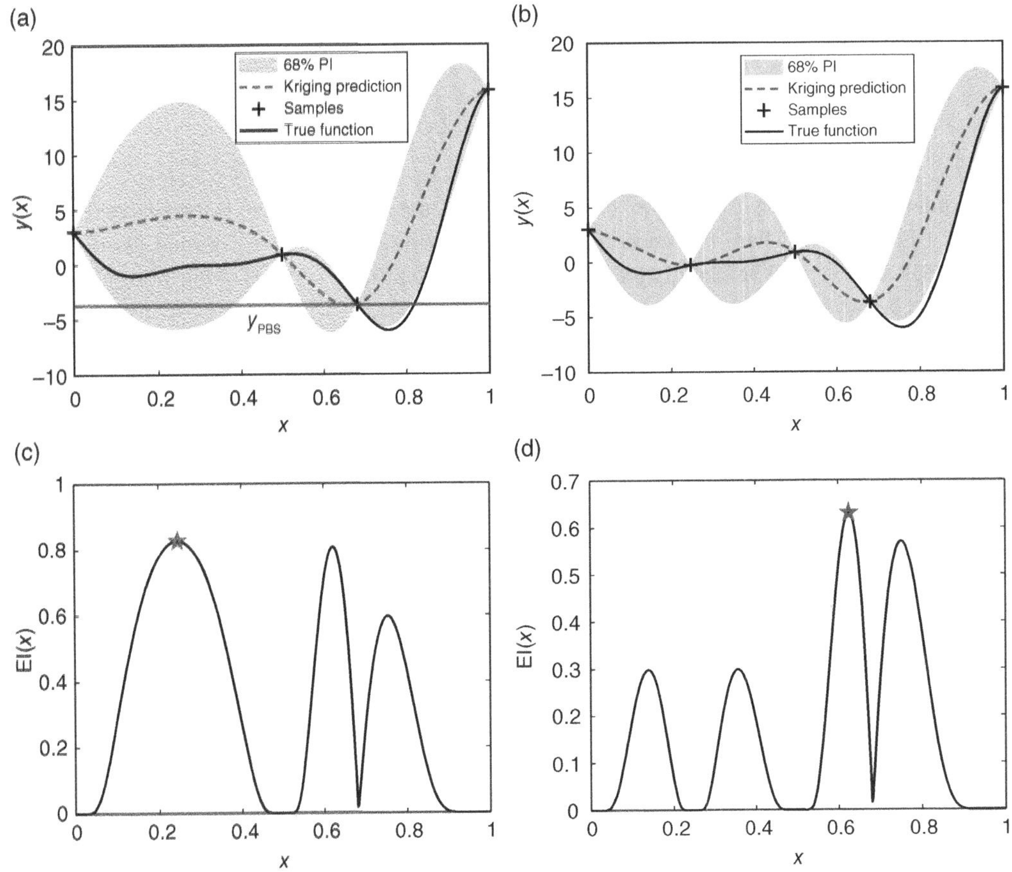


**Figure 10.3** Illustration of the EGO process using EI. (a) Predictions and uncertainty with four samples, (b) predictions and uncertainty with five samples, (c) EI with four samples, and (d) EI with five samples.

sample is added at $x = 0.246$ and the Kriging surrogate is built with five samples, whose predictions are shown in Figure 10.3(b). Although $x = 0.246$ has the maximum EI, it does not mean that the sample at this point becomes a new $y_{PBS}$. However, it is clear that the new sample significantly reduces prediction uncertainty in the range of $0 < x < 0.5$. In addition, the prediction accuracy of the new Kriging surrogate is better than that of the previous surrogate. The plot of EI in Figure 10.3(d) shows that $\mathrm{EI}(0.246) = 0$, which is expected. However, it has a maximum of $\mathrm{EI} = 0.631$ at $x = 0.625$. Considering $y_{PBS}$ is located at $x = 0.68$, this location is obviously an exploitation. Near $y_{PBS}$, EI varies rapidly, as shown in Figure 10.3(d) because it has a high probability of reducing $y_{PBS}$ further. It is noted that during the process of adding samples, the present best sample $y_{PBS}$ did not change, and it is far away from the global minimum of $y_{opt} = -6.02$ at $x_{opt} = 0.757$. However, the process of adding a sample at the maximum EI location can eventually find the global optimum.

**Example 10.1 Adding Additional Samples Using Expected Improvement**
For the function $y(x) = (6x-2)^2 \sin(12x-4)$ shown in Figure 10.3, perform three more EGO iterations to find the global optimum. Compare the accuracy of the surrogate using the three error metrics at each iteration.

**Solution:**

Fitting a Kriging surrogate can be achieved using the `Kriging` Matlab function in Chapter 7. Once prediction $\hat{y}(x)$ and prediction standard deviation $s_y(x)$ are available at prediction points, they can be used to calculate EI($x$). Instead of using an optimization algorithm to find the maximum EI($x$), we use 500 equally spaced prediction points to find the maximum EI($x$). The following Matlab script is used.

```
myfun = @(x) (6*x-2).^2.*sin(12*x-4);                          % In-line function
xs=[0 0.5 0.68 1]';                                            % Sample locations
%xs=[0 0.246 0.5 0.68 1]';                                     % Sample locations
%xs=[0 0.246 0.5 0.625 0.68 1]';                               % Sample locations
%xs=[0 0.246 0.5 0.625 0.68 0.743 1]';                         % Sample locations
%xs=[0 0.246 0.5 0.625 0.68 0.743 0.762 1]';                   % Sample locations
%xs=[0 0.246 0.5 0.625 0.68 0.743 0.758 0.762 1]';             % Sample locations
ys=myfun(xs);                                                    % output samples
xp=linspace(0,1,501)';                                          % Prediction points
[yp, ysig, nb]=Kriging(xs,ys,xp,0,0);                     % Build Kriging surrogate
ny=size(ys,1);                                                 % Number of samples
cilb = yp + tinv(0.16,ny-nb)*ysig;       % Lower-bounds of prediction intervals
ciub = yp + tinv(0.84,ny-nb)*ysig;       % Upper-bounds of prediction intervals
ytrue=myfun(xp);                  % True function values at prediction points
figure; hold on;                                               % Plotting results
xShaded=[xp;sort(xp,'descend')]; yShaded=[cilb;ciub(end:-1:1)];
patch(xShaded,yShaded,[.24 .73 .89],'EdgeColor','none');
plot(xp,yp,'r-',xs,ys,'k+',xp,ytrue,'k-','Markersize',10,'Linewidth',1.5)
xlabel('x');ylabel('y(x)');
legend('68% PI','Kriging prediction','Samples','True function');
%
[xopt, yopt]=fminsearch(@myfun,0.6);                   % Optimum of true function
[yPBS,I]=min(ys);xPBS=xs(I);                               % Present best sample
u=(yPBS-yp)./ysig;                                            % Normalized zPBS
EI=ysig.*(u.*cdf('Normal',u,0,1)+pdf('Normal',u,0,1));%Expected
improvement
[EImax, I]=max(EI); xEImax=xp(I);                % Maximum expected improvement
fprintf('      xPBS       yPBS     xEImax      EImax\n')
fprintf('%8.3g %8.3g %8.3g %8.3g\n',xPBS,yPBS,xEImax,EImax);
figure;
plot(xp,EI,'k',xEImax,EImax,'pentagram','Markersize',10,
'Linewidth',1.5);
xlabel('x');ylabel('EI(x)');
eRMS=sqrt(sum((ytrue-yp).^2)/501);                  % RMS error at N grid points
eav=sum(abs(ytrue-yp))/501;                     % Average error at NxN grid points
emax=max(abs(ytrue-yp));                        % Maximum error at NxN grid points
fprintf('RMS error:%8.3g\nAve error:%8.3g\nMax
error:%8.3g\n',eRMS,eav,emax);
```

It is noted that there are several lines to define variable `xs`, which is the vector of input sample locations. It is necessary to remove the comment symbol `%` one at a time to fit a Kriging surrogate with different samples. These additional sample locations are obtained from `xEImax` variable, which stores the sample location that maximizes the EI.

**Table 10.1** Performance of the iterative EGO process.

| $n_y$ | $x_{PBS}$ | $y_{PBS}$ | $x_{EI}$ | maxEI | $e_{RMS}$ | $e_{av}$ | $e_{max}$ |
|---|---|---|---|---|---|---|---|
| 4 | 0.68 | −3.68 | 0.246 | 0.827 | 4.08 | 3.51 | 7.94 |
| 5 | 0.68 | −3.68 | 0.625 | 0.631 | 3.05 | 2.16 | 7.61 |
| 6 | 0.68 | −3.68 | 0.743 | 1.15 | 1.87 | 1.32 | 4.55 |
| 7 | 0.743 | −5.92 | 0.762 | 0.15 | 1.21 | 0.85 | 2.64 |
| 8 | 0.762 | −6.01 | 0.758 | 0.0116 | 0.769 | 0.463 | 2.01 |

Table 10.1 summarizes the outcomes from the iterative EGO process. When the number of samples $n_y = 4$, the maximum EI occurs at $x_{EI} = 0.246$, as shown in Figure 10.3(c). Then, in the next iteration, this point is added as an additional sample, and a new Kriging surrogate is fitted with five samples, whose results are shown in the row of $n_y = 5$. It is obvious that the fifth sample ($x_5 = 0.246$) is for exploration, while all other samples are used for exploitation. It is noted that for the first three iterations, $y_{PBS} = -3.68$ remains unchanged because samples were added mostly to reduce the prediction uncertainty. A new improved best sample is obtained at the seventh sample, where $y_{PBS} = -5.92$, and it is further improved at the eighth sample with $y_{PBS} = -6.01$. Considering the global optimum is at $x_{opt} = 0.757$ with $y_{opt} = -6.02$, this result is close enough to stop the iterative EGO process. Figure 10.4(a) and (c) show the Kriging predictions and EI with six samples, while Figure 10.4(b) and (d) show the same with seven samples.

An interesting observation can be drawn from the previous example. The maximum EI initially fluctuates, but it decreases as $y_{PBS}$ is close to the global optimum, and the prediction uncertainty is reduced. Therefore, the magnitude of the maximum EI can be used as a stopping criterion of the iterative EGO process.

### 10.2.2 Probability of Improvement

The EI in the previous section is the most popular criterion for selecting the next sample location. It focuses on how to improve the present best sample. Alternatively, we could iteratively add a sample in an effort to maximize the probability of improving upon a target value $y_T$. The target value normally has a lower value than the present best sample, $y_T \le y_{PBS}$. It is often suggested that $y_T$ is about 25% less than the $y_{PBS}$. This is an attractive strategy if a user has a good idea of achieving the target of the objective function. Given a target value, the improvement in Eq. (10.1) is modified to

$$I_T(\mathbf{x}) = \max(y_T - Y(\mathbf{x}), 0)$$

This is the distribution of $Y(\mathbf{x})$ below $y_T$. Then, the probability of improving the present best solution beyond $y_T$ is

$$PI(\mathbf{x}) = \Pr[I_T(\mathbf{x})] = \Phi\left(\frac{y_T - \hat{y}(\mathbf{x})}{s_y(\mathbf{x})}\right) = \Phi(z_T) \tag{10.5}$$

where $z_T = (y_T - \hat{y}(\mathbf{x}))/s_y(\mathbf{x})$ is the normalized target value. Similar to the EI, a new sample location is determined to maximize the PI as

$$\mathbf{x}_{new} = \arg\max_{\mathbf{x}} \mathrm{PI}(\mathbf{x}) \tag{10.6}$$

Once the new sample location is determined, the remaining process is identical to the case of the EI.

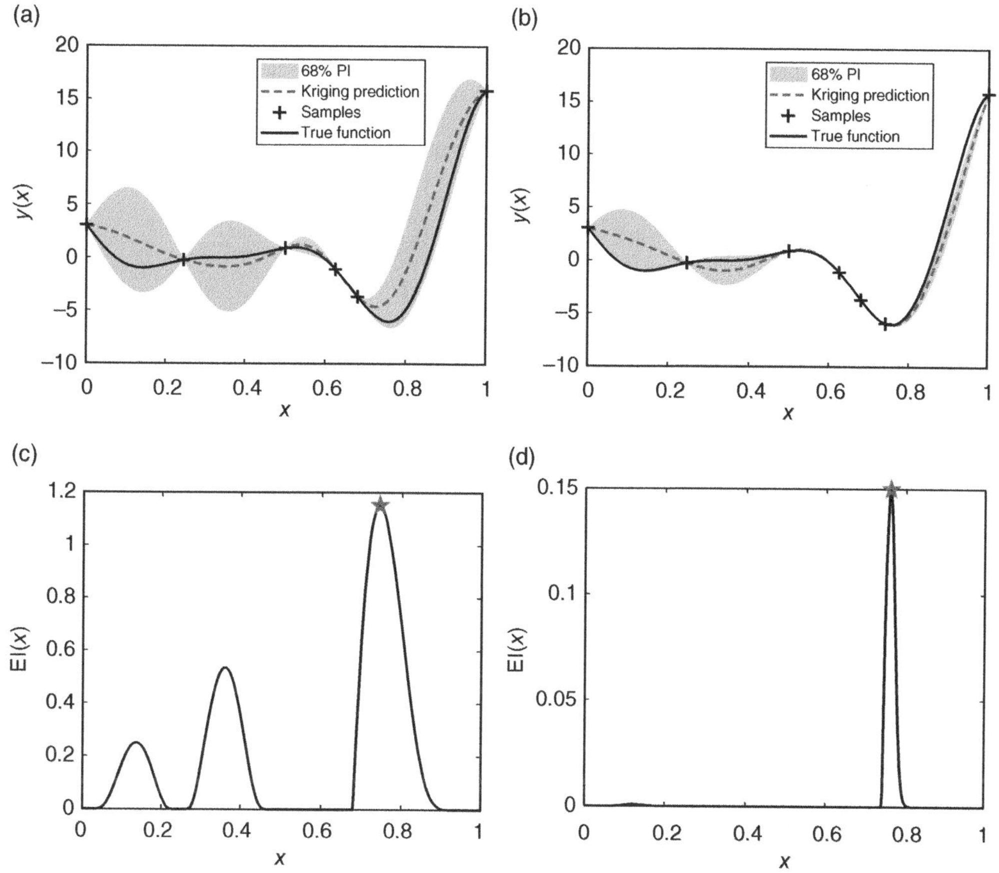


**Figure 10.4** Illustration of the EGO process: (a) predictions and uncertainty with six samples, (b) predictions and uncertainty with seven samples, (c) expected improvement with six samples, and (d) expected improvement with seven samples.

Figure 10.5(a) shows the initial Kriging surrogate with four samples from the true function $y(x) = (6x - 2)^2 \sin(12x - 4)$. Based on the present best sample (0.68, −3.68), the target value $y_T =$ −4.6 is chosen as 25% below the present best sample (dashed-dotted line). Figure 10.5(c) shows the plot of the PI. In the case of EI criterion, the next sample location was $x_{\text{new}} = 0.246$ in order to explore the design space. In the case of PI, the next sample location is chosen based on exploitation at $x_{\text{new}} = 0.633$, which is close to the present best sample. A similar level of PI can be found near $x = 0.75$, but due to a large surrogate prediction error, $x_{\text{new}} = 0.633$ has a larger probability.

Figure 10.5(b) shows the initial Kriging surrogate with the target value $y_T = -6.5$, which is a more aggressive target than that of Figure 10.5(a). In this case, the plot of the PI has its maximum value at $x_{\text{new}} = 0.244$, which is similar to the case of the EI. Therefore, in this case, the new sample location was chosen for the purpose of exploration. If the target value is conservative, the EGO process tends to choose near the present best sample in order to fine-tune the optimum design. On the other hand, if the target value is aggressive, the EGO process tends to explore the design space for possible significant improvement due to large uncertainty. In reality, however, it is difficult to estimate the target value that is close to the true optimum design.

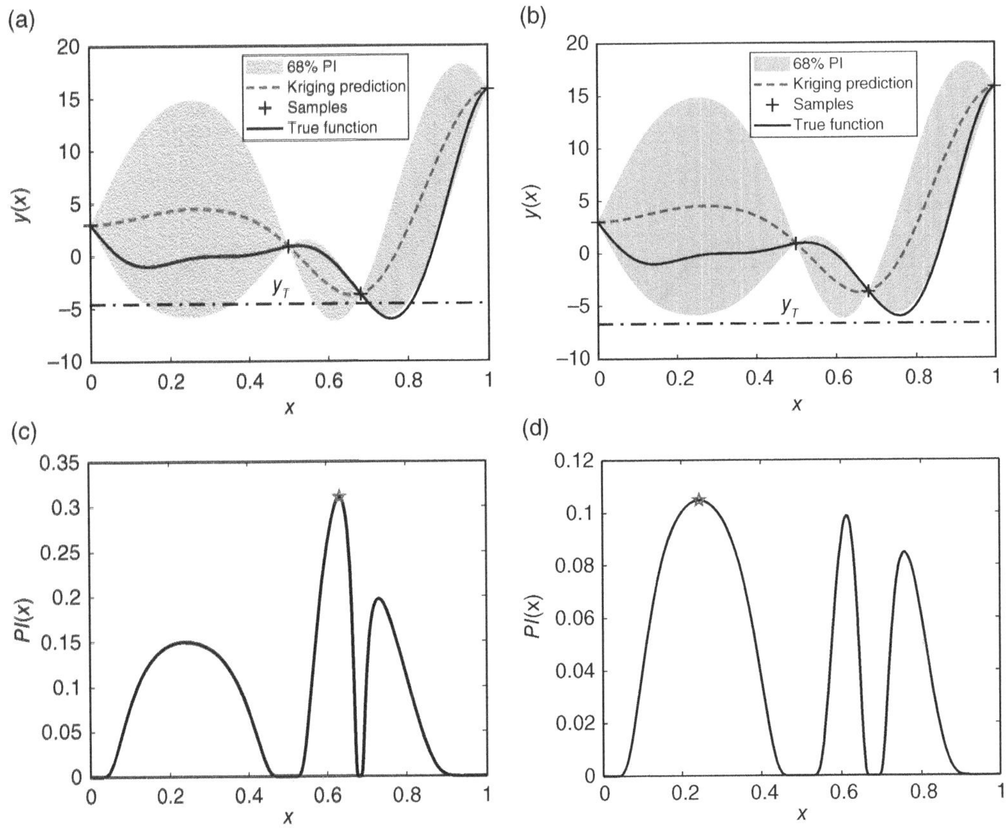


**Figure 10.5** Kriging surrogate predictions and the probability of improvement. (a) Predictions and uncertainty with $y_T = -4.6$, (b) predictions and uncertainty with $y_T = -6.5$, (c) PI with $y_T = -4.6$, and (d) PI with $y_T = -6.5$.

### Example 10.2 EGO with the Probability of Improvement

For the function $y(x) = (6x - 2)^2 \sin(12x - 4)$ shown in Figure 10.5(a), perform five EGO iterations using the PI to find the global optimum. Choose the target value as 25% below the present best sample. Compare the accuracy of the surrogate using the three error metrics at each iteration.

**Solution:**

The following Matlab script is used to perform EGO iterations with the PI criterion. The Matlab function `Kriging` in Chapter 7 is used for fitting the Kriging surrogate. Once prediction $\hat{y}(x)$ and prediction standard deviation $s_y(x)$ are calculated at prediction points, they can be used to calculate the PI($x$). Instead of using an optimization algorithm to find the maximum PI($x$), we use 500 equally spaced prediction points to find the maximum PI($x$).

```
myfun = @(x) (6*x-2).^2.*sin(12*x-4);                    % In-line function
xs=[0 0.5 0.68 1]';                                        % Input samples
%xs=[0 0.5 0.633 0.68 1]';
%xs=[0 0.5 0.633 0.68 0.723 1]';
%xs=[0 0.244 0.5 0.633 0.68 0.723 1]';
```

```
%xs=[0 0.12 0.244 0.5 0.633 0.68 0.723 1]';
%xs=[0 0.12 0.244 0.5 0.633 0.68 0.723 0.794 1]';
ys=myfun(xs);                                          % output samples
xp=linspace(0,1,501)';                             % Prediction points
[yp, ysig, nb]=Kriging(xs,ys,xp,0,0);          % Build Kriging surrogate
ny=size(ys,1);                                      % Number of samples
cilb = yp + tinv(0.16,ny-nb)*ysig;    % Lower-bounds of prediction intervals
ciub = yp + tinv(0.84,ny-nb)*ysig;    % Upper-bounds of prediction intervals
ytrue=myfun(xp);                  % True function values at predictio points
figure; hold on;                                     % Plotting results
xShaded=[xp;sort(xp,'descend')]; yShaded=[cilb;ciub(end:-1:1)];
patch(xShaded,yShaded,[.24 .73 .89],'EdgeColor','none');
plot(xp,yp,'r-',xs,ys,'k+',xp,ytrue,'k-','Markersize',10,'Linewidth',1.5)
xlabel('x');ylabel('y(x)');
legend('68% PI','Kriging prediction','Samples','True function');
%
[yPBS,I]=min(ys);xPBS=xs(I);                          % Present best sample
ytarget=yPBS-0.25*abs(yPBS);                              % Target value
ztarget=(ytarget-yp)./ysig;                     % Normalized target value
PI=cdf('Normal',ztarget,0,1);                 % Probability of improvement
[PImax, I]=max(PI); xPImax=xp(I);       % Max. of probability of improvement
fprintf('    xPBS      yPBS    xPImax     PImax  ytarget\n')
fprintf('%8.3g %8.3g %8.3g %8.3g %8.3g\n',xPBS,yPBS,xPImax,PImax,ytarget);
figure;
plot(xp,PI,'k',xPImax,PImax,'pentagram','Markersize',10,'Linewidth',1.5);
xlabel('x');ylabel('PI(x)');
eRMS=sqrt(sum((ytrue-yp).^2)/501);             % RMS error at N grid points
eav=sum(abs(ytrue-yp))/501;                % Average error at NxN grid points
emax=max(abs(ytrue-yp));                   % Maximum error at NxN grid points
fprintf('RMS error:%8.3g\nAve error:%8.3g\nMax
error:%8.3g\n',eRMS,eav,emax);
```

It is noted that there are several lines to define variable `xs`, which is the vector of input sample locations. It is necessary to remove the comment symbol `%` one at a time to fit a Kriging surrogate with different samples. These additional sample locations are obtained from `xPImax` variable, which stores the sample location that maximizes the PI. Table 10.2 summarizes the outcomes from the iterative EGO process using the PI. When the number of samples $n_y = 4$, the present best sample is $y_{\text{PBS}} = -3.68$ and the target value is $y_T = -4.6$. The maximum PI occurs at $x_{\text{PI}} = 0.633$, as shown in Figure 10.5(c). Then, in the next iteration, this point is added as an additional sample, and a new Kriging surrogate is fitted with five samples, whose results are shown in the row of $n_y = 5$. It is obvious that the fifth sample ($x_5 = 0.633$) is for exploitation as it is close to the present best sample. Figure 10.6(a) shows Kriging predictions with five samples.

When the next sample ($x_6 = 0.723$) is added, the present best sample is changed to $y_{\text{PBS}} = -5.46$, and accordingly, the target value is modified to $y_T = -6.83$. Since the true function has its minimum at $y_{\text{opt}} = -6.02$, it is impossible to attain the new target value. A difficulty in the PI criterion occurs when the target value is too ambitious to attain. As shown in Table 10.2, the PI decreases rapidly, and the algorithm fails to place a sample at the global optimum point. Figure 10.6(c) shows Kriging

**Table 10.2** Performance of the iterative EGO process with probability of improvement.

| $n_y$ | $x_{PBS}$ | $y_{PBS}$ | $x_{PI}$ | maxPI | $y_T$ | $e_{RMS}$ | $e_{av}$ | $e_{max}$ |
|---|---|---|---|---|---|---|---|---|
| 4 | 0.68 | −3.68 | 0.633 | 0.311 | −4.6 | 4.08 | 3.51 | 7.94 |
| 5 | 0.68 | −3.68 | 0.723 | 0.539 | −4.6 | 2.52 | 1.9 | 4.48 |
| 6 | 0.723 | −5.46 | 0.244 | 0.133 | −6.83 | 1.68 | 1.09 | 3.93 |
| 7 | 0.723 | −5.46 | 0.12 | 0.000593 | −6.83 | 1.14 | 0.802 | 2.5 |
| 8 | 0.723 | −5.46 | 0.794 | 3.37E-5 | −6.83 | 0.535 | 0.292 | 1.72 |
| 9 | 0.723 | −5.46 | 0.353 | 1.8E-31 | −6.83 | 0.418 | 0.262 | 1.27 |

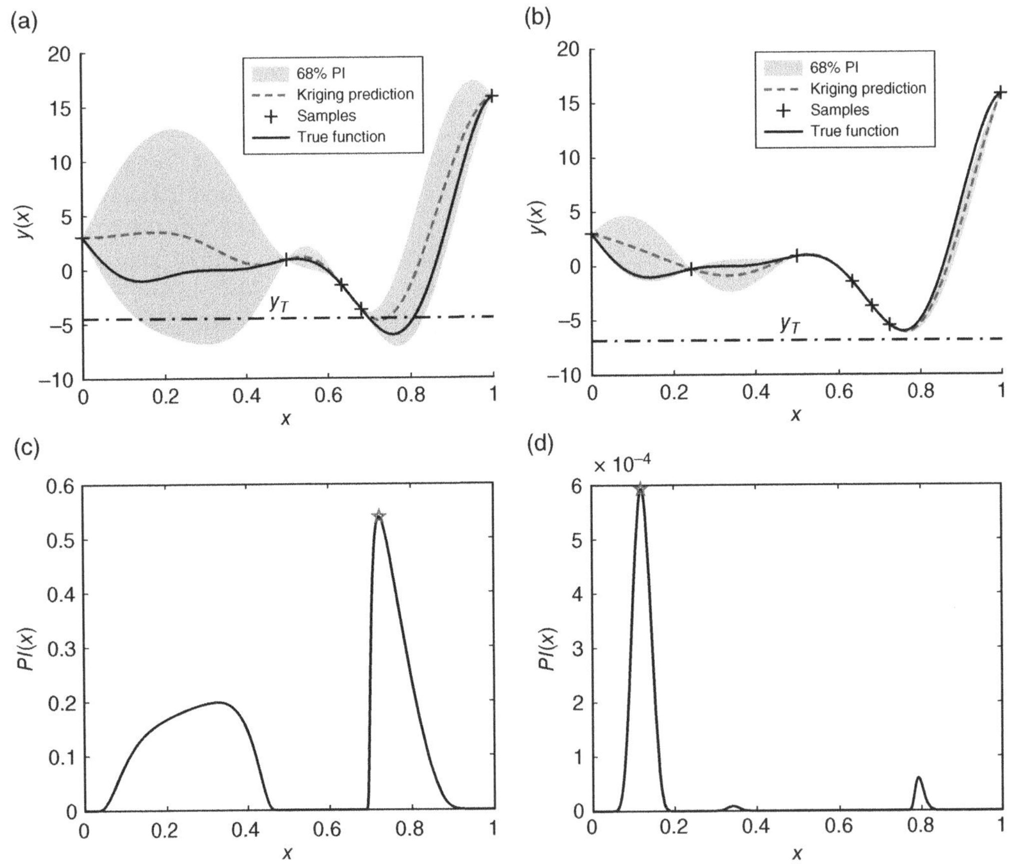


**Figure 10.6** Illustration of the EGO process using the probability of improvement. (a) Kriging predictions with $n_y$ = 5, (b) Kriging predictions with $n_y$ = 7, (c) probability of improvement with $n_y$ = 5, and (d) probability of improvement with $n_y$ = 7.

predictions with seven samples. As the PI in Figure 10.6(d) shows, the probability of achieving the target value of $y_T = -6.83$ is very small. Therefore, when the maximum PI is too small, it means that the target value is too ambitious. Then, it would be necessary to make the target closer to the present best sample.

**Table 10.3** EGO process with the target value of the global optimum.

| $n_y$ | $x_{PBS}$ | $y_{PBS}$ | $x_{PI}$ | maxPI | $y_T$ | $e_{RMS}$ | $e_{av}$ | $e_{max}$ |
|---|---|---|---|---|---|---|---|---|
| 4 | 0.68 | −3.68 | 0.617 | 0.136 | −6.02 | 4.08 | 3.51 | 7.94 |
| 5 | 0.68 | −3.68 | 0.754 | 0.218 | −6.02 | 2.56 | 1.92 | 4.75 |
| 6 | 0.754 | −6.02 | 0.758 | 0.611 | −6.02 | 1.58 | 1.01 | 3.78 |

**Example 10.3 Probability of Improvement with Optimum Target**

In the previous example, it is shown that the EGO process is sensitive to the target value. For the function $y(x) = (6x - 2)^2 \sin(12x - 4)$ shown in Figure 10.3, perform three EGO iterations to find the global optimum. Use the target value as the global optimum $y_T = -6.02$. Compare the accuracy of the surrogate using the three error metrics at each iteration.

**Solution:**

Table 10.3 shows the iterations of the EGO process with the target value of $y_T = -6.02$. It is interesting to note that the EGO process focuses on exploitation, and the global optimum point is found in the third iteration. Figure 10.7(a) and (c) show the Kriging predictions and the PI. The first

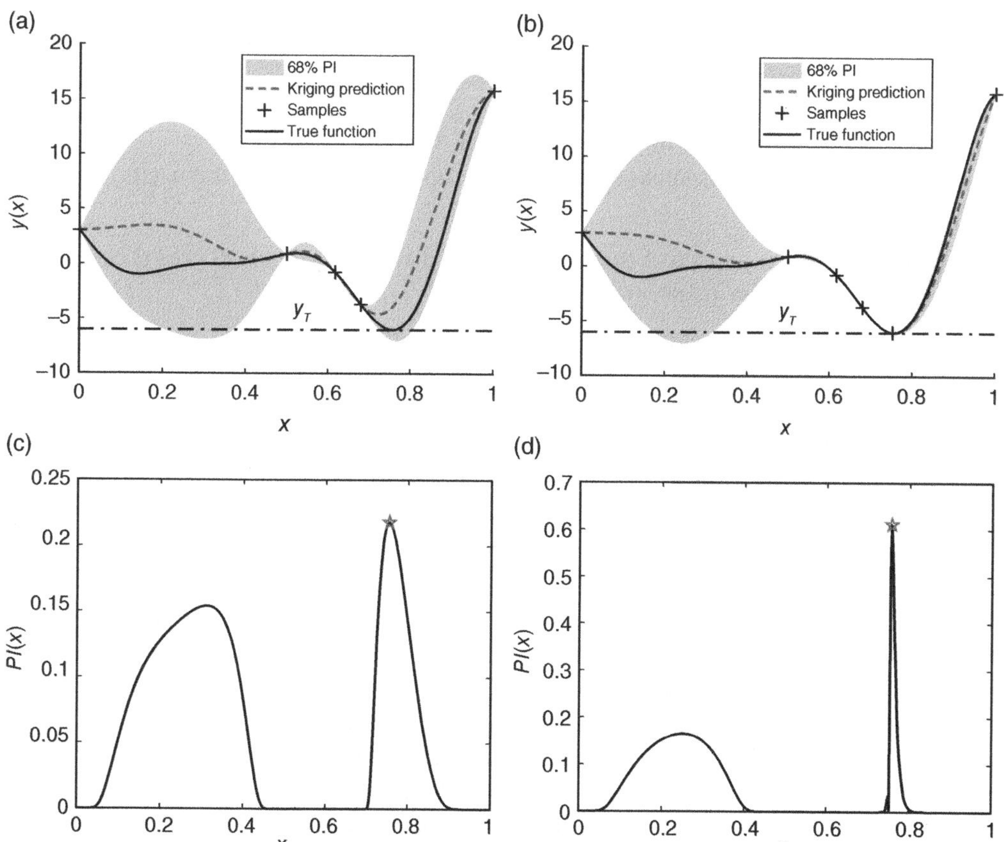


**Figure 10.7** EGO process with the optimum target value. (a) Kriging predictions with $n_y = 5$, (b) probability of improvement with $n_y = 5$, (c) Kriging predictions with $n_y = 6$, and (d) PI with $n_y = 6$.

sample is added at $x = 6.17$, which does not change $y_{PBS}$. However, the second sample at $x = 0.754$ makes the sample close enough to the global optimum. The third sample would be added at $x = 0.758$, but that would not change the global optimum.

### 10.2.3 Adaptive Target for Probability of Improvement

As shown in the previous two examples, the target value plays a critically important role in the EGO process with the PI. If the value of the global optimum is known, this process can find the global optimum point quickly. However, it is rare that users know the global optimum in advance. Therefore, this process does not have much benefit as it is. An important observation from Example 10.2 is that the PI decreases rapidly when the target value is unattainable. Therefore, when the maximum PI is below a threshold, it would be necessary to reduce the target value close to the present best sample. Therefore, with $y_{T_0} = y_{PBS} - 0.25|y_{PBS}|$ as the initial target value, the following adaptive logic can be introduced to adjust the target value based on the PI.

$$y_{T_{k+1}} = \begin{cases} y_{T_k} & \text{if } \max(\text{PI}) > 0.05 \\ \dfrac{y_{T_k} + y_{PBS}}{2} & \text{otherwise} \end{cases} \tag{10.7}$$

The threshold value 0.05 can be adjusted depending on how aggressive the process can be. If the threshold increases, the EGO tends to emphasize exploitation, while if the threshold decreases, more exploration is sought. At each EGO iteration, the updating process in Eq. (10.7) needs to be repeated until max(PI) > 0.05 before adding an additional sample.

**Example 10.4 Probability of Improvement with Adaptive Target**

Repeat the EGO process in Example 10.2 using the adaptive target strategy in Eq. (10.7).

**Solution:**

In the Matlab script in Example 10.2, those lines that calculate the maximum PI need to be replaced with the following adaptive target strategy:

```
ytarget=yPBS-0.25*abs(yPBS);
while 1
 ztarget=(ytarget-yp)./ysig;
 PI=cdf('Normal',ztarget,0,1);
 [PImax, I]=max(PI); xPImax=xp(I);
 if PImax<0.05, ytarget=(ytarget+yPBS)/2; else, break; end
end
```

The same optimization problem with Example 10.2 is solved with the adaptive target strategy. Table 10.4 summarizes the performance of the EGO process. The process is identical to that in Example 10.2 for the first three iterations. When adding the seventh sample, EGO figured out that the target value of −6.83 is unattainable, and thus, the adaptive target strategy is activated to reduce the target value to −6.15, with which the next sample location was changed to $x_{new} = 0.764$. As the EGO process approaches the global optimum, the adjusted target value converges to the global optimum, which is shown at the eighth sample. Figure 10.8 shows the nine samples and Kriging predictions after the EGO process converges. It turns out that there remains a significant amount of prediction uncertainty in the nonoptimal region, while prediction uncertainty almost disappears near the global optimum region. That means the EGO process does not improve the prediction

**Table 10.4** Performance of the iterative EGO process with adaptive target.

| $n_y$ | $x_{PBS}$ | $y_{PBS}$ | $x_{PI}$ | maxPI | $y_T$ | $e_{RMS}$ | $e_{av}$ | $e_{max}$ |
|---|---|---|---|---|---|---|---|---|
| 4 | 0.68 | −3.68 | 0.633 | 0.311 | −4.6 | 4.08 | 3.51 | 7.94 |
| 5 | 0.68 | −3.68 | 0.723 | 0.539 | −4.6 | 2.52 | 1.9 | 4.48 |
| 6 | 0.723 | −5.46 | 0.244 | 0.133 | −6.83 | 1.68 | 1.09 | 3.93 |
| 7 | 0.723 | −5.46 | 0.764 | 0.35 | −6.15 | 1.14 | 0.802 | 2.5 |
| 8 | 0.764 | −6 | 0.758 | 0.366 | −6.02 | 0.766 | 0.465 | 1.98 |
| 9 | 0.758 | −6.02 | 0.758 | 1.0 | −6.02 | 1.1 | 0.689 | 2.97 |

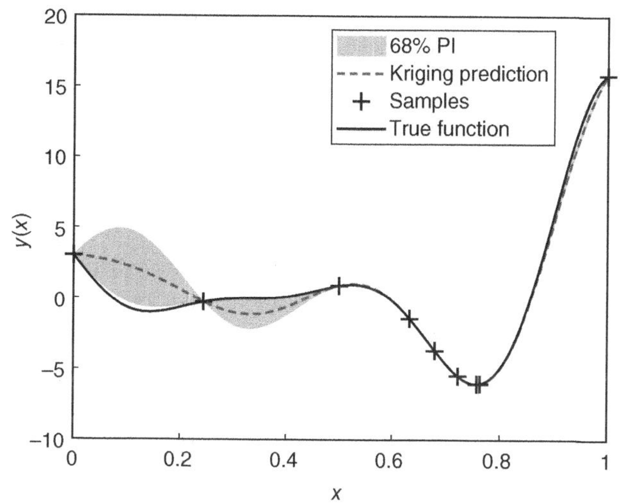


**Figure 10.8** Kriging surrogate after the EGO process converges using adaptive target strategy.

accuracy everywhere. This is confirmed by the three error metrics in Table 10.4. All three error metrics actually increase at the final iteration.

### 10.2.4 Expected Feasibility

So far, EGO criteria focus on improving an objective function over the design domain. However, as shown in Chapter 4, many engineering optimization problems include various constraints. Minimizing an objective function without considering constraints may not satisfy design requirements. Therefore, many engineering optimization problems require surrogate models for objective and constraint functions simultaneously. Although the standard form of optimization problems is defined in Section 4.3, we consider a simple optimization problem with an inequality constraint as

$$\begin{aligned} &\text{minimize}\, y(\mathbf{x}) \\ &\text{such that}\, g(\mathbf{x}) \le \bar{g} \end{aligned} \tag{10.8}$$

In order to solve the above optimization problem, we need one surrogate for the objective function $\hat{y}(\mathbf{x})$ and one surrogate for the constraint function $\hat{g}(\mathbf{x})$.

When we build two surrogate models, it makes sense to use the same sample locations because the objective and constraint functions might be evaluated using the same computer simulation. The difficulty is that the goal of the objective function is to find its minimum point, while the goal of the constraint function is to find its boundary. Therefore, the EI or the PI in the previous section are customized for the objective function, but not for the constraint function. In addition, when we evaluate the constraint, we do not mind having poor accuracy when the constraint is far away from its critical value. Instead, accurate prediction of the constraint boundary is important. The region to satisfy the constraint is called the feasible set, as defined in Chapter 4. The goal of constraint surrogate is to find the boundary of the feasible set accurately, rather than predicting the constraint function accurately.

In this section, a new criterion, expected feasibility (EF), is proposed for the EGO process with constraint surrogate. When the constraint surrogate is modeled using Kriging or PRS, it is considered a distribution, as with the objective surrogate. Since Kriging is a stochastic process, the feasibility at a point $\mathbf{x}$ is defined as

$$F(\mathbf{x}) = \max(\epsilon(\mathbf{x}) - |\overline{g} - G(\mathbf{x})|, 0) \tag{10.9}$$

where $G(\mathbf{x})$ is an uncertain variable of surrogate predictions. As discussed in Chapters 2 and 7, surrogate prediction is normally distributed as $G(\mathbf{x}) \sim N\left(\hat{g}(\mathbf{x}), s_g^2(\mathbf{x})\right)$, where $\hat{g}(\mathbf{x})$ is the mean prediction and $s_g^2(\mathbf{x})$ is the prediction variance of the constraint surrogate. In the definition of feasibility, $\epsilon(\mathbf{x})$ represents uncertainty in the surrogate. Since this uncertainty is proportional to prediction variance, for example, it is possible that $\epsilon(\mathbf{x}) = 2s_g(\mathbf{x})$. Therefore, the feasibility function is positive when the predicted constraint boundary is less than $\epsilon(\mathbf{x})$. This is similar to $\epsilon$-active constraint set in Eq. (5.36), but now it is not a constant. The feasibility function becomes zero when the constraint is inactive, i.e., $g(\mathbf{x}) \le \overline{g} - \epsilon(\mathbf{x})$, or violated too much, i.e., $g(\mathbf{x}) \ge \overline{g} + \epsilon(\mathbf{x})$. It has a positive value only when the constraint is within the following range: $\overline{g} - \epsilon(\mathbf{x}) \le g(\mathbf{x}) \le \overline{g} + \epsilon(\mathbf{x})$.

EF is the expectation of $F(\mathbf{x})$. It is the first moment of $Y(\mathbf{x})$ from the negative infinity to $y_{\text{PBS}}$. Mathematically, EI can be defined as

$$\text{EF}(\mathbf{x}) = \mathbb{E}[F(\mathbf{x})] = \int_{\overline{g}-\epsilon}^{\overline{g}+\epsilon} [\epsilon(\mathbf{x}) - |\overline{g} - g(\mathbf{x})|] f_G(g)dg \tag{10.10}$$

In order to make the calculation convenient, the uncertain variable is normalized using $Z = (G - \hat{g}(\mathbf{x}))/s_g(\mathbf{x})$. Then, the uncertain variable $Z$ follows the standard normal distribution, whose PDF is $\phi(z)$ and CDF is $\Phi(z)$. In the above equation, since $\epsilon(\mathbf{x})$ is independent of the uncertainty variable $G$, it can come out of the integral, and it becomes $\epsilon(\mathbf{x})[\Phi(\overline{z}^+) - \Phi(\overline{z}^-)]$, where $\overline{z}^{\pm} = (\overline{g} \pm \epsilon - \hat{g}(\mathbf{x}))/s_g$ is the normalized constraint range. Also, the normalized variable at the constraint boundary is defined as $\overline{z} = (\overline{g} - \hat{g}(\mathbf{x}))/s_g$. Similar to the case of EI, the remaining integrands can be split into two ranges after adding and subtracting $\hat{g}(\mathbf{x})$ as

$$\begin{aligned}\text{EF}(\mathbf{x}) = &\int_{\overline{g}}^{\overline{g}+\epsilon} [(\hat{g}(\mathbf{x}) - \overline{g}) + (g(\mathbf{x}) - \hat{g}(\mathbf{x}))] f_G(g)dg \\ &- \int_{\overline{g}-\epsilon}^{\overline{g}} [(\hat{g}(\mathbf{x}) - \overline{g}) + (g(\mathbf{x}) - \hat{g}(\mathbf{x}))] f_G(g)dg \\ &+ \epsilon(\mathbf{x})[\Phi(z^+) - \Phi(z^-)]\end{aligned}$$

It is noted that the term $(\hat{g}(\mathbf{x}) - \overline{g})$ can also come out of the integral. Therefore, the two integrals of $(\hat{g}(\mathbf{x}) - \overline{g})$ term become $(\hat{g}(\mathbf{x}) - \overline{g})[\Phi(\overline{z}) - \Phi(\overline{z}^+) + \Phi(\overline{z}) - \Phi(\overline{z}^-)]$. The remaining integrand can be normalized by $z = (g(\mathbf{x}) - \hat{g}(\mathbf{x}))/s_g(\mathbf{x})$. Using the formula, $\phi'(z) = -z\phi(z)$, the integrals of $(g(\mathbf{x}) - \hat{g}(\mathbf{x}))$ become $-s_g(\mathbf{x})[\phi(\bar{z}) - \phi(\bar{z}^+) + \phi(\bar{z}) - \phi(\bar{z}^-)]$. Finally, the EF can be written as

$$\begin{aligned} \mathrm{EF}(\mathbf{x}) &= (\hat{g}(\mathbf{x}) - \bar{g})[2\Phi(\bar{z}) - \Phi(\bar{z}^+) - \Phi(\bar{z}^-)] \\ &\quad - s_g(\mathbf{x})[2\phi(\bar{z}) - \phi(\bar{z}^+) - \phi(\bar{z}^-)] \\ &\quad + \epsilon(\mathbf{x})[\Phi(\bar{z}^+) - \Phi(\bar{z}^-)] \end{aligned} \tag{10.11}$$

The term $(\hat{g}(\mathbf{x}) - \bar{g})$ is the predicted difference of the constraint boundary at $\mathbf{x}$. This term is penalized by the probability within the constraint range. Therefore, this term becomes large when $\hat{g}(\mathbf{x})$ is close to $\bar{g}$, which promotes exploitation. On the other hand, the negative of the second term becomes large when the prediction uncertainty $s_g(\mathbf{x})$ is large, which promotes exploration. Then, the new sample location is determined by maximizing the EF as

$$\mathbf{x}_{\text{new}} = \arg\max_{\mathbf{x}} \mathrm{EF}(\mathbf{x}) \tag{10.12}$$

Different strategies can be implemented between EI and EF. If samples of $y(\mathbf{x})$ and $g(\mathbf{x})$ can be obtained from a single simulation, it would be beneficial to obtain them at the same sample location. In such a case, two samples can be added during the EGO process: one for maximizing EI and the other for maximizing EF. On the other hand, when $y(\mathbf{x})$ and $g(\mathbf{x})$ are obtained from different simulations, there is no reason to calculate them at the same sample location. In such a case, EI is used to determine the sample location of $y(\mathbf{x})$ and EF is used to determine that of $g(\mathbf{x})$.

## 10.3 Efficient Global Optimization Using Polynomial Response Surface

Although the concept of EGO can be applicable to any surrogate models, it has mostly been developed using Kriging surrogate in Chapter 7 partly because Kriging assumes that samples are accurate, and also Kriging provides prediction uncertainty. It is also possible to use a polynomial response surface (PRS) in Chapter 2 for the purpose of EGO because PRS also provides prediction uncertainty. However, it is important to consider that PRS predictions do not interpolate samples, and the accuracy of PRS depends on the model form. In this section, the Matlab function EGOPRS is introduced, which uses a PRS surrogate for the purpose of EGO. We will only consider the EGO process with EI. We leave the EGO process with the PI as an exercise problem (see Exercise Problem 10.1).

Although PRS is developed in Chapter 2, we summarize important equations here for the purpose of explaining the Matlab code. First, PRS predictions can be written as

$$\hat{y}(\mathbf{x}) = \xi(\mathbf{x})^T\mathbf{b} \tag{10.13}$$

where $\xi(\mathbf{x})$ is the vector of basis functions, and $\mathbf{b}$ is the vector of coefficients. The vector of coefficients is determined using the following regression equation:

$$(\mathbf{X}^T\mathbf{X})\mathbf{b} = \mathbf{X}^T\mathbf{y} \tag{10.14}$$

where $\mathbf{X}$ is the design matrix whose row is the vector of basis functions at a sample location, and $\mathbf{y}$ is the vector of output samples. The prediction standard deviation of PRS can be written as

$$s_y(\mathbf{x}) = \hat{\sigma}\sqrt{\xi(\mathbf{x})^T(\mathbf{X}^T\mathbf{X})^{-1}\xi(\mathbf{x})} \tag{10.15}$$

where the estimated noise standard deviation can be written as

$$\hat{\sigma} = \sqrt{\frac{(\mathbf{y}-\mathbf{Xb})^T(\mathbf{y}-\mathbf{Xb})}{n_y - n_\beta}} \tag{10.16}$$

where $n_y$ is the number of samples, and $n_\beta$ is the number of polynomial coefficients. The mean prediction in Eq. (10.13) and the prediction standard deviation in Eq. (10.15) are used to calculate the EI or the PI.

In the previous examples, the maximum of the EI or the PI is found based on a fine grid of prediction points. In a multidimensional surrogate, this is inefficient, if not impossible, as it requires numerous surrogate evaluations. Instead, optimization algorithms can be used to find the maximum. The EI and PI are multimodal functions, as shown in the previous examples. For multimodal functions, gradient-based optimization algorithms tend to find a local optimum near the initial point. Therefore, global search algorithms, such as genetic algorithms, are preferred to find the maximum point of EI. In literature, the differential evolution algorithm is popular for this purpose. We will use the genetic algorithm implemented in Matlab. However, interesting readers are encouraged to try different algorithms.

In order to call the objective function to generate a new sample, it is necessary to define the objective function in a separate file. For example, the following `EGOfn.m` needs to be defined first.

```
function f=EGOfn(x)
 f = sin(x);
end
```

Since Matlab can call many evaluation points simultaneously, it is necessary to vectorize `EGOfn` such that it can calculate a vector of output `f` with a vector of input `x`. This function is used to generate initial samples and additional samples during EGO iteration. If an output at the sample location is calculated using a computer simulation, this function needs to be modified to launch a simulation, such as finite element analysis or computational fluid dynamics, and extract the output and return it.

The following Matlab function `EGOPRS` performs EGO using a polynomial response surface. The Matlab code is composed of three functions: `EGOPRS`, `exImp`, and `basis`. These three functions can be saved in a single file `EGOPRS.m` or in three separate files. The `EGOPRS` function takes initial sample pairs (`xs`, `ys`), prediction points `xp`, the maximum number of samples `nmax`, the range of design space `range`, and the order of polynomials `p` as input parameters. Upon completing the EGO process, the function returns the updated sample pairs (`xs`, `ys`), surrogate predictions `yp` at prediction points, prediction standard deviation `ysig`, and the number of polynomial coefficients `nb`.

```
function [xs, ys, yp, ysig, nb]=EGOPRS(xs,ys,xp,nmax,range,p)
% INPUTS  (xs, ys): initial sample pairs
%          xp: np x NDV prediction points
%          nmax: maximum number of samples
%          range: range of design domain
%          p: order or PRS
% OUTPUTS (xs, ys): samples pairs included added samples
%          yp: np x 1 mean predictions
%          ysig: np x 1 prediction standard deviations
%          nb: # of coefficients in trend function
%
[ny,NDV]=size(xs);np=size(xp,1);                        % Size of samples
```

```
meanx=mean(xs);stdx=std(xs);xs=(xs-meanx)./stdx; % Scaling sample locations
meany=mean(ys); stdy=std(ys);ys=(ys-meany)/stdy;          % Scaling samples
xp=(xp-meanx)./stdx;                             % Scaling prediction points
range=(range-meanx)./stdx;                            % Scaling design space
nb=factorial(NDV+p)/(factorial(NDV)*factorial(p));    % No. of coefficients
%
fprintf(" ny   yPBS   maxEI   xNew    yNew\n")                % Table legend
for loop=ny:nmax                                                % EGO loop
 X=basis(xs,NDV,p);                                          % Design matrix
 b=(X'*X)\(X'*ys);                                             % Fitting PRS
 %
 % Optimization using EI to find a new sample location
 par=struct('b',b,'nb',nb,'p',p,'sigma2',sum((ys-X*b).^2)/(ny-nb),...
            'R',inv(X'*X),'yPBS',min(ys));      % Structure of database
 opts=optimset('Display','off','TolFun',1e-6,'DiffMinChange',0.001);
 xNew=ga(@(x) exImp(x,par),NDV,[],[],[],[],range(1,:),range(2,:),[],opts);
 % Stopping criteria
 if (loop==nmax||max(min(abs(xNew-xs)))<1E-3||exImp(xNew,par)>-1E-3)
 fprintf('%4d %8.3g\n',loop,par.yPBS.*stdy+meany); break;
 end
 %
 fprintf('%4d %8.3g %8.3g',loop,par.yPBS.*stdy+meany,-exImp(xNew,par));
 xs=[xs;xNew]; ny=ny+1;                            % Adding a new sample
 yNew=(EGOfn(xNew.*stdx+meanx)-meany)/stdy; ys=[ys;yNew];
 fprintf(' %8.3g',xNew.*stdx+meanx);fprintf(' %8.3g\n',yNew*stdy+meany);
end
xi=basis(xp,NDV,p);                     % Basis vector at prediction points
yp=(xi*b)*stdy+meany;                                  % PRS predictions
ysig=zeros(np,1);                                       % Prediction STD
for i=1:np, ysig(i)=sqrt(par.sigma2.*xi(i,:)*par.R*xi(i,:)')*stdy; end
xs=xs.*stdx+meanx;ys=ys*stdy+meany;        % Denormalizing samples
end
%
function EI=exImp(x,par)
 xi = basis(x,size(x,2),par.p);                           % Basis vector
 yhat=xi*par.b;                                        % PRS predictions
 ysig=sqrt(par.sigma2*xi*par.R*xi');                    % Prediction STD
 u  = (par.yPBS-yhat)./ysig;                            % Normalized PBS
 EI = -(par.yPBS-yhat).*normcdf(u,0,1)-ysig.*normpdf(u,0,1); % Negative EI
end
%
function X=basis(x, n, p)
% Calculating polynomial basis functions with n variables and p orders
 ny=size(x,1);
 X(:,1) = ones(ny,1);
 if p > 0
  X=[X x];
  X1=x;
  n_loc=1:n;
  n_loc1=1;
  for i=2:p
   nc=size(X1,2);
```

```
65      X2=[];
66      ctr=1;
67      for k=1:n
68        l_ctr=0;
69        for j=n_loc(k):nc
70          X2(:,ctr)=x(:,k).*X1(:,j);
71          ctr=ctr+1;
72          l_ctr=l_ctr+1;
73        end
74        n_loc1(k+1)=l_ctr+n_loc1(k);
75      end
76      X=[X X2];
77      X1=X2;
78      n_loc=n_loc1;
79    end
80  end
81 end
```

Function `EGOPRS`: The initial number of samples and the number of design variables are determined using the size of input samples: `ny` samples with `NDV` variables (Line 12). In order to make the optimization process standardized, all samples are normalized using their mean and standard deviation, including the lower and upper bounds of the design domain (Lines 13~16). Line 17 calculates the number of polynomial coefficients using the order of the polynomial and the number of design variables. EGO iteration is in the `for` loop between Lines 20 and 44. First, the PRS surrogate is fitted using the design matrix in Line 21 and calculating the regression coefficients in Line 22. In order to calculate the EI, the fitting parameters need to be stored in the parameter structure `par` in Lines 25 and 26. The genetic optimization algorithm is used to find the sample location that maximizes the EI in Line 28. During optimization, the `exImp` function is used to calculate the EI at different points. Upon optimization, variable `xNew` stores the new sample location that maximizes the EI. Line 30 determines if the EGO loop continues or stops. Three stopping criteria are used. First, the EGO iteration stops when the total number of samples becomes `nmax`. If the new sample location is too close to the old sample location, it means that optimization has converged, and the loop stops. Also, as we discussed in Example 10.1, if the value of EI is too small, it indicates that there is not much probability to improve, and therefore, the algorithm converges. The EGO iteration repeats until any of these three criteria is satisfied. If the EGO process does not converge, additional iteration is conducted by adding `xNew` as a new sample (Line 35), and `EGOfn` is used to evaluate the sample value at the new point (Line 36). Then, the process repeats.

Once the EGO process is converged, the prediction points `xp` are used to evaluate surrogate predictions. First, the basis vectors are evaluated at prediction points (Line 39), and then, they are multiplied by coefficients to calculate the surrogate predictions (Line 40). Also, the prediction standard deviation is calculated using Eq. (10.15) (Line 42). It is important to denormalize all samples and predictions (Lines 40, 42, and 43).

Function `exImp`: Function `exImp` calculates the EI at a given point `x` using the parameter structure `par`. First, it calculates the basis function at `x` (Line 47) and then calculates the prediction (Line 48) and the prediction standard deviation (Line 49). With these two quantities, the normalized present best sample is calculated at Line 50, and the negative EI at Line 51. Since the optimization algorithm minimizes the objective function, the negative EI is minimized instead of maximizing the EI.

Function `basis`: It is straightforward to define polynomial basis functions when the design problem has a single variable. However, when the number of variables and the order of polynomials are arbitrary, it is not straightforward to determine the monomial basis functions. Function `basis` calculates the number of terms for the basis functions and calculates the basis functions at given point x. In the case of three-dimensional input variables, for example, the orders of monomials are $b_0$, $b_1x_1$, $b_2x_2$, $b_3x_3$, $b_4x_1^2$, $b_5x_1x_2$, $b_6x_1x_3$, $b_7x_2^2$, $b_8x_2x_3$, $b_9x_3^3$, ....

**Example 10.5 EGO with PRS for 1D Function**

Using EGOPRS, find the global minimum of function $y(x) = (6x-2)^2 \sin(12x-4)$ in the range of $x \in [0, 1]$ using EGO with quartic PRS. Start with five initial samples $\mathbf{x}_s = [0.0\ 0.25\ 0.5\ 0.68\ 1.0]$. Show the PRS predictions and 95% prediction intervals along with samples at each EGO iteration.

**Solution:**

In order to use EGOPRS, the following Matlab script is used. As explained previously, the EGOfn function should return $y(x) = (6x-2)^2 \sin(12x-4)$. The following Matlab script is used to use EGOPRS to find the global optimum.

```
rng('default');                         % Initializing random number generator
xs=[0 0.25 0.5 0.68 1]';ys=EGOfn(xs);                % Initial sample pairs
N=100;xp=linspace(0,1,N)';                              % Prediction points
[xs, ys, yp, ysig, nb]=EGOPRS(xs,ys,xp,20,[0; 1],4);        % Calling EGO
ny=size(ys,1);                                % Number of updated samples
cilb = yp + tinv(0.025,ny-nb)*ysig;  % Lower-bounds of prediction intervals
ciub = yp + tinv(0.975,ny-nb)*ysig;  % Upper-bounds of prediction intervals
ytrue=EGOfn(xp);                % True function values at prediction points
figure; hold on;                                        % Plotting results
xShaded=[xp;sort(xp,'descend')]; yShaded=[cilb;ciub(end:-1:1)];
patch(xShaded,yShaded,[.24 .73 .89],'EdgeColor','none');
plot(xp,yp,'r-',xs,ys,'k+',xp,ytrue,'k-','Markersize',10,'Linewidth',1.5)
xlabel('x');ylabel('y(x)');
legend('95% PI','Kriging prediction','Samples','True function');
[yPBS,I]=min(ys);
fprintf('Opt at x=');fprintf('%8.3g',xs(I,:));fprintf(',
yPBS=%8.3g\n',yPBS);
eRMS=sqrt(sum((ytrue-yp).^2)/N);               % RMS error at NxN grid points
eav=sum(abs(ytrue-yp))/N;                  % Average error at NxN grid points
emax=max(abs(ytrue-yp));                   % Maximum error at NxN grid points
fprintf('RMS error:%8.3g\nAve error:%8.3g\nMax
error:%8.3g\n',eRMS,eav,emax);
```

First, the random number generator is initialized (Line 1). Random numbers are used for the genetic optimization algorithm. Therefore, this is necessary to keep the results the same. The EGOfn function is used to generate initial samples (Line 2). A total of 100 equally spaced grid points is used for prediction points (Line 3). EGOPRS is called with 20 maximum samples and quartic PRS (Line 4). EGOPRS converged with nine samples. Therefore, four additional samples are added during EGO iteration.

Table 10.5 show the results of EGO iterations. When $n_y = 5$, the prediction uncertainty cannot be calculated because the number of coefficients of a quartic PRS is $n_\beta = 5$. For this example, EGO iteration determines to add samples for the purpose of exploitation. Knowing the global optimum point is $x_{opt} = 0.757$, all additional samples are placed near the optimum point. The EGO process concludes that the global optimum is at $x = 0.77$ with $y = -5.93$, which is slightly different from the true global optimum of $y_{opt} = -6.02$. This is because of a model-form error. A more complicated

**Table 10.5** Performance of the iterative EGO process with PRS.

| $n_y$ | $x_{PBS}$ | $y_{PBS}$ | $x_{EI}$ | $y_{EI}$ | maxEI | $e_{RMS}$ | $e_{av}$ | $e_{max}$ |
|---|---|---|---|---|---|---|---|---|
| 5 | 0.68 | −3.68 | 0.693 | −4.30 | — | 2.01 | 1.32 | 6.19 |
| 6 | 0.693 | −4.30 | 0.77 | −5.93 | 0.121 | 2.08 | 1.36 | 6.38 |
| 7 | 0.77 | −5.93 | 0.78 | −5.74 | 0.00317 | 2.26 | 1.49 | 6.82 |
| 8 | 0.77 | −5.93 | 0.779 | −5.76 | 0.00101 | 2.28 | 1.5 | 6.86 |
| 9 | 0.77 | −5.93 | — | — | — | 2.29 | 1.61 | 6.88 |

form of a polynomial model is required to accurately model the behavior of the true function. It is noted that the maximum EI gradually decreases as more samples are added. However, it is interesting to note that the three error metrics are not improved with more samples. This is because EGO does not improve the overall performance of the surrogate. Instead, it focuses on near the global optimum point.

Figure 10.9 shows predictions and prediction intervals during EGO iteration. As the number of samples increases, the overall prediction uncertainty is reduced, but the errors at $0 < x < 0.5$ and $0.8 < x < 1.0$ are not improved.

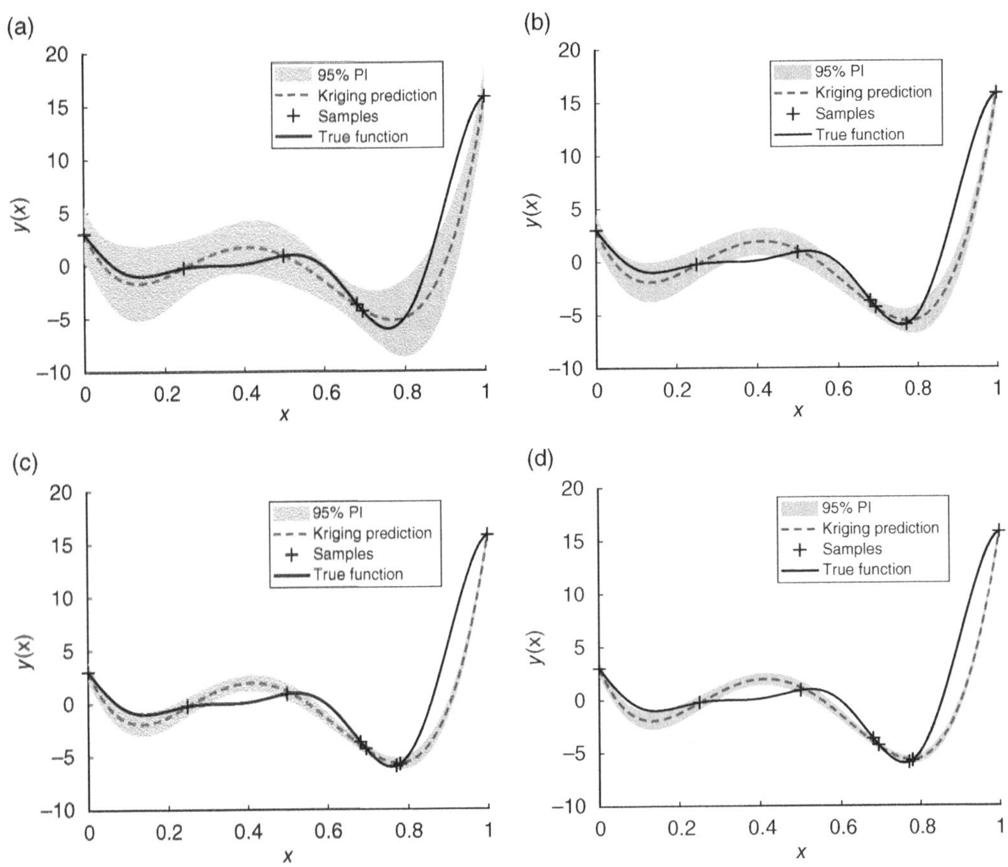


**Figure 10.9** PRS predictions and prediction intervals during EGO iteration. (a) $n_y = 6$, (b) $n_y = 7$, (c) $n_y = 8$, and (d) $n_y = 9$.

**Example 10.6 EGO with PRS for 2D Function**

Using EGOPRS, find the global minimum of function $y(x_1, x_2) = \sin(x_1)\cos(x_2)$ in the range of $x_1, x_2 \in [0, 2\pi]$ using EGO. Use cubic polynomials for PRS and maximum 40 samples. Start with 20 initial LHS samples and evaluate the accuracy of the surrogate using $21 \times 21$ equally spaced grid points.

**Solution:**

The function has three global minima at $(x_1, x_2) = (\pi/2, \pi), (3\pi/2, 0), (3\pi/2, 2\pi)$. The EGO algorithm will not find all three global minima but one. The choice of which global minimum it will find depends on the initial sample locations, surrogate model-form, and prediction uncertainty.

In order to call the objective function to generate a new sample, it is necessary to define the objective function as a separate file. For example, the following EGOfn.m needs to be defined first:

```
function f=EGOfn(x)
 f = sin(x(:,1)).*cos(x(:,2));
end
```

Then, the following Matlab script can be used to perform EGO process.

```
rng('default'); N=21; dN=2*pi/(N-1); ny=20;
xs=2*pi*lhsdesign(ny,2,'iteration',5000);                  % 20 LHS samples
ys=EGOfn(xs);                                     % Samples from true function
[XX,YY]=meshgrid(-0:dN:2*pi, 0:dN:2*pi);           % NxN grid of predictions
xp=[reshape(XX,N^2,1) reshape(YY,N^2,1)];                % Prediction points
figure(1); scatter(xs(:,1),xs(:,2),'filled');hold on  % Plot sample locations
axis([0 2*pi 0 2*pi]); xlabel('x_1');ylabel('x_2');
[xs, ys, yp, ysig, nb]=EGOPRS(xs,ys,xp,40,[0 0; 2*pi 2*pi],3);
YP=reshape(yp,N,N);                                    % Predictions at grid
YT=reshape(EGOfn(xp),N,N);                           % True function at grid
scatter(xs(ny+1:end,1),xs(ny+1:end,2),'d');       % Plot new sample locations
axis([0 2*pi 0 2*pi]); xlabel('x_1');ylabel('x_2');
figure(2); hold on; surf(XX,YY,YP);                    % Plot of predictions
plot3(xs(:,1),xs(:,2),ys,'ro');                            % Plot of samples
axis([0 2*pi 0 2*pi]); xlabel('x_1');ylabel('x_2');zlabel('y(x)');
figure(3); hold on; surf(XX,YY,YT)                   % PLot of true function
plot3(xs(:,1),xs(:,2),ys,'ro');                            % Plot of samples
axis([0 2*pi 0 2*pi]); xlabel('x_1');ylabel('x_2');zlabel('y(x)');
[yPBS,I]=min(ys);
fprintf('Opt at x=');fprintf('%8.3g',xs(I,:));fprintf(', yPBS=%8.3g\n',yPBS);
eRMS=sqrt(sum(sum((YT-YP).^2))/N^2);           % RMS error at NxN grid points
eav=sum(sum(abs(YT-YP)))/N^2;              % Average error at NxN grid points
emax=max(max(abs(YT-YP)));                 % Maximum error at NxN grid points
fprintf('RMS error:%8.3g\nAve error:%8.3g\nMax
error:%8.3g\n',eRMS,eav,emax);
```

The initial sample locations are determined using Latin hypercube sampling (Line 2). Since lhsdesign function in Matlab generates LHS samples in the range of [0, 1], it is scaled to $[0, 2\pi]$. Then, EGOfn is called with the initial sample locations to calculate output samples (Line 3). The $21 \times 21$ prediction points are reshaped to have a column vector (Line 5), and the initial sample locations are plotted (Line 6) with filled circle markers.

The EGO process is called with maximum 40 samples and cubic PRS (Line 8). Once the EGO process is converged, the added samples during the EGO iteration are plotted with diamond-shaped markers. The predictions and the true functions are plotted at $21 \times 21$ grid points. Finally, the three error metrics are calculated using these grid points.

The EGO iteration stopped after adding 10 more samples. Table 10.6 shows the present best sample, maximum EI and sample location during EGO iteration. Figure 10.10(a) shows the initial 20 samples in filled circular markers and the added 10 samples in diamond-shaped markers. Among 10 added samples, 4 of them are used to reduce uncertainty on the boundary, which belongs to

**Table 10.6** Performance of the iterative EGO process with PRS for 2D example.

| $n_y$ | $y_{PBS}$ | maxEI | $x_{1new}$ | $x_{2new}$ | $y_{new}$ |
|---|---|---|---|---|---|
| 20 | −0.922 | 4.65 | 6.28 | 6.28 | 0.0 |
| 21 | −0.922 | 0.655 | 0.0 | 2.66 | 0.0 |
| 22 | −0.922 | 1.29 | 6.28 | 0.0 | 0.0 |
| 23 | −0.922 | 0.233 | 4.47 | 0.0 | −0.97 |
| 24 | −0.97 | 0.174 | 4.52 | 0.0 | −0.982 |
| 25 | −0.982 | 0.142 | 4.55 | 0.0 | −0.987 |
| 26 | −0.987 | 0.0427 | 0.0 | 4.16 | 0.0 |
| 27 | −0.987 | 0.0969 | 4.51 | 0.0 | −0.98 |
| 28 | −0.987 | 0.0877 | 4.53 | 0.0 | −0.983 |
| 29 | −0.987 | 0.0804 | 4.54 | 0.0 | −0.985 |
| 30 | −0.987 | — | — | — | — |

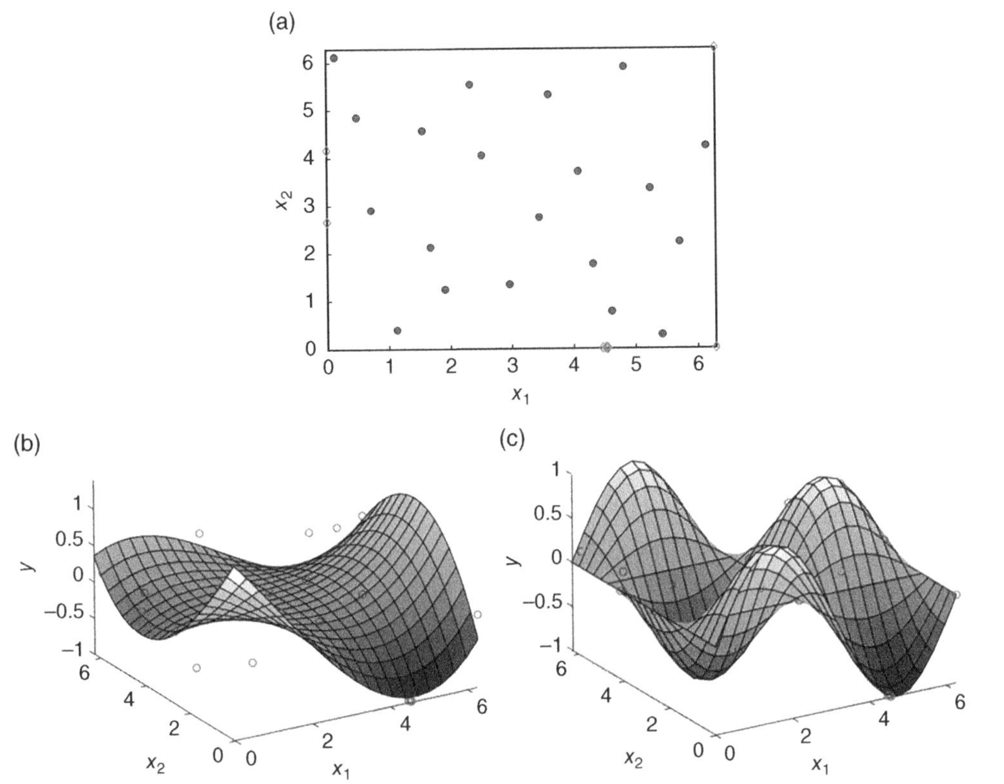


**Figure 10.10** PRS surrogate during EGO iteration. (a) Sample locations, (b) PRS predictions, and (c) true function.

the extrapolation region. Then, the remaining six added samples are placed near the global optimum $(3\pi/2, 0)$, which is exploitation. The PRS identifies the global optimum at (4.55,0) with the optimum value of −0.987. The algorithm cannot locate the true global optimum, which is $x_{opt} = (4.71,0)$ with the optimum value of $y_{opt} = -1.0$. Same as the previous example, this is due to the model-form error of a cubic PRS. A more interesting observation is the surrogate predictions shown in Figure 10.10(b). Compared to the plot of the true function in Figure 10.10(c), the PRS predictions have a significant error. Since EGO process finds the optimum point using samples, however, the optimum point has less than 1.3% error. This is a great performance considering that the root-mean-squared error of PRS surrogate is 0.45, which is another important advantage of EGO process.

## 10.4 Efficient Global Optimization Using Kriging Surrogate

In this section, a Matlab function EGO is presented, which performs EGO iterations to find the global optimum design using Kriging surrogate. The required functions are EGO, exImp, fitKRG, and trend. We only present EGO and exImp in this section, and fitKRG and trend functions are available in Chapter 7. All functions are combined into a single file EGO.m, or they can be separated and located in the Matlab path, including the current folder. The Matlab function EGO can support a multidimensional design space.

For the input parameters, the function EGO takes initial sample pairs (xs, ys). The matrix of input sample locations xs is $n_y \times n$ dimensions with $n_y$ being the number of samples and $n$ being the dimension of input variables. Similarly, the output samples ys is a column vector $(n_y \times 1)$. Upon converging EGO iterations, the function calculates Kriging predictions at prediction points xp. The order of polynomials of the trend function is stored in the trPRS variable, and the maximum number of samples is stored in nmax. The range matrix stores the lower and upper bounds of the design domain. When samples include random noise, use noise = 1. Otherwise, use noise = 0.

Once EGO iterations are converged, the function returns the updated sample pairs (xs, ys), Kriging predictions yp at prediction point xp, prediction standard deviation ysig, and the number of coefficients of the trend function nb.

```
function [xs, ys, yp, ysig, nb]=EGO(xs,ys,xp,trPRS,nmax,range,noise)
% INPUTS  xs: ny x NDV sample points
%         ys: ny x 1 samples
%         xp: np x NDV prediction points
%         trPRS: polynomial order of trend function (0, 1, 2)
%         nmax: maximum number of samples
%         range: range of design domain
%         noise: 1 for noisy samples, 0 for no noise
% OUTPUTS (xs, ys): samples pairs included added samples
%         yp: np x 1 mean predictions
%         ysig: np x 1 prediction standard deviations
%         nb: # of coefficients in trend function
%
[ny,NDV]=size(xs);nvar=NDV+1;np=size(xp,1);                  % Size of samples
meanx=mean(xs);stdx=std(xs);xs=(xs-meanx)./stdx;     % Scaling sample locations
meany=mean(ys); stdy=std(ys);ys=(ys-meany)/stdy;              % Scaling samples
xp=(xp-meanx)./stdx;                                % Scaling prediction points
range=(range-meanx)./stdx;                              % Scaling design space
%
fprintf("   ny      yPBS     Theta      maxEI      xNew      yNew\n")
```

```
for loop=ny:nmax                                                    % EGO loop
 nij=ny*(ny-1)/2; l=0;                   % # of upper-triangular component of R
 diff=zeros(nij,NDV);                             % Distance between samples
 ij=zeros(nij,2);                               % Index of correlation matrix
 for i=1:ny-1
  l=l(end)+(1:ny-i);
  ij(l,:)=[repmat(i,ny-i,1),(i+1:ny)'];
  diff(l,:)=repmat(xs(i,:),ny-i,1)-xs(i+1:ny,:);
 end
 %
 % Estimating hyper-parameters using max. likelihood estimate
 par=struct('xs',xs,'ys',ys,'d',diff,'ij',ij,'trPRS',min(trPRS,2));
 hyp0= [ny^(-1/4)*ones(NDV,1); min(0.1,noise)];                 % Initial value
 lb=[0.01*ones(NDV,1);0];ub=[6*ones(NDV,1);noise];       % Lower- & upper-bound
 opts=optimset('Display','off','TolFun',1e-6,'DiffMinChange',0.001);
 %hypOpt=fmincon(@(hyp) fitKRG(hyp,par),hyp0,[],[],[],[],lb,ub,[],opts);
 hypOpt = ga(@(hyp) fitKRG(hyp,par),nvar,[],[],[],[],lb,ub,[],opts)';
 %hypOpt = simulannealbnd(@(hyp) fitKRG(hyp,par),hyp0,lb,ub,opts);
 hypmle=hypOpt(1:size(xs,2)); tau=max(0,hypOpt(end));        % Opt. hypereparam.
 %
  par.hyp=hypmle; par.yPBS=min(ys);     % Add hyperparameter & PBS to database
 [~,fit]=fitKRG(hypOpt,par);                           % Fitting Kriging surrogate
 %
 % Optimization using EI to find a new sample location
 xNew=ga(@(x) exImp(x,fit,par),NDV,[],[],[],[],range(1,:),range(2,:),[],opts);
 % Stopping criteria
 if (loop==nmax||max(min(abs(xNew-xs)))<1E-3||exImp(xNew,fit,par)>-1E-3)
  fprintf('%4d %8.3g\n',loop,par.yPBS.*stdy+meany); break;
 end
 %
 fprintf('%4d %8.3g',loop,par.yPBS.*stdy+meany);
 fprintf(" %8.3g %8.3g",hypmle,-exImp(xNew,fit,par));
 xs=[xs;xNew]; ny=ny+1;                                    % Adding a new sample
 yNew=(EGOfn(xNew.*stdx+meanx)-meany)/stdy; ys=[ys;yNew];
 fprintf(' %8.3g',xNew.*stdx+meanx); fprintf(' %8.3g\n',yNew*stdy+meany);
 %
end
xi = trend(xp,trPRS); nb=size(xi,2);            % Trend basis vector at np points
xrep=repmat(xp',ny,1); xrep=transpose(reshape(xrep,NDV,np*ny));
d=xrep-repmat(xs,np,1);                     % Distance b/w predictions & samples
 r = (1-tau)*exp(sum(-(d./hypmle').^2,2));     % Correlation b/w preds & samples
 r = reshape(r,ny,np);
yp=(xi*fit.beta + (fit.alpha*r)')*stdy+meany;                % Kriging predictions
%
rt=fit.C\r;
q=fit.G\(xi'-fit.Ft'*rt);
sigma2=fit.sigma2.*stdy.^2;                                     % Process variance
ysig=sqrt(sigma2.*(1-sum(rt.^2,1)+sum(q.^2,1))');                 % Prediction STD
 xs=xs.*stdx+meanx;ys=ys*stdy+meany;                       % Denormalizing samples
end
%
function EI=exImp(x,fit,par)
 xi = trend(x,par.trPRS);                 % Trend basis vector at prediction points
 d=x-par.xs;                                % Distance b/w predictions & samples
```

```
75  r = exp(sum(-(d./par.hyp').^2,2));             % Correlation b/w preds & samples
76  yhat=xi*fit.beta + (fit.alpha*r)';                     % Kriging predictions
77  %
78  rt=fit.C\r;
79  q=fit.G\(xi'-fit.Ft'*rt);
80  sigma2=fit.sigma2;                                        % Process variance
81  ysig=sqrt(sigma2.*(1-sum(rt.^2,1)+sum(q.^2,1))');          % Prediction STD
82  u = (par.yPBS-yhat)./ysig;              % Calculating negative expected improvement
83  EI  =  -(par.yPBS-yhat).*normcdf(u,0,1)-ysig.*normpdf(u,0,1);
84 end
```

Function EGO: Function EGO performs EGO iterations using Kriging surrogate based on the EI. The function requires fitKRG and trend functions from Chapter 7. The initial number of samples and the number of design variables are determined using the size of input samples: ny samples with NDV variables (Line 14). In order to make the optimization process standardized, all samples are normalized using their mean and standard deviation, including the lower and upper bounds of the design domain (Lines 15–18).

EGO iteration is conducted in the for loop between Lines 21 and 57. First, the distances between samples are calculated along with the indices in the correlation matrix in Lines 23–29. With parameter structure par (Line 32) and lower and upper bounds (Line 34), the hyperparameters of Kriging surrogate are found using optimization algorithms (Lines 33 –39). The optimization algorithm calls the fitKRG function that calculates the negative log-likelihood and finds the optimum hyperparameters that minimize it. Different optimization algorithms can be applied by removing the comment symbol %. Once the optimum hyperparameters are found, they are stored in the parameter structure par. The present best sample is also stored in the same structure (Line 41). With the optimum hyperparameters, the Kriging surrogate is fitted and stored in the Kriging structure fit (Line 42).

Once the Kriging surrogate model with the current samples is available, it is used to calculate the maximum of EI. This requires another optimization algorithm that can find the maximum location of EI (Line 45). As mentioned earlier, since the EI is a multimodal function, a global search algorithm, such as genetic optimization, is preferred to gradient-based algorithms. During optimization, the exImp function is used to calculate the EI at different points. Upon optimization, variable xNew stores the new sample location that maximizes the EI.

Lines 47–49 determine if the EGO loop continues or stops. Three stopping criteria are used. First, the EGO iteration stops when the total number of samples becomes nmax. If the new sample location is too close to the old sample location, it means that optimization has converged. Since the EI is zero at a sample location, not much improvement is expected at the new sample, and the loop stops. Also, as we discussed in Example 10.1, if the value of EI is small, it indicates that the algorithm converges. The EGO iteration repeats until any of these three criteria is satisfied. If the EGO process does not converge, additional iteration is conducted by adding xNew as a new sample (Line 53), and EGOfn is used to evaluate the sample value at the new point (Line 54). Then, the process repeats.

Once the EGO process is converged, the prediction points xp are used to evaluate surrogate predictions. First, the basis vectors of the trend function are evaluated at prediction points (Line 58), and then, the distances between the prediction points and sample locations are calculated in d (Lie 60), which is used to calculate the correlation between prediction points and samples (Line 61). The Kriging predictions are calculated in Line 63. In order to calculate the prediction standard deviation, the same process as with Kriging surrogate in Chapter 7 is performed in Lines 65–68. It is noted that the Kriging predictions and standard deviations are denormalized before returning to the main function (Lines 63, 67). In addition, all samples, including the initial samples and added samples, need to be denormalized as well (Line 69).

Function `exImp`: In the EGO process with PRS, the same name of function is used to calculate the negative of EI. This function calculates the negative EI using Kriging surrogate. It takes the evaluation point `x`, the fitted Kriging structure `fit`, and Kriging parameter structure `par` as inputs, and it returns the negative EI as an output (Line 72). For the given evaluation point `x`, the trend basis function is calculated first (Line 73). Using the distance between the evaluation point and samples (Line 74), the correlation vector is calculated (Line 75). Since the fitted Kriging parameters are stored in the `fit` structure, the correlation vector is used to calculate the Kriging prediction (Line 76). Similarly, the process variance (Line 80) is used to calculate the prediction standard deviation (Line 81). With Kriging prediction and standard deviation, the normalized $z_{PBS}$ is calculated at Line 82, and the negative EI is calculated at Line 83.

**Example 10.7 EGO with Kriging for 1D Example**

Find the global optimum of the function in Example 10.5 using EGO with Kriging surrogate. Instead of five samples, start with four samples: $\mathbf{x}_s = [0, 0.5, 0.68, 1.0]$.

**Solution:**

Most of the Matlab script in Example 10.5 remains the same. The only differences are the initial sample locations and calling `EGO` instead of `EGOPRS`. Lines 2 and 4 need to be replaced by

```
xs=[0 0.5 0.68 1]';ys=EGOfn(xs);                % Initial sample pairs
[xs, ys, yp, ysig, nb]=EGO(xs,ys,xp,0,20,[0; 1],0);
```

In EGO, we used a constant trend function, 20 maximum number of samples, and a noise-free sample option. Table 10.7 summarizes EGO iterations. With initial four samples, $y_{PBS} = -3.68$ at $x_{PBS} = 0.68$. With this initial surrogate, the EI has a maximum of 0.099 at $x_{new} = 0.246$. Therefore, the fifth sample is added at this location. During this exploration step, the function yields $y_{new} = -0.238$, which is much higher than the present best sample. Therefore, $y_{PBS} = -3.68$ remains the same. Figure 10.11(a) shows Kriging predictions with prediction intervals with five samples.

The Kriging surrogate with five samples estimates the maximum EI of 0.0755 at $x_{new} = 0.626$. This point is close to the present best sample location $x_{PBS} = 0.68$. Therefore, it is considered as exploitation to further improve $y_{PBS} = -3.68$. However, the actual function value $y_{new} = -1.11$ at this point turns out to be higher than the present best sample, but it significantly reduces uncertainty near the present best sample, as shown in Figure 10.11(b). Again, $y_{PBS} = -3.68$ remains the same at this iteration.

After adding the sixth sample, the Kriging surrogate estimates the maximum EI of 0.138 at $x_{new} = 0.743$. This point is close to the present best sample location $x_{PBS} = 0.68$. Therefore, it is considered as exploitation to further improve $y_{PBS}$. In fact, the actual function value at this point turns out to be $y_{new} = -5.92$, which is lower than the present best sample. Therefore, this new sample becomes the present best sample. At this point, prediction uncertainty near the global optimum almost disappears, as shown in Figure 10.11(c).

**Table 10.7** Performance of the iterative EGO process with Kriging.

| $n_y$ | $x_{PBS}$ | $y_{PBS}$ | $x_{EI}$ | $y_{EI}$ | maxEI | $e_{RMS}$ | $e_{av}$ | $e_{max}$ |
|---|---|---|---|---|---|---|---|---|
| 4 | 0.68 | −3.68 | 0.246 | −0.238 | 0.099 | 4.07 | 3.49 | 7.94 |
| 5 | 0.68 | −3.68 | 0.626 | −1.11 | 0.0755 | 3.04 | 2.15 | 7.6 |
| 6 | 0.68 | −3.68 | 0.743 | −5.92 | 0.138 | 1.86 | 1.31 | 4.54 |
| 7 | 0.743 | −5.92 | 0.761 | −6.01 | 0.0172 | 1.2 | 0.845 | 2.64 |
| 8 | 0.761 | −6.01 | — | — | — | 0.767 | 0.46 | 2.01 |

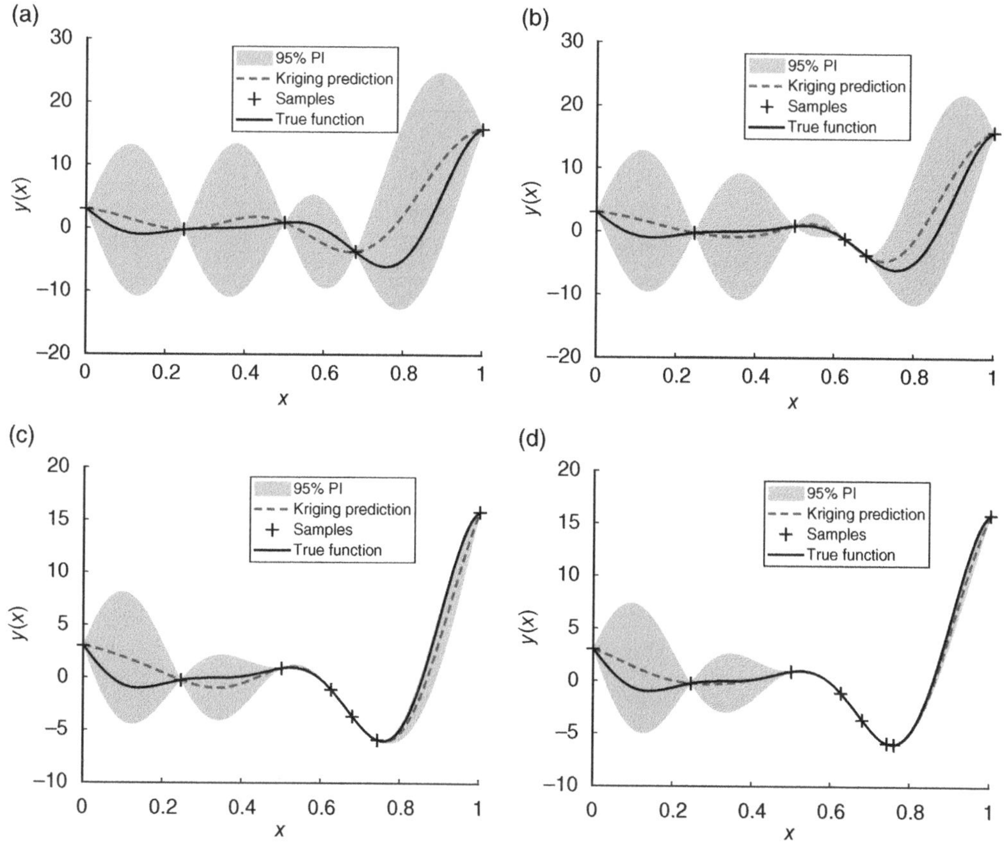


**Figure 10.11** Kriging predictions and prediction intervals during EGO iteration: (a) $n_y = 5$, (b) $n_y = 6$, (c) $n_y = 7$, and (d) $n_y = 8$.

After adding the seventh sample, the EGO process identifies the next sample location at $x_{\text{new}} = 0.761$, which is another exploitation step. This new point further reduces the present best sample to $y_{\text{PBS}} = -6.01$, and at this iteration, the EGO satisfies the stopping criterion. Figure 10.11 (d) shows the Kriging predictions and prediction intervals at the final iteration. It is noted that the uncertainty in the non-optimum region is still high, but added samples remove uncertainty near the global optimum.

**Example 10.8 EGO with Kriging for 2D Example**
Find the global optimum of the function in Example 10.6 using EGO with Kriging surrogate.

**Solution:**
Most of the Matlab script in Example 10.6 remains the same. The only difference is the command line calling `EGO` instead of `EGOPRS`. Line 8 needs to be replaced by

```
[xs, ys, yp, ysig, nb]=EGO(xs,ys,xp,0,40,[0 0; 2*pi 2*pi],0);
```

In EGO, we used a constant trend function, 40 maximum number of samples, and a noise-free sample option. Table 10.8 summarizes EGO iterations. The EGO process converged with

**Table 10.8** Performance of the iterative EGO process with Kriging for 2D example.

| $n_y$ | $y_{PBS}$ | maxEI | $x_{1new}$ | $x_{2new}$ | $y_{new}$ |
|---|---|---|---|---|---|
| 20 | −0.922 | 0.155 | 4.5 | 0.0 | −0.977 |
| 21 | −0.977 | 0.117 | 1.66 | 3.14 | −0.996 |
| 22 | −0.996 | 0.0836 | 4.58 | 8.28 | −0.999 |
| 23 | −0.999 | 0.0629 | 6.28 | 6.28 | 0.0 |
| 24 | −0.999 | 0.0075 | 4.76 | 0.0 | −0.999 |
| 25 | −0.999 | 0.00103 | 1.77 | 3.09 | −0.978 |
| 26 | −0.999 | 0.0019 | 4.7 | 0.0 | −1.0 |
| 27 | −1.0 | 0.00611 | 4.53 | 6.28 | −0.983 |
| 28 | −1.0 | — | — | — | — |

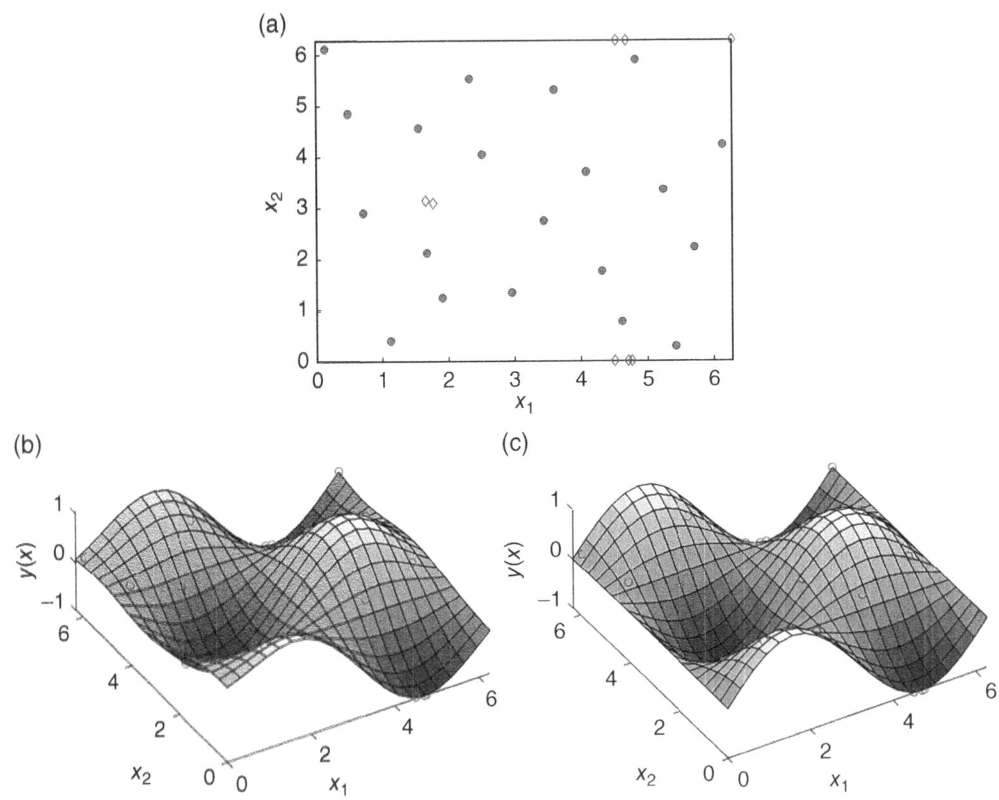


**Figure 10.12** Kriging surrogate during EGO iteration: (a) sample locations, (b) Kriging predictions, and (c) true function.

28 samples, which means eight additional samples. Among the eight, only a single sample is used for exploration at (6.28, 6.28). All other samples are used for exploitation. The true function has three global minima at $(x_1, x_2) = (\pi/2, \pi)$, $(3\pi/2, 0)$, $(3\pi/2, 2\pi)$. The EGO process added samples near the three global minima. As shown in Figure 10.12(a), two samples are added near $(\pi/2, \pi)$

and $(3\pi/2, 2\pi)$, while three samples are added near $(3\pi/2, 0)$. Since Kriging interpolates samples, it is efficient in terms of finding the global minimum, compared to the performance of the PRS surrogate in Example 10.6. This also confirms using the root-mean-squared error. In the case of Kriging, $e_{RMS} = 0.0802$ with 28 samples, while the PRS has $e_{RMS} = 0.45$ with 30 samples. The surrogate predictions shown in Figure 10.12(b) are much closer than the predictions from PRS, compared to the true function shown in Figure 10.12(c).

## Exercise

**10.1** The EGO process using PRS in Section 10.3 is based on maximizing EI. Modify `EGOPRS` function to use the PI with the adaptive target strategy. Test Examples 10.5 and 10.6 and compare the performance with the EGO process using EI.

**10.2** The Branin–Hu function is a popular two-dimensional analytical function that is used for testing surrogate models. The functional form is defined in $x_1 \in [-5, 10]$, and $x_2 \in [0, 15]$ as

$$f(x_1, x_2) = \left(x_2 - \frac{5.1}{4\pi^2}x_1^2 + \frac{5}{\pi}x_1 - 6\right)^2 + 10\left(1 - \frac{1}{8\pi}\right)\cos(x_1) + 10$$

Generate 10 initial samples using Latin hypercube sampling and perform the EGO process using Kriging with maximum 30 samples to find the global minimum. Evaluate the accuracy of the final Kriging surrogate against the true function at $51 \times 51$ uniform grids in terms of root-mean-square error, average error, and maximum error. Plot the Kriging surrogate and the true function.

**10.3** Repeat Problem 10.2 using EGO with PRS. Use a cubic polynomial to approximate the function using samples.

**10.4** Find the global minimum of the Rosenbrock function using EGO with PRS and compare it against the true global minimum. The Rosenbrock function is given as

$$y(x_1, x_2) = 100\left(x_1^2 - x_2\right)^2 + (1 - x_1)^2$$

The range of input data is given as $x_1 \in [-2, 2]$ and $x_2 \in [-1, 3]$. Generate 15 initial samples using Latin hypercube sampling and add additional samples up to 30 using EGO. Use a quartic polynomials to approximate the function.

**10.5** Find the global minimum of the Rosenbrock function in Problem 10.4 using EGO with Kriging and compare it against the true global minimum.

**10.6** Find the global minimum of the Sasena function using EGO with PRS and compare it against the true global minimum. The Sasena function is given as

$$y(x_1, x_2) = 2 + 0.01\left(x_2 - x_1^2\right)^2 + (1 - x_1)^2 + 2(2 - x_2)^2 + 7\sin(0.5x_1)\sin(0.7x_1x_2)$$

The range of input data is given as $x_1$, $x_2 \in [0.0, 5.0]$. Generate 15 initial samples using Latin hypercube sampling and add additional samples up to 30 using EGO. Use a quartic polynomial to approximate the function.

**10.7** Find the global minimum of the Sasena function in Problem 10.6 using EGO with Kriging and compare it against the true global minimum.

**10.8** Find the global minimum of the Hartmann3 function using EGO with Kriging and compare it against the true global minimum. The Hartmann3 function is given as

$$y(\mathbf{x}) = -\sum_{i=1}^{4} a_i \exp\left(-\sum_{j=1}^{3} B_{ij}\left(x_j - D_{ij}\right)^2\right)$$

where $\mathbf{a} = \{1.0, 1.2, 3.0, 3.2\}$ is the vector of coefficients, and the range of input data is given as $x_j \in [0.0, 1.0]$. Generate 20 initial samples using Latin hypercube sampling and add additional samples up to 40 using EGO. The matrices $\mathbf{B}$ and $\mathbf{D}$ are defined as

$$\mathbf{B} = \begin{bmatrix} 3 & 10 & 30 \\ 0.1 & 10 & 35 \\ 3 & 10 & 30 \\ 0.1 & 10 & 35 \end{bmatrix}, \quad \mathbf{D} = \begin{bmatrix} 0.3689 & 0.1170 & 0.2673 \\ 0.4699 & 0.4387 & 0.7470 \\ 0.1091 & 0.8732 & 0.5547 \\ 0.0382 & 0.5743 & 0.8828 \end{bmatrix}$$

**10.9** Find the global minimum of Himmelblau's function using EGO with Kriging. The true Himmelblau function is defined as

$$f(x_1, x_2) = \left(x_1^2 + x_2 - 11\right)^2 + \left(x_1 + x_2^2 - 7\right)^2$$

The function has one local maximum at $\mathbf{x} = (-0.270845, -0.923939)$ with $f_{\max} = 181.617$ and four identical local minima at $\mathbf{x} = (3.0, 2.0)$, $(-2.805118, 3.131312)$, $(-3.779310, -3.283186)$, $(3.584428, -1.848126)$. Use 10 initial samples from LHS in the domain $x_1, x_2 \in [-4, 4]$. Add maximum of 20 more samples to find the global optimum.

# References


1 Hughes, K.T., Balachandar, S., Diggs, A. et al. (2020). Simulation-driven design of experiments examining the large-scale, explosive dispersal of particles. *Shock Waves* 30 (4): 325–347.

2 Zhang, F., Frost, D.L., Thibault, P.A., and Murray, S.B. (2001). Explosive dispersal of solid particles. *Shock Waves* 10 (6): 431–443.

3 Myers, R.H. and Montgomery, D.C. (2002). *Response Surface Methodology: Process and Product Optimization Using Designed Experiments*, 2e. Wiley.

4 Blanning, R.W. (1975). The construction and implementation of metamodels. *Simulation* 24 (6): 177–184.

5 Willcox, K. and Peraire, J. (2002). Balanced model reduction via the proper orthogonal decomposition. *AIAA Journal* 40 (11): 2323–2330.

6 O'Hagan, A. (2006). Bayesian analysis of computer code outputs: a tutorial. *Reliability Engineering & System Safety* 91 (10): 1290–1300.

7 Bieker, H.P., Slupphaug, O., Johansen, T.A. et al. (2007). Real-time production optimization of oil and gas production systems: a technology survey. *SPE Production & Operations* 22 (4): 382–391.

8 Robinson, T., Eldred, M., Willcox, K., and Haimes, R. (2008). Surrogate-based optimization using multi-fidelity models with variable parameterization and corrected space mapping. *AIAA Journal* 46 (11): 2814–2822.

9 Goel, T., Haftka, R.T., Shyy, W., and Queipo, N.V. (2007). Ensemble of multiple surrogates. *Structural and Multidisciplinary Optimization* 33 (3): 199–216.

10 Galton, F. (1886). Regression towards mediocrity in hereditary stature. *The Journal of the Anthropological Institute of Great Britain and Ireland* 15: 246–263.

11 Freedman, D.A. (2009). *Statistical Models: Theory and Practice*, 26. Cambridge University Press.

12 Rencher, A.C. and Christensen, W.F. (2012). *Methods of Multivariate Analysis, Wiley Series in Probability and Statistics*, 3e, vol. 709, 19. Wiley. ISBN 9781118391679.

13 Trefethen, L.N. and Bau, D. III (1997). *Numerical Linear Algebra*. Philadelphia, PA: Society for Industrial and Applied Mathematics. ISBN 978-0-898713-61-9.

14 Kim, N.H., Wang, H., and Queipo, N.V. (2006). Adaptive reduction of design variables using global sensitivity in reliability-based optimization. *International Journal of Reliability and Safety* 1 ((1–2)): 102–119.

15 Gogu, C., Le Riche, R., Molimard, J., and Haftka, R.T. (2010). Effect of approximation fidelity on vibration-based elastic constants identification. *Structural and Multidisciplinary Optimization* 42: 293–304.

16 Hotelling, H. (1933). Analysis of a complex of statistical variables into principal components. *Journal of Educational Psychology* 24: 417, 498–441, 520.


*Surrogate Modeling and Optimization: Theories, Applications, and Limitations*, First Edition. Nam-Ho Kim.

Published 2026 by John Wiley & Sons, Inc.
Companion website: www.wiley.com/go/kim/surrogate_modeling

**17** Casella, G. (2002). *Statistical Inference*, (Second ed.).e, 556. Pacific Grove: Duxbury/Thomson Learning. ISBN 9788131503942.

**18** Cawley, G.C. and Talbot, N.L.C. (2010). On over-fitting in model selection and subsequent selection bias in performance evaluation. *Journal of Machine Learning Research* 11: 2079–2107.

**19** Grossman, R., Seni, G., Elder, J. et al. (2010). Ensemble methods in data mining: improving accuracy through combining predictions. In: *Synthesis Lectures on Data Mining and Knowledge Discovery*, vol. 2 (ed. J. Han, L. Getoor, and J. Gehrke), 1–126. Morgan & Claypool.

**20** Tarpey, T. (2000). A note on the prediction sum of squares statistic for restricted least squares. *The American Statistician* 54 (2): 116–118.

**21** Picheny, V., Wagner, T., and Ginsbourger, D. (2013). A benchmark of kriging-based infill criteria for noisy optimization. *Structural and Multidisciplinary Optimization* 48: 607–626.

**22** Fisher, R. (1971 [1935]). *The Design of Experiments*, 9e. Macmillan. ISBN 0-02-844690-9.

**23** Box, G.E.P. and Hunter, J.S. (1957). Multifactor experimental designs for exploring response surfaces. *The Annals of Mathematical Statistics* 28: 195–241.

**24** Box, G.E.P. and Behnken, D.W. (1960). Some new three-level designs for the study of quantitative variables. *Technometrics* 2: 455–475. https://doi.org/10.1080/00401706.1960.10489912.

**25** Atkinson, A.C., Donev, A.N., and Tobias, R.D. (2007). *Optimum Experimental Designs, with SAS*, 511 +xvi. Oxford University Press. ISBN 978-0-19-929660-6.

**26** Jones, B., Aleen-Moyer, K., and Goos, P. (2021). A-optimal versus D-optimal design of screening experiments. *Journal of Quality Technology* 53 (4): 369–382. https://doi.org/10.1080/00224065.2020.1757391.

**28** Johnson, M., Moore, L., and Ylvisaker, D. (1990). Minimax and maximin distance designs. *Journal of Statistical Planning and Inference* 26: 131–148.

**29** Iman, R. and Conover, W. (1982). A distribution-free approach to inducing rank correlation among input variables. *Communications in Statistics Part B: Simulation and Computation* 11: 311–334.

**27** Box, G.E.P. and Draper, N.R. (1959). A basis for the selection of a response surface design. *Journal of the American Statistical Association* 136: 3221–3230.

**30** Hedayat, A., Sloane, N., and Stufken, J. (1999). Orthogonal arrays: theory and applications. In: *Springer, Series in Statistics*. Berlin: Springer.

**31** McKay, M., Conover, W., and Beckman, R. (1979). A comparison of three methods for selecting values of input variables in the analysis of output from a computer code. *Technometrics* 21: 239–245.

**32** Ye, K. (1998). Orthogonal column latin hypercubes and their application in computer experiments. *Journal of the American Statistical Association* 93: 1430–1439.

**33** Palmer, K. and Tsui, K. (2001). A minimum bias latin hypercube design. *IIE Transactions* 33: 793–808.

**34** Leary, S., Bhaskar, A., and Keane, A. (2003). Optimal orthogonal arraybased latin hypercubes. *Journal of Applied Statistics* 30: 585–598.

**35** Box, G.E.P., Hunter, W.G., and Hunter, J.S. (1978). *Statistics for Experimenters: An Introduction to Design, Data Analysis, and Model Building*. Wiley.

**36** Owen, A.B. (1992). Orthogonal arrays for computer experiments, integration and visualization. *Statistica Sinica* 2 (2): 439–452. http://www.jstor.org/stable/24304869.

**37** Hedayat, A.S., Sloane, N.J.A., and Stufken, J. (1999). *Orthogonal Arrays: Theory and Applications*. New York: Springer-Verlag.

**38** Queipo, N.V., Haftka, R.T., Shyy, W. et al. (2005). Surrogate-based analysis and optimization. *Progress in Aerospace Sciences* 41: 1–28.

**39** Arora, J.S. (2017). *Introduction to Optimum Design*, 4e. Academic Press.

**40** Haftka, R.T. and Gurdal, Z. *Elements of Structural Optimization*. Springer Dordrecht.

41 Svanberg, K. (1987). The method of moving asymptotes – a new method for structural optimization. *International Journal for Numerical Methods in Engineering* 24: 359–373.

42 Protter, M.H. and Morrey, C.B. (1977). The boundedness and extreme–value theorems. In: *A First Course in Real Analysis* (ed. M.H. Protter and C.B. Murrey), 71–73. New York: Springer. ISBN 0-387-90215-5.

43 Kuhn, H. W. and Tucker, A. W. (1951). Nonlinear programming. *Proceedings of 2nd Berkeley Symposium*, 481–492. Berkeley: University of California Press (31 July–12 August 1950).

44 Karush, W. (1939). Minima of functions of several variables with inequalities as side constraints. M.Sc. thesis. Dept. of Mathematics, Univ. of Chicago, Chicago, IL.

45 Berke, L. and Khot, N.S. (1987). Structural optimization using optimality criteria. In: *Computer Aided Optimal Design: Structural and Mechanical Systems*, vol. 27 (ed. C.A.M. Soares), 271–311. Berlin, Heidelberg: Springer Berlin Heidelberg.

46 Patnaik, S. N., Guptill, J. D., and Berke, L. (1993). Merits and limitations of optimality criteria method for structural optimization. *NASA Technical Paper 3373*. National Aeronautics and Space Administration: Cleveland, OH.

47 Sigmund, O. (2016). A 99-line topology optimization code written Matlab. *Structural and Multidisciplinary Optimization* 21: 120–127.

48 Kim, N.H., Dong, T., Weinberg, D., and Dalidd, J. (2021). Generalized optimality criteria method for topology optimization. *Applied Sciences* 11: 3175.

49 Choi, K.K. and Kim, N.H. (2004). *Structural Sensitivity Analysis and Optimization I: Linear Systems*. NY: Springer.

50 Fletcher, R. (1976). Conjugate gradient methods for indefinite systems. In: *Numerical Analysis* (ed. G. Watson), 73–89. Berlin, Heidelberg: Springer. https://doi.org/10.1007/BFb0080116.

51 Fletcher, R. and Powell, M.J.D. (1963). A rapidly convergent descent method for minimization. *The Computer Journal* 6: 163–180.

52 Fletcher, R. and Reeves, R.M. (1964). Function minimization by conjugate gradients. *The Computer Journal* 7: 149–160.

53 Knuth, D. (1974). Postscript about NP-hard problems. *ACM SIGACT News* 6 (2): 15–16.

54 Wolpert, D.H. and Macready, W.G. (1997). No free lunch theorems for optimization. *IEEE Transactions on Evolutionary Computation* 1: 67–82.

55 Weise, T. (2009). Global optimization algorithms – theory and application. http://www.it-weise.de/projects/book.pdf (accessed 12 September 2025).

56 Spendley, W., Hext, G.R., and Himsworth, F.R. (1962). Sequential application of simplex designs in optimization and evolutionary operation. *Technometrics* 4 (4): 441–461.

57 Nelder, J.A. and Mead, R. (1965). A simplex method for function minimization. *The Computer Journal* 7: 308–313.

58 Lagarias, J.C., Reeds, J.A., Wright, M.H., and Wright, P.E. (1998). Convergence properties of the Nelder-Mead simplex method in low dimensions. *SIAM Journal on Optimization* 9 (1): 112–147.

59 Jones, D.R., Perttunen, C.D., and Stuckman, B.E. (1993). Lipschitzian optimization without the Lipschitz constant. *Journal of Optimization Theory and Applications* 79: 157–181.

60 Shubert, B.O. (1972). A sequential method seeking the global maximum of a function. *SIAM Journal on Numerical Analysis* 9 (3): 379–388.

61 Cox, S.E., Haftka, R.T., Baker, C.A. et al. (2001). A comparison of global optimization methods for the design of a high-speed civil transport. *Journal of Global Optimization* 21: 415–433.

62 Holland, J. (1992). *Adaptation in Natural and Artificial Systems*. Cambridge, MA: MIT Press. ISBN 978-0262581110.

63 Glover, F. and Kochenberger, G.A. (2003). *Handbook of Metaheuristics*, vol. 57. Springer, International Series in Operations Research & Management Science.

64 Kennedy, J. and Eberhart, R. (1995). Particle swarm optimization. *Proceedings of IEEE International Conference on Neural Networks* IV: 1942–1948.

65 Mezura-Montes, E. and Coello Coello, C.A. (2011). Constraint-handling in nature-inspired numerical optimization: past, present and future. *Swarm and Evolutionary Computation* 1: 173–194.

66 Pedersen, M.E. (2010). *Good Parameters for Particle Swarm Optimization*. Luxembourg: Hvass Laboratories.

67 Metropolis, N., Rosenbluth, A.W., Rosenbluth, M.N. et al. (1953). Equation of state calculations by fast computing machines. *The Journal of Chemical Physics* 21 (6): 1087–1092.

68 Kirkpatrick, S., Gelatt, C.D. Jr., and Vecchi, M.P. (1983). Optimization by simulated annealing. *Science* 220 (4598): 671–680.

69 Cerny, V. (1985). Thermodynamical approach to the traveling salesman problem: an efficient simulation algorithm. *Journal of Optimization Theory and Applications* 45: 41–52.

70 Rutenbar, R.A. (1989). Simulated annealing algorithms: an overview. *IEEE Circuits and Devices Magazine* 5: 19–26.

71 Johnson, D.S., Aragon, C.R., McGeoch, L.A., and Schevon, C. (1990). Optimization by simulated annealing: an experimental evaluation. Part I. Graph partitioning. *Operations Research* 37: 865–893.

72 Aarts, E. and Korst, J. (1989). *Simulated Annealing and Boltzmann Machines: A Stochastic Approach to Combinatorial Optimization and Neural Computing*. Wiley.

73 Nahar, S., Sahni, S., and Shragowithz, E. V. (1985). Experiments with Simulated Annealing. *Conference: 22*nd *ACM/IEEE Design Automation Conference*, Las Vegas, Nevada, USA (23–26 June 1985), 748–752.

74 Elperin, T. (1988). Monte Carlo structural optimization in discrete variables with annealing algorithm. *International Journal for Numerical Methods in Engineering* 26: 815–821.

75 Kincaid, R. K., and Padula, S. L., (1990). Minimizing distortion and internal forces in truss structures by simulated annealing, *Proceedings of the AIAA/ASME/ASCE/AHS/ASC 31st Structures, Structural Dynamics, and Materials Conference*, Long Beach, CA (02–04 April 1990). Part 1, 327–333.

76 Balling, R. J. and May, S. A. 1990. Large-scale discrete structural optimization: simulated annealing, branch-and-bound, and other techniques, *AIAA/ASME/ASCE/AHS/ASC 32nd Structures, Structural Dynamics, and Materials Conference*, Long Beach, CA.

77 Chen, G.-S., Bruno, R.J., and Salama, M. (1991). Optimal placement of active/passive members in structures using simulated annealing. *AIAA Journal* 29 (8): 1327–1334.

78 Krige, D.G. (1951). A statistical approach to some basic mine valuation problems on the Witwatersrand. *Journal of the Chemical, Metallurgical and Mining Society of South Africa* 52 (6): 119–139.

79 Matheron, G. (1963). Principles of geostatistics. *Economic Geology* 58: 1246–1266.

80 Breitkopf, P., Naceur, H., Rassineux, A., and Villon, P. Moving least squares response surface approximation: formulation and metal forming applications. *Computers and Structures* 83: 1411–1428.

81 Broomhead, D.S. and Lowe, D. (1988). Multivariable functional interpolation and adaptive networks. *Complex Systems* 2: 321–355.

82 Rasmussen, C.E. and Williams, C.K.I. (2006). *Gaussian Processes for Machine Learning*. Cambridge, MA: MIT Press.

83 Xiong, Y., Chen, W., Apley, D., and Ding, X. (2007). A non-stationary covariance-based Kriging method for metamodelling in engineering design. *International Journal for Numerical Methods in Engineering* 71 (6): 733–756.

84 Chen, J.S. and Kim, N.H. (2007). Meshfree method and application for shape optimization. In: *Optimization of Structural and Mechanical Systems* (ed. J.S. Arora), 389–414. World Scientific Publishing/Imperial College Press, Chapter 14.

85 Toal, D.J.J., Bressloff, N.W., and Keane, A.J. (2008). Kriging hyperparameter tuning strategies. *AIAA Journal* 46 (5): 1240–1252.

86 Lophaven, S.N., Nielsen, H.B., and Sondergaard, J. (2002). DACE – A Matlab Kriging Toolbox, *Technical report IMM-TR-2002-12, Technical University of Denmark.*

87 McCulloch, W.S. and Pitts, W. (1943). A logical calculus of the ideas immanent in nervous activity. *The Bulletin of Mathematical Biophysics* 5 (4): 115–133.

88 Hebb, D.O. (2002). *The Organization of Behavior: A Neuropsychological Theory*. Psychology Press.

89 Rosenblatt, F. (1958). The perceptron: a probabilistic model for information storage and organization in the brain. *Psychological Review* 65 (6): 386–408.

90 Minsky, M. and Papert, S. (1969). *Perceptrons: An Introduction to Computational Geometry*. MIT Press.

91 Hopfield, J.J. (1982). Neural networks and physical systems with emergent collective computational abilities. *Proceedings of the National Academy of Sciences* 79 (8): 2554–2558.

92 Rumelhart, D.E., Hinton, G.E., and Williams, R.J. (1986). Learning representations by back-propagating errors. *Nature* 323 (6088): 533–536.

93 Zell, A. (1994). *Simulation Neuronaler Netze [Simulation of Neural Networks]* (in German), 1e, 73. Addison-Wesley.

94 Billings, S.A. (2013). *Nonlinear System Identification: NARMAX Methods in the Time, Frequency, and Spatio-Temporal Domains*. Wiley.

95 Miljanovic, M. (2012). Comparative analysis of recurrent and finite impulse response neural networks in time series prediction. *Indian Journal of Computer and Engineering* 3 (1): 180–191.

96 Liu, P. and Li, H. (2004). *Fuzzy Neural Network Theory and Application*. Singapore: World Scientific.

97 He, Y., Tan, Y., and Sun, Y. (2004). Wavelet neural network approach for fault diagnosis of analogue circuits. *IEE Proceedings – Circuits, Devices and Systems* 151 (4): 379–384.

98 Bicciato, S., Pandin, M., Didonè, G., and Bello, C. D. 2001. Analysis of an associative memory neural network for pattern identification in gene expression data, Biokdd01, *Work-shop on Data Mining in Bioinformatics*, 22–30, San Francisco, CA (26 August 2001).

99 Happel, B.L.M. and Murre, J.M.J. (1994). The design and evolution of modular neural net-work architectures. *Neural Networks* 7: 985–1004.

100 Zhang, G. and Yuen, K.K.F. (2013). Toward a hybrid approach of primitive cognitive network process and particle swarm optimization neural network for forecasting. *Procedia Computer Science* 17: 441–448.

101 Rovithakis, G.A., Maniadakis, M., and Zervakis, M. (2004). A hybrid neural network/genetic algorithm approach to optimizing feature extraction for signal classification. *IEEE Transactions on Systems, Man, and Cybernetics, Part B: Cybernetics* 34 (1): 695–702.

102 Duch, W. and Jankowski, N. (1999). Survey of neural transfer functions. *Neural Computing Surveys* 2: 163–212.

103 Yu, H. and Wilamowski, B.M. (2011). Levenberg–Marquardt training. *Industrial Electronics Handbook* 5 (12): 1–16.

104 Beale, M.H., Hagan, M.T., and Demuth, H.B. (2015). *Neural Network Toolbox™ 8.4. User's Guide*. Natick: The MathWorks, Inc.

105 Orr, G. and Schraudolph, N. (2016). *Momentum and learning rate adaptation*. CS-449: Neural Networks.

106 Coppe, A., Haftka, R.T., Kim, N.H., and Yuan, F.G. (2010). Uncertainty reduction of damage growth properties using structural health monitoring. *Journal of Aircraft* 47 (6): 2030–2038.

**107** Fernández-Godino, M.G., Park, C., Kim, N.H., and Haftka, R.T. (2019). Issues in deciding whether to use multifidelity surrogates. *AIAA Journal* 57 (5): 2039–2054.

**108** Park, C., Haftka, R.T., and Kim, N.H. (2017). Remarks on multi-fidelity surrogates. *Structural and Multidisciplinary Optimization* 55 (3): 1029–1050.

**109** Bandler, J.W., Biernacki, R.M., Chen, S.H. et al. (1994). Space mapping technique for electromagnetic optimization. *IEEE Transactions on Microwave Theory and Techniques* 42 (12): 2536–2544.

**110** Park, C., Haftka, R.T., and Kim, N.H. (2018). Low-fidelity scale factor improves Bayesian multi-fidelity prediction by reducing bumpiness of discrepancy function. *Structural and Multidisciplinary Optimization* 58 (2): 399–414.

**111** Zhang, Y., Kim, N.H., Park, C., and Haftka, R.T. (2018). Multi-fidelity surrogate based on single linear regression. *AIAA Journal* 56 (12): 4944–4952.

**112** Choi, W., Radhakrishnan, K., Kim, N.H., and Park, J.S. (2021). Multi-fidelity surrogate models for predicting averaged heat transfer coefficients on endwall of turbine blades. *Energies* 14 (2): 482.

**113** Balabanov, V., Haftka, R. T., Grossman, B., Mason, W. H., and Watson, L. T. (1998). Multifidelity response surface model for HSCT wing bending material weight. *Proceedings of 7th AIAA/USAF/NASA/ISSMO Symposium on Multidisciplinary Analysis and Optimization*, St. Louis, MO, USA (02–04 September 1998).

**114** Kennedy, M.C. and O'Hagan, A. (2000). Predicting the output from a complex computer code when fast approximations are available. *Biometrika* 87 (1): 1–13.

**115** Jin, R., Chen, W., and Sudjianto, A. (2005). An efficient algorithm for constructing optimal design of computer experiments. *Journal of Statistical Planning and Inference* 134 (1): 268–287.

**116** Morris, M.D., Mitchell, T.J., and Ylvisaker, D. (1993). Bayesian design and analysis of computer experiments: use of derivatives in surface prediction. *Technometrics* 35 (3): 243–255.

**117** Xiong, S., Qian, P.Z., and Wu, C.J. (2013). Sequential design and analysis of high-accuracy and low-accuracy computer codes. *Technometrics* 55 (1): 37–46.

**118** Zhang, Y., Schutte, J.F., Seneviratne, W.P. et al. (2017). Sampling by exploration and replication for estimating experimental strength of composite structures. *AIAA Journal* 55 (10): 3594–3602.

**119** Guo, Z., Song, L., Park, C. et al. (2018). Analysis of dataset selection for multi-fidelity surrogates for a turbine problem. *Structural and Multidisciplinary Optimization* 57 (6): 2127–2142.

**120** Jones, D., Schonlau, M., and Welch, W. (1998). Efficient global optimization of expensive black-box functions. *Journal of Global Optimization* 13 (4): 455–492.

**121** Huang, D., Allen, T.T., Notz, W.I., and Zeng, N. (2006). Global optimization of stochastic black-box systems via sequential kriging meta-models. *Journal of Global Optimization* 34 (3): 441–466.

# Index



*Surrogate Modeling and Optimization: Theories, Applications, and Limitations*, First Edition. Nam-Ho Kim.

Published 2026 by John Wiley & Sons, Inc.
Companion website: www.wiley.com/go/kim/surrogate_modeling

Printed and bound by CPI Group (UK) Ltd, Croydon, CR0 4YY

17/09/2026

14964315-0006